1994

Probability

Classics in Applied Mathematics

Editor: Robert E. O'Malley, Jr.
University of Washington
Seattle, Washington

Classics in Applied Mathematics is a series of textbooks and research monographs that were once declared out of print. SIAM is publishing this series as a professional service to foster a better understanding of applied mathematics.

Classics in Applied Mathematics

Lin, C. C. and Segel, L. A., Mathematics Applied to Deterministic Problems in the Natural Sciences

Belinfante, Johan G. F. and Kolman, Bernard, A Survey of Lie Groups and Lie Algebras with Applications and Computational Methods

Ortega, James M., Numerical Analysis: A Second Course

Fiacco, Anthony V. and McCormick, Garth P., Nonlinear Programming: Sequential Unconstrained Minimization Techniques

Clarke, F. H., Optimization and Nonsmooth Analysis

Carrier, George F. and Pearson, Carl E., Ordinary Differential Equations

Breiman, Leo, Probability

Probability

Leo Breiman
University of California, Berkeley

siam.
Society for Industrial and Applied Mathematics
Philadelphia

Library of Congress Cataloging-in-Publication Data

Breiman, Leo
Probability / Leo Breiman.
p. cm. -- (Classics in applied mathematics ; 7)
Originally published: Reading, Mass. : Addison-Wesley Pub. Co., 1968. (Addison-Wesley series in statistics)
Includes bibliographical references and index.
ISBN 0-89871-296-3
1. Probabilities. I. Title. II. Series.
QA273.B864 1992
519.2--dc20 92-1381

This SIAM edition is an unabridged, corrected republication of the work first published by Addison-Wesley Publishing Company, Inc., Reading, Massachusetts, 1968.

Preface

A few years ago I started a book by first writing a very extensive preface. I never finished that book and resolved that in the future I would write first the book and then the preface. Having followed this resolution I note that the result is a desire to be as brief as possible.

This text developed from an introductory graduate course and seminar in probability theory at UCLA. A prerequisite is some knowledge of real variable theory, such as the ideas of measure, measurable functions, and so on. Roughly, the first seven chapters of *Measure Theory* by Paul Halmos [64] is sufficient background. There is an appendix which lists the essential definitions and theorems. This should be taken as a rapid review or outline for study rather than as an exposition. No prior knowledge of probability is assumed, but browsing through an elementary book such as the one by William Feller [59, Vol. I], with its diverse and vivid examples, gives an excellent feeling for the subject.

Probability theory has a right and a left hand. On the right is the rigorous foundational work using the tools of measure theory. The left hand "thinks probabilistically," reduces problems to gambling situations, coin-tossing, motions of a physical particle. I am grateful to Michel Loève for teaching me the first side, and to David Blackwell, who gave me the flavor of the other.

David Freedman read through the entire manuscript. His suggestions resulted in many substantial revisions, and the book has been considerably improved by his efforts. Charles Stone worked hard to convince me of the importance of analytic methods in probability. The presence of Chapter 10 is largely due to his influence, and I am further in his debt for reading parts of the manuscript and for some illuminating conversations on diffusion theory.

Of course, in preparing my lectures, I borrowed heavily from the existing books in the field and the finished product reflects this. In particular, the books by M. Loève [108], J. L. Doob [39], E. B. Dynkin [43], and K. Ito and H. P. McKean [76] were significant contributors.

Two students, Carl Maltz and Frank Kontrovich, read parts of the manuscript and provided lists of mistakes and unreadable portions. Also, I was blessed by having two fine typists, Louise Gaines and Ruth Goldstein, who rose above mere patience when faced with my numerous revisions of the "final draft." Finally, I am grateful to my many nonmathematician friends who continually asked when I was going to finish "that thing," in voices that could not be interminably denied.

Leo Breiman
Topanga, California
January, 1968

Preface to the Classic Edition

This is the first of four books I have written; the one I worked the hardest on; and the one I am fondest of. It marked my goodbye to mathematics and probability theory. About the time the book was written, I left UCLA to go into the world of applied statistics and computing as a full-time freelance consultant.

The book went out of print well over ten years ago, but before it did a generation of statisticians, engineers, and mathematicians learned graduate probability theory from its pages. Since the book became unavailable, I have received many calls asking where it could be bought and then for permission to copy part or all of it for use in graduate probability courses.

These reminders that the book was not forgotten saddened me and I was delighted when SIAM offered to republish it in their Classics Series. The present edition is the same as the original except for the correction of a few misprints and errors, mainly minor.

After the book was out for a few years it became commonplace for a younger participant at some professional meeting to lean over toward me and confide that he or she had studied probability out of my book. Lately, this has become rarer and the confiders older. With republication, I hope that the age and frequency trends will reverse direction.

Leo Breiman
University of California, Berkeley
January, 1992

Contents

Chapter 5 Martingales

Chapter 6 Stationary Processes and the Ergodic Theorem

Chapter 7 Markov Chains

Chapter 8 Convergence in Distribution and the Tools Thereof

Chapter 12 Stochastic Processes and Brownian Motion

Chapter 13 Invariance Theorems

Chapter 14 Martingales and Processes with Stationary, Independent Increments

Chapter 15 Markov Processes, Introduction and Pure Jump Case

Chapter 16 Diffusions

To my mother and father
and Tuesday's children

CHAPTER 1

INTRODUCTION

A good deal of probability theory consists of the study of limit theorems. These limit theorems come in two categories which we call strong and weak. To illustrate and also to dip into history we begin with a study of coin-tossing and a discussion of the two most famous prototypes of weak and strong limit theorems.

1. n INDEPENDENT TOSSES OF A FAIR COIN

These words put us immediately into difficulty. What meaning can be assigned to the words, *coin, fair, independent?* Take a pragmatic attitude—all computations involving n tosses of a fair coin are based on two givens:

a) There are 2^n possible outcomes, namely, all sequences n-long of the two letters H and T (Heads and Tails).

b) Each sequence has probability 2^{-n}.

Nothing else is given. All computations regarding odds, and so forth, in fair coin-tossing are based on (a) and (b) above. Hence we take (a) and (b) as being the complete definition of n independent tosses of a fair coin.

2. THE "LAW OF AVERAGES"

Vaguely, almost everyone believes that for large n, the number of heads is *about the same* as the number of tails. That is, if you toss a fair coin a large number of times, then about half the tosses result in heads.

How to make this mathematics? All we have at our disposal to mathematize the "law of averages" are (a) and (b) above. So if there is anything at all corresponding to the law of averages, it must come out of (a) and (b) with no extra added ingredients.

Analyze the 2^n sequences of H and T. In how many of these sequences do exactly k heads appear? This is a combinatorial problem which clearly can be rephrased as: Given n squares, in how many different ways can we distribute k crosses on them? (See Fig. 1.1.) For example, if $n = 3$, $k = 2$, then we have the result shown in Fig. 1.2, and the answer is 3.

To get the answer in general, take the k crosses and subscript them so they become different from each other, that is, $+_1, +_2, \ldots, +_k$. Now we

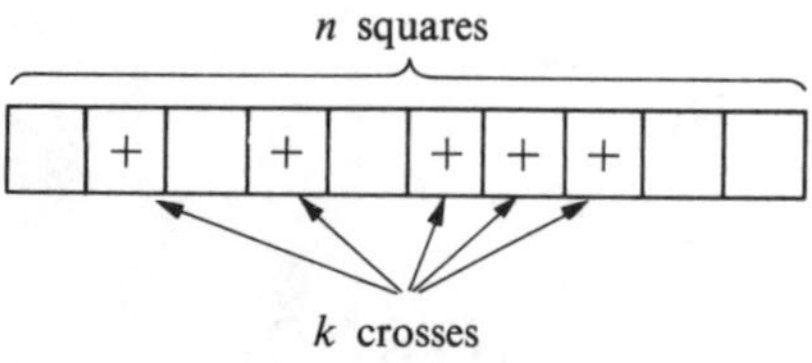

Figure 1.1

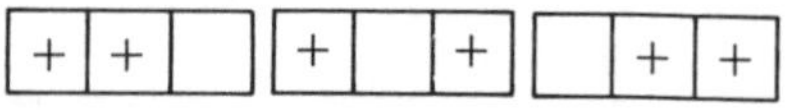

Figure 1.2

may place these latter crosses in n squares in $n(n-1)\cdots(n-k+1)$ ways [$+_1$ may be put down in n ways, then $+_2$ in $(n-1)$ ways, and so forth]. But any permutation of the k subscripted crosses among the boxes they occupy gives rise to exactly the same distribution of unsubscripted crosses. There are $k!$ permutations. Hence

Proposition 1.1. *There are exactly*

$$ {}_nC_k = \frac{n!}{k!\,(n-k)!} $$

sequences of H, T, n-long in which k heads appear.

Simple computations show that if n is even, ${}_nC_k$ is a maximum for $k = n/2$ and if n is odd, ${}_nC_k$ has its maximum value at $k = (n-1)/2$ and $k = (n+1)/2$.

Stirling's Approximation [59, Vol. I, pp. 50 ff.]

(1.2) $$ n! = e^{-n}n^n\sqrt{2\pi n}(1 + \epsilon_n), $$

where $\epsilon_n \to 0$ *as* $n \to \infty$.

We use this to get

(1.3) $$ {}_{2n}C_n = \frac{(2n)!}{n!\,n!} = \frac{e^{-2n}(2n)^{2n}\sqrt{4\pi n}}{e^{-2n}n^{2n}(2\pi n)}(1 + \delta_n) $$

$$ = 2^{2n}\cdot\frac{1}{\sqrt{\pi n}}(1 + \delta_n), $$

where $\delta_n \to 0$ as $n \to \infty$.

In $2n$ trials there are 2^{2n} possible sequences of outcomes H, T. Thus (1.3) implies that $k = n$ for only a fraction of about $1/\sqrt{\pi n}$ of the sequences. Equivalently, the probability that the number of heads equals the number of tails is about $1/\sqrt{\pi n}$ for n large (see Fig. 1.3).

Conclusion. As n becomes large, the proportion of sequences such that heads comes up exactly $n/2$ times goes to zero (see Fig. 1.3).

Whatever the "law of averages" may say, it is certainly not reasonable in a thousand tosses of a fair coin to expect exactly 500 heads. It is not

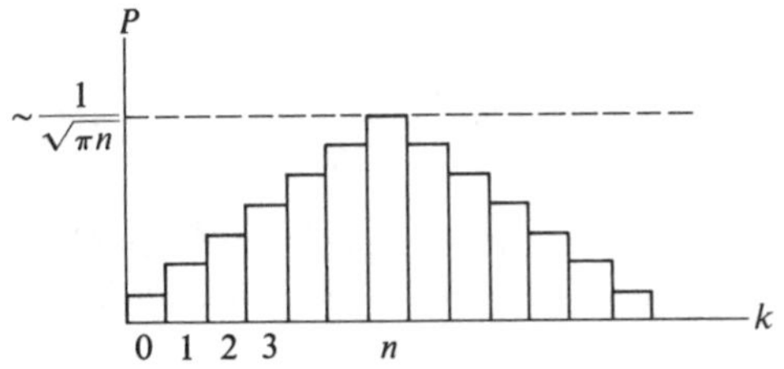

Figure 1.3 Probability of exactly k heads in $2n$ tosses.

possible to fix a number M such that for n large most of the sequences have the property that the number of heads in the sequence is within M of $n/2$. For $2n$ tosses this fraction of the sequences is easily seen to be less than $2M/\sqrt{\pi n}$ (forgetting δ_n) and so becomes smaller and smaller.

To be more reasonable, perhaps the best we can get is that usually the *proportion* of heads in n tosses is close to $\frac{1}{2}$. More precisely—

Question. Given any $\epsilon > 0$, for how many sequences does the proportion of heads differ from $\frac{1}{2}$ by less than ϵ?

The answer to this question is one of the earliest and most famous of the limit theorems of probability. Let $N(n, \epsilon)$ be the number of sequences n-long satisfying the condition of the above question.

Theorem 1.4. $\lim_n 2^{-n} N(n, \epsilon) = 1$.

In other words, the fraction of sequences such that the proportion of heads differs from $\frac{1}{2}$ by less than ϵ goes to one as n increases for any $\epsilon > 0$.

This theorem is called *the weak law of large numbers for fair coin tossing.* To prove this theorem we need to show that

(1.5) $$\lim_n \left[\frac{1}{2^n} \sum_{k; |k/n - 1/2| < \epsilon} {}_nC_k\right] = 1.$$

Theorem 1.4 states that most of the time, if you toss a coin n times, the proportion of heads will be close to $\frac{1}{2}$. Is this what is intuitively meant by *the law of averages*? Not quite—the abiding faith seems to be that no matter how badly you have done on the first n tosses, eventually things will settle down and smooth out *if you keep tossing the coin.*

Ignore this faith for the moment. Let us go back and establish some notation and machinery so we can give Theorem 1.4 an interesting proof. One proof is simply to establish (1.5) by direct computation. It was done this way originally, but the following proof is simpler.

Definition 1.6

a) *Let Ω_n be the space consisting of all sequences n-long of H, T. Denote these sequences by $\omega_1, \omega_2, \ldots, \omega_N$, $N = 2^n$.*

b) *Let A, B, C, and so forth, denote subsets of Ω_n. The probability P(A) of any subset A is defined as the sum of the probabilities of all sequences in A, that is,*

$$P(A) = 2^{-n} \text{ (number of sequences in } A),$$

equivalently, P(A) is the fraction of the total number of sequences that are in A.

For example, one interesting subset of Ω_n is the set A_1 of all sequences such that the first member is H. This set can be described as "the first toss results in heads." We should certainly have, if (b) above makes sense, $P(A_1) = \frac{1}{2}$. This is so, because there are exactly 2^{n-1} members of Ω_n whose first member is H.

c) *Let $\mathsf{X}(\omega)$ be any real-valued function on Ω_n. Define the expected value of X as*

$$E\mathsf{X} = \sum_{\omega \in \Omega_n} \mathsf{X}(\omega) \cdot \frac{1}{2^n} \cdot$$

Note that the expected value of X is just its average weighted by the probability. Suppose $\mathsf{X}(\omega)$ takes the value x_1 on the set of sequences A_1, x_2 on A_2, and so forth; then, of course,

$$E\mathsf{X} = \sum_i x_i P(A_i).$$

And also note that $E\mathsf{X}$ is an integral, that is,

$$E(\alpha \mathsf{X} + \beta \mathsf{Y}) = \alpha E\mathsf{X} + \beta E\mathsf{Y},$$

where α, β are real numbers, and $E\mathsf{X} \geq 0$, for $\mathsf{X} \geq 0$. Also, in the future we will denote by $\{\omega;\ \cdots\}$ the subset of Ω_n satisfying the conditions following the semicolon.

The proof of 1.4 will be based on the important Chebyshev inequality.

Proposition 1.7. *For $\mathsf{X}(\omega)$ any function on Ω_n and any $\epsilon > 0$,*

$$P(\omega; |\mathsf{X}(\omega)| \geq \epsilon) \leq \frac{1}{\epsilon^2} E(\mathsf{X}^2).$$

Proof

$$P(\omega; |\mathsf{X}| \geq \epsilon) = \frac{1}{2^n} (\text{number of } \omega; |\mathsf{X}(\omega)| \geq \epsilon) = \sum_{\{\omega;\, |\mathsf{X}| \geq \epsilon\}} \frac{1}{2^n}$$

$$\leq \sum_{\{\omega;\, |\mathsf{X}| \geq \epsilon\}} \frac{\mathsf{X}^2(\omega)}{\epsilon^2} \frac{1}{2^n} \leq \frac{1}{\epsilon^2} \sum_{\omega \in \Omega_n} \mathsf{X}^2(\omega) \cdot \frac{1}{2^n} = \frac{1}{\epsilon^2} E\mathsf{X}^2.$$

Define functions $\mathsf{X}_1(\omega), \ldots, \mathsf{X}_n(\omega), \mathsf{S}_n(\omega)$ on Ω_n by

(1.8)
$$\mathsf{X}_j(\omega) = \begin{cases} 1 & \text{if } j\text{th member of } \omega \text{ is } H, \\ 0 & \text{if } j\text{th member of } \omega \text{ is } T, \end{cases}$$
$$\mathsf{S}_n(\omega) = \mathsf{X}_1(\omega) + \cdots + \mathsf{X}_n(\omega),$$

so that $\mathsf{S}_n(\omega)$ is exactly the number of heads in the sequence ω. For practice, note that

$$EX_1 = 0 \cdot P(\omega;\ \text{first toss} = T) + 1 \cdot P(\omega;\ \text{first toss} = H) = \tfrac{1}{2},$$
$$EX_1X_2 = 0 \cdot P(\omega;\ \text{either first toss or second toss} = T)$$
$$+ 1 \cdot P(\omega;\ \text{both first toss and second toss} = H) = \tfrac{1}{4}$$

(since there are 2^{n-2} sequences beginning with HH). Similarly, check that if $i \neq j$, then

$$(\mathsf{X}_i - \tfrac{1}{2})(\mathsf{X}_j - \tfrac{1}{2}) = \begin{cases} \tfrac{1}{4} & \text{on } 2^{n-1} \text{ sequences}, \\ -\tfrac{1}{4} & \text{on } 2^{n-1} \text{ sequences}, \end{cases}$$

so that

$$E(\mathsf{X}_i - \tfrac{1}{2})(\mathsf{X}_j - \tfrac{1}{2}) = 0, \quad i \neq j.$$

Also,

$$(\mathsf{X}_i - \tfrac{1}{2})^2 \equiv \tfrac{1}{4}, \quad \Rightarrow E(\mathsf{X}_i - \tfrac{1}{2})^2 = \tfrac{1}{4}.$$

Finally, write

$$\mathsf{S}_n - \frac{n}{2} = \sum_{j=1}^{n} (\mathsf{X}_j - \tfrac{1}{2})$$

so that

(1.9)
$$E\left(\frac{\mathsf{S}_n}{n} - \frac{1}{2}\right)^2 = \frac{1}{n^2} E\left(\sum_{i,j} (\mathsf{X}_i - \tfrac{1}{2})(\mathsf{X}_j - \tfrac{1}{2})\right)$$
$$= \frac{1}{n} E(\mathsf{X}_1 - \tfrac{1}{2})^2 = \frac{1}{4n}.$$

Proof of Theorem 1.4. By Chebyshev's inequality,

$$P\left(\omega; \left|\frac{\mathsf{S}_n(\omega)}{n} - \frac{1}{2}\right| \geq \epsilon\right) \leq \frac{E(\mathsf{S}_n/n - \tfrac{1}{2})^2}{\epsilon^2}.$$

Use (1.9) now to get

(1.10)
$$P\left(\omega; \left|\frac{\mathsf{S}_n(\omega)}{n} - \frac{1}{2}\right| \geq \epsilon\right) \leq \frac{1}{4n\epsilon^2},$$

implying

$$\lim_n P\left(\omega; \left|\frac{\mathsf{S}_n(\omega)}{n} - \frac{1}{2}\right| \geq \epsilon\right) = 0.$$

Since $P(\Omega_n) = 1$, this completes the proof.

Definition 1.11. *Consider n independent tosses of a biased coin with probability p of heads. This is defined by*

a) *there are* 2^n *possible outcomes consisting of all sequences in* Ω_n.

b) *the probability* $P(\omega)$ *of any sequence* ω *is given by*

$$P(\omega) = (p^{\#\text{ of } H \text{ in } \omega})(q^{\#\text{ of } T \text{ in } \omega}), \quad q = 1 - p.$$

As before, define $P(A)$, $A \subset \Omega_n$, *by* $P(A) = \sum_{\omega \in A} P(\omega)$. *For* $\mathsf{X}(\omega)$ *any real valued function on* Ω_n, *define* $E\mathsf{X} = \sum_{\omega \in \Omega_n} \mathsf{X}(\omega)\, P(\omega)$.

The following problems concern biased coin-tossing.

Problems

1. Show that $P(\Omega_n) = 1$, $P(A \cup B) \leq P(A) + P(B)$ with equality if A and B are disjoint.

2. Show that

$$P(\omega;\ \mathsf{S}_n(\omega) = k) = {}_nC_k p^k q^{n-k}.$$

3. Show that Chebyshev's inequality 1.7 remains true for biased coin-tossing.

4. Prove the weak law of large numbers in the form: for any $\epsilon > 0$,

$$\lim_n P\left(\omega; \left|\frac{\mathsf{S}_n(\omega)}{n} - p\right| \geq \epsilon\right) = 0.$$

5. Using Stirling's approximation, find an approximation to the value of $P(\mathsf{S}_{2n} = n)$.

Definition 1.12. *For* $\omega \in \Omega_n$, $\omega = (\omega_1, \ldots, \omega_n)$, *where* $\omega_i \in \{H, T\}$, *call* ω_i *the ith coordinate of* ω *or the outcome of the ith toss. Any subset* $A \subset \Omega_n$ *will be referred to as an event. An event* $A \subset \Omega_n$ *will be said to depend only on the* $i_1, \ldots, i_k$*th tosses if it is of the form*

$$A = \{\omega;\ (\omega_{i_1}, \ldots, \omega_{i_k}) \in E\}, \qquad E \subset \Omega_k,\ \ k \leq n.$$

Problems

6. If A is of the form above, show that $P(A) = P'(E)$, where $P'(\omega)$ is defined on Ω_k by 1.11(b).

7. If $A, B \subset \Omega_n$ are such that A depends on the $i_1, \ldots, i_k$ tosses and B on the $j_1, \ldots, j_m$ tosses and the sets $(i_1, \ldots, i_k)$, $(j_1, \ldots, j_m)$ have no common member, then

$$P(A \cap B) = P(A)P(B).$$

[*Hint:* On Problems 6 and 7 above induction works.]

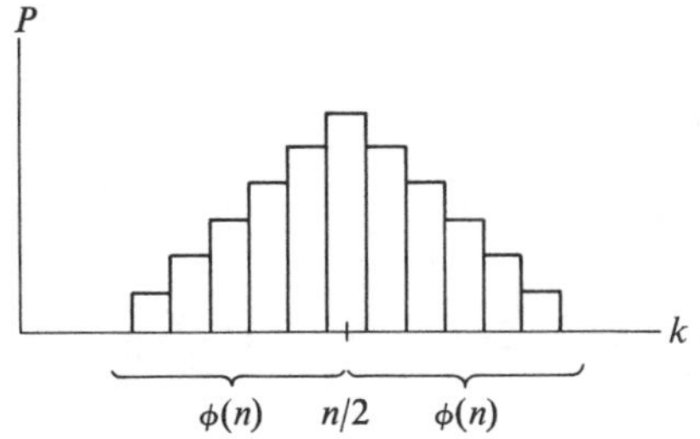

Figure 1.4 Probability of k heads in n tosses.

3. THE BELL-SHAPED CURVE ENTERS (Fluctuation Theory)

For large n, the weak law of large numbers says that most outcomes have about $n/2$ heads, more precisely, that the number of heads falls in the range $(n/2)(1 \pm \epsilon)$ with probability almost one for n large. Pose the question, how large a fluctuation about $n/2$ is nonsurprising? For instance, if you get 60 H in 100 tosses, will you strongly suspect the coin of being biased? If you get 54? 43? and so on. Look at the graph in Fig. 1.4. What we want is a function $\varphi(n)$ increasing with n such that

$$P(|S_n - n/2| < \varphi(n)) \to \alpha, \quad 0 < \alpha < 1. \tag{1.13}$$

There are useful things we know about $\varphi(n)$. As the maximum height of the graph is order $1/\sqrt{n}$ we suspect that we will have to go about $\sqrt{n}$ steps on either side. Certainly if we put $\varphi(n) = x_n \sqrt{n}/2$ (the $\frac{1}{2}$ factor to make things work out later on), then $\underline{\lim}\, x_n > 0$, otherwise the limit in (1.13) would be zero. By Chebyshev's inequality,

$$P(|S_n - n/2| \geq x_n\sqrt{n}/2) \leq n/nx_n^2 = 1/x_n^2.$$

So $\overline{\lim}\, x_n < \infty$, otherwise α in (1.13) would have to be one. These two bounds lead to the immediate suspicion that we could take $x_n \to x$, $0 < x < \infty$. But there is no reason, then, not to try $x_n \equiv x$, all n. First, examine the case for n even. We want to evaluate

$$P_n(x) = P(|S_{2n} - n| < x(\sqrt{2n}/2)).$$

This is given by

$$\sum_{|k-n| < x\sqrt{n/2}} 2^{-2n}\, {}_{2n}C_k.$$

Put $k = n + j$, to get

$$P_n(x) = \sum_{|j| < x\sqrt{n/2}} 2^{-2n}\, {}_{2n}C_{n+j}.$$

$${}_{2n}C_{n+j} = \frac{(2n)!}{(n+j)!\,(n-j)!},$$

$$(n+j)! = (n+j)\cdots(n+1)n!,$$

$$(n-j)! = n!/n \cdot (n-1)\cdots(n-j+1).$$

Put

$$P_n = P(S_{2n} = n) = 2^{-2n}(2n)!/n!\,n!$$

and write

$$2^{-2n}\,{}_{2n}C_{n+j} = P_n \cdot \frac{n(n-1)\cdots(n-j+1)}{(n+j)\cdots(n+1)}.$$

Let $D_{j,n}$ be the second factor above,

$$D_{j,n} = \frac{1}{(1 + j/n)(1 + j/(n-1))\cdots(1 + j/(n-j+1))}$$

and

$$\log D_{j,n} = -\sum_{k=0}^{j-1} \log\bigl(1 + j/(n-k)\bigr).$$

Use the expansion $\log(1 + x) = x\bigl(1 + \epsilon(x)\bigr)$, where $\lim_{x\to 0} \epsilon(x) = 0$.

$$\log D_{j,n} = -\sum_{k=0}^{j-1} \frac{j}{n-k}\left(1 + \epsilon\left(\frac{j}{n-k}\right)\right).$$

Note that j is restricted to the range $R_n = \{j;\ |j| < x\sqrt{n/2}\}$, so that if we write

$$\log D_{j,n} = -(1 + \epsilon_{j,n})\sum_{k=0}^{j-1} \frac{j}{n-k},$$

then $\sup_{j\in R_n} \epsilon_{j,n} \to 0$. Writing

$$\frac{j}{n-k} = \frac{j}{n}\cdot\frac{1}{1-k/n},$$

since $0 \le k < j$, we find that

$$\log D_{j,n} = -(1 + \epsilon'_{j,n})\sum_{k=0}^{j-1} \frac{j}{n} = -(1 + \epsilon'_{j,n})\frac{j^2}{n},$$

where again $\sup_{j\in R_n} \epsilon'_{j,n} \to 0$. Also for $j \in R_n$, $j^2/n < x^2/2$, so that

$$D_{j,n} = e^{-j^2/n}(1 + \Delta_{j,n}),\quad j \in R_n,$$

where $\sup_{j\in R_n} \Delta_{j,n} \to 0$. By (1.3),

$$P_n(x) = (1 + \delta_n)\sum_{j\in R_n} \frac{1}{\sqrt{\pi n}}\, e^{-j^2/n},\quad \lim_n \delta_n = 0.$$

Make the changes of variable, $t_j = j\sqrt{2/n}$, $\Delta t = t_{j+1} - t_j = \sqrt{2/n}$, so the condition $j \in R_n$ becomes

$$-x < t_j < x.$$

Now the end is near:

$$P_n(x) = (1 + \delta_n) \sum_{-x<t_j<x} \frac{1}{\sqrt{2\pi}} e^{-t_j^2/2} \Delta t.$$

The factor on the right is graciously just the approximating sum for an integral, that is, we have now shown that

$$\textbf{(1.14)} \qquad \lim_n P_n(x) = \frac{1}{\sqrt{2\pi}} \int_{-x}^{+x} e^{-t^2/2}\, dt.$$

To get the odd values of n take $h > 0$ and note that for n sufficiently large

$$\left\{\omega; |S_{2n} - n| < \frac{(x-h)}{2}\sqrt{2n}\right\} \subset \left\{\omega; \left| S_{2n+1} - \frac{2n+1}{2}\right| < \frac{x}{2}\sqrt{2n+1}\right\}$$
$$\subset \left\{\omega; |S_{2n} - n| < \frac{(x+h)}{2}\sqrt{2n}\right\},$$

yielding

$$\lim_n P_n(x-h) \le \underline{\lim}\, P\left(\left| S_{2n+1} - \frac{2n+1}{2}\right| < \frac{x}{2}\sqrt{2n+1}\right)$$
$$\le \overline{\lim}\, P\left(\left| S_{2n+1} - \frac{2n+1}{2}\right| < \frac{x}{2}\sqrt{2n+1}\right)$$
$$\le \lim_n P_n(x+h).$$

Thus we have proved, as done originally (more or less), a special case of the famous central limit theorem, which along with the law of large numbers shares the throne in probability theory.

Theorem 1.15

$$\lim_n P\left(\left| S_n - \frac{n}{2}\right| < \frac{x}{2}\sqrt{n}\right) = \frac{1}{\sqrt{2\pi}} \int_{-x}^{+x} e^{-t^2/2}\, dt.$$

There is a more standard form for this theorem: Let

$$\textbf{(1.16)} \qquad \Phi(x) = \frac{1}{\sqrt{2\pi}} \int_{-\infty}^{x} e^{-t^2/2}\, dt$$

and $Z_n = 2S_n - n$, that is, Z_n is the excess of heads over tails in n tosses, or if

$$Y_j = \begin{cases} +1 & \text{if } \omega_j = H, \\ -1 & \text{if } \omega_j = T, \end{cases}$$

then $Z_n = \sum_1^n Y_j$. From 1.15

$$\lim_n P\left(\frac{|Z_n|}{\sqrt{n}} < x\right) = \Phi(x) - \Phi(-x).$$

By symmetry,

$$P\left(0 \le \frac{Z_n}{\sqrt{n}} < x\right) = \tfrac{1}{2}P\left(\frac{|Z_n|}{\sqrt{n}} < x\right) + \tfrac{1}{2}P(Z_n = 0),$$

$$P(Z_n < 0) = \tfrac{1}{2} - \tfrac{1}{2}P(Z_n = 0),$$

giving

$$\lim_n P\left(\frac{Z_n}{\sqrt{n}} < x\right) = \frac{\Phi(x) + 1 - \Phi(-x)}{2}.$$

But $\Phi(+\infty) = 1$, so $1 - \Phi(-x) = \Phi(x)$, and therefore,

Theorem 1.17

$$\lim_n P\left(\frac{Z_n}{\sqrt{n}} < x\right) = \Phi(x).$$

Thus, the asymptotic distribution of the deviation of the number of heads from $n/2$ is governed by $\Phi(x)$. That $\Phi(x)$, the normal curve, should be singled out from among all other limiting distributions is one of the most magical and puzzling results in probability. Why $\Phi(x)$? The above proof gives very little insight as to what properties of $\Phi(x)$ cause its sudden appearance against the simple backdrop of fair coin-tossing. We return to this later.

Problems

8. Using Theorem 1.15 as an approximation device, find the smallest integer N such that in 1600 tosses of a fair coin, there is probability at least 0.99 that the number of heads will fall in the range $800 \pm N$.

9. Show that for j even,

$$P(Z_{2n} = j) = \frac{e^{-j^2/4n}}{\sqrt{\pi n}}(1 + \epsilon_n),$$

where $\epsilon_n \to 0$ as $n \to 0$.

10. Show that

$$\textbf{(1.18)} \qquad \sup_j |\sqrt{\pi n}\, P(Z_{2n} = j) - e^{-j^2/4n}| \to 0,$$

where the sup is over all even j.

11. Consider a sequence of experiments such that on the nth experiment a coin is tossed independently n times with probability of heads p_n. If $\lim_n np_n = \lambda$, $0 < \lambda < \infty$, then letting S_n be the number of heads occurring in the nth experiment, show that

$$\lim_n P(S_n = j) = \frac{\lambda^j}{j!}e^{-\lambda}, \quad j = 0, 1, 2, \ldots$$

4. STRONG FORM OF THE "LAW OF AVERAGES"

Definition 1.19. *Let Ω be the space consisting of all infinite sequences of H's and T's. Denote a point in Ω by ω. Let ω_j be the* jth *letter in ω, that is, $\omega = (\omega_1, \omega_2, \ldots)$. Define functions on Ω as follows:*

$$\mathsf{X}_j(\omega) = \begin{cases} 1 & \text{if } \omega_j = H, \\ 0 & \text{if } \omega_j = T, \end{cases}$$

$$\mathsf{S}_n(\omega) = \sum_{j=1}^{n} \mathsf{X}_j(\omega).$$

We are concerned with the behavior of $\mathsf{S}_n(\omega)/n$ for large n. The intuitive notion of the law of averages would be that

$$\text{(1.20)} \qquad \lim_n \frac{\mathsf{S}_n(\omega)}{n} = \frac{1}{2}$$

for all $\omega \in \Omega$. This is obviously false; the limit need not exist and even if it does, certainly does not need to equal $\frac{1}{2}$. [Consider $\omega = (H, H, H, \ldots)$.] The most we can ask is whether for *almost all* sequences in Ω, (1.20) holds.

What about the set E of ω such that (1.20) does not hold? We would like to say that in some sense the probability of this exceptional set is small. Here we encounter an essential difficulty. We know how to assign probabilities to sets of the form

$$(\omega;\ \omega_1 = H, \ldots, \omega_n = T).$$

Indeed, if $A \subset \Omega_n$, we know how to assign probabilities to all sets of the form

$$(\omega;\ (\omega_1, \ldots, \omega_n) \in A),$$

that is, simply as before

$$P(\omega; (\omega_1, \ldots, \omega_n) \in \mathrm{A}) = \frac{1}{2^n}(\text{number of sequences in A}).$$

But the exceptional set E is the set such that $\mathsf{S}_n(\omega)/n \not\rightarrow \frac{1}{2}$ and so does not depend on $\omega_1, \ldots, \omega_n$ for any n, but rather on the asymptotic distribution of heads and tails in the sequence ω.

But anyhow, let's try to push through a proof and then see what is wrong with it and what needs to be fixed up.

Theorem 1.21 *(Strong law of large numbers). The probability of the set of sequences E such that $\mathsf{S}_n(\omega)/n \not\rightarrow \frac{1}{2}$ is zero.*

Proof. First note that

$$\lim_n \frac{\mathsf{S}_n}{n} = \frac{1}{2} \Leftrightarrow \lim_m \frac{\mathsf{S}_{m^2}}{m^2} = \frac{1}{2},$$

since for any n, take m such that

$$m^2 \le n < (m+1)^2 \qquad \text{or} \qquad 0 \le n - m^2 \le 2m.$$

For this m,

$$\left|\frac{\mathsf{S}_n}{n} - \frac{\mathsf{S}_{m^2}}{m^2}\right| = \left|\frac{\mathsf{S}_n}{m^2} - \frac{\mathsf{S}_{m^2}}{m^2} + \left(\frac{1}{n} - \frac{1}{m^2}\right)\mathsf{S}_n\right|$$

$$\le \frac{|n - m^2|}{m^2} + n\left|\frac{1}{n} - \frac{1}{m^2}\right| \le \frac{2}{m} + \frac{2}{m} = \frac{4}{m}.$$

Fix $\epsilon > 0$ and let E_ϵ be the set $\{\omega;\ \overline{\lim}\,|S_{m^2}/m^2 - \frac{1}{2}| > \epsilon\}$. Look at the set $E_{m_0,m_1} \subset \Omega$ of sequences such that the inequality $|S_{m^2}/m^2 - \frac{1}{2}| > \epsilon$ occurs at least once for $m_0 \le m \le m_1$. That is,

$$E_{m_0,m_1} = \bigcup_{m=m_0}^{m_1} \left\{\omega;\ \left|\frac{\mathsf{S}_{m^2}}{m^2} - \frac{1}{2}\right| > \epsilon\right\}.$$

The set E_{m_0,m_1} is a set of sequences that depends only on the coordinates $\omega_1, \ldots, \omega_{m_1^2}$. We know how to assign probability to such sets, and applying the result of Problem 1,

$$P(E_{m_0,m_1}) \le \sum_{m=m_0}^{m_1} P\left(\omega;\ \left|\frac{\mathsf{S}_{m^2}}{m^2} - \frac{1}{2}\right| > \epsilon\right).$$

Using Chebyshev's inequality in the form (1.10) we get

$$P(E_{m_0,m_1}) \le \frac{1}{4\epsilon^2} \sum_{m=m_0}^{m_1} \frac{1}{m^2}.$$

Let m_1 go to infinity and note that

$$E_{m_0} = \lim_{m_1} E_{m_0,m_1} = \bigcup_{m=m_0}^{\infty} \left\{\omega;\ \left|\frac{\mathsf{S}_{m^2}}{m^2} - \frac{1}{2}\right| > \epsilon\right\}$$

is the set of all sequences such that the inequality $|\mathsf{S}_{m^2}/m^2 - \frac{1}{2}| > \epsilon$ occurs at least once for $m \ge m_0$. Also note that the $\{E_{m_0,m_1}\}$ are an increasing sequence of sets in m_1 for m_0 fixed. *If we could make a vital transition and say that*

$$P(E_{m_0}) = \lim_{m_1} P(E_{m_0,m_1}),$$

then it would follow that

$$P(E_{m_0}) \le \frac{1}{4\epsilon^2} \sum_{m=m_0}^{\infty} \frac{1}{m^2}.$$

Now $\overline{\lim}\,|\mathsf{S}_{m^2}/m^2 - \frac{1}{2}| > \epsilon$ if and only if for any m_0, $\exists m > m_0$ such $|\mathsf{S}_{m^2}/m^2 - \frac{1}{2}| > \epsilon$. From this, $E_\epsilon = \lim_{m_0} E_{m_0}$, where the sets E_{m_0} are decreasing in m_0. (The limits of increasing or decreasing sequences of sets are well defined, for example, $\lim_{m_0} E_{m_0} = \bigcap_{m_0=1}^{\infty} E_{m_0}$, and so forth.) *If we could again assert as above that*

$$P(E_\epsilon) = \lim_{m_0} P(E_{m_0}),$$

then

$$P(E_\epsilon) \leq \lim_{m_0} \left[\frac{1}{4\epsilon^2} \sum_{m=m_0}^{\infty} \frac{1}{m^2}\right] = 0.$$

By definition, E is the set $\{\omega;\ \overline{\lim}\ |\mathsf{S}_{m^2}/m^2 - \tfrac{1}{2}| > 0\}$, so $E = \lim_k E_{1/k}$, k running through the positive integers, and the sets $E_{1/k}$ increasing in k.

Once more, if we assert that

$$P(E) = \lim_k P(E_{1/k}),$$

then since $P(E_{1/k}) = 0$, all $k > 0$, consequently $P(E) = 0$ and the theorem is proven. *Q.E.D*???

The real question is one of how may probability be assigned to subsets of Ω. What we need for the above proof is an assignment of probability $P(\cdot)$ on a class of subsets $\mathcal{F}$ of Ω such that $\mathcal{F}$ contains all the sets that appear in the above proof and such that $P(\cdot)$ in some way corresponds to a fair coin-tossing probability. More concretely, what we want are the statements

(1.22)

i) $\mathcal{F}$ contains all subsets depending only on a finite number of tosses, that is, all sets of the form $\{\omega;\ (\omega_1, \ldots, \omega_n) \in A\}$, $A \subset \Omega_n$, $n \geq 1$, and $P(\cdot)$ is defined on these sets by

$$P(\omega;\ (\omega_1, \ldots, \omega_n) \in A) = P_n(A),$$

where P_n is the probability defined previously on Ω_n;

ii) if A_n is any monotone sequence of sets in $\mathcal{F}$, then $\lim_n A_n$ is also in $\mathcal{F}$;

iii) if the A_n are as in (ii) above, then

$$\lim_n P(A_n) = P\left(\lim_n A_n\right);$$

iv) if $A, B \in \mathcal{F}$ are disjoint, then

$$P(A \cup B) = P(A) + P(B).$$

Of these four, one is simply the requirement that the assignment be consistent with our previous definition of independent coin-tossing. Two and three are exactly the statement of what is needed to make the transitions in the proof of the law of large numbers valid. Four is that the assignment $P(\cdot)$ continue to have on Ω the property that the probability assignment has on Ω_n and whose absence would seem intuitively most offensive, namely, that if two sets of outcomes are disjoint, then the probability of getting into either one or the other is the sum of the probabilities of each one.

Also, is the assignment of $P(\cdot)$ unique in any sense? If it is not, then we are in real difficulty. We can put the above questions into more amenable form. Let $\mathcal{F}_0$ be the class of all subsets of Ω depending on only a finite number of tosses, then

Proposition 1.23. *$\mathcal{F}_0$ is a field, where*

Definition 1.24. *A class of subsets $\mathcal{C}$ of a space Ω is a field if it is closed under finite unions, intersections, and complementation. The complement of Ω is the empty set $\emptyset$.*

The proof of (1.23) is a direct verification. For economy, take $\mathcal{F}$ to be the smallest class of sets containing $\mathcal{F}_0$ such that $\mathcal{F}$ has property (1.22ii). That such a smallest class exists can be established by considering the sets common to every class of sets containing $\mathcal{F}_0$ satisfying (1.22ii). But (see Appendix A), these properties imply

Proposition 1.25. *$\mathcal{F}$ is the smallest σ-field containing $\mathcal{F}_0$, where*

Definition 1.26. *A class of subsets $\mathcal{F}$ of Ω is a σ-field if it is closed under complementation, and countable intersections and unions. For any class $\mathcal{C}$ of subsets of Ω, denote by $\mathcal{F}(\mathcal{C})$ the smallest σ-field containing $\mathcal{C}$.*

Also

Proposition 1.27. *$P(\cdot)$ on $\mathcal{F}$ satisfies* (1.22) *iff $P(\cdot)$ is a probability measure, where*

Definition 1.28. *A nonnegative set function $P(\cdot)$ defined on a σ-field $\mathcal{F}$ of subsets of Ω is a probability measure if*

i) *(normalization)* $P(\Omega) = 1$;
ii) *(σ-additivity) for every finite or countable collection $\{B_k\}$ of sets in $\mathcal{F}$ such that B_k is disjoint from B_j, $k \neq j$,*

$$P\left(\bigcup_k B_k\right) = \sum_k P(B_k).$$

Proof of 1.27. If $P(\cdot)$ satisfies (1.22), then by finite induction on (iv)

$$P\left(\bigcup_1^n B_k\right) = \sum_1^n P(B_k).$$

Let $A_n = \bigcup_1^n B_k$, then the A_n are a monotone sequence of sets, $\lim A_n = \bigcup_k B_k$. By (1.22iii),

$$P\left(\bigcup_k B_k\right) = \lim_n \left(\sum_1^k P(B_k)\right) = \sum_k P(B_k).$$

Conversely, if $P(\cdot)$ is σ-additive, it implies (1.22). For if the $\{A_n\}$ are a monotone sequence, say $A_n \subset A_{n+1}$, we can let $B_k = A_k - A_{k-1}$, $k > 1$, $B_1 = A_1$. The $\{B_k\}$ are disjoint, and $\lim_n A_n = \bigcup_k B_k$. Thus σ-additivity gives

$$P\left(\lim_n A_n\right) = \sum_k P(B_k).$$

Use $\bigcup_1^n B_k = A_n$, $P(A_n) = \sum_1^n P(B_k)$ to get

$$\lim_n P(A_n) = \lim_n \left(\sum_1^n P(B_k)\right) = \sum_k P(B_k).$$

The starting point is a set function with the following properties:

Definition 1.29. *A nonnegative set function P on a field $\mathcal{F}_0$ is a finite probability measure if*

i) $P(\Omega) = 1$;

ii) *for $A, B \in \mathcal{F}_0$, and disjoint,*

$$P(A \cup B) = P(A) + P(B).$$

Now the original question can be restated in more standard form: Given the finite probability measure $P(\cdot)$ defined on $\mathcal{F}_0$ by (1.22i), does there exist a probability measure defined on $\mathcal{F}$ and agreeing with $P(\cdot)$ on $\mathcal{F}_0$? And in what sense is the measure unique? The problem is seen to be one of extension—given $P(\cdot)$ on $\mathcal{F}_0$, is it possible to extend the domain of definition of $P(\cdot)$ to $\mathcal{F}$ such that it is σ-additive? But this is a standard measure theoretical question. The surprise is that the attempt to patch up the strong law of large numbers has led directly to this well-known problem (see Appendix A.9).

5. AN ANALYTIC MODEL FOR COIN-TOSSING

The fact that the sequence $X_1(\omega), X_2(\omega), \ldots$ comprised functions depending on consecutive independent tosses of a fair coin was to some extent immaterial in the proof of the strong law. For example, produce functions on a different space Ω' this way: Toss a well-balanced, six-sided die independently, let Ω' be the space of all infinite sequences $\omega' = (\omega_1', \omega_2', \ldots)$, where ω_k' takes values in $(1, \ldots, 6)$. Define $X_n'(\omega')$ to be one if the nth throw results in an even face, zero if in an odd face. The sequence $X_1'(\omega'), X_2'(\omega'), \ldots$ has the same probabilistic structure as $X_1(\omega), \ldots$ in the sense that the probability of any sequence n-long of zeros and ones is $1/2^n$ in both models (with the appropriate definition of independent throws of a well-balanced die). But this assignment of probabilities is the important information, rather than the exact nature of the underlying space. For example, the same argument leading to the strong law of large numbers holds for the variables $X_1', X_2', \ldots$. Therefore, in general, we will consider as a model for fair coin-tossing any set of functions $X_1, X_2, \ldots$, with values zero or one defined on a space Ω of points ω such that probability $1/2^n$ is assigned to all sets of the form

$$\{\omega;\ X_1 = s_1, \ldots, X_n = s_n\}$$

for $s_1, \ldots, s_n$ any sequence of zeros and ones.

An interesting analytic model can be constructed on the half-open unit interval $\Omega = [0, 1)$. It can be shown that every number x in $[0, 1)$ has a unique binary expansion containing an infinite number of zeros. The latter restriction takes care of binary rational points which have two expansions, that is

$$\frac{1}{2} = 0.1000 \cdots = \frac{1}{2} + \frac{0}{2^2} + \frac{0}{2^3} + \cdots$$

$$= 0.0111 \cdots = \frac{0}{2} + \frac{1}{2^2} + \frac{1}{2^3} + \cdots$$

Now for any $x \in [0, 1)$ write down this expansion $x = .x_1x_2 \cdots$ and define

(1.30) $$X_n(x) = x_n.$$

That is, $X_n(x)$ is the nth digit in the expansion of x (see Fig. 1.5).

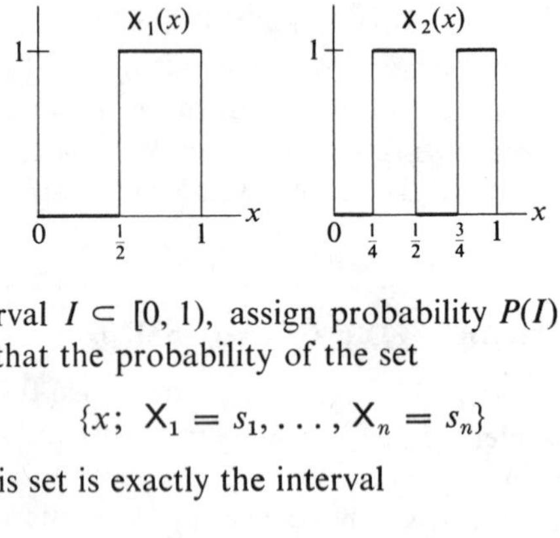

Figure 1.5

To every interval $I \subset [0, 1)$, assign probability $P(I) = \|I\|$, the length of I. Now check that the probability of the set

$$\{x;\ X_1 = s_1, \ldots, X_n = s_n\}$$

is $1/2^n$, because this set is exactly the interval

$$\left\{x;\ \frac{s_1}{2} + \frac{s_2}{2^2} + \cdots + \frac{s_n}{2^n} \le x < \frac{s_1}{2} + \frac{s_2}{2^2} + \cdots + \frac{s_n}{2^n} + \frac{1}{2^n}\right\}.$$

Thus, $X_1(x), X_2(x), \ldots$ on $[0, 1)$ with the given probability is a model for fair coin-tossing.

The interest in this particular model is that the extension of P is a classical result. The smallest σ-field containing all the intervals

$$\{x;\ X_1(x) = s_1, \ldots, X_n(x) = s_n\}$$

is the Borel field $\mathcal{B}_1([0, 1))$ of subsets of $[0, 1)$, and there is a unique extension of P to a probability on this field, namely, Lebesgue measure. The theorem establishing the existence of Lebesgue measure makes the proof of the strong law of large numbers rigorous for the analytic model. The statement in this context is:

Theorem 1.31. *For almost all* $x \in [0, 1)$ *with respect to Lebesgue measure, the asymptotic proportions of zero's and one's in the binary expansion of* x *is* $\frac{1}{2}$.

The existence of Lebesgue measure in the analytic model for coin-tossing makes it plausible that there is an extended probability in the original model. This is true, but we defer the proof to the next chapter.

Another way of looking at the analytic model is to say that the binary expansion of a number in $[0, 1)$ produces independent zeros and ones with probability $\frac{1}{2}$ each with respect to Lebesgue measure. Thus, as Theorem 1.31 illustrates, any results established for fair coin-tossing can be written as theorems concerning functions and numbers on and in $[0, 1)$ with respect to Lebesgue measure. Denote Lebesgue measure from now on by dx or $l(dx)$.

Problem 12. (The Rademacher Functions). Let

$$\gamma_i(x) = \begin{cases} -1, & x_i = 1, \\ +1, & x_i = 0, \end{cases}$$

where $x = .x_1x_2 \cdots$ is the unique binary expansion of x containing an infinite number of zeros. Show from the properties of coin-tossing that if $i_k \neq i_j$ for $j \neq k$,

$$\int_0^1 \prod_1^m \gamma_{i_k}(x)\, dx = 0.$$

Graph $\gamma_3(x)$, $\gamma_4(x)$. Show that the sequence of functions $\{\gamma_i(x)\}$ is orthonormal with respect to Lebesgue measure.

6. CONCLUSIONS

The strong law of large numbers and the central limit theorem illustrate the two main types of limit theorems in probability.

Strong limit theorems. *Given a sequence of functions* $\mathsf{Y}_1(\omega), \mathsf{Y}_2(\omega), \ldots$ *there is a limit function* $\mathsf{Y}(\omega)$ *such that* $P(\omega; \lim_n \mathsf{Y}_n(\omega) = \mathsf{Y}(\omega)) = 1$.

Weak limit theorems. *Given a sequence of functions* $\mathsf{Y}_1(\omega), \mathsf{Y}_2(\omega), \ldots$ *show that*

$$\lim_n P(\omega; \mathsf{Y}_n(\omega) < x)$$

exists for every x.

There is a great difference between strong and weak theorems which will become more apparent. We will show later, for instance, that $Z_n/\sqrt{n}$ has no limit in any reasonable way. A more dramatic example of this is: on $([0, 1), \mathcal{B}_1([0, 1)))$ with P being Lebesgue measure, define

$$Y_n(y) = \begin{cases} 0, & y < \frac{1}{2}, \\ 1, & \frac{1}{2} \le y < 1, \end{cases}$$

for n even. For n odd,

$$Y_n(y) = \begin{cases} 1, & y < \frac{1}{2}, \\ 0, & \frac{1}{2} \le y < 1. \end{cases}$$

For all n, $P(y;\ Y_n(y) < x) = P(y;\ Y_1(y) < x)$. But for every $y \in [0, 1)$

$$\overline{\lim_n}\, Y_n(y) = 1, \qquad \underline{\lim_n}\, Y_n(y) = 0.$$

To begin with we concentrate on strong limit theorems. But to do this we need a more firmly constructed measure theoretic foundation.

NOTES

To get some of the fascinating interplay between probability and number theory, refer to Mark Kac's monograph [83].

Although there will be very little subsequent work with combinatorics in this text, they occupy an honored and powerful place in probability theory. First, for many of the more important theorems, the original version was for independent fair coin-tossing. Even outside of this, there are some strong theorems in probability for which the most interesting proofs are combinatorial. A good source for these uses are Feller's books [59].

An elegant approach to the measure theoretic aspects of probability can be found in Neveu's book [113].

CHAPTER 2

MATHEMATICAL FRAMEWORK

1. INTRODUCTION

The context that is necessary for the strong limit theorems we want to prove is:

Definition 2.1. *A probability space consists of a triple* $(\Omega, \mathcal{F}, P)$ *where*

i) Ω *is a space of points* ω, *called the sample space and sample points.*
ii) $\mathcal{F}$ *is a* σ*-field of subsets of* Ω. *These subsets are called events.*
iii) $P(\cdot)$ *is a probability measure on* $\mathcal{F}$; *henceforth refer to* P *as simply a probability.*

On Ω there is defined a sequence of real-valued functions $X_1(\omega), X_2(\omega), \ldots$ which are random variables in the sense of

Definition 2.2. *A function* $X(\omega)$ *defined on* Ω *is called a random variable if for every Borel set* B *in the real line* $R^{(1)}$, *the set* $\{\omega; X(\omega) \in B\}$ *is in* $\mathcal{F}$. ($X(\omega)$ *is a measurable function on* $(\Omega, \mathcal{F})$.)

Whether a given function is a random variable, of course, depends on the pair $(\Omega, \mathcal{F})$. The reason underlying 2.2 is that we want probability assigned to all sets of the form $\{\omega; X(\omega) \in I\}$, where I is some interval. It will follow from 2.29 that if $\{\omega; X(\omega) \in I\}$ is in $\mathcal{F}$ for all intervals I, then X must be a random variable.

Definition 2.3. *A countable stochastic process, or process, is a sequence of random variables* $X_1, X_2, \ldots$ *defined on a common probability space* $(\Omega, \mathcal{F}, P)$.

But in a probabilistic model arising in gambling or science the given data are usually an assignment of probability to a much smaller class of sets. For example, if all the variables $X_1, X_2, \ldots$ take values in some countable set F, the probability of all sets of the form

(2.4) $$\{\omega; X_1 = s_1, \ldots, X_n = s_n\}, \quad s_k \in F, \quad k = 1, \ldots, n$$

is usually given. If the $X_1, X_2, \ldots$ are not discrete, then often the specification is for all sets of the form

(2.5) $$\{\omega; X_1 \in I_1, \ldots, X_n \in I_n\},$$

where $I_1, \ldots, I_n$ are intervals.

To justify the use of a probability space as a framework for probability theory it is really necessary to show that a reasonable assignment of probabilities to a small class of sets has a unique extension to a probability P on a probability space $(\Omega, \mathcal{F}, P)$. There are fairly general results to this effect. We defer this until we have explored some of the measure-theoretic properties of processes.

2. RANDOM VECTORS

Given two spaces Ω and R, let $\mathbf{X}$ be a function on Ω to R, $\mathbf{X}\colon \Omega \to R$. The inverse image under $\mathbf{X}$ of a set $B \subset R$ is $\{\omega;\ \mathbf{X} \in B\}$. We abbreviate this by $\{\mathbf{X} \in B\}$.

Proposition 2.6. *Set operations are preserved under inverse mappings, that is,*

$$\left\{\mathbf{X} \in \bigcup_\lambda B_\lambda\right\} = \bigcup_\lambda \{\mathbf{X} \in B_\lambda\},$$

$$\left\{\mathbf{X} \in \bigcap_\lambda B_\lambda\right\} = \bigcap_\lambda \{\mathbf{X} \in B_\lambda\},$$

$$\{\mathbf{X} \in B^c\} = \{\mathbf{X} \in B\}^c.$$

(B^c denotes the complement of the set B.)

Proof. By definition.

This quickly gives

Proposition 2.7. *If $\mathbf{X}\colon \Omega \to R$, and $\mathcal{B}$ is a σ-field in R, the class of sets $\{\mathbf{X} \in B\}$, $B \in \mathcal{B}$, is a σ-field. If $\mathcal{F}$ is a σ-field in Ω, then the class of subsets B in R such that $\{\mathbf{X} \in B\} \in \mathcal{F}$ is a σ-field.*

Proof. Both assertions are obvious from 2.6.

Definition 2.8. *If there are σ-fields $\mathcal{F}$ and $\mathcal{B}$, in Ω, R respectively, $\mathbf{X}\colon \Omega \to R$ is called a random vector if $\{\mathbf{X} \in B\} \in \mathcal{F}$, for all $B \in \mathcal{B}$. ($\mathbf{X}$ is a measurable map from $(\Omega, \mathcal{F})$ to $(R, \mathcal{B})$.)*

We will sometimes refer to $(R, \mathcal{B})$ as the range space of $\mathbf{X}$. But the *range* of $\mathbf{X}$ is the direct image under $\mathbf{X}$ of Ω, that is, the union of all points $\mathbf{X}(\omega)$, $\omega \in \Omega$.

Denote by $\mathcal{F}(\mathbf{X})$ the σ-field of all sets of the form $\{\mathbf{X} \in B\}$, $B \in \mathcal{B}$.

Definition 2.9. *If $\mathcal{A}$ is a σ-field contained in $\mathcal{F}$, call $\mathbf{X}$ $\mathcal{A}$-measurable if $\mathcal{F}(\mathbf{X}) \subset \mathcal{A}$.*

If there is a probability space $(\Omega, \mathcal{F}, P)$ and $\mathbf{X}$ is a random vector with range space $(R, \mathcal{B})$, then $\hat{P}$ can be naturally defined on $\mathcal{B}$ by

$$\hat{P}(B) = P(\mathbf{X} \in B).$$

It is easy to check that $\hat{P}$ defined this way is a probability on $\mathcal{B}$.

Definition 2.10. *$\hat{P}$ is called the probability distribution of the random vector* $\mathbf{X}$.

Conversely, suppose $\mathbf{X}$ is a random vector on $(\Omega, \mathcal{F})$ to $(R, \mathcal{B})$ and there is a probability distribution $\hat{P}$ defined on $\mathcal{B}$. Since every set in $\mathcal{F}(\mathbf{X})$ is of the form $\{\mathbf{X} \in B\}$, $B \in \mathcal{B}$, can P be defined on $\mathcal{F}(\mathbf{X})$ by

$$\textbf{(2.11)} \qquad P(\mathbf{X} \in B) = \hat{P}(B)?$$

The answer, in general, is no! The difficulty is that the same set $A \in \mathcal{F}(\mathbf{X})$ may be represented in two different ways as $\{\mathbf{X} \in B_1\}$ and $\{\mathbf{X} \in B_2\}$, and there is no guarantee that $\hat{P}(B_1) = \hat{P}(B_2)$. What is true is

Proposition 2.12. *Let F be the range of* $\mathbf{X}$. *If $B \in \mathcal{B}$, $B \subset F^c$ implies $\hat{P}(B) = 0$, then P is uniquely defined on $\mathcal{F}(\mathbf{X})$ by* 2.11, *and is a probability.*

Proof. If $A = \{\mathbf{X} \in B_1\} = \{\mathbf{X} \in B_2\}$, then $B_1 - B_2$ and $B_2 - B_1$ are both in F^c. Hence $\hat{P}(B_1) = \hat{P}(B_2)$. The σ-additivity is quickly verified.

Problem 1. Use 2.12 and the existence of the analytic model for coin-tossing to prove the existence of the desired extension of P in the original model.

3. THE DISTRIBUTION OF PROCESSES

Denote by $R^{(\infty)}$ the space consisting of all infinite sequences $(x_1, x_2, \ldots)$ of real numbers. In $R^{(\infty)}$ an n-dimensional rectangle is a set of the form

$$\{\mathbf{x} \in R^{(\infty)};\ x_1 \in I_1, \ldots, x_n \in I_n\},$$

where $I_1, \ldots, I_n$ are finite or infinite intervals. Take the Borel field $\mathcal{B}_\infty$ to be the smallest σ-field of subsets of $R^{(\infty)}$ containing all finite-dimensional rectangles.

If each component of $\mathbf{X} = (\mathsf{X}_1, \ldots)$ is a random variable, then it follows that the vector $\mathbf{X}$ is a measurable mapping to $(R^{(\infty)}, \mathcal{B}_\infty)$. In other words,

Proposition 2.13. *If $\mathsf{X}_1, \mathsf{X}_2, \ldots$ are random variables on $(\Omega, \mathcal{F})$, then for $\mathbf{X} = (\mathsf{X}_1, \mathsf{X}_2, \ldots)$ and every $B \in \mathcal{B}_\infty$, $\{\mathbf{X} \in B\} \in \mathcal{F}$.*

Proof. Let S be a finite-dimensional rectangle

$$S = \{\mathbf{x} \in R^{(\infty)};\ x_1 \in I_1, \ldots, x_n \in I_n\}.$$

Then

$$\{\mathbf{X} \in S\} = \{\mathsf{X}_1 \in I_1\} \cap \{\mathsf{X}_2 \in I_2\} \cap \cdots \cap \{\mathsf{X}_n \in I_n\}.$$

This is certainly in $\mathcal{F}$. Now let $\mathcal{C}$ be the class of sets C in $\mathcal{B}_\infty$ such that $\{\mathbf{X} \in C\} \in \mathcal{F}$. By 2.7 $\mathcal{C}$ is a σ-field. Since $\mathcal{C}$ contains all rectangles, $\mathcal{C} = \mathcal{B}_\infty$.

If all that we observe are the values of a process $\mathsf{X}_1(\omega), \mathsf{X}_2(\omega), \ldots$ the underlying probability space is certainly not uniquely determined. As an example, suppose that in one room a fair coin is being tossed independently, and calls zero or one are being made for tails or heads respectively. In another room a well-balanced die is being cast independently and zero or one called as the resulting face is odd or even. There is, however, no way of discriminating between these two experiments on the basis of the calls.

Denote $\mathbf{X} = (X_1, \ldots)$. From an observational point of view, the thing that really interests us is not the space $(\Omega, \mathcal{F}, P)$, but the distribution of the values of $\mathbf{X}$. If two processes, $\mathbf{X}$ on $(\Omega, \mathcal{F}, P)$, $\mathbf{X}'$ on $(\Omega', \mathcal{F}', P')$ have the same probability distribution,

(2.14) $$\hat{P}(B) = P(\mathbf{X} \in B) = P'(\mathbf{X}' \in B), \qquad \text{all } B \in \mathcal{B}_\infty,$$

then there is no way of distinguishing between the processes by observing them.

Definition 2.15. *Two processes* $\{X_n\}$ *on* $(\Omega, \mathcal{F}, P)$ *and* $\{X_n'\}$ *on* $(\Omega', \mathcal{F}', P')$ *will be said to have the same distribution if* (2.14) *holds.*

The distribution of a process contains all the information which is relevant to probability theory. All theorems we will prove depend only on the distribution of the process, and hence hold for all processes having that distribution. Among all processes having a given distribution $\hat{P}$ on $\mathcal{B}_\infty$, there is one which has some claim to being the simplest.

Definition 2.16. *For any given distribution* $\hat{P}$ *define random variables* $\hat{X}_1, \hat{X}_2, \ldots$ *on* $(R^{(\infty)}, \mathcal{B}_\infty, \hat{P})$ *by*

$$\hat{X}_n(x_1, x_2, \ldots) = x_n.$$

This process is called the coordinate representation process and has the same distribution as the original process.

This last assertion is immediate since for any $B \in \mathcal{B}_\infty$,

$$\hat{P}(\mathbf{x};\ (\hat{X}_1, \hat{X}_2, \ldots) \in B) = \hat{P}(\mathbf{x};\ \mathbf{x} \in B) = \hat{P}(B).$$

This construction also leads to the observation that given any probability $\hat{P}$ on $\mathcal{B}_\infty$, there exists a process $\mathbf{X}$ such that $P(\mathbf{X} \in B) = \hat{P}(B)$.

Define the Borel field $\mathcal{B}_n$ in $R^{(n)}$ as the smallest σ-field containing all rectangles $\{(x_1, \ldots, x_n);\ x_1 \in I_1, \ldots, x_n \in I_n\}$, $I_1, \ldots, I_n$ intervals.

Definition 2.17. *An n-dimensional cylinder set in* $R^{(\infty)}$ *is any set of the form* $\{\mathbf{x} \in R^{(\infty)};\ (x_1, \ldots, x_n) \in B\}$, $B \in \mathcal{B}_n$.

Problems

2. Show that the class of all finite-dimensional cylinder sets is a field, but not a σ-field.

3. Let F be a countable set, $F = \{r_j\}$. Denote by $F^{(\infty)}$ the set of all infinite sequences with coordinates in F. Show that $X\colon \Omega \to F$ is a random variable with respect to $(\Omega, \mathcal{F})$ iff $\{\omega;\ X(\omega) = r_j\} \in \mathcal{F}$, all j.

4. Given two processes $\mathbf{X}, \mathbf{X}'$ such that both take values in $F^{(\infty)}$, show that they have the same distribution iff

$$P(X_1 = t_1, \ldots, X_n = t_n) = P'(X_1' = t_1, \ldots, X_n' = t_n)$$

for every n-sequence $(t_1, \ldots, t_n)$, $t_k \in \{r_j\}$, $k = 1, \ldots, n$.

4. EXTENSION IN SEQUENCE SPACE

Given the concept of the distribution of a process, the extension problem can be looked at in a different way. The given data is a specification of values $P(\mathbf{X} \in B)$ for a class of sets in $\mathcal{B}_\infty$. That is, a set function $\hat{P}$ is defined for a class of sets $\mathcal{C} \subset \mathcal{B}_\infty$, and $\hat{P}(B)$ is the probability that the observed values of the process fall in $B \in \mathcal{C}$. Now ask: Does there exist *any* process whose distribution agrees with $\hat{P}$ on $\mathcal{C}$? Alternatively—construct a process $\mathbf{X}$ such that $P(\mathbf{X} \in B) = \hat{P}(B)$ for all $B \in \mathcal{C}$. This is equivalent to the question of whether $\hat{P}$ on $\mathcal{C}$ can be extended to a probability on $\mathcal{B}_\infty$. Because if so, the coordinate representation process has the desired distribution. As far as the original sample space is concerned, once $\hat{P}$ on $\mathcal{B}_\infty$ is gotten, 2.12 can be used to get an extension of P to $\mathcal{F}(\mathbf{X})$, if $\hat{P}$ assigns probability zero to sets $B \in \mathcal{B}_\infty$ falling in the complement of the range of $\mathbf{X}$.

Besides this, another reason for looking at the extension problem on $(R^{(\infty)}, \mathcal{B}_\infty)$ is that this is the smoothest space on which we can always put a process having any given distribution. It has some topological properties which allow nice extension results to be proved.

The basic extension theorem we use is the analog in $\mathcal{B}_\infty$ of the extension of measures on the real line from their values on intervals. Let $\mathcal{C}$ be the class of all finite-dimensional rectangles, and assume that $\hat{P}$ is defined on $\mathcal{C}$. A finite-dimensional rectangle may be written as a disjoint union of finite-dimensional rectangles, for instance,

$$\{\mathbf{x} \in R^{(\infty)};\ x_1 \in [0, 1)\} = \{\mathbf{x} \in R^{(\infty)};\ x_1 \in [0, \tfrac{1}{3})\} \cup \{\mathbf{x} \in R^{(\infty)};\ x_1 \in [\tfrac{1}{3}, 1)\}.$$

Of course, we will insist that if a rectangle S is a finite union $\bigcup_j S_j$ of disjoint rectangles, then $\hat{P}(S) = \Sigma_j \hat{P}(S_j)$. But an additional regularity condition is required, simply because the class of finite probabilities is much larger than the class of probabilities.

Extension Theorem 2.18. *Let $\hat{P}$ be defined on the class $\mathcal{C}$ of all finite-dimensional rectangles and have the properties:*

a) $\hat{P}(\cdot) \geq 0$, $\hat{P}(R^{(\infty)}) = 1$;

b) *if $\{S_j\}$, $j = 1, \ldots, m$, are disjoint n-dimensional rectangles and*

$$S = \bigcup_{j=1}^{m} S_j$$

is a rectangle, then

$$\hat{P}(S) = \sum_{j=1}^{m} \hat{P}(S_j);$$

c) *if* $\{S_j\}$ *are a nondecreasing sequence of n-dimensional rectangles, and* $S_j \uparrow S$,

$$\lim_j \hat{P}(S_j) = \hat{P}(S).$$

Then there is a unique extension of $\hat{P}$ *to a probability on* $\mathcal{B}_\infty$.

Proof. As this result belongs largely to the realm of measure theory rather than probability, we relegate its proof to Appendix A.48.

Theorem 2.18 translates into probability language as: If probability is assigned in a reasonable way to rectangles, then there exists a process $X_1, X_2, \ldots$ such that $P(X_1 \in I, \ldots, X_n \in I_n)$ has the specified values.

If the probabilities are assigned to rectangles, then in order to be well defined, the assignment must be *consistent.* This means here that since an n-dimensional rectangle is also an $(n + 1)$-dimensional rectangle (take $I_{n+1} = R^{(1)}$), its assignment as an $(n + 1)$-dimensional rectangle must agree with its assignment as an n-dimensional rectangle.

Now consider the situation in which the probability distributions of all finite collections of random variables in a process are specified. Specifically, probabilities $\hat{P}_n$ on $\mathcal{B}_n$, $n = 1, 2, \ldots$, are given and $\hat{P}$ is defined on the class of all finite-dimensional cylinder sets (2.17) by

$$\hat{P}(\mathbf{x} \in R^{(\infty)}; (x_1, \ldots, x_n) \in B) = \hat{P}_n(B), \quad B \in \mathcal{B}_n.$$

In order for $\hat{P}$ to be well-defined, the $\hat{P}_n$ must be consistent—every n-dimensional cylinder set is also an $(n + 1)$-dimensional cylinder set and must be given the same probability by $\hat{P}_n$ and $\hat{P}_{n+1}$.

Corollary 2.19. (*Kolmogorov extension theorem*). *There is a unique extension of* $\hat{P}$ *to a probability on* $\mathcal{B}_\infty$.

Proof. $\hat{P}$ is defined on the class $\mathcal{C}$ of all finite-dimensional rectangles and is certainly a finite probability on $\mathcal{C}$. Let S^*, S_j^* be rectangles in $R^{(n)}$, $S_j^* \uparrow S^*$, then

$$\hat{P}(\mathbf{x} \in R^{(\infty)}; \quad (x_1, \ldots, x_n) \in S_j^*) = \hat{P}_n(S_j^*).$$

Since $\hat{P}_n$ is a probability on $\mathcal{B}_n$, it is well behaved under monotone limits (1.27). Hence $\hat{P}_n(S_j^*) \uparrow \hat{P}_n(S^*)$, and Theorem 2.18 is in force.

The extension requirements become particularly simple when the required process takes only values in a countable set.

Corollary 2.20. *Let* $F \subset R^{(1)}$ *be a countable set. If* $p(s_1, \ldots, s_n)$ *is specified for all finite sequences of elements of F and satisfies*

$$p(\cdot) \geq 0,$$

$$\sum_{s_{n+1} \in F} p(s_1, \ldots, s_{n+1}) = p(s_1, \ldots, s_n),$$

$$\sum_{s_1 \in F} p(s_1) = 1,$$

then there exists a process $X_1, X_2, \ldots$ *such that*

$$P(X_1 = s_1, \ldots, X_n = s_n) = p(s_1, \ldots s_n).$$

Proof. Let s denote an n-tuple with coordinates in F. For any $B \in \mathcal{B}_n$, define

$$\hat{P}_n(B) = \sum_{s \in B} p(s).$$

It is easy to check that $\hat{P}_n$ is a finite probability on $\mathcal{B}_n$. Furthermore, $B_k \uparrow B$ implies

$$\lim_k \hat{P}_n(B_k) = \hat{P}_n(B).$$

Thus we conclude (see 1.27) that $\hat{P}_n$ is a probability on $\mathcal{B}_n$. The $\hat{P}_n$ are clearly consistent. Now apply 2.19 to get the result.

The extension results in this section are a bit disquieting, because even though the results are purely measure-theoretic, the proofs in the space $(R^{(\infty)}, \mathcal{B}_\infty)$ depend essentially on the topological properties of Euclidean spaces. This is in the nature of the problem. For example, if one has an infinite product $(\Omega^{(\infty)}, \mathcal{F}_\infty)$ of $(\Omega, \mathcal{F})$ spaces, that is:

$$\Omega^{(\infty)} = \{(\omega_1, \ldots); \omega_k \in \Omega\};$$

$\mathcal{F}_\infty$ the smallest σ-field containing all sets of the form

$$\{(\omega_1, \ldots); \ \omega_1 \in A_1, \ldots, \omega_n \in A_n\}, \quad A_1, \ldots, A_n \in \mathcal{F};$$

P_n a probability on all n-dimensional cylinder sets; and the set $\{P_n\}$ consistent; then a probability P on $\mathcal{F}_\infty$ agreeing with P_n on n-dimensional cylinder sets may or may not exist. For a counter-example, see Jessen and Andersen [77] or Neveu [113, p. 84].

Problem 5. Take $\Omega = (0, 1]$,

$$X_n(x) = \begin{cases} 1, & 0 < x < 1/n, \\ 0, & 1/n < x \leq 1. \end{cases}$$

Let $\mathcal{F}_0 = \bigcup_{n \geq 1} \mathcal{F}(X_1, \ldots, X_n)$. Characterize the sets in $\mathcal{F}_0$, $\mathcal{F}(\mathcal{F}_0)$. For $B \in \mathcal{B}_n$, assign

$$P((X_1, \ldots, X_n) \in B) = \begin{cases} 1, & \text{if } (1, 1, \ldots, 1) \in B, \\ 0, & \text{otherwise.} \end{cases}$$

Prove that P is additive on $\mathcal{F}_0$, but that there is no extension of P to a probability on $\mathcal{F}(\mathcal{F}_0)$.

5. DISTRIBUTION FUNCTIONS

What is needed to ensure that two processes have the same distribution?

Definition 2.21. *Given a process* $\{X_n\}$ *on* $(\Omega, \mathcal{F}, P)$, *define the n-dimensional distribution functions by*

$$F_n(x_1, \ldots; x_n) = P(X_1 < x_1, \ldots, X_n < x_n).$$

The functions $F_n(\cdot)$ are real-valued functions defined on $R^{(n)}$. Denote these at times by $F_n(\mathbf{x}_n)$ or, to make their dependence on the random variables explicit, by $F_{X_1\ldots X_n}(x_1, \ldots, x_n)$ or $F_{\mathbf{X}_n}(\mathbf{x}_n)$.

Theorem 2.22. *Two processes have the same distribution iff all their distribution functions are equal.*

The proof of 2.22 follows from a more general result that we want on the record.

Proposition 2.23. *Let Q, Q' be two probabilities on $(\Omega, \mathcal{F})$. Let $\mathcal{C}$ be a class of sets such that $A, B \in \mathcal{C} \Rightarrow A \cap B \in \mathcal{C}$, and $\mathcal{F} = \mathcal{F}(\mathcal{C})$. Then $Q = Q'$ on $\mathcal{C}$ implies that $Q = Q'$ on $\mathcal{F}$.*

There seems to be a common belief that 2.23 is true without the hypothesis that $\mathcal{C}$ be closed under $\cap$. To disprove this, let $\Omega = \{a, b, c, d\}$, $Q_1(a) = Q_1(d) = Q_2(b) = Q_2(c) = \frac{1}{6}$, and $Q_1(b) = Q_1(c) = Q_2(a) = Q_2(d) = \frac{1}{3}$. $\mathcal{F}$ is the class of all subsets of Ω, and

$$\mathcal{C} = \{a \cup b, d \cup c, a \cup c, b \cup d\}.$$

Proof. Let $\mathcal{F}_0(\mathcal{C})$ be the smallest field containing $\mathcal{C}$. By the unique extension theorem it suffices to show that $Q = Q'$ on $\mathcal{F}_0(\mathcal{C})$. Let $\mathcal{D}$ be the smallest class of sets such that

i) $\mathcal{C} \subset \mathcal{D}$,
ii) $\Omega \in \mathcal{D}$,
iii) $A, B \in \mathcal{D}$, $B \subset A \Rightarrow A - B \in \mathcal{D}$.

Then $\mathcal{D} = \mathcal{F}_0(\mathcal{C})$. To see this, let $\mathcal{U}$ be the class of sets A in $\mathcal{D}$ such that $A \cap C \in \mathcal{D}$ for all $C \in \mathcal{C}$. Then notice that $\mathcal{U}$ satisfies (i), (ii), (iii) above, so $\mathcal{U} = \mathcal{D}$. This implies that $A \cap C \in \mathcal{D}$ for all $A \in \mathcal{D}$, $C \in \mathcal{C}$. Now let $\mathcal{E}$ be the class of sets E in $\mathcal{D}$ such that $A \cap E \in \mathcal{D}$, all $A \in \mathcal{D}$. Similarly $\mathcal{E}$ satisfies (i), (ii), (iii), so $\mathcal{E} = \mathcal{D}$. This yields $\mathcal{D}$ closed under $\cap$, but by (ii), (iii), $\mathcal{D}$ is also closed under complementation, proving the assertion. Let $\mathcal{G}$ be the class of sets G in $\mathcal{F}$ such that $Q(G) = Q'(G)$. Then $\mathcal{G}$ satisfies (i), (ii), (iii) $\Rightarrow \mathcal{D} \subset \mathcal{G}$ or $\mathcal{F}_0(\mathcal{C}) \subset \mathcal{G}$.

Returning to the proof of 2.22. Let $\hat{P}$, $\hat{P}'$ be defined on $\mathcal{B}_\infty$ by $P(\mathbf{X} \in B)$, $P'(\mathbf{X}' \in B)$, respectively. Let $\mathcal{C} \subset \mathcal{B}_\infty$ be the class of all sets of the form $C = \{\mathbf{x};\ x_1 < y_1, \ldots, x_n < y_n\}$. Then clearly $\mathcal{C}$ is closed under $\cap$, and $\mathcal{F}(\mathcal{C}) = \mathcal{B}_\infty$. Now $\hat{P}(C) = F_{\mathbf{X}_n}(y_1, \ldots, y_n)$ and $\hat{P}'(C) = F_{\mathbf{X}'_n}(y_1, \ldots, y_n)$, so that $\hat{P} = \hat{P}'$ on $\mathcal{C}$ by hypothesis. By 2.23 $\hat{P} = \hat{P}'$ on $\mathcal{B}_\infty$.

Another proof of 2.22 which makes it more transparent is as follows: For any function $G(x_1, \ldots, x_n)$ on $R^{(n)}$ and I an interval $[a, b)$, $a < b$, $\mathbf{x} = (x_1, \ldots, x_n)$, write

$$\underset{k}{\Delta}_I G(\mathbf{x}) = G(x_1, \ldots, x_{k-1}, b, x_{k+1}, \ldots, x_n) - G(x_1, \ldots, x_{k-1}, a, x_{k+1}, \ldots, x_n).$$

By definition, since

$$F_n(x_1, \ldots, x_n) = P(\mathsf{X}_1 < x_1, \ldots, \mathsf{X}_n < x_n),$$

the probability of any rectangle $\{\mathsf{X}_1 \in I_1, \ldots, \mathsf{X}_n \in I_n\}$ with $I_1, \ldots, I_n$ left closed, right open, can be expressed in terms of F_n by

$$\textbf{(2.24)} \qquad P(\mathsf{X}_1 \in I_1, \ldots, \mathsf{X}_n \in I_n) = \Delta^{(1)}_{I_1} \cdots \Delta^{(n)}_{I_n} F_n(\mathbf{x})$$

because, for $I_n = [a_n, b_n)$,

$$\begin{aligned} \Delta^{(n)}_{I_n} F_n(\mathbf{x}) &= P(\mathsf{X}_1 < x_1, \ldots, \mathsf{X}_n < b_n) - P(\mathsf{X}_1 < x_1, \ldots, \mathsf{X}_n < a_n) \\ &= P(\mathsf{X}_1 < x_1, \ldots, \mathsf{X}_{n-1} < x_{n-1}, \mathsf{X}_n \in I_n). \end{aligned}$$

By taking limits, we can now get the probabilities of all rectangles. From the extension theorem 2.18 we know that specifying P on rectangles uniquely determines it.

Frequently, the distribution of a process is specified by giving a set of distribution functions $\{F_n(\mathbf{x})\}$, $n = 1, 2, \ldots$ But in order that $\{F_n(\mathbf{x})\}$ be derived from a process $\{\mathsf{X}_n\}$ on a probability space $(\Omega, \mathcal{F}, P)$, they must have certain essential properties.

Proposition 2.25. *The distribution functions $F_n(\mathbf{x})$ satisfy the conditions:*

i) *Non-negativity. For finite intervals $I_k = [a_k, b_k)$, $k = 1, \ldots, n$,*

$$\Delta^{(1)}_{I_1} \cdots \Delta^{(n)}_{I_n} F_n(\mathbf{x}) \geq 0.$$

ii) *Continuity from below. If $\mathbf{x}^{(k)} = (x_1^{(k)}, \ldots, x_n^{(k)})$ and $x_j^{(k)} \uparrow x_j$, $j = 1, \ldots, n$, then*

$$F_n(\mathbf{x}^{(k)}) \uparrow F_n(\mathbf{x}).$$

iii) *Normalization. All limits of F_n exist as*

$$x_j \uparrow +\infty \qquad \text{or} \qquad x_j \downarrow -\infty.$$

If $x_j \downarrow -\infty$, then $F_n(\mathbf{x}) \to 0$. If all x_j, $j = 1, \ldots, n \uparrow +\infty$, then $F_n(\mathbf{x}) \to 1$.

The set of distribution functions are connected by

iv) *Consistency*

$$\lim_{x_n \uparrow +\infty} F_n(x_1, \ldots, x_n) = F_{n-1}(x_1, \ldots, x_{n-1}), \quad n > 1.$$

Proof. The proof of (i) follows from (2.24). To prove (ii), note that

$$\{\omega;\ \mathsf{X}_1 < x_1^{(k)}, \ldots, \mathsf{X}_n < x_n^{(k)}\} \uparrow \{\omega;\ \mathsf{X}_1 < x_1, \ldots, \mathsf{X}_n < x_n\}.$$

Use the essential fact that probabilities behave nicely under monotone limits to get (ii). Use this same fact to prove (iii) and (iv); e.g. if $x_j \downarrow -\infty$, then $\{\omega;\ \mathsf{X}_1 < x_1, \ldots, \mathsf{X}_n < x_n\} \downarrow \emptyset$. If all $x_1, \ldots, x_n \uparrow +\infty$, then

$$\{\omega;\ \mathsf{X}_1 < x_1, \ldots, \mathsf{X}_n < x_n\} \uparrow \Omega.$$

Another important construction theorem verifies that the conditions of 2.25 characterize the distribution functions of a process.

Theorem 2.26. *Given a set of functions* $\{F_n(\mathbf{x})\}$ *satisfying* 2.25 (*i*), (*ii*), (*iii*), (*iv*), *there is a process* $\{\mathsf{X}_n\}$ *on* $(\Omega, \mathcal{F}, P)$ *such that*

$$P(\mathsf{X}_1 < x_1, \ldots, \mathsf{X}_n < x_n) = F_n(x_1, \ldots, x_n).$$

Proof. The idea of how the proof should go is simple. Use $\Omega = R^{(\infty)}$, $\mathcal{F} = \mathcal{B}_\infty$, and use the coordinate representation process $\hat{\mathsf{X}}_1, \hat{\mathsf{X}}_2, \ldots$ We want to construct $\hat{P}$ on $\mathcal{B}_\infty$ such that if $S \in \mathcal{B}_\infty$ is a semi-infinite rectangle of the form

$$\{\mathbf{x} \in R^{(\infty)}; x_1 < y_1, \ldots, x_n < y_n\},$$

then

$$\hat{P}(S) = F_n(y_1, \ldots, y_n).$$

To construct $\hat{P}$ starting from F_n, define $\hat{P}$ on rectangles whose sides are left closed, right open, intervals $I_1, \ldots, I_n$ by

$$\hat{P}(\mathbf{x}; x_1 \in I_1, \ldots, x_n \in I_n) = \Delta^{(1)}_{I_1} \cdots \Delta^{(n)}_{I_n} F_n(\mathbf{x}).$$

Extend this to all rectangles by taking limits. The consistency 2.25 (iv) guarantees that $\hat{P}$ is well defined on all rectangles. All that is necessary to do now is to verify the conditions of 2.18. If S_j, S are left closed, right open rectangles, and $S_j \uparrow S$, then the continuity from below of F_n, 2.25 (ii), yields

$$\lim_j \hat{P}(S_j) = \hat{P}(S).$$

To verify the above for general rectangles, use the fact that their probabilities can be defined as limits of probabilities of left closed, right open rectangles.

The complication is in showing additivity of $\hat{P}$ on rectangles. It is sufficient to show that

$$\hat{P}(S) = \sum_j \hat{P}(S_j)$$

for left closed, right open, disjoint rectangles $S_1, \ldots, S_k$ whose union is a rectangle S. In one dimension the statement $\hat{P}(S) = \sum_j \hat{P}(S_j)$ follows from the obvious fact that for $a_0 < a_1 < \cdots < a_k$,

$$\Delta_{[a_0, a_k)} F_1(x_1) = \sum_{j=1}^{k} \Delta_{[a_{j-1}, a_j)} F_1(x_1).$$

The general result is a standard theorem in the theory of the Stieltjes integral (McShane [111a, pp. 245–246]).

If a function $F(x_1, \ldots, x_n)$ satisfies only the first three conditions of 2.25 then Theorem 2.26 implies the following.

Corollary 2.27. *There are random variables* $\mathsf{X}_1, \ldots, \mathsf{X}_n$ *on a space* $(\Omega, \mathcal{F}, P)$ *such that*

$$P(\mathsf{X}_1 < x_1, \ldots, \mathsf{X}_n < x_n) = F(x_1, \ldots, x_n).$$

Hence, any such function will be called an n-dimensional distribution function.

If a set $\{F_n\}$, $n = 1, 2, \ldots,$ of n-dimensional distribution functions satisfies 2.25 (iv), call them *consistent.* The specification of a consistent set of $\{F_n\}$ is pretty much the minimum amount of data needed to completely specify the distribution of a process in the general case.

Problems

6. For any random variable X, let $F_{\mathsf{X}}(x) = P(\mathsf{X} < x)$. The function $F_{\mathsf{X}}(x)$ is called the distribution function of the variable X.
Prove that $F_{\mathsf{X}}(x)$ satisfies

i) $x \le x' \Rightarrow F_{\mathsf{X}}(x) \le F_{\mathsf{X}}(x')$,

ii) $\lim_{x \uparrow +\infty} F_{\mathsf{X}}(x) = 1, \quad \lim_{x \downarrow -\infty} F_{\mathsf{X}}(x) = 0,$

iii) $\lim_{x_k \uparrow x} F_{\mathsf{X}}(x_k) = F_{\mathsf{X}}(x).$

7. If a function $F(x_1, \ldots, x_n)$ is nondecreasing in each variable separately, does this imply $\Delta^{(1)}_{I_1} \cdots \Delta^{(n)}_{I_n} F_n(\mathbf{x}) > 0$? Give an example of a function $F(x, y)$ such that

i) $F(x, y) \ge 0$,

ii) $x' > x, y' > y \Rightarrow F(x', y) \ge F(x, y), F(x, y') \ge F(x, y).$

iii) There are finite intervals I_1, I_2 such that

$$\Delta^{(1)}_{I_1} \Delta^{(2)}_{I_2} F(x, y) < 0.$$

8. Let $F_1(x), F_2(x), \ldots$ be functions satisfying the conditions of Problem 6. Prove that the functions

$$F_n(x_1, \ldots, x_n) = \prod_1^n F_k(x_k)$$

form a consistent set of distribution functions.

6. RANDOM VARIABLES

From now on, for reasons sufficient and necessary, we study random variables defined on a probability space. The sufficient reason is that the extension theorems state that given a fairly reasonable assignment of probabilities, a process can be constructed fitting the specified data. The necessity is that most strong limit theorems require this kind of an environment. Now we record a few facts regarding random variables and probability spaces.

Proposition 2.28. *Let $\mathcal{C}$ be a class of Borel sets such that $\mathcal{F}(\mathcal{C}) = \mathcal{B}_1$, X a real-valued function on Ω. If $\{\mathsf{X} \in C\} \in \mathcal{F}$, all $C \in \mathcal{C}$, then X is a random variable on $(\Omega, \mathcal{F})$.*

Proof. Let $\mathcal{D} \subset \mathcal{B}_1$ be the class of all Borel sets D such that $\{\mathsf{X} \in D\} \in \mathcal{F}$. $\mathcal{D}$ is a σ-field. $\mathcal{C} \subset \mathcal{D} \Rightarrow \mathcal{D} = \mathcal{B}_1$.

Corollary 2.29. *If $\{\mathsf{X} \in I\} \in \mathcal{F}$ for all intervals I, then X is a random variable.*

At times functions come up which may be infinite on some parts of Ω but which are random variables on subsets where they are finite.

Definition 2.30. *An extended random variable X on $(\Omega, \mathcal{F})$ may assume the values $\pm\infty$, but $\{\mathsf{X} \in B\} \in \mathcal{F}$, for all $B \in \mathcal{B}_1$.*

Proposition 2.31. *Let $\mathbf{X}$ be a random vector to $(R, \mathcal{B})$. If $\varphi(\mathbf{x})$ is a random variable on $(R, \mathcal{B})$, then $\varphi(\mathbf{X})$ is a random variable on $(\Omega, \mathcal{F})$, measurable $\mathcal{F}(\mathbf{X})$.*

Proof. Write, for $A \in \mathcal{B}_1$,

$$\{\varphi(\mathbf{X}) \in A\} = \{\mathbf{X} \in \varphi^{-1}(A)\} \in \mathcal{F}(\mathbf{X}),$$

$\varphi^{-1}(A)$ here denoting the inverse image of A under φ.

Definition 2.32. *For random variables $\mathsf{X}_1, \mathsf{X}_2, \ldots$ on Ω, the σ-field of all events depending on the first n outcomes is the class of sets $\{(\mathsf{X}_1, \ldots, \mathsf{X}_n) \in B\}$, $B \in \mathcal{B}_n$. Denote this by $\mathcal{F}(\mathsf{X}_1, \ldots, \mathsf{X}_n)$. The class of sets depending on only a finite number of outcomes is*

$$\mathcal{F}_0 = \bigcup_{n \geq 1} \mathcal{F}(\mathsf{X}_1, \ldots, \mathsf{X}_n).$$

In general, $\mathcal{F}_0$ is a field, but not a σ-field. But the fact that $\mathcal{F}(\mathbf{X}) = \mathcal{F}(\mathcal{F}_0)$ follows immediately from the definitions.

Proposition 2.33. *Given a process $\mathsf{X}_1, \mathsf{X}_2, \ldots$. For every set $A_1 \in \mathcal{F}(\mathbf{X})$ and $\epsilon > 0$, there is a set A_2 in some $\mathcal{F}(\mathsf{X}_1, \ldots, \mathsf{X}_n)$ such that*

$$P(A_1 \triangle A_2) \leq \epsilon.$$

($A_1 \triangle A_2$ is the symmetric set difference $(A_1 - A_2) \cup (A_2 - A_1)$.)

Proof. The proof of this is one of the standard results which cluster around the construction used in the Carathéodory extension theorem. The statement is that if P on $\mathcal{F}(\mathcal{F}_0)$ is an extension of P on $\mathcal{F}_0$, then for every set $A_1 \in \mathcal{F}(\mathcal{F}_0)$ and $\epsilon > 0$, there is a set A_2 in the field $\mathcal{F}_0$ such that $P(A_2 \triangle A_1) \leq \epsilon$ (see Appendix A.12). Then 2.33 follows because $\mathcal{F}(\mathbf{X})$ is the smallest σ-field containing

$$\bigcup_{n \geq 1} \mathcal{F}(\mathsf{X}_1, \ldots, \mathsf{X}_n).$$

If all the random variables in a process $\mathsf{X}_1, \mathsf{X}_2, \ldots$ take values in a Borel set $E \in \mathcal{B}_1$, it may be more convenient to use the range space $(E^{(\infty)}, \mathcal{B}_\infty(E))$, where $\mathcal{B}_\infty(E)$ consists of all sets in $\mathcal{B}_\infty$ which are subsets of $E^{(\infty)}$. For example, if $\mathsf{X}_1, \mathsf{X}_2, \ldots$ are coin-tossing variables, then each one takes values in $\{0, 1\}$, and the relevant R, $\mathcal{B}$ for the process is

$$(\{0, 1\}^{(\infty)}, \mathcal{B}_\infty(\{0, 1\})).$$

If a random variable X has distribution function $F(x)$, then $P(\mathsf{X} \in B)$ is a probability measure on $\mathcal{B}_1$ which is an extension of the measure on intervals $[a, b)$ given by $F(b) - F(a)$. Thus, use the notation:

Definition 2.34. *For X a random variable, denote by $P(\mathsf{X} \in dx)$ or $F(dx)$ the probability measure $P(\mathsf{X} \in B)$ on $\mathcal{B}_1$. Refer to $F(dx)$ as the distribution of X.*

Definition 2.35. *A sequence $\mathsf{X}_1, \mathsf{X}_2, \ldots$ of random variables all having the same distribution $F(dx)$ are called identically distributed. Similarly, call random vectors $\mathbf{X}_1, \mathbf{X}_2, \ldots$ with the same range space $(R, \mathcal{B})$ identically distributed if they have the common distribution*

$$\hat{P}(d\mathbf{x}) = P(\mathbf{X}_n \in d\mathbf{x}).$$

Problems

9. Show that $\mathcal{B}_\infty(\{0, 1\})$ is the smallest σ-field containing all sets of the form

$$\{\mathbf{x} \in \{0, 1\}^{(\infty)}; x_1 = s_1, \ldots, x_n = s_n\},$$

where $s_1, \ldots, s_n$ is any sequence n-long of zeros and ones, $n = 1, 2, \ldots$

10. Given a process $\mathsf{X}_1, \mathsf{X}_2, \ldots$ on $(\Omega, \mathcal{F}, P)$. Let $m_1, m_2, \ldots$ be positive integer-valued random variables on $(\Omega, \mathcal{F}, P)$. Prove that the sequence $\mathsf{X}_{m_1}, \mathsf{X}_{m_2}, \ldots$ is a process on $(\Omega, \mathcal{F}, P)$.

7. EXPECTATIONS OF RANDOM VARIABLES

Definition 2.36. *Let X be a random variable on $(\Omega, \mathcal{F}, P)$. Define the expectation of X, denoted $E\mathsf{X}$, by $\int \mathsf{X}(\omega)\, dP(\omega)$. This is well defined if $E|\mathsf{X}| < \infty$. Alternative notations for the integrals are*

$$\int \mathsf{X}\, dP, \qquad \int \mathsf{X}(\omega) P(d\omega).$$

Definition 2.37. *For any probability space $(\Omega, \mathcal{F}, P)$ define*

i) *if $A \in \mathcal{F}$, the set indicator $\chi_A(\omega)$ is the random variable*

$$\chi_A(\omega) = \begin{cases} 1, & \omega \in A, \\ 0, & \omega \in A^c. \end{cases}$$

ii) *If* X *is a random variable, then* X^+, X^- *are the random variables*

$$X^+(\omega) = \begin{cases} X(\omega), & \omega \in \{X \geq 0\}, \\ 0, & \text{otherwise;} \end{cases}$$

$$X^-(\omega) = \begin{cases} -X(\omega), & \omega \in \{X \leq 0\}, \\ 0, & \text{otherwise.} \end{cases}$$

A number of results we prove in this and later sections depend on a principle we state as

Proposition 2.38. *Consider a class* $\mathfrak{L}$ *of random variables having the properties*

i) $X, Y \in \mathfrak{L}$, $\alpha, \beta \geq 0 \Rightarrow \alpha X + \beta Y \in \mathfrak{L}$.

ii) $X_n \in \mathfrak{L}$, $X_n(\omega) \uparrow X(\omega)$ *for every* $\omega \Rightarrow X \in \mathfrak{L}$.

iii) For every set $A \in \mathcal{F}$, $\chi_A(\omega) \in \mathfrak{L}$.

Then $\mathfrak{L}$ *includes all nonnegative random variables on* $(\Omega, \mathcal{F}, P)$.

Proof. See Appendix A.22.

This is used to prove

Proposition 2.39. *Let the processes* $\mathbf{X}$ *on* $(\Omega, \mathcal{F}, P)$, $\mathbf{X}'$ *on* $(\Omega', \mathcal{F}', P')$ *have the same distribution. Then if* $\varphi(\mathbf{x})$ *is measurable* $(R^{(\infty)}, \mathcal{B}_\infty)$,

$$\textbf{(2.40)} \qquad \int_\Omega \varphi(\mathbf{X}(\omega))P(d\omega) = \int_{\Omega'} \varphi(\mathbf{X}'(\omega'))P'(d\omega'),$$

in the sense that if either side is well defined, so is the other, and the two are equal.

Proof. Consider all φ for which (2.40) is true. This class satisfies (i) and (ii) of 2.38. Further, let $B \in \mathcal{B}_\infty$ and let $\varphi(\mathbf{x}) = \chi_B(\mathbf{x})$. Then the two sides of (2.40) become $P(\mathbf{X} \in B)$ and $P'(\mathbf{X}' \in B)$, respectively. But these are equal since the processes have the same distribution. Hence (2.40) holds for all nonnegative φ. Thus, for any φ, it holds true for $|\varphi|$, φ^+, φ^-.

Corollary 2.41. *Define* $\hat{P}(\cdot)$ *on* $\mathcal{B}_\infty$ *by* $\hat{P}(B) = P(\mathbf{X} \in B)$. *Then if* φ *is measurable* $(R^{(\infty)}, \mathcal{B}_\infty)$ *and* $E\,|\varphi(\mathbf{X})| < \infty$,

$$\int \varphi(\mathbf{X})\, dP = \int \varphi(\mathbf{x})\hat{P}(d\mathbf{x}).$$

Proof. $\{X_n\}$ on $(\Omega, \mathcal{F}, P)$ and $\{\hat{X}_n\}$ on $(R^{(\infty)}, \mathcal{B}_\infty, \hat{P})$, have the same distribution, where $\hat{X}_n$ is the coordinate representation process. Thus

$$\int \varphi(X_1, X_2, \ldots)\, dP = \int \varphi(\hat{X}_1, \hat{X}_2, \ldots)\hat{P}(d\mathbf{x}),$$

but $\hat{X}_n(\mathbf{x}) = x_n$.

8. CONVERGENCE OF RANDOM VARIABLES

Given a sequence of random variables $\{X_n\}$, there are various modes of strong convergence of X_n to a limiting random variable X.

Definition 2.42

i) X_n *converges to* X *almost surely* (a.s.) *if*

$$P(\omega; \lim X_n(\omega) = X(\omega)) = 1.$$

Denote this by $X_n \xrightarrow{\text{a.s.}} X$ *or* $X_n \to X$ a.s.

ii) X_n *converges to* X *in* rth *mean, for* $r > 0$, *if* $E|X_n - X|^r \to 0$. *Denote this by* $X_n \xrightarrow{\text{r}} X$.

iii) X_n *converges in probability to* X *if for every* $\epsilon > 0$,

$$P(|X_n - X| > \epsilon) \to 0 \text{ as } n \to \infty.$$

Denote this by $X_n \xrightarrow{\text{P}} X$.

The important things to notice are:

First, all these convergences are "probabilistic." That is, if $X, X_1, \ldots$ has the same distribution as $X', X_1', \ldots$, then $X_n \to X$ in any of the above senses implies that $X_n' \to X'$ in the same sense. This is obvious for $\xrightarrow{\text{r}}, \xrightarrow{\text{P}}$. See Problem 12 for $\xrightarrow{\text{a.s.}}$.

Secondly, Cauchy convergence in any one of these senses gives convergence.

Proposition 2.43. *If* $X_m - X_n \to 0$ *in any of the above ways as* $m, n \to \infty$ *in any way, then there is a random variable* X *such that* $X_n \to X$ *in the same way.*

Proof. Do a.s. convergence first. For all ω such that $X_n(\omega)$ is Cauchy convergent, $\lim_n X_n(\omega)$ exists. Hence $P(\omega;\ \lim_n X_n(\omega) \text{ exists}) = 1$. Let $X(\omega) = \lim_n X_n(\omega)$ for all ω such that the limit exists, otherwise put it equal to zero, then $X_n \xrightarrow{\text{a.s.}} X$. For the other modes of Cauchy convergence the proof is deferred until Section 3 of the next chapter (Problems 6 and 7).

Thirdly, of these various kinds of convergences $\xrightarrow{\text{a.s.}}$ is usually the hardest to establish and more or less the strongest. To get from a.s. convergence to $\xrightarrow{\text{r}}$, some sort of boundedness condition is necessary. Recall

Theorem 2.44. (*Lebesgue bounded convergence theorem*). *If* $Y_n \xrightarrow{\text{a.s.}} Y$ *and if there is a random variable* $Z \geq 0$ *such that* $EZ < \infty$, *and* $|Y_n| \leq Z$ *for all* n, *then* $EY_n \to EY$. (See Appendix A.28).

Hence, using $Y_n = |X_n - X|^r$ in 2.44, we get

(2.45) $$X_n \xrightarrow{\text{a.s.}} X,\ |X_n|^r \le Z,\ EZ < \infty \Rightarrow X_n \xrightarrow{r} X.$$

Convergence in probability is the weakest. The implications go

(2.46) $$\text{i) } \xrightarrow{\text{a.s.}} \Rightarrow \xrightarrow{P}$$
$$\text{ii) } \xrightarrow{r} \Rightarrow \xrightarrow{P}.$$

Problems

11. Prove (2.46i and ii). [Use a generalization of Chebyshev's inequality on (ii).]

12. Let $\{X_n\}$, $\{X'_n\}$ have the same distribution. Prove that if $X_n \to X$ a.s., there is a random variable X' such that $X'_n \to X'$ a.s.

13. For a process $\{X_n\}$ prove that the set

$$\left\{\omega;\ \lim_n X_n(\omega) \text{ does not exist}\right\}$$

is an event (i.e., is in $\mathcal{F}$).

14. Prove that for X a random variable with $E|X| < \infty$, then $A_n \in \mathcal{F}$, $A_n \downarrow \emptyset$, implies

$$\lim_n \int_{A_n} X\, dP = 0.$$

NOTES

The use of a probability space $(\Omega, \mathcal{F}, P)$ as a context for probability theory was formalized by Kolmogorov [98], in a monograph published in 1933. But, as Kolmogorov pointed out, the concept had already been current for some time.

Subsequent work in probability theory has proceeded, almost without exception, from this framework. There has been controversy about the correspondence between the axioms for a probability space and more primitive intuitive notions of probability. A different approach in which the probability of an event is defined as its asymptotic frequency is given by von Mises [112]. The argument can go on at several levels. At the top is the contention that although it seems reasonable to assume that P is a finite probability, there are no strong intuitive grounds for assuming it σ-additive. Thus, in their recent book [40] Dubins and Savage assume only finite additivity, and even within this weaker framework prove interesting limit theorems. One level down is the question of whether a probability measure P need be additive at all. The more basic property is argued to be: $A \subset B \Rightarrow P(A) \le P(B)$. But, as always, with weaker assumptions fewer nontrivial theorems can be proven.

At the other end, it happens that some σ-fields have so many sets in them that examples occur which disturb one's intuitive concept of probability. Thus, there has been some work in the direction of restricting the type of σ-field to be considered. An interesting article on this is by Blackwell [8].

For a more detailed treatment of measure theory than is possible in Appendix A, we recommend the books of Neveu [113], Loève [108], and Halmos [64].

CHAPTER 3

INDEPENDENCE

Independence, or some form of it, is one of the central concepts of probability, and it is largely responsible for the distinctive character of probability theory.

1. BASIC DEFINITIONS AND RESULTS

Definition 3.1

(a) *Given random variables* $X_1, \ldots, X_n$, *on* $(\Omega, \mathcal{F}, P)$, *they are said to be independent if for any sets* $B_1, \ldots, B_n \in \mathcal{B}_1$,

$$P(X_1 \in B_1, \ldots, X_n \in B_n) = \prod_1^n P(X_k \in B_k).$$

(b) *Given a probability space* $(\Omega, \mathcal{F}, P)$ *and* σ*-fields* $\mathcal{F}_1, \ldots, \mathcal{F}_n$ *contained in* $\mathcal{F}$, *they are said to be independent if for any sets* $A_1 \in \mathcal{F}_1, \ldots, A_n \in \mathcal{F}_n$

$$P(A_1 \cap \cdots \cap A_n) = \prod_1^n P(A_k).$$

Obviously, $X_1, \ldots, X_n$ *are independent random variables iff* $\mathcal{F}(X_1), \ldots, \mathcal{F}(X_n)$ *are independent* σ*-fields.*

These definitions have immediate generalizations to random vectors.

Definition 3.2. *Random vectors* $\mathbf{X}_1, \mathbf{X}_2, \ldots, \mathbf{X}_n$ *are said to be independent if the* σ*-fields* $\mathcal{F}(\mathbf{X}_1), \mathcal{F}(\mathbf{X}_2), \ldots, \mathcal{F}(\mathbf{X}_n)$, *are independent.*

Virtually all the results of this section stated for independent random variables hold for independent random vectors. But as the generalization is so apparent, we usually omit it.

Definition 3.3. *The random variables* $X_1, X_2, \ldots$ *are called independent if for every* $n \geq 2$, *the random variables* $X_1, X_2, \ldots, X_n$, *are independent.*

Proposition 3.4. *Let* $X_1, X_2, \ldots$ *be independent random variables and* $B_1, B_2, \ldots$ *any sets in* $\mathcal{B}_1$. *Then*

$$P(X_1 \in B_1, X_2 \in B_2, \ldots) = \prod_1^\infty P(X_k \in B_k).$$

Proof. Let $A_n = \{X_1 \in B_1, \ldots, X_n \in B_n\}$. Then A_n are decreasing, hence

$$P(\lim A_n) = \lim P(A_n).$$

But,

$$\lim A_n = \{X_1 \in B_1, X_2 \in B_2, \ldots\}$$

and

$$\lim P(A_n) = \lim_n \prod_1^n P(X_k \in B_k) = \prod_1^\infty P(X_k \in B_k).$$

Note that the same result holds for independent σ-fields.

Proposition 3.5. *Let $X_1, X_2, \ldots$ be independent random variables, $(i_1, i_2, \ldots)$, $(j_1, j_2, \ldots)$ disjoint sets of integers. Then the fields*

$$\mathcal{F}_1 = \mathcal{F}(X_{i_1}, X_{i_2}, \ldots), \mathcal{F}_2 = \mathcal{F}(X_{j_1}, X_{j_2}, \ldots)$$

are independent.

Proof. Consider any set $D \in \mathcal{F}_2$ of the form $D = \{X_{j_1} \in B_1, \ldots, X_{j_m} \in B_m\}$, $B_k \in \mathcal{B}_1, k = 1, \ldots, m$. Define two measures Q_1 and Q_1' on $\mathcal{F}_1$ by, for $A \in \mathcal{F}_1$,

$$Q_1(A) = P(A \cap D), \qquad Q_1'(A) = P(A)P(D).$$

Consider the class of sets $\mathcal{C} \subset \mathcal{F}_1$ of the form

$$C = \{X_{i_1} \in E_1, \ldots, X_{i_n} \in E_n\}, E_l \in \mathcal{B}_1, l = 1, \ldots, n.$$

Note that

$$\begin{aligned} Q_1(C) &= P\left(\bigcap_{l=1}^n \{X_{i_l} \in E_l\} \bigcap_{k=1}^m \{X_{j_k} \in B_k\}\right) \\ &= \prod_{l=1}^n P(X_{i_l} \in E_l) \prod_{k=1}^m P(X_{j_k} \in B_k) \\ &= P(C)P(D) = Q_1'(C). \end{aligned}$$

Thus $Q_1 = Q_1'$ on $\mathcal{C}$, $\mathcal{C}$ is closed under $\cap$, $\mathcal{F}(\mathcal{C}) = \mathcal{F}_1 \Rightarrow Q_1 = Q_1'$ on $\mathcal{F}_1$ (see 2.23). Now repeat the argument. Fix $A \in \mathcal{F}_1$ and define $Q_2(\cdot)$, $Q_2'(\cdot)$ on $\mathcal{F}_2$ by $P(A \cap \cdot)$, $P(A)P(\cdot)$. By the preceding, for any D of the form given above, $Q_2(D) = Q_2'(D)$, implying $Q_2 = Q_2'$ on $\mathcal{F}_2$ and thus for any $A_1 \in \mathcal{F}_1$, $A_2 \in \mathcal{F}_2$,

$$P(A_1 \cap A_2) = P(A_1)P(A_2).$$

Corollary 3.6. *Let $X_1, X_2, \ldots$ be independent random variables, $J_1, J_2, \ldots$ disjoint sets of integers. Then the σ-fields $\mathcal{F}_k = \mathcal{F}(\{X_j\}, j \in J_k)$ are independent.*

Proof. Assume that $\mathcal{F}_1, \ldots, \mathcal{F}_n$ are independent. Let

$$J = \bigcup_1^n J_k, \qquad \text{and} \qquad \mathcal{F}' = \mathcal{F}(\{X_j\}, j \in J).$$

Since J and J_{n+1} are disjoint, $\mathcal{F}'$ and $\mathcal{F}_{n+1}$ satisfy the conditions of Proposition 3.5 above. Let $A_1 \in \mathcal{F}_1, \ldots, A_n \in \mathcal{F}_n$. Then since $\mathcal{F}_k \subset \mathcal{F}'$, $k = 1, \ldots, n$,

$$A_1 \cap \cdots \cap A_n \in \mathcal{F}'.$$

Let $A' = A_1 \cap \cdots \cap A_n$ and $A_{n+1} \in \mathcal{F}_{n+1}$. By 3.5,

$$P(A_{n+1} \cap A') = P(A_{n+1})P(A') = P(A_{n+1})P(A_n) \cdots P(A_1)$$

by the induction hypothesis.

From 3.5 and 3.6 we extract more concrete and interesting consequences. For instance, 3.5 implies that the fields $\mathcal{F}(X_1, \ldots, X_n)$ and $\mathcal{F}(X_{n+1}, \ldots)$ are independent. As another example, if $\varphi_1, \varphi_2, \ldots$ are measurable $(R^{(m)}, \mathcal{B}_m)$, then the random variables

$$Z_1 = \varphi_1(X_1, \ldots, X_m), \quad Z_2 = \varphi_2(X_{m+1}, \ldots, X_{2m}), \ldots$$

are independent. Another way of stating 3.6 is to say that the random vectors $\mathbf{X}_k = (\{X_j\}, j \in J_k)$, are independent.

How and when do we get independent random variables?

Theorem 3.7. *A necessary and sufficient condition for* $X_1, X_2, \ldots,$ *to be independent random variables is that for every n, and n-tuple* $(x_1, \ldots, x_n)$

$$F_{\mathbf{X}_n}(x_1, \ldots, x_n) = F_{X_1}(x_1) \cdots F_{X_n}(x_n).$$

Proof. It is obviously necessary—consider the sets $\{X_1 < x_1, \ldots, X_n < x_n\}$. To go the other way, we want to show that for arbitrary $B_1, \ldots, B_n \in \mathcal{B}_1$,

$$P(X_1 \in B_1, \ldots, X_n \in B_n) = \prod_1^n P(X_k \in B_k).$$

Fix $x_2, \ldots, x_n$ and define two σ-additive measures Q and Q' on $\mathcal{B}_1$ by

$$Q_1(B) = P(X_1 \in B, X_2 < x_2, \ldots, X_n < x_n),$$

$$Q_1'(B) = P(X_1 \in B)P(X_2 < x_2, \ldots, X_n < x_n).$$

Now on all sets of the form $(-\infty, x)$, Q and Q' agree, implying that $Q = Q'$ on $\mathcal{B}_1$. Repeat this by fixing $B_1 \in \mathcal{B}_1, x_3, \ldots, x_n$ and defining

$$Q_2(B) = P(X_1 \in B_1, X_2 \in B, X_3 < x_3, \ldots, X_n < x_n),$$

$$Q_2'(B) = P(X_1 \in B_1)P(X_2 \in B)P(X_3 < x_3, \ldots, X_n < x_n),$$

so $Q_2 = Q_2'$ on the sets $(-\infty, x)$, hence $Q_2 = Q_2'$ on $\mathcal{B}_1$, and continue on down.

The implication of this theorem is that if we have any one-dimensional distribution functions $F_1(x), F_2(x), \ldots$ and we form the consistent set of distribution functions (see Problem 8, Chapter 2) $F_1(x_1) \cdots F_n(x_n)$, then any resulting process $X_1, X_2, \ldots$ having these distribution functions consists of independent random variables.

Proposition 3.8. *Let* X *and* Y *be independent random variables, f and g* $\mathcal{B}_1$-*measurable functions such that* $E\,|f(X)| < \infty$, $E\,|g(Y)| < \infty$, *then*

$$E\,|f(X)g(Y)| < \infty$$

and

$$E\,[f(X)g(Y)] = [Ef(X)][Eg(Y)].$$

Proof. For any set $A \in \mathcal{B}_1$, take $f(x) = \chi_A(x)$; and consider the class $\mathcal{L}$ of nonnegative functions $g(y)$ for which the equality in 3.8 holds. $\mathcal{L}$ is closed under linear combinations. By the Lebesgue monotone convergence theorem applied to both sides of the equation in 3.8, $g_n \in \mathcal{L}$, $g_n \uparrow g \Rightarrow g \in \mathcal{L}$. If $B \in \mathcal{B}_1$ and $g(y) = \chi_B(y)$, then the equation becomes

$$P(X \in A, Y \in B) = P(X \in A)P(Y \in B),$$

which holds by independence of X and Y. By 2.38, $\mathcal{L}$ includes all nonnegative g-measurable $\mathcal{B}_1$. Now fix g and apply to f to conclude that 3.8 is valid for all nonnegative g and f.

For general g and f, note that 3.8 holding for nonnegative g and f implies $E\,|f(X)g(Y)| = [E\,|f(X)|][E\,|g(Y)|]$, so integrability of $f(X)$ and $g(Y)$ implies that of $f(X)g(Y)$. By writing $f = f^+ - f^-$, $g = g^+ - g^-$ we obtain the general result.

Note that if X and Y are independent, then so are the random variables $f(X)$ and $g(Y)$. So actually the above proposition is no more general than the statement: *Let* X *and* Y *be independent random variables,* $E\,|X| < \infty$, $E\,|Y| < \infty$, *then* $E\,|XY| < \infty$ *and* $EXY = EX \cdot EY$.

By induction, we get

Corollary 3.9. *Let* $X_1, \ldots, X_n$ *be independent random variables such that* $E\,|X_k| < \infty$, $k = 1, \ldots, n$. *Then*

$$E\,|X_1 \cdots X_n| < \infty \qquad \text{and} \qquad E\,(X_1 \cdots X_n) = \prod_1^n EX_k.$$

Proof. Follows from 3.8 by induction.

Problems

1. Let $\mathcal{F}_1$, $\mathcal{F}_2$ be independent σ-fields. Show that if a set A is both in $\mathcal{F}_1$ and $\mathcal{F}_2$, then $P(A) = 0$ or 1.

2. Use Fubini's theorem (Appendix A.37) to show that for X and Y independent random variables

a) for any $B \in \mathcal{B}_1$, $P(X \in B - y)$ is a $\mathcal{B}_1$-measurable function of y,

b) $P(X + Y \in B) = \int P(X \in B - y)P(Y \in dy)$.

2. TAIL EVENTS AND THE KOLMOGOROV ZERO-ONE LAW

Consider the set E again, on which $S_n/n \not\to \frac{1}{2}$ for fair coin-tossing. As pointed out, this set has the odd property that whether or not $\omega \in E$ does not depend on the first n coordinates of ω no matter how large n is. Sets which have this fascinating property we call tail events.

Definition 3.10. *Let* $X_1, X_2, \ldots$ *be any process. A set* $E \in \mathcal{F}(\mathbf{X})$ *will be called a tail event if* $E \in \mathcal{F}(X_n, X_{n+1}, \ldots)$, *all* n. *Equivalently, let* $\mathfrak{T}$ *be the* σ*-field* $\bigcap_{n=1}^{\infty} \mathcal{F}(X_n, X_{n+1}, \ldots)$, *then* $\mathfrak{T}$ *is called the tail* σ*-field and any set* $E \in \mathfrak{T}$ *is called a tail event.*

This definition may seem formidable, but it captures formally the sense in which certain events do not depend on any finite number of their coordinates. For example,

$$E = \left\{\omega; \frac{X_1 + \cdots + X_n}{n} \not\to \frac{1}{2}\right\}$$

is a tail event. Because for any $k \geq 1$,

$$E = \left\{\omega; \frac{X_k + \cdots + X_n}{n} \not\to \frac{1}{2}\right\},$$

hence $E \in \mathcal{F}(X_k, X_{k+1}, \ldots)$ for all $k \geq 1$, $\Rightarrow E \in \mathfrak{T}$.

An important class of tail events is given as follows:

Definition 3.11. *Let* $X_1, X_2, \ldots$ *be any process,* $B_1, B_2, \ldots$ *Borel sets. The set* X_n *in* B_n *infinitely often, denoted* $\{X_n \in B_n \text{ i.o.}\}$ *is the set* $\{\omega;\ {}_nX(\omega) \in B_n$ *occurs for an infinite number of* $n\}$. *Equivalently,*

$$\{X_n \in B_n \text{ i.o.}\} = \lim_{m \uparrow \infty} \bigcup_{n=m}^{\infty} \{X_n \in B_n\}.$$

It is fairly apparent that for many strong limit theorems the events involved will be tail. Hence it is most gratifying that the following theorem is in force.

Theorem 3.12. *(Kolmogorov zero-one law). Let* $X_1, X_2, \ldots$ *be independent random variables. Then if* $E \in \mathfrak{T}$, $P(E)$ *is either zero or one.*

Proof. $E \in \mathcal{F}(\mathbf{X})$. By 2.33, there are sets $E_n \in \mathcal{F}(X_1, \ldots, X_n)$ such that $P(E_n \,\triangle\, E) \to 0$. This implies $P(E_n) \to P(E)$, and $P(E_n \cap E) \to P(E)$. But $E \in \mathcal{F}(X_{n+1}, X_{n+2}, \ldots)$, hence E and E_n are in independent σ-fields. Thus $P(E_n \cap E) = P(E_n)P(E)$. Taking limits in this latter equation gives

$$P(E) = [P(E)]^2.$$

The only solutions of $x = x^2$ are $x = 0$ or 1. Q.E.D.

This is really a heart-warming result. It puts us into secure business with strong limit theorems for independent random variables involving tail events. Either the theorem holds true for almost all $\omega \in \Omega$ or it fails almost surely.

Problems

3. Show that $\{X_n \in B_n \text{ i.o.}\}$ is a tail event.

4. In the coin-tossing game let $\mathbf{s}$ be any sequence m-long of zeros or ones. Let Z_n be the vector $(X_{n+1}, \ldots, X_{n+m})$, and F the set $\{Z_n = \mathbf{s} \text{ i.o.}\}$. Show that $F \in \mathfrak{J}$.

5. (the random signs problem). Let c_n be any sequence of real numbers. In the fair coin-tossing game let $Y_n = \pm 1$ as the nth toss is H or T. Let $D = \{\omega;\ \Sigma\, c_n Y_n \text{ converges}\}$; show that $D \in \mathfrak{J}$.

3. THE BOREL-CANTELLI LEMMA

Every tail event has probability zero or one. Now the important question is: how to decide which is which. The Borel-Cantelli lemma is a most important step in that direction. It applies to a class of events which includes many tail-events, but it also has other interesting applications.

Definition 3.13. *In $(\Omega, \mathcal{F}, P)$, let $A_n \in \mathcal{F}$. The set $\{A_n \text{ i.o.}\}$ is defined as $\{\omega;\ \omega \in A_n$ for an infinite number of $n\}$, or equivalently*

$$\{A_n \text{ i.o.}\} = \lim_m \bigcup_{n>m} A_n.$$

Borel-Cantelli Lemma 3.14

I. *The direct half. If $A_n \in \mathcal{F}$, then $\sum_1^\infty P(A_n) < \infty$* implies $P(A_n \text{ i.o.}) = 0$.

To state the second part of the Borel-Cantelli lemma we need

Definition 3.15. *Events $A_1, A_2, \ldots,$ in $(\Omega, \mathcal{F}, P)$ will be called independent events if the random variables $\chi_{A_1}, \chi_{A_2}, \ldots$ are independent* (*see Problem* 8).

II. *The converse half. If $A_n \in \mathcal{F}$ are independent events then $\sum_1^\infty P(A_n) = \infty$* implies $P(A_n \text{ i.o.}) = 1$.

Proof of I

$$P(A_n \text{ i.o.}) = P\left(\lim_m \bigcup_m^\infty A_n\right) = \lim_m P\left(\bigcup_m^\infty A_n\right) \le \lim_m \left(\sum_m^\infty P(A_n)\right).$$

But obviously $\sum_1^\infty P(A_n) < \infty$ implies that $\sum_m^\infty P(A_n) \to 0$, as $m \to \infty$.

Proof of II

$P\left(\bigcup_m^\infty A_n\right) = 1 - P\left(\bigcap_m^\infty A_n^c\right)$. We show that $P\left(\bigcap_m^\infty A_n^c\right) = 0$. Because the events $\{A_n\}$ are independent,

$$P\left(\bigcap_m^\infty A_n^c\right) = \prod_m^\infty P(A_n^c) = \prod_m^\infty (1 - P(A_n)).$$

Use the inequality $\log(1 - x) \le -x$ to get

$$\log\left(\prod_m^\infty (1 - P(A_n))\right) = \sum_m^\infty \log(1 - P(A_n)) \le -\sum_m^\infty P(A_n) = -\infty.$$

Application 1. In coin-tossing, let $\mathbf{s}$ be any sequence k-long of H, T.

$$A_n = \{\omega;\ (\omega_n, \ldots, \omega_{n+k-1}) = \mathbf{s}\},\quad 0 < P(\text{Heads}) < 1.$$

Proposition 3.16. $P(A_n \text{ i.o.}) = 1$.

Proof. Let $B_1 = \{\omega;\ (\omega_1, \ldots, \omega_k) = \mathbf{s}\}$, $B_2 = \{\omega;\ (\omega_{k+1}, \ldots, \omega_{2k}) = \mathbf{s}\}, \ldots$ The difficulty is that the A_n are not independent events because of the overlap, for instance, between A_1 and A_2, but the B_n are independent, and $\{A_n \text{ i.o.}\} \supset \{B_n \text{ i.o.}\}$. Now $P(B_n) = P(B_1) > 0$, so $\sum_1^\infty P(B_n) = \infty$, implying by 3.14(II) that $P(B_n \text{ i.o.}) = 1$.

Another way of putting this proposition is that in coin-tossing (biased or not), given any finite sequence of H, T's, this sequence will occur an infinite number of times as the tossing continues, except on a set of sequences of probability zero.

Application 2. Again, in coin-tossing, let $\mathsf{Y}_i = \pm 1$, as ith toss is H or T, $\mathsf{Z}_n = \mathsf{Y}_1 + \cdots + \mathsf{Y}_n$. If $\mathsf{Z}_n = 0$, we say that an equalization (or return to the origin) takes place at time n. Let $A_n = \{\mathsf{Z}_n = 0\}$. Then $\{A_n \text{ i.o.}\} = \{\omega;$ an infinite number of equalizations occur$\}$.

Proposition 3.17. *If $P(Heads) \neq \frac{1}{2}$, then* $P(\mathsf{Z}_n = 0 \text{ i.o.}) = 0$.

Proof. Immediate, from the Borel-Cantelli lemma and the asymptotic expression for $P(\mathsf{Z}_n = 0)$.

Another statement of 3.17 is that in biased coin-tossing, as we continue tossing, we eventually come to a last equalization and past this toss there are no more equalizations. What if the coin is fair?

Theorem 3.18. *For a fair coin,* $P(\mathsf{Z}_n = 0 \text{ i.o.}) = 1$.

Proof. The difficulty, of course, is that the events $A_n = \{\mathsf{Z}_n = 0\}$ are not independent, so 3.14 is not directly applicable. In order to get around this, we manufacture a most pedestrian proof, which is typical of the way in which the Borel-Cantelli lemma is stretched out to cover cases of nonindependent events.

The idea of the proof is this; we want to apply the converse part of the Borel-Cantelli lemma, but in order to do this we can look only at the random variables X_k related to disjoint stretches of tosses. That is, if we consider a subsequence $n_1 < n_2 < n_3 < \cdots$ of the integers, then any events $\{C_k\}$ such that each C_k depends only on $\{\mathsf{Y}_{n_k+1}, \mathsf{Y}_{n_k+2}, \ldots, \mathsf{Y}_{n_{k+1}}\}$ are independent events to which the Borel-Cantelli lemma applies. Suppose, for instance, that we select $n_k < m_k < n_{k+1}$ and define

$$C_k = \{\mathsf{Y}_{n_k+1} + \cdots + \mathsf{Y}_{m_k} \leq -n_k\} \cap \{\mathsf{Y}_{m_k+1} + \cdots + \mathsf{Y}_{n_{k+1}} \geq m_k\}.$$

The purpose of defining C_k this way is that we know

$$\mathsf{Z}_{n_k} = \mathsf{Y}_1 + \cdots + \mathsf{Y}_{n_k} \leq n_k,$$

because each Y_i is ± 1. Hence $\omega \in C_k \Rightarrow \mathsf{Z}_{m_k} \leq 0$. Again $\mathsf{Z}_{m_k} \geq -m_k$, so, in addition,

$$\{\omega \in C_k\} \Rightarrow \mathsf{Z}_{n_{k+1}} \geq 0.$$

Therefore $\{\omega \in C_k\} \Rightarrow \{\mathsf{Z}_n = 0$ at least once for $n_k + 1 \leq n \leq n_{k+1}\}$. We have used here a standard trick in probability theory of considering stretches $n_1, n_2, \ldots$ so far apart that the effect of what happened previously to n_k is small as compared to the amount that Z_n can change between n_k and n_{k+1}. Now

$$\{C_k \text{ i.o.}\} \subset \{\mathsf{Z}_n = 0 \text{ i.o.}\}.$$

So we need only to prove now that the n_k, m_k can be selected in such a way that $\sum_1^\infty P(C_k) = \infty$.

Assertion: Given any number α, $0 < \alpha < 1$, *and integer* $k \geq 1$, $\exists$ *an integer* $\varphi(k) \geq 1$ *such that*

$$P(|\mathsf{Z}_{\varphi(k)}| < k) \leq \alpha.$$

Proof. We know that for any fixed j,

$$P(\mathsf{Z}_n = j) \to 0.$$

Hence for k fixed, as $n \to \infty$,

$$\sum_{|j|<k} P(\mathsf{Z}_n = j) \to 0.$$

Simply take $\varphi(k)$ sufficiently large so that

$$\sum_{|j|<k} P(\mathsf{Z}_{\varphi(k)} = j) \leq \alpha.$$

Define n_k, m_k as follows: $n_1 = 1$, $m_k = n_k + \varphi(n_k)$, $n_{k+1} = m_k + \varphi(m_k)$. Compute $P(C_k)$ as follows:

$$P(C_k) = P(\mathsf{Y}_{n_k+1} + \cdots + \mathsf{Y}_{m_k} \leq -n_k)P(\mathsf{Y}_{m_k+1} + \cdots + \mathsf{Y}_{n_{k+1}} \geq m_k).$$

By symmetry,

$$P(C_k) = \tfrac{1}{4}P(|\mathsf{Y}_{n_k+1} + \cdots + \mathsf{Y}_{m_k}| \geq n_k)P(|\mathsf{Y}_{m_k+1} + \cdots + \mathsf{Y}_{n_{k+1}}| \geq m_k).$$

Thus, since the distribution of the vector $(Y_{i+1}, \ldots, Y_{i+j})$ is the same as that of $(Y_1, \ldots, Y_j)$,

$$\begin{aligned} P(C_k) &= \tfrac{1}{4}P(|Y_1 + \cdots + Y_{m_k - n_k}| \geq n_k)P(|Y_1 + \cdots + Y_{n_{k+1}-m_k}| \geq m_k) \\ &= \tfrac{1}{4}P(|Y_1 + \cdots + Y_{\varphi(n_k)}| \geq n_k)P(|Y_1 + \cdots + Y_{\varphi(m_k)}| \geq m_k) \\ &\geq \tfrac{1}{4}(1 - \alpha)^2. \end{aligned}$$

Q.E.D.

This proof is a bit of a mess. Now let me suggest a much more exciting possibility. Suppose we can prove that $P(Z_n = 0$ at least once$) = 1$. Now every time there is an equalization, everything starts all over again. That is, if $Z_n = 0$, then the game starts from the $(n + 1)$ toss as though it were beginning at $n = 0$. Consequently, we are sure now to have at least one more equalization. Continue this argument now ad infinitum to conclude that $P(Z_n = 0$ at least once$) = 1 \Rightarrow P(Z_n = 0$ i.o.$) = 1$. We make this argument hold water when 3.18 is generalized in Section 7, and generalize it again in Chapter 7.

Problems

6. Show, by using 3.14, that $X_n \xrightarrow{P} X \Rightarrow \exists$ a subsequence X_{n_k} such that $X_{n_k} \xrightarrow{\text{a.s.}} X$.

7. Show, using 3.14, that if $X_n - X_m \xrightarrow{P} 0$, $\exists$ a random variable X such that $X_n \xrightarrow{P} X$. [Hint: Take $\epsilon_k \downarrow 0$ and n_k such that for $m, n \geq n_k$,

$$P(|X_n - X_m| > \epsilon_k) \leq \frac{1}{2^k}.$$

Now prove that there is a random variable X such that $X_{n_k} \xrightarrow{\text{a.s.}} X$.]

8. In order that events $A_1, A_2, \ldots$ be independent, show it is sufficient that

$$P(A_{i_1} \cap \cdots A_{i_m}) = P(A_{i_1}) \cdots P(A_{i_m})$$

for every finite subcollection $A_{i_1}, \ldots, A_{i_m}$. [One interesting approach to the required proof is: Let $\mathfrak{D}$ be the smallest field containing $A_1, \ldots, A_N$. Define Q on $\mathfrak{D}$ by $Q(B_1 \cap \cdots \cap B_N) = P(B_1) \cdots P(B_N)$, where the sets B_k are equal to A_k or A_k^c. Use $P(A_{i_1} \cap \cdots \cap A_{i_m}) = P(A_{i_1}) \cdots P(A_{i_m})$ to show that $P = Q$ on a class of sets to which 2.23 can be applied. Conclude that $P = Q$ on $\mathfrak{D}$.]

9. Use the strong law of large numbers in the form $S_n/n \to p$ a.s. to prove 3.17.

10. Let $X_1, X_2, \ldots$ be independent identically distributed random variables. Prove that $E|X_1| < \infty$ if and only if

$$P(|X_n)| > n \text{ i.o.}) = 0.$$

(See Loève [108, p. 239].)

4. THE RANDOM SIGNS PROBLEM

In Problem 5, it is shown for $Y_1, Y_2, \ldots,$ independent $+1$ or -1 with probability $\frac{1}{2}$, that the set $\{\omega; \sum_1^n c_k Y_k \text{ converges}\}$ is a tail event. Therefore it has probability zero or one. The question now is to characterize the sequences $\{c_n\}$ such that $\sum_1^n c_k Y_k$ converges a.s.

This question is naturally arrived at when you look at the sequence $1/n$, that is, $\Sigma\, 1/n$ diverges, but $\Sigma\, (-1)^n 1/n$ converges. Now what happens if the signs are chosen at random? In general, look at the consecutive sums $\sum_1^n X_k$ of any sequence $X_1, X_2, \ldots$ of independent random variables. The convergence set is again a tail event. When does it have probability one?

The basic result here is that in this situation convergence in probability implies the much stronger convergence almost surely.

Theorem 3.19. *For $X_1, X_2, \ldots$ independent random variables,*

$$\sum_1^n X_k \xrightarrow{P} \Leftrightarrow \sum_1^n X_k \xrightarrow{\text{a.s.}} \tag{3.20}$$

Proof. Proceeds by an important lemma which is due to Skorokhod, [125].

Lemma 3.21. *Let $S_1, \ldots, S_N$ be successive sums of independent random variables such that* $\sup_{j \le N} P(|S_N - S_j| > \alpha) = c < 1$. *Then*

$$P\left(\sup_{j \le N} |S_j| > 2\alpha\right) \le \frac{1}{1-c} P(|S_N| > \alpha).$$

Proof. Let $j^*(\omega) = \{\text{first } j \text{ such that } |S_j| > 2\alpha\}$. Then

$$P\left(|S_N| > \alpha,\quad \sup_{j \le N} |S_j| > 2\alpha\right) = \sum_{j=1}^N P(|S_N| > \alpha,\quad j^* = j)$$

$$\ge \sum_{j=1}^N P(|S_N - S_j| \le \alpha,\quad j^* = j).$$

The set $\{j^* = j\}$ is in $\mathcal{F}(X_1, \ldots, X_j)$, and $S_N - S_j$ is measurable $\mathcal{F}(X_{j+1}, \ldots, X_N)$, so the last sum on the right above equals

$$\sum_{j=1}^N P(|S_N - S_j| \le \alpha) P(j^* = j).$$

Use $P(|S_N - S_j| \le \alpha) \ge 1 - c$ to get

$$P\left(|S_N| > \alpha,\quad \sup_{j \le N} |S_j| > 2\alpha\right) \ge (1-c) \sum_{j=1}^N P(j^* = j)$$

$$= (1-c) P\left(\sup_{j \le N} |S_j| > 2\alpha\right).$$

The observation that

$$P(|S_N| > \alpha, \sup_{j \le N} |S_j| > 2\alpha) \le P(|S_N| > \alpha)$$

completes the proof.

To finish the proof of 3.19: If a sequence s_n of real numbers does not converge, then there exists an $\epsilon > 0$ such that for every m,

$$\sup_{n; n > m} |s_n - s_m| > \epsilon.$$

So if $\sum_1^n X_k$ diverges with positive probability then there exists an $\epsilon > 0$ and $\delta > 0$ such that for every m fixed,

$$P\left(\sup_{n>m} \left|\sum_{m+1}^n X_k\right| > \epsilon\right) \ge \delta.$$

By (3.21), keeping m, N fixed

$$P\left(\sup_{m<n\le N} \left|\sum_{m+1}^n X_k\right| > \epsilon\right) \le \frac{1}{1 - C_{m,N}} P\left(\left|\sum_{m+1}^N X_k\right| > \frac{\epsilon}{2}\right),$$

where

$$C_{m,N} = \sup_{m<n\le N} P\left(\left|\sum_n^N X_k\right| > \frac{\epsilon}{2}\right).$$

If $\sum_1^n X_k$ is convergent in probability, then $\sum_1^N X_k - \sum_1^m X_k \xrightarrow{P} 0$ as $m,N \to \infty$. Hence, as $m,N \to \infty$,

$$P\left(\left|\sum_{m+1}^N X_k\right| > \frac{\epsilon}{2}\right) \to 0,$$

so we find that

$$C_{m,N} \to 0.$$

Taking first $N \to \infty$, conclude

$$\lim_m P\left(\sup_{n>m} \left|\sum_{m+1}^n X_k\right| > \epsilon\right) = 0.$$

This contradiction proves the theorem.

We can use convergence in second mean to get an immediate criterion.

***Corollary* 3.22.** *If* $EX_k = 0$, *all* k, *and* $\sum_1^\infty EX_k^2 < \infty$, *then the sums* $\sum_1^n X_k$ *converge* a.s.

In particular, for the random signs problem, mentioned at the beginning of this section, the following corollary holds.

Corollary 3.23. *A sufficient condition for the sums* $\sum_1^n c_k Y_k$ *to converge* a.s. *is* $\sum_1^\infty c_k^2 < \infty$.

The open question is necessity. Marvelously enough, the converse of 3.23 is true, so that $\Sigma\, c_k Y_k$ converges if and only if $\Sigma\, c_k^2 < \infty$. In fact, a partial converse of 3.22 holds.

Theorem 3.24. *Let* $X_1, X_2, \ldots$ *be independent random variables such that* $EX_k = 0$, *and* $|X_k| \le \alpha < \infty$, *all* k. *Then* $\sum_1^n X_k$ *converges* a.s. *implies* $\sum_1^\infty EX_k^2 < \infty$.

Proof. For any $\lambda > 0$, define $n^*(\omega)$ by

$$n^*(\omega) = \begin{cases} \text{1st } n \text{ such that } \left|\sum_1^n X_k\right| > \lambda, \\ +\infty \text{ if no such } n \text{ exists,} \end{cases}$$

where n^* is an extended random variable. For any integers $j < N$, look at

$$\int_{\{n^*=j\}} \left(\sum_1^N X_k\right)^2 dP.$$

Since $\{n^* = j\} \in \mathcal{F}(X_1, \ldots, X_j)$, then by independence, and $EX_k = 0$, all k,

$$\int_{\{n^*=j\}} \left(\sum_1^N X_k\right)^2 dP = \int_{\{n^*=j\}} \left(\sum_1^j X_k\right)^2 dP + \sum_{j+1}^N \int_{\{n^*=j\}} X_k^2\, dP.$$

On $\{n^* = j\}$, $\left|\sum_1^{j-1} X_k\right| \le \lambda$. Hence since $|X_j| \le \alpha$, $\left|\sum_1^j X_k\right| \le \lambda + \alpha$. And, by independence, for $k > j$,

$$\int_{\{n^*=j\}} X_k^2\, dP = P(n^* = j) EX_k^2.$$

Using these,

$$\int_{\{n^*=j\}} \left(\sum_1^N X_k\right)^2 dP \le (\lambda + \alpha)^2 P(n^* = j) + P(n^* = j) \sum_1^N EX_k^2.$$

Sum on j from 1 up to N to get

$$\int_{\{n^*\le N\}} \left(\sum_1^N X_k\right)^2 dP \le (\lambda + \alpha)^2 P(n^* \le N) + P(n^* \le N) \cdot \sum_1^N EX_k^2.$$

Also,

$$\int_{\{n^*>N\}} \left(\sum_1^N X_k\right)^2 dP \le \lambda^2 P(n^* > N) \le (\lambda + \alpha)^2 P(n^* > N).$$

Adding this to the above inequality we get

$$\sum_1^N EX_k^2 \leq (\lambda + \alpha)^2 + P(n^* \leq N) \sum_1^N EX_k^2$$

or

$$P(n^* > N) \sum_1^N EX_k^2 \leq (\lambda + \alpha)^2.$$

Letting $N \to \infty$, we find that

$$P(n^* = \infty) \cdot \sum_1^\infty EX_k^2 \leq (\lambda + \alpha)^2.$$

But, since $\sum_1^n X_k$ converges a.s., then there must exist a λ such that

$$P\left(\sup_n \left| \sum_1^n X_k \right| \leq \lambda\right) > 0,$$

implying $P(n^* = \infty) > 0$ and $\sum_1^\infty EX_k^2 < \infty$.

The results of 3.24 can be considerably sharpened. But why bother; elegant necessary and sufficient conditions exist for the convergence of sums $\sum_1^n X_k$ where the only assumption made is that the X_k are independent. This is the "three-series" theorem of Kolmogorov (see Loève [108, p. 237]). More on this will appear in Chapter 9. Kac [82] has interesting analytic proofs of 3.23 and its converse.

Problems

11. Let $X_1, X_2, \ldots$ be independent, and $X_k \geq 0$. If for some δ, $0 < \delta < 1$, there exists an x such that

$$\int_{\{X_k > x\}} X_k \, dP \leq \delta EX_k,$$

for all k, show that

$$\sum_1^\infty X_k < \infty \text{ a.s.} \Rightarrow \sum_1^\infty EX_k < \infty.$$

Give an example to show that in general, $X_1, X_2, \ldots$ independent non-negative random variables and $\sum_1^\infty X_k < \infty$ a.s. does not imply that

$$\sum_1^\infty EX_k < \infty.$$

12. Let $Y_1, Y_2, \ldots$ be a process. We will say that the integer-valued random variables $m_1, m_2, \ldots$ are *optional skipping* variables if

i) $1 < m_1 < m_2 < \cdots$,
ii) $\{m_k = j\} \in \mathcal{F}(Y_1, \ldots, Y_{j-1})$

(i.e., the decision as to which game to play next depends only on the previous outcomes). Denote $\check{Y}_k = Y_{m_k}$. Show that

a) If the $Y_1, Y_2, \ldots$ are independent and identically distributed then the sequence $\check{Y}_1, \check{Y}_2, \ldots$ has the same distribution as $Y_1, Y_2, \ldots$
b) For $Y_1, Y_2, \ldots$ as in (a), show that the sequence $Y_{m_1}, Y_{m_1+1}, Y_{m_1+2}, \ldots$ has the same distribution as $Y_1, Y_2, \ldots$
c) Give an example where the $Y_1, Y_2, \ldots$ are independent, but the $\check{Y}_1, \check{Y}_2, \ldots$ are not independent.

5. THE LAW OF PURE TYPES

Suppose that $X_1, X_2, \ldots$ are independent and $\sum_1^n X_k$ converges a.s. What can be said about the distribution of the limit $X = \sum_1^\infty X_k$? In general, very little! In the nature of things the distribution of X can be anything. For example, let $X_k = 0$, $k > 1$, then $X = X_1$. There is one result available here which is an application of the Kolmogorov zero-one law and remarkable for its simplicity and elegance.

Definition 3.25. *A random variable X is said to have a distribution of pure type if either*

i) *There is a countable set D such that $P(X \in D) = 1$,*
ii) *$P(X = x) = 0$ for every $x \in R^{(1)}$, but there is a set $D \in \mathcal{B}_1$ of Lebesgue measure zero such that $P(X \in D) = 1$, or*
iii) *$P(X \in dx) \ll l(dx)$(Lebesgue measure)*

[Recall that $\mu \ll \nu$ for two measures μ, ν denotes μ absolutely continuous with respect to ν; see Appendix A.29].

Theorem 3.26 *(Jessen-Wintner law of pure types* [78]). *Let $X_1, X_2, \ldots$ be independent random variables such that*

i) $\sum_1^n X_k \to X$ a.s.,
ii) *For each k, there is a countable set F_k such that $P(X_k \in F_k) = 1$.*

Then the distribution of X is of pure type.

Proof. Let $F = \bigcup_{k \geq 1} F_k$. Take G to be the smallest additive group in $R^{(1)}$ containing F. G consists of all numbers of the form $m_1 x_1 + \cdots + m_j x_j$,

$x_1, \ldots, x_j \in F$ and $m_1, \ldots, m_j$ integers. F is countable, hence G is countable. For any set $B \subset R^{(1)}$ write

$$G \oplus B = \{x \in R^{(1)};\ x = x_1 + x_2, x_1 \in G, x_2 \in B\}.$$

Note that

i) B countable $\Rightarrow G \oplus B$ countable,
ii) $B \in \mathcal{B}_1 \Rightarrow G \oplus B \in \mathcal{B}_1$,
iii) $B \in \mathcal{B}_1$, $l(B) = 0 \Rightarrow l(G \oplus B) = 0$.

For $B \in \mathcal{B}_1$, and $C = \{\omega, \sum_1^n \mathsf{X}_k \text{ converges}\}$, consider the event

$$A = \{\mathsf{X} \in G \oplus B\} \cap C.$$

The point is that A is a tail event. Because if $x_1 - x_2 \in G$, then

$$x_1 \in G \oplus B \Leftrightarrow x_2 \in G \oplus B.$$

But $\mathsf{X} - \sum_n^\infty \mathsf{X}_k \in G$ for all ω in C. Hence

$$A = \left\{\sum_n^\infty \mathsf{X}_k \in G \oplus B\right\} \cap C,\ n = 1, 2, \ldots$$

By the zero-one law $P(A) = 0$ or 1. This gives the alternatives:

a) Either there is a countable set D such that $P(\mathsf{X} \in D) = 1$ or $P(\mathsf{X} \in G \oplus B) = 0$, hence $P(\mathsf{X} \in B) = 0$, for all countable sets B.
b) If the latter in (a) holds, then either there is a set $D \in \mathcal{B}_1$ such that $l(D) = 0$ and $P(\mathsf{X} \in D) = 1$ or $P(\mathsf{X} \in G \oplus B) = 0$, for all $B \in \mathcal{B}_1$ such that $l(B) = 0$.
c) In this latter case $B \in \mathcal{B}_1$, $l(B) = 0 \Rightarrow P(\mathsf{X} \in B) = 0$, that is, the distribution is absolutely continuous with respect to $l(dx)$.

Theorem 3.26 gives no help as to which type the distribution of the limit random variable belongs to. In particular, for $\mathsf{Y}_1, \ldots$ independent ± 1 with probability $\frac{1}{2}$, the question of the type of the distribution of the sums $\sum_1^\infty c_n \mathsf{Y}_n$ is open. Some important special cases are given in the following problems.

Problems

13. Show that $P\left(\sum_1^\infty \mathsf{Y}_k/2^k \in dx\right)$ is Lebesgue measure on [0, 1]. [Recall the analytic model for coin-tossing.]

14. If X and Y are independent random variables, use Problem 2 to show that $P(\mathsf{X} \in dx) \ll l(dx) \Rightarrow P(\mathsf{X} + \mathsf{Y} \in dx) \ll l(dx)$.

15. Use Problems 13 and 14 to show that the distribution of

$$\sum_1^\infty \mathsf{Y}_k/k \quad \text{is} \quad \ll l(dx).$$

16. Show that if independent random variables $\mathsf{X}_1, \mathsf{X}_2, \ldots$ take values in a countable set F, and if there are constants $\alpha_n \in F$ such that

$$\sum_1^\infty P(\mathsf{X}_n \neq \alpha_n) < \infty \quad \text{and} \quad \sum_1^\infty \alpha_n < \infty,$$

then the sum $\mathsf{X} = \sum_1^\infty \mathsf{X}_n$ has distribution concentrated on a countable number of points. (The converse is also true; see Lévy [101].)

6. THE LAW OF LARGE NUMBERS FOR INDEPENDENT RANDOM VARIABLES

From the random signs problem, by some juggling, we can generalize the law of large numbers for independent random variables.

Theorem 3.27. *Let $\mathsf{X}_1, \mathsf{X}_2, \ldots$ be independent random variables, $E\mathsf{X}_k = 0$, $E\mathsf{X}_k^2 < \infty$. Let $b_n \geq 0$ converge up to $+\infty$. If $\sum_1^\infty E\mathsf{X}_k^2/b_k^2 < \infty$, then*

$$\frac{\mathsf{X}_1 + \cdots + \mathsf{X}_n}{b_n} \xrightarrow{\text{a.s.}} 0.$$

Proof. To prove this we need:

Kronecker's Lemma 3.28. *Let $x_1, x_2, \ldots$ be a sequence of real numbers such that $\sum_1^n x_k \to s$ finite. Take $b_n \uparrow \infty$, then*

$$\frac{1}{b_n}\sum_1^n b_k x_k \to 0.$$

Proof. Let $r_n = \sum_{n+1}^\infty x_k$, $r_0 = s$; then $x_n = r_{n-1} - r_n$, $n = 1, 2, \ldots$, and

$$\sum_1^n b_k x_k = \sum_1^n b_k(r_{k-1} - r_k) = \sum_0^{n-1} b_{k+1} r_k - \sum_1^n b_k r_k$$
$$= \sum_1^{n-1} (b_{k+1} - b_k) r_k + b_1 s - b_n r_n$$

Thus

$$\left|\sum_1^n b_k x_k\right| \leq \sum_1^{n-1} (b_{k+1} - b_k)\,|r_k| + b_1\,|s| + b_n\,|r_n|. \tag{3.29}$$

For any $\epsilon > 0$, take N such that $|r_k| \le \epsilon$ for $k \ge N$. Then letting $\bar{r} = \max_{n\ge 1} |r_n|$ we get

$$\sum_1^{n-1} (b_{k+1} - b_k)\,|r_k| \le \sum_{k=1}^{N-1} (b_{k+1} - b_k)\,|r_k| + \epsilon \sum_N^{n-1} (b_{k+1} - b_k)$$
$$\le \bar{r}(b_N - b_1) + \epsilon(b_n - b_N).$$

Divide (3.29) by b_n, and take lim sup, noting that $b_N/b_n \to 0$, $|r_n| \to 0$, to get

$$\overline{\lim_n} \left| \frac{1}{b_n} \sum_1^n b_k x_k \right| \le \epsilon.$$

To prove 3.27, by Kronecker's lemma, if $\sum_1^n (\mathsf{X}_k/b_k)$ converges a.s., then

$$\frac{1}{b_n} \sum \mathsf{X}_k \to 0 \text{ a.s.}$$

By 3.22, it is sufficient that

$$\sum_1^\infty E(\mathsf{X}_k/b_k)^2 = \sum_1^\infty E\mathsf{X}_k^2/b_k^2 < \infty.$$

Q.E.D.

As a consequence of 3.27, if the $\mathsf{X}_1, \mathsf{X}_2, \ldots$ are identically distributed, $E\mathsf{X}_1 = 0$, and $E\mathsf{X}_1^2 = \sigma^2 < \infty$, then

$$\sum_1^\infty \frac{1}{b_k^2} < \infty \Rightarrow \frac{\mathsf{X}_1 + \cdots + \mathsf{X}_n}{b_n} \to 0 \text{ a.s.}$$

This is stronger than 1.21, which gives the same conclusion for $b_n = n$. For example, we could take $b_n = n^{1/2+\epsilon}$, any $\epsilon > 0$; or $b_n = \sqrt{n} \log n$. But the strong law of large numbers is basically a first-moment theorem.

Theorem 3.30. *Let $\mathsf{X}_1, \mathsf{X}_2, \ldots$ be independent and identically distributed random variables; if $E\,|\mathsf{X}_1| < \infty$ then*

$$\frac{\mathsf{X}_1 + \cdots + \mathsf{X}_n}{n} \xrightarrow{\text{a.s.}} E\mathsf{X}_1;$$

if $E\,|\mathsf{X}_1| = \infty$, then the above averages diverge almost everywhere.

Proof. In order to apply 3.27 define truncated random variables $\tilde{\mathsf{X}}_n$ by

$$\tilde{\mathsf{X}}_n = \begin{cases} \mathsf{X}_n, & \text{if } |\mathsf{X}_n| \le n, \\ 0, & |\mathsf{X}_n| > n. \end{cases}$$

By Problem 10, $P(|\mathsf{X}_n| > n \text{ i.o.}) = 0$. Hence (3.30) is equivalent to

$$\frac{\tilde{\mathsf{X}}_1 + \cdots + \tilde{\mathsf{X}}_n}{n} \xrightarrow{\text{a.s.}} E\mathsf{X}_1.$$

But $E\tilde{X}_n \to EX_1$, so (3.30) will follow if 3.27 can be applied to show that

$$\frac{1}{n}\sum_1^n (\tilde{X}_k - E\tilde{X}_k) \to 0.$$

Since $E(\tilde{X}_k - E\tilde{X}_k)^2 \le E\tilde{X}_k^2$, it is sufficient to show that

$$\sum_1^\infty \frac{1}{n^2} E\tilde{X}_n^2 = \sum_1^\infty \frac{1}{n^2}\int_{|x|\le n} x^2 F(dx) < \infty.$$

This follows from writing the right-hand side as

$$\sum_{n=1}^\infty \sum_{k=1}^n \frac{1}{n^2} \int_{k-1<|x|\le k} x^2 F(dx).$$

Interchange order of summation, and use $\sum_k^\infty 1/n^2 < 2/k$, $k \ge 1$, to get

$$\sum_1^\infty \frac{1}{n^2} E\tilde{X}_k^2 \le \sum_{k=1}^\infty \frac{2}{k} \int_{k-1<|x|\le k} x^2 F(dx) \le 2E\,|X_1|.$$

For the converse, suppose that S_n/n converges on a set of positive probability. Then it converges a.s. The contradiction is that

$$\frac{X_n}{n} = \frac{S_n}{n} - \left(\frac{n-1}{n}\right)\frac{S_{n-1}}{n-1}$$

must converge a.s. to zero, implying $P(|X_n| > n \text{ i.o.}) = 0$. This is impossible by Problem 10.

7. RECURRENCE OF SUMS

Through this section let $X_1, X_2, \ldots$ be a sequence of independent, identically distributed random variables. Form the successive sums $S_1, S_2, \ldots$

Definition 3.31. *For $x \in R^{(1)}$, call x a recurrent state if for every neighborhood I of x, $P(S_n \in I \text{ i.o.}) = 1$.*

The problem is to characterize the set of recurrent states. In coin-tossing, with $X_1, X_2, \ldots$ equaling ± 1 with probability p, q, if $p \ne \frac{1}{2}$, then $P(S_n = 0 \text{ i.o.}) = 0$. In fact, the strong law of large numbers implies that for any state j, $P(S_n = j \text{ i.o.}) = 0$—no states are recurrent. For fair coin-tossing, every time S_n returns to zero, the probability of entering the state j is the same as it was at the start when $n = 0$. It is natural to surmise that in this case $P(S_n = j \text{ i.o.}) = 1$ for all j. But we can use this kind of reasoning for any distribution, that is, if there is any recurrent state, then all states should be recurrent.

Definition 3.32. *Say that a random variable* X *is distributed on the lattice* $L_d = \{nd\}, n = 0, \pm 1, \ldots, d$ *any real number* >0, *if* $\Sigma_n P(\mathsf{X} = nd) = 1$ *and there is no smaller lattice having this property. If* X *is not distributed on any lattice, it is called nonlattice. In this case, say that it is distributed on* L_0, *where* $L_0 = R^{(1)}$.

Theorem 3.33. *If* $\mathsf{X}_1, \mathsf{X}_2, \ldots$ *are distributed on* L_d, $d \geq 0$, *then either every state in* L_d *is recurrent, or no states are recurrent.*

Proof. Let G be the set of recurrent points. Then G is closed. Because $x_n \in G$, $x_n \to x$ implies that for every neighborhood I of x, $x_n \in I$ for n sufficiently large. Hence $P(\mathsf{S}_n \in I \text{ i.o.}) = 1$.

Define y to be a *possible* state if for every neighborhood I of y, $\exists\, k$ such that $P(\mathsf{S}_k \in I) > 0$. I assert that

$$x \text{ recurrent}, \quad y \text{ possible} \Rightarrow x - y \text{ recurrent}.$$

To show this, take any $\epsilon > 0$, and k such that $P(|\mathsf{S}_k - y| < \epsilon) > 0$. Then

$$\begin{aligned} &P(|\mathsf{S}_n - x| < \epsilon \text{ finitely often}) \\ &\qquad \geq P(|\mathsf{S}_k - y| < \epsilon, |\mathsf{S}_{k+n} - \mathsf{S}_k - (x - y)| < 2\epsilon \text{ finitely often}) \\ &\qquad = P(|\mathsf{S}_k - y| < \epsilon) P(|\mathsf{S}_n - (x - y)| < 2\epsilon \text{ finitely often}). \end{aligned}$$

The left side is zero, implying

$$P(|\mathsf{S}_n - (x - y)| < 2\epsilon \text{ finitely often}) = 0.$$

If G is not empty, it contains at least one state x. Since every recurrent state is a possible state, $x - x = 0 \in G$. Hence G is a group, and therefore is a closed subgroup of $R^{(1)}$. But the only closed subgroups of $R^{(1)}$ are the lattices L_d, $d \geq 0$. For every possible state y, $0 - y \in G \Rightarrow y \in G$. For $d > 0$, this implies $L_d \subset G$, hence $L_d = G$. If X_1 is non-lattice, G cannot be a lattice, so $G = R^{(1)}$.

A criterion for which alternative holds is established in the following theorem.

Theorem 3.34. *Let* $\mathsf{X}_1, \mathsf{X}_2, \ldots$ *be distributed on* L_d, $d \geq 0$. *If there is a finite interval* J, $J \cap L_d \neq \emptyset$ *such that* $\sum_1^\infty P(\mathsf{S}_n \in J) < \infty$, *then no states are recurrent. If there is a finite interval* J *such that* $\sum_1^\infty P(S_n = J) = \infty$, *then all states in* L_d *are recurrent.*

Proof. If $\sum_1^\infty P(\mathsf{S}_n \in J) < \infty$, use the Borel-Cantelli lemma to get $P(\mathsf{S}_n \in J \text{ i.o.}) = 0$. There is at least one state in L_d that is not recurrent, hence none are. To go the other way, we come up against the same difficulty as in proving 3.18, the successive sums $\mathsf{S}_1, \mathsf{S}_2, \ldots$ are not independent. Now we make essential use of the idea that every time one of the sums

$S_1, S_2, \ldots$ comes back to the neighborhood of a state x, the subsequent process behaves nearly as if we had started off at the state x at $n = 0$. If $\sum_1^\infty P(S_n \in J) = \infty$, for any $\epsilon > 0$ and less than half the length of J, there is a subinterval $I = (x - \epsilon, x + \epsilon) \subset J$ such that $\sum_1^\infty P(S_n \in I) = \infty$. Define sets

$$A_k = \begin{cases} \{S_k \in I, S_{n+k} \notin I, n = 1, 2, \ldots\}, & k \geq 1, \\ \{S_n \notin I, n = 1, 2, \ldots\}, & k = 0. \end{cases}$$

A_k is the set on which the last visit to I occurred at the kth trial. Then

$$\{S_n \in I \quad \text{finitely often}\} = \bigcup_0^\infty A_k.$$

The A_k are disjoint, hence

$$P(S_n \in I \text{ finitely often}) = \sum_0^\infty P(A_k).$$

For $k \geq 1$,

$$P(A_k) \geq P(S_k \in I, |S_{n+k} - S_k| \geq 2\epsilon, n = 1, 2, \ldots).$$

Use independence, then identical distribution of the $\{X_n\}$ to get

$$\begin{aligned} P(A_k) &\geq P(S_k \in I)P(|S_{n+k} - S_k| \geq 2\epsilon, \quad n = 1, 2, \ldots) \\ &= P(S_k \in I)P(|S_n| \geq 2\epsilon, \quad n = 1, 2, \ldots). \end{aligned}$$

This inequality holds for all $k \geq 1$. Thus

$$P(S_n \in I \text{ finitely often}) \geq P(|S_n| \geq 2\epsilon, n = 1, 2, \ldots) \sum_1^\infty P(S_k \in I).$$

Since $\sum_1^\infty P(S_k \in I) = \infty$, we conclude that for every $\epsilon > 0$

(3.35) $$P(|S_n| \geq 2\epsilon, \quad n = 1, 2, \ldots) = 0.$$

Now take $I = (-\epsilon, +\epsilon)$, and define the sets A_k as above. Denote $I_\delta = (-\delta, +\delta)$, so that

$$A_k = \lim_{\delta \uparrow \epsilon} \{S_k \in I_\delta, S_{n+k} \notin I, n = 1, 2, \ldots\}, \quad k \geq 1.$$

Since the sequence of sets is monotone,

$$P(A_k) = \lim_{\delta \uparrow \epsilon} P(S_k \in I_\delta, S_{n+k} \notin I, n = 1, 2, \ldots), \quad k \geq 1.$$

Now use (3.35) in getting

$$\begin{aligned} &P(S_k \in I_\delta, S_{n+k} \notin I, n = 1, 2, \ldots) \\ &\leq P(S_k \in I_\delta, |S_{n+k} - S_k| \geq \epsilon - \delta, n = 1, \ldots) \\ &= P(S_k \in I_\delta)P(|S_n| \geq \epsilon - \delta, n = 1, \ldots) = 0 \end{aligned}$$

to conclude $P(A_k) = 0$, $k \geq 1$. Use (3.35) directly to establish

$$P(A_0) = P(|S_n| \geq \epsilon, n = 1, 2, \ldots) = 0.$$

Therefore $P(S_n \in I \text{ finitely often}) = 0$, and the sums S_n enter every neighborhood of the origin infinitely often with probability one. So the origin is a recurrent state, and consequently all states in L_d are recurrent. Q.E.D.

Look again at the statement of this theorem. An immediate application of the Borel-Cantelli lemma gives a zero-one property:

Corollary 3.36. *Either*

$$P(S_n \in I \text{ i.o.}) = 1$$

for all finite intervals I such that $L_d \cap I \neq \emptyset$; or

$$P(S_n \in I \text{ i.o.}) = 0$$

for all such I.

Definition 3.37. *If the first alternative in 3.36 holds, call the process $S_1, S_2, \ldots$ recurrent. If the second holds, call it transient.*

A quick corollary of 3.34 is a proof that fair coin-tossing is recurrent. Simply use the estimate $P(S_{2n} = 0) \sim 1/\sqrt{\pi n}$ to deduce that

$$\sum_1^\infty P(S_n = 0)$$

diverges.

The criterion for recurrence given in 3.34 is difficult to check directly in terms of the distribution of $X_1, X_2, \ldots$ A slightly more workable expression will be developed in Chapter 8. There is one important general result, however. If

$$E|X_1| < \infty \quad \text{and} \quad EX_1 \neq 0,$$

then by the law of large numbers, the sums are transient. If $EX_1 = 0$, the issue is in doubt. All that is known is that $S_n = o(n)$ [$o(n)$ denoting small order of n]. There is no reason why the sums should behave as regularly as the sums in coin-tossing. (See Problem 17 for a particularly badly-behaved example of successive sums with zero means.) But, at any rate,

Theorem 3.38. *If $EX_1 = 0$, then the sums $S_1, S_2, \ldots$ are recurrent.*

Proof. First, we need to prove

Proposition 3.39. *If I is any interval of length a, then*

$$\sum_1^\infty P(S_n \in I) \leq 1 + \sum_1^\infty P(|S_n| \leq a).$$

Proof. Denote $N = \sum_1^\infty \chi_I(\mathsf{S}_n)$, so that N counts the number of times that the sums enter the interval I. Define an extended random variable n^* by

$$\mathsf{n}^* = \begin{cases} \text{1st } n \text{ such that } \mathsf{S}_n \in I, \\ \infty \quad \text{if} \quad \mathsf{S}_n \notin I, \ n = 1, 2, \ldots \end{cases}$$

$$EN = \sum_1^\infty \int_{\{\mathsf{n}^* = k\}} N \, dP.$$

On $\{\mathsf{n}^* = k\}$, $\chi_I(\mathsf{S}_n) = 0$, $n < k$, and $\chi_I(\mathsf{S}_k) = 1$. Thus, on $\{\mathsf{n}^* = k\}$, denoting by $I - y$ the interval I shifted left a distance y,

$$\begin{aligned} N &\leq 1 + \sum_{n=1}^\infty \chi_{I - \mathsf{S}_k}(\mathsf{S}_{n+k} - \mathsf{S}_k) \\ &\leq 1 + \sum_{n=1}^\infty \chi_{[-a, +a]}(\mathsf{S}_{n+k} - \mathsf{S}_k). \end{aligned}$$

Since $\{\mathsf{n}^* = k\} \in \mathcal{F}_1(\mathsf{X}_1, \ldots, \mathsf{X}_k)$,

$$EN \leq P(\mathsf{n}^* < \infty)\left[1 + \sum_1^\infty P(|\mathsf{S}_n| \leq a)\right].$$

We use 3.39 to prove 3.38 as follows: For any positive integer M,

$$\begin{aligned} \sum_1^\infty P(-M < \mathsf{S}_n \leq M) &= \sum_{k=-M}^{M-1} \sum_1^\infty P(\mathsf{S}_n \in (k, k+1)) \\ &\leq 2M\left(1 + \sum_1^\infty P(|\mathsf{S}_n| \leq 1)\right). \end{aligned}$$

Hence

$$\overline{\lim} \frac{1}{2M} \sum_1^\infty P(|\mathsf{S}_n| < M) \leq 1 + \sum_1^\infty P(|\mathsf{S}_n| \leq 1). \tag{3.40}$$

The strong law of large numbers implies the weaker result $\mathsf{S}_n/n \xrightarrow{P} 0$, or $P(|\mathsf{S}_n| < \epsilon n) \to 1$ for every $\epsilon > 0$. Fix ϵ, and take m so that $P(|\mathsf{S}_n| < \epsilon n) > \frac{1}{2}$, $n > m$. Then $P(|\mathsf{S}_n| < M) > \frac{1}{2}$, $m \leq n < M/\epsilon$, which gives

$$\sum_1^\infty P(|\mathsf{S}_n| < M) \geq \tfrac{1}{2}(M/\epsilon - m),$$

$$(1/2M) \sum_1^\infty P(|\mathsf{S}_n| < M) \geq 1/4\epsilon - m/4M.$$

Substituting this into (3.40), we get

$$1 + \sum_1^\infty P(|\mathsf{S}_n| \leq 1) \geq 1/4\epsilon.$$

Since ϵ is arbitrary, conclude that

$$\sum_1^\infty P(|S_n| \le 1) = \infty.$$

By 3.34 the sums are recurrent. Q.E.D.

Problems

17. Unfavorable "fair" game, due to Feller [59, Vol. I, p. 246]. Let $X_1, X_2, \ldots$ be independent and identically distributed, and take values in $\{0, 2, 2^2, 2^3, \ldots\}$ so that

$$P(X_1 = 2^k) = \frac{1}{2^k k(k+1)},$$

and define $P(X_1 = 0)$ to make the sum unity. Now $EX_1 = 1$, but show that for every $\epsilon > 0$,

$$P\left(S_n - n < -\frac{(1-\epsilon)n}{\log_2 n}\right) \to 1.$$

18. Consider k fair coin-tossing games being carried on independently of each other, giving rise to the sequence of random variables

$$\begin{array}{l} Y_1^{(1)}, Y_2^{(1)}, \ldots, \\ Y_1^{(2)}, Y_2^{(2)}, \ldots, \\ \vdots \\ Y_1^{(k)}, Y_2^{(k)}, \ldots, \end{array}$$

where $Y_n^{(j)}$ is ± 1 as the nth outcome of the jth game is H or T. Let $Z_n^{(j)} = Y_1^{(j)} + \cdots + Y_n^{(j)}$, and plot the progress of each game on one axis of $R^{(k)}$. The point described is $\mathbf{Z}_n = (Z_n^{(1)}, \ldots, Z_n^{(k)}) = \mathbf{Y}_1 + \cdots + \mathbf{Y}_n$, where $\mathbf{Y}_1$ takes any one of the values $(\pm 1, \ldots, \pm 1)$ with probability $1/2^k$. Denote $\mathbf{0} = (0, 0, \ldots, 0)$. Now $\mathbf{Z}_n = \mathbf{0}$ only if equalization takes place in all k games simultaneously. Show that

$$P(\mathbf{Z}_n = \mathbf{0} \text{ i.o.}) = \begin{cases} 1, & k = 1, 2, \\ 0, & k \ge 3. \end{cases}$$

8. STOPPING TIMES AND EQUIDISTRIBUTION OF SUMS

Among the many nice applications of the law of large numbers, I am going to pick one. Suppose that $X_1, X_2, \ldots$ are distributed on the lattice L_d and the sums are recurrent. For any interval I such that $I \cap L_d \ne \emptyset$, the number of $S_1, \ldots, S_n$ falling into I goes to ∞. Denote this number by $N_n(I)$. Then

N_n/n, the average number of landings in I per unit time, goes to zero in general (see Problem 20).

An interesting result is that the points $\mathsf{S}_1, \ldots, \mathsf{S}_n$ become a.s. uniformly distributed in the sense that for any two finite intervals

$$\frac{\mathsf{N}_n(I_1)}{\mathsf{N}_n(I_2)} \xrightarrow{\text{a.s.}} \frac{\|I_1\|}{\|I_2\|}.$$

(In the lattice case, define $\|I\|$ as the number of points of L_d in $I \cap L_d$). This equidistribution is clearly a strong property. The general proof is not elementary (see Harris and Robbins [69]). But there is an interesting proof in the lattice case which introduces some useful concepts. The idea is to look at the number of landings of the S_n sequence in I between successive zeros of S_n.

Definition 3.41. *A positive, integer-valued random variable* n^* *is called a stopping time for the sums* $\mathsf{S}_1, \mathsf{S}_2, \ldots$ *if*

$$\{\mathsf{n}^* = k\} \in \mathcal{F}(\mathsf{S}_1, \ldots, \mathsf{S}_k).$$

The field of events $\mathcal{F}(\mathsf{S}_k, k \leq \mathsf{n}^*)$ *depending on* S_n *up to time of stopping consists of all* $A \in \mathcal{F}(\mathbf{X})$ *such that*

$$A \cap \{\mathsf{n}^* = k\} \in \mathcal{F}(\mathsf{S}_1, \ldots, \mathsf{S}_k).$$

For example, in recurrent case, $d = 1$, let n^* be the first entry of the sums $\{\mathsf{S}_n\}$ into the state j. This is a stopping time. More important, once at state j, the process continues by adding independent random variables, so that $\mathsf{S}_{\mathsf{n}^*+k} - \mathsf{S}_{\mathsf{n}^*}$ should have the same distribution as S_k and be independent of anything that happened up to time n^*.

Proposition 3.42. *If* n^* *is a stopping time, then the process* $\breve{\mathsf{S}}_k = \mathsf{S}_{\mathsf{n}^*+k} - \mathsf{S}_{\mathsf{n}^*}$, $k = 1, \ldots$ *has the same distribution as* S_k, $k = 1, \ldots$ *and* $\mathcal{F}(\breve{\mathsf{S}}_1, \breve{\mathsf{S}}_2, \ldots)$ *is independent of* $\mathcal{F}(\mathsf{S}_k, k \leq \mathsf{n}^*)$.

Proof. Let $A \in \mathcal{F}(\mathsf{S}_k, k \leq \mathsf{n}^*)$, $B \in \mathcal{B}_\infty$, and write

$$P(\breve{\mathbf{S}} \in B, A) = \sum_{n=1}^{\infty} P(\breve{\mathbf{S}} \in B, A, \mathsf{n}^* = n).$$

On the set $\{\mathsf{n}^* = n\}$, the $\breve{\mathsf{S}}_k$ process is equal to the $\mathsf{S}_{k+n} - \mathsf{S}_n$ process, and $A \cap \{\mathsf{n}^* = n\} \in \mathcal{F}(\mathsf{S}_1, \ldots, \mathsf{S}_n)$. Since $\mathsf{S}_{k+n} - \mathsf{S}_n, k = 1, 2, \ldots$ has the same distribution as S_k, $k = 1, 2, \ldots$

$$P(\breve{\mathbf{S}} \in B, A) = P(\mathbf{S} \in B) \sum_{n=1}^{\infty} P(A, \mathsf{n}^* = n) = P(\mathbf{S} \in B) P(A).$$

Note that n^* itself is measurable $\mathcal{F}(\mathsf{S}_k, k \leq \mathsf{n}^*)$.

Definition 3.43. *The times of the zeros of* S_n *are defined by*

$$\begin{aligned} \mathsf{R}_1 &= \min\,\{n;\ \mathsf{S}_n = 0, n > 0\}, \\ \mathsf{R}_2 &= \min\,\{n;\ \mathsf{S}_n = 0, n > \mathsf{R}_1\}, \\ &\ \ \vdots \\ \mathsf{R}_k &= \min\,\{n;\ \mathsf{S}_n = 0, n > \mathsf{R}_{k-1}\}. \end{aligned}$$

R_k *is called the* kth *occurrence time of* $\{\mathsf{S}_n = 0\}$. *The times between zeros are defined by* $\mathsf{T}_1 = \mathsf{R}_1$, $\mathsf{T}_2 = \mathsf{R}_2 - \mathsf{R}_1, \ldots$

The usefulness of these random variables is partially accounted for by

Proposition 3.44. *If* $P(\mathsf{S}_n = 0 \text{ i.o.}) = 1$, *then the* $\mathsf{T}_1, \mathsf{T}_2, \ldots$ *are independent and identically distributed random variables.*

Proof. T_1 is certainly a stopping time. By 3.42, $\breve{\mathsf{S}}_k = \mathsf{S}_{k+T_1} - \mathsf{S}_{T_1}$ has the same distribution as S_k, but this process is independent of T_1. Thus, T_2, which is the first equalization time for the $\breve{\mathsf{S}}_k$ process, is independent of T_1 and has the same distribution. Repeat this argument for $k = 3, \ldots$.

Theorem 3.45. *Let* $\mathsf{X}_1, \mathsf{X}_2, \ldots$ *have lattice distance one, and* $P(\mathsf{S}_n = 0 \text{ i.o.}) = 1$. *Then, for any two states,* j, l,

$$\frac{\mathsf{N}_n(j)}{\mathsf{N}_n(l)} \xrightarrow{\text{a.s.}} 1.$$

Proof. Let $\mathsf{R}_1, \mathsf{R}_2, \ldots$ be the times of successive returns to the origin, $\mathsf{T}_1, \mathsf{T}_2, \ldots$ the times between return. The $\mathsf{T}_1, \mathsf{T}_2, \ldots$ are independent and identically distributed. Let $\mathsf{M}_1(j)$ be the number of landings in j before the first return to the origin, $\mathsf{M}_2(j)$, the number between the first and second returns, etc. The $\mathsf{M}_1, \mathsf{M}_2, \ldots$ are similarly independent and identically distributed (see Problem 22). The law of large numbers could be applied to $(\mathsf{M}_1 + \cdots + \mathsf{M}_k)/k$ if we knew something about $E\mathsf{M}_1(j)$. Denote $\pi(j) = E\mathsf{M}_1(j)$, and assume for the moment that for all $j \in L_1$, $0 < \pi(j) < \infty$. Since $\mathsf{M}_1(j) + \cdots + \mathsf{M}_k(j) = \mathsf{N}_{\mathsf{R}_k}(j)$,

$$\frac{\mathsf{N}_{\mathsf{R}_k}(j)}{\mathsf{N}_{\mathsf{R}_k}(l)} \xrightarrow{\text{a.s.}} \frac{\pi(j)}{\pi(l)}, \qquad j \neq 0, \quad l \neq 0.$$

This gives convergence of $\mathsf{N}_n(j)/\mathsf{N}_n(l)$ along a random subsequence. To get convergence over the full sequence, write

$$\max_{0 \le m \le \mathsf{T}_{k+1}} \left| \frac{\mathsf{N}_{\mathsf{R}_k+m}(j)}{\mathsf{N}_{\mathsf{R}_k+m}(l)} - \frac{\mathsf{N}_{\mathsf{R}_k}(j)}{\mathsf{N}_{\mathsf{R}_k}(l)} \right| \le \frac{\mathsf{N}_{\mathsf{R}_{k+1}}(j) - \mathsf{N}_{\mathsf{R}_k}(j)}{\mathsf{N}_{\mathsf{R}_k}(l)} + \frac{\mathsf{N}_{\mathsf{R}_k}(j)[\mathsf{N}_{\mathsf{R}_{k+1}}(l) - \mathsf{N}_{\mathsf{R}_k}(l)]}{[\mathsf{N}_{\mathsf{R}_k}(l)]^2}.$$

Dividing the top and bottom of the first term on the right by k we have that $\overline{\lim}_k$ of that term is given by

$$\frac{1}{E\mathsf{M}_1(l)}\overline{\lim_k}\left(\frac{\mathsf{M}_{k+1}(j)}{k}\right).$$

By Problem 10, this term is a.s. zero. The second term is treated similarly to get

$$\frac{\mathsf{N}_n(j)}{\mathsf{N}_n(l)} \xrightarrow{\text{a.s.}} \frac{\pi(j)}{\pi(l)}. \tag{3.46}$$

Given that we have landed in j for the first time, let λ_j be the probability, starting from j, that we return to the origin before another landing in j. This must occur with positive probability, otherwise $P(\mathsf{S}_n = 0 \text{ i.o.}) < 1$. So whether or not another landing occurs before return is decided by tossing a coin with probability λ_j of failure. The expected number of additional landings past the first is given by the expected number of trials until failure. This is given by $\sum_1^\infty m(1 - \lambda_j)^m\lambda_j < \infty$, hence $\pi(j)$ is finite. Add the convention $\pi(0) = 1$, then (3.46) holds for all states, j and l. Let n^* be the first time that state l is entered. By (3.46)

$$\frac{\mathsf{N}_{\mathsf{n}^*+n}(j)}{\mathsf{N}_{\mathsf{n}^*+n}(l)} \xrightarrow{\text{a.s.}} \frac{\pi(j)}{\pi(l)}.$$

But $\mathsf{N}_{\mathsf{n}^*+n}(l)$ is the number of times that the sums $\mathsf{S}_{k+\mathsf{n}^*} - \mathsf{S}_{\mathsf{n}^*}$, $k = 1, \ldots, n$ land in state zero, and $\mathsf{N}_{\mathsf{n}^*+n}(j)$ is the number of times that $\mathsf{S}_{k+\mathsf{n}^*} - \mathsf{S}_{\mathsf{n}^*}$, $k = 1, \ldots, n$ land in $j - l$, plus the number of landings in j by the sums S_k up to time n^*. Therefore, $\pi(j)/\pi(l) = \pi(j - l)/\pi(0)$, or

$$\pi(j) = \pi(l)\pi(j - l).$$

This is the exponential equation on the integers—the only solutions are $\pi(j) = r^j$. Consider any sequence of states $m_1, \ldots, m_{n-1}, 0$ terminating at zero. I assert that

$$\begin{aligned} P(\mathsf{S}_1 &= m_1, \ldots, \mathsf{S}_{n-1} = m_{n-1}, \mathsf{S}_n = 0) \\ &= P(\mathsf{S}_1 = -m_{n-1}, \mathsf{S}_2 = -m_{n-2}, \ldots, \mathsf{S}_{n-1} = -m_1, \mathsf{S}_n = 0). \end{aligned}$$

The first probability is $P(\mathsf{X}_1 = m_1, \ldots, \mathsf{X}_n = -m_{n-1})$. The fact that $\mathsf{X}_1, \ldots, \mathsf{X}_n$ are identically distributed lets us equate this to

$$P(\mathsf{X}_1 = -m_{n-1}, \ldots, \mathsf{X}_n = m_1),$$

which is the second probability. This implies that $\pi(j) = \pi(-j)$, hence $\pi(j) \equiv 1$.

Problems

19. For fair coin-tossing equalizations $P(\mathsf{T}_1 = 2n) = 1/n({}_{2n-2}C_{n-1})2^{-2n+1}$. Use this result to show that

a) $P(\mathsf{T}_1 = 2n) \sim \dfrac{1}{2\sqrt{\pi}n^{3/2}}$.

b) From (a) conclude $E\mathsf{T}_1 = \infty$.

c) Using (a) again, show that $P(\mathsf{T}_1 \geq 2n) \sim \dfrac{1}{\sqrt{\pi}n^{1/2}}$.

d) Use (c) to show that $P(\mathsf{T}_k > 2k^2 \text{ i.o.}) = 1$.

e) Conclude from (d) that $P\left(\overline{\lim}_n \dfrac{\mathsf{R}_n}{n} = \infty\right) = 1$.

(There are a number of ways of deriving the exact expression above for $P(\mathsf{T}_1 = 2n)$; see Feller [59, Vol. I, pp. 74–75].)

20. For fair coin-tossing, use $P(\mathsf{S}_{2n} = 0) \sim 1/\sqrt{\pi n}$ to show that

$$E[\mathsf{N}_n(0)]^2 \sim cn.$$

Use an argument similar to the proof of the strong law of large numbers for fair-coin tossing, Theorem 1.21, to prove

$$\frac{\mathsf{N}_n(0)}{n} \xrightarrow{\text{a.s.}} 0.$$

21. Define

$$\mathsf{n}_1^* = \min\{n; \mathsf{S}_n > 0\},$$
$$\mathsf{n}_2^* = \min\{n; \mathsf{S}_{n+\mathsf{n}_1^*} > \mathsf{S}_{\mathsf{n}_1^*}\}, \text{ and so forth,}$$

$$\mathsf{Y}_k = \begin{cases} \mathsf{S}_{\mathsf{n}_1^* + \cdots + \mathsf{n}_k^*} - \mathsf{S}_{\mathsf{n}_1^* + \cdots + \mathsf{n}_{k-1}^*}, & k > 1, \\ \mathsf{S}_{\mathsf{n}_1^*}, \quad k = 1. \end{cases}$$

Show that the sequence $(\mathsf{n}_k^*, \mathsf{Y}_k)$ are independent, identically distributed vectors. Use the law of large numbers to prove that if $E|\mathsf{X}_1| < \infty$, $E\mathsf{X}_1 > 0$, then if one of $E\mathsf{Y}_1$, $E\mathsf{n}_1^*$ is finite, so is the other, and

$$(E\mathsf{n}_1^*)(E\mathsf{X}_1) = E\mathsf{Y}_1.$$

Show by using the sequence

$$\tilde{\mathsf{X}}_k = \begin{cases} \mathsf{X}_k, & \mathsf{X}_k \leq \lambda, \\ 0, & \mathsf{X}_k > \lambda, \end{cases}$$

for $\lambda > 0$, that $E\mathsf{Y}_1 < \infty$, $E\mathsf{n}_1^* < \infty$. (See Blackwell [7].)

22. For sums $S_1, S_2, \ldots$ such that $P(S_n = 0 \text{ i.o.}) = 1$, and $R_1, R_2, \ldots$ the occurrence times of $\{S_n = 0\}$, define the vectors $\mathbf{Z}_k$ by

$$\mathbf{Z}_k = (S_{R_k+1}, \ldots, S_{R_{k+1}}).$$

Define the appropriate range space $(R, \mathcal{B})$ for each of these vectors and show that they are independent and identically distributed.

9. HEWITT-SAVAGE ZERO-ONE LAW

Section 7 proved that for any interval I, $P(S_n \in I \text{ i.o.})$ is zero or one. But these sets are not tail events; whether $S_n = 0$ an infinite number of times depends strongly on X_1. However, there is another zero-one law in operation, formulated recently by Hewitt and Savage [71] which covers a variety of non-tail events.

Definition 3.47. *For a process $X_1, X_2, \ldots$, $A \in \mathcal{F}(\mathbf{X})$ is said to be symmetric if for any finite permutation $\{i_1, i_2, \ldots\}$ of $\{1, 2, \ldots\}$, there is a set $B \in \mathcal{B}_\infty$ such that*

$$A = \{(X_1, X_2, \ldots) \in B\} = \{(X_{i_1}, X_{i_2}, \ldots) \in B\}.$$

Theorem 3.48. *For $X_1, X_2, \ldots$ independent and identically distributed, every symmetric set has probability zero or one.*

Proof. The short proof we give here is due to Feller [59, Vol. II]. Take $A_n \in \mathcal{F}(X_1, \ldots, X_n)$ so that $P(A \,\triangle\, A_n) \to 0$. A_n can be written

$$\{(X_1, \ldots, X_n) \in B_n\}, \qquad B_n \in \mathcal{B}_n.$$

Because the $\tilde{\mathbf{X}} = (X_{i_1}, X_{i_2}, \ldots)$ process, $i_1, i_2, \ldots$ any sequence of distinct integers, has the same distribution as $\mathbf{X}$,

$$P(\tilde{\mathbf{X}} \in C) = P(\mathbf{X} \in C), \qquad C \in \mathcal{B}_\infty.$$

Hence, for any $B \in \mathcal{B}_\infty$,

$$\textbf{(3.49)} \qquad P(\{\tilde{\mathbf{X}} \in B\} \,\triangle\, \{\tilde{\mathbf{X}} \in B_n\}) = P(\{\mathbf{X} \in B\} \,\triangle\, \{\mathbf{X} \in B_n\}).$$

Take $(i_1, i_2, \ldots) = (2n, 2n-1, \ldots, 1, 2n+1, \ldots)$.

$$\tilde{A}_n = \{\tilde{\mathbf{X}} \in B_n\} = \{(X_{2n}, \ldots, X_{n+1}) \in B_n\}.$$

By (3.49), taking $B \in \mathcal{B}_\infty$ such that $\{\tilde{\mathbf{X}} \in B\} = \{\mathbf{X} \in B\} = A$,

$$P(A \,\triangle\, \tilde{A}_n) = P(A \,\triangle\, A_n) \to 0 \Rightarrow P(A_n \,\triangle\, \tilde{A}_n) \to 0 \Rightarrow P(A_n \cap \tilde{A}_n) \to P(A).$$

But A_n and $\tilde{A}_n$ are independent, thus

$$P(A_n \cap \tilde{A}_n) = P(A_n)P(\tilde{A}_n) \to P(A)^2.$$

Again, as in the Kolmogorov zero-one law, we wind up with

$$P(A) = P(A)^2. \qquad \text{Q.E.D.}$$

Corollary 3.50. *For* $X_1, X_2, \ldots$ *independent and identically distributed,* $\{S_n\}$ *the sums, every tail event on the* $S_1, S_2, \ldots$ *process has probability zero or one.*

Proof. For A a tail event, if $\{i_1, i_2, \ldots\}$ permutes only the first n indices, take $B \in \mathcal{B}_\infty$ such that

$$A = \{(S_{n+1}, \ldots) \in B\}.$$

Thus A is a symmetric set.

This result leads to the mention of another famous strong limit theorem which will be proved much later. If $EX_k = 0$, $EX_k^2 < \infty$ for independent, identically distributed random variables, then the form 3.27 of the strong law of large numbers implies

$$\frac{X_1 + \cdots + X_n}{\sqrt{n} \log n} \xrightarrow{\text{a.s.}} 0.$$

On the other hand, it is not hard to show that

$$\text{(3.51)} \qquad \begin{aligned} \underline{\lim} \frac{X_1 + \cdots + X_n}{\sqrt{n}} &= -\infty \text{ a.s.} \\ \overline{\lim} \frac{X_1 + \cdots + X_n}{\sqrt{n}} &= +\infty \text{ a.s.} \end{aligned}$$

Therefore fluctuations of S_n should be somewhere in the range $\sqrt{n}$ to $\sqrt{n} \log n$. For any function $h(n) \uparrow \infty$, the random variable $\overline{\lim}\, |S_n|/h(n)$ is a tail random variable, hence a.s constant. The famous law of the iterated logarithm is

Theorem 3.52

$$\overline{\lim} \frac{|S_n|}{\sigma\sqrt{2n \log (\log n)}} = 1 \text{ a.s.}$$

This is equivalent to: For every $\epsilon > 0$,

$$P(|S_n| \geq (1 - \epsilon)h(n) \text{ i.o.}) = 1,$$
$$P(|S_n| \geq (1 + \epsilon)h(n) \text{ i.o.}) = 0,$$

with $h(n) = \sigma \sqrt{2n \log (\log n)}$.

Therefore, a more general version of the law of the iterated logarithm would be a separation of all nondecreasing $h(n)$ into two classes

$$P(|S_n| > h(n) \text{ i.o.}) = 0 \quad \text{or} \quad 1.$$

The latter dichotomy holds because of 3.50. The proof of 3.52 is quite tricky, to say nothing of the more general version. The simplest proof around for coin-tossing is in Feller [59, Vol. I].

Actually, this theorem is an oddity. Because, even though it is a strong limit theorem, it is a second-moment theorem and its proof consists of ingenious uses of the Borel-Cantelli lemma combined with the central limit theorem. We give an illuminating proof in Chapter 13.

Problem 23. Use the Kolmogorov zero-one law and the central limit theorem to prove (3.51) for fair coin-tossing.

Remark. The important theorems for independence come up over and over again as their contents are generalized. In particular, the random signs problem connects with martingales (Chapter 5), the strong law of large numbers generalizes into the ergodic theorem (Chapter 6), and the notions of recurrence of sums comes up again in Markov processes (Chapter 7).

NOTES

The strong law of large numbers was proven for fair coin-tossing by Borel in 1909. The forms of the strong law given in this chapter were proved by Kolmogorov in 1930, 1933 [92], [98]. The general solution of the random signs problem is due to Khintchine and Kolmogorov [91] in 1925. A special case of the law of the iterated logarithm was proven by Khintchine [88, 1924].

The work on recurrence is more contemporary. The theorems of Section 7 are due to Chung and Fuchs [18, 1951], but the neat proof given that $EX = 0$ implies recurrence was found by Chung and Ornstein [20, 1962]. But this work was preceded by some intriguing examples due to Polya [116, 1921]. (These will be given in Chapter 7). The work of Harris and Robbins (loc. cit.) on the equidistribution of sums appeared in 1953.

There is a bound for sums of independent random variables with zero means which is much more well-known than the Skorokhod's inequality, that is,

$$P\left(\max_{k\leq n} |S_k| > \epsilon\right) \leq \frac{1}{\epsilon^2}\sum_1^n EX_k^2,$$

$S_k = X_1 + \cdots + X_k$. This is due to Kolmogorov. Compare it with the Chebyshev bound for $P(|S_n| > \epsilon)$. A generalization is proved in Chapter 5.

The strong law for identically distributed random variables depends essentially on $E|X| < \infty$. One might expect that even if $E|X| = \infty$, there would be another normalization $N_n \uparrow \infty$ such that the normed sums

$$\frac{X_1 + \cdots + X_n}{N_n}$$

converge a.s. One answer is trivial; you can always take N_n increasing so rapidly that a.s. convergence to zero follows. But Chow and Robbins [14] have obtained the strong result that if $E|\mathsf{X}| = \infty$, there is no normalization $N_n \uparrow \infty$ such that one gets a.s. convergence to anything but zero.

If $E|\mathsf{X}_1| < \infty$, then if the sums are nonrecurrent, either $\mathsf{S}_n \to +\infty$ a.s. or $\mathsf{S}_n \to -\infty$ a.s. But if $E|\mathsf{X}_1| = \infty$, the sums can be transient and still change sign an infinite number of times; in fact, one can get $\overline{\lim}\, \mathsf{S}_n = +\infty$ a.s., $\underline{\lim}\, \mathsf{S}_n = -\infty$ a.s. Examples of this occur when $\mathsf{X}_1, \mathsf{X}_2, \ldots$ have one of the symmetric stable distributions discussed in Chapter 9.

Strassen [135] has shown that the law of the iterated logarithm is a second-moment theorem in the sense that if $E\mathsf{X}_1 = 0$, and

$$\overline{\lim}\, \frac{\mathsf{X}_1 + \cdots + \mathsf{X}_n}{\sqrt{n \log(\log n)}} < \infty \text{ a.s.},$$

then $E\mathsf{X}_1^2 < \infty$. There is some work on other forms of this law when $E\mathsf{X}_1^2 = \infty$, but the results (Feller [54]) are very specialized.

For more extensive work with independent random variables the most interesting source remains Paul Lévy's book [103]. Loève's book has a good deal of the classical material. For an elegant and interesting development of the ideas of recurrence see Spitzer [130].

CHAPTER 4

CONDITIONAL PROBABILITY AND CONDITIONAL EXPECTATION

1. INTRODUCTION

More general tools need to be developed to handle relationships between dependent random variables. The concept of conditional probability—the distribution of one set of random variables given information concerning the observed values of another set—will turn out to be a most useful tool.

First consider the problem: What is the probability of an event B given that A has occurred? If we know that $\omega \in A$, then our new sample space is A. The probability of B is proportional to the probability of that part of it lying in A. Hence

Definition 4.1. *Given $(\Omega, \mathcal{F}, P)$, for sets $A, B \in \mathcal{F}$, such that $P(A) > 0$, the conditional probability of B given that A has occurred is defined as*

$$\frac{P(B \cap A)}{P(A)},$$

and is denoted by $P(B \mid A)$.

This extends immediately to conditioning by random variables taking only a countable number of values.

Definition 4.2. *If X takes values in $\{x_k\}$, the conditional probability of A given $\mathsf{X} = x_k$ is defined by*

$$P(A \mid \mathsf{X} = x_k) = \frac{P(A, \mathsf{X} = x_k)}{P(\mathsf{X} = x_k)}$$

if $P(\mathsf{X} = x_k) > 0$ and arbitrarily defined as zero if $P(\mathsf{X} = x_k) = 0$.

Note that there is probability zero that X takes values in the set where the conditional probability was not defined by the ratio. $P(A \mid \mathsf{X} = x_k)$ is a probability on $\mathcal{F}$, and the natural definition of the conditional expectation of a random variable Y given $\mathsf{X} = x_k$ is

$$E(\mathsf{Y} \mid \mathsf{X} = x_k) = \int \mathsf{Y}(\omega) P(d\omega \mid \mathsf{X} = x_k) \tag{4.3}$$

if the integral exists.

What needs to be done is to generalize the definition so as to be able to handle random variables taking on nondenumerably many values. Look at the simplest case of this: Suppose there is a random variable X on $(\Omega, \mathcal{F}, P)$, and let $A \in \mathcal{F}$. If $B \in \mathcal{B}_1$ is such that $P(\mathsf{X} \in B) > 0$, then as above, the conditional probability of A given $\mathsf{X} \in B$, is defined by

$$P(A \mid \mathsf{X} \in B) = \frac{P(A, \mathsf{X} \in B)}{P(\mathsf{X} \in B)}.$$

But suppose we want to give meaning to the conditional probability of A given $\mathsf{X}(\omega) = x$. Of course, if $P(\mathsf{X} = x) > 0$, then we have no trouble and proceed as in 4.2. But many of the interesting random variables have the property that $P(\mathsf{X} = x) = 0$ for all x. This causes a fuss. An obvious thing to try is taking limits, i.e., to try defining

(4.4) $$P(A \mid \mathsf{X} = x) = \lim_{h \downarrow 0} \frac{P\big(A, \mathsf{X} \in (x - h, x + h)\big)}{P\big(\mathsf{X} \in (x - h, x + h)\big)}.$$

In general, this is no good. If $P(\mathsf{X} = x_0) = 0$, then there is no guarantee, unless we put more restrictive conditions on P and X, that the limit above will exist for

$$x = x_0.$$

So either we add these restrictions (very unpleasant), or we look at the problem a different way. Look at the limit in (4.4) globally as a function of x. Intuitively, it looks as though we are trying to take the derivative of one measure with respect to another. This has a familiar ring; we look back to see what can be done.

On $\mathcal{B}_1$ define two measures as follows: Let

(4.5) $$\hat{Q}(B) = P(A, \mathsf{X} \in B),$$
$$\hat{P}(B) = P(\mathsf{X} \in B).$$

Note that $0 \le \hat{Q}(B) \le \hat{P}(B)$ so that $\hat{Q}$ is absolutely continuous with respect to $\hat{P}$. By the Radon-Nikodym theorem (Appendix A.30) we can define the derivative of $\hat{Q}$ with respect to $\hat{P}$, which is exactly what we are trying to do with limits in (4.4). But we must pay a price for taking this elegant route. Namely, recall that $d\hat{Q}/d\hat{P}$ is defined as any $\mathcal{B}_1$-measurable function $\varphi(x)$ satisfying

(4.6) $$\hat{Q}(B) = \int_B \varphi(x)\hat{P}(dx), \quad \text{all } B \in \mathcal{B}_1.$$

If φ satisfies (4.6) so does φ' if $\varphi = \varphi'$ a.s. $\hat{P}$. Hence this approach, defining $P(A \mid \mathsf{X} = x)$ as any function satisfying (4.6) leads to an arbitrary selection of one function from among a class of functions equivalent (a.s. equal) under $\hat{P}$. This is a lesser evil.

Definition 4.7. *The conditional probability* $P(A \mid \mathsf{X} = x)$ *is defined as any* $\mathcal{B}_1$*-measurable function satisfying*

$$P(A, \mathsf{X} \in B) = \int_B P(A \mid \mathsf{X} = x)\hat{P}(dx), \quad \text{all } B \in \mathcal{B}_1.$$

In 4.7 above, $P(A \mid \mathsf{X} = x)$ is defined as a $\mathcal{B}_1$-measurable function $\varphi(x)$, unique up to equivalence under $\hat{P}$. For many purposes it is useful to consider the conditional probability as a random variable on the original $(\Omega, \mathcal{F}, P)$ space, rather than the version above which resembles going into representation space. The natural way to do this is to define

$$P(A \mid \mathsf{X}(\omega)) = \varphi(\mathsf{X}(\omega)).$$

Since φ is $\mathcal{B}_1$-measurable, then $\varphi(\mathsf{X}(\omega))$ is a random variable on $(\Omega, \mathcal{F})$. Since any two versions of φ are equivalent under $\hat{P}$, any two versions of $P(A \mid \mathsf{X}(\omega))$ obtained in this way are equivalent under P. But there is a more direct way to get to $P(A \mid \mathsf{X})$, analogous to 4.7. Actually, what is done is just transform 4.7 to $(\Omega, \mathcal{F}, P)$.

Definition 4.8. *The conditional probability of* A *given* $\mathsf{X}(\omega)$, *is defined as any random variable on* Ω, *measurable* $\mathcal{F}(\mathsf{X})$, *and satisfying*

$$P(A, \mathsf{X} \in B) = \int_{\{\mathsf{X} \in B\}} P(A \mid \mathsf{X})\, dP, \quad \text{all } B \in \mathcal{B}_1.$$

Any two versions of $P(A \mid \mathsf{X})$ *differ on a set of probability zero.*

This gives the same $P(A \mid \mathsf{X})$ as starting from 4.7 to get $\varphi(\mathsf{X}(\omega))$, where $\varphi(x) = P(A \mid \mathsf{X} = x)$. To see this, apply 2.41 to 4.7 and compare the result with 4.8. A proof that is a bit more interesting utilizes a converse of 2.31.

Proposition 4.9. *Let* $\mathbf{X}$ *be a random vector on* $(\Omega, \mathcal{F})$ *taking values in* $(R, \mathcal{B})$. *If* Z *is a random variable on* $(\Omega, \mathcal{F})$, *measurable* $\mathcal{F}(\mathbf{X})$, *then there is a random variable* $\theta(\mathbf{x})$ *on* $(R, \mathcal{B})$ *such that*

$$\mathsf{Z} = \theta(\mathbf{X}).$$

Proof. See Appendix A.21.

The fact that $P(A \mid \mathsf{X})$ is $\mathcal{F}(\mathsf{X})$-measurable implies by this proposition that $P(A \mid \mathsf{X}) = \theta(\mathsf{X})$, where $\theta(x)$ is $\mathcal{B}_1$-measurable. But $\theta(\mathsf{X})$ satisfies

$$P(A, \mathsf{X} \in B) = \int_{\{\mathsf{X} \in B\}} \theta(\mathsf{X}(\omega))\, dP = \int_B \theta(x)\hat{P}(dx)$$

(this last by 2.41). Hence $\theta = \varphi$ a.s. $\hat{P}$.

We can put 4.8 into a form which shows up a seemingly curious phenomenon. Since $\mathcal{F}(\mathsf{X})$ is the class of all sets $\{\mathsf{X} \in B\}$, $B \in \mathcal{B}_1$, $P(A \mid \mathsf{X})$ is any random variable satisfying

$$\textbf{(4.10)} \qquad P(A \cap D) = \int_D P(A \mid \mathsf{X})\, dP, \quad \text{all } D \in \mathcal{F}(\mathsf{X}).$$

From this, make the observation that if X_1 and X_2 are two random variables which contain the same information in the sense that $\mathcal{F}(\mathsf{X}_1) = \mathcal{F}(\mathsf{X}_2)$, then

$$\textbf{(4.11)} \qquad P(A \mid \mathsf{X}_1) = P(A \mid \mathsf{X}_2) \quad \text{a.s.}$$

In a way this is not surprising, because $\mathcal{F}(\mathsf{X}_1) = \mathcal{F}(\mathsf{X}_2)$ implies that X_1 and X_2 are functions of each other, that is, from 4.9,

$$\mathsf{X}_2(\omega) = \theta_1(\mathsf{X}_1(\omega)), \qquad \mathsf{X}_1(\omega) = \theta_2(\mathsf{X}_2(\omega)).$$

The idea here is that $P(A \mid \mathsf{X})$ does not depend on the values of X, but rather on the sets in $\mathcal{F}$ that X discriminates between.

The same course can be followed in defining the conditional expectation of one random variable, given the value of another. Let X, Y be random variables on $(\Omega, \mathcal{F}, P)$. What we wish to define is the conditional expectation of Y given $\mathsf{X} = x$, in symbols, $E(\mathsf{Y} \mid \mathsf{X} = x)$. If $B \in \mathcal{B}$ were such that $P(\mathsf{X} \in B) > 0$, intuitively $E(\mathsf{Y} \mid \mathsf{X} \in B)$ should be defined as $\int \mathsf{Y}(\omega) P(d\omega \mid \mathsf{X} \in B)$, where $P(\cdot \mid \mathsf{X} \in B)$ is the probability on $\mathcal{F}$ defined as

$$P(A \mid \mathsf{X} \in B) = P(A, \mathsf{X} \in B)/P(\mathsf{X} \in B).$$

Again, we could take $B = (x - h, x + h)$, let $h \downarrow 0$, and hope the limit exists. More explicitly, we write the ratio

$$\frac{\int \mathsf{Y}(\omega) P(d\omega, \mathsf{X} \in B)}{P(\mathsf{X} \in B)}$$

and hope that as $P(\mathsf{X} \in B) \to 0$, the limiting ratio exists. Again the derivative of one set function with respect to another is coming up. What to do is similar: Define

$$\hat{Q}(B) = \int \mathsf{Y}(\omega) P(d\omega, \mathsf{X} \in B),$$

$$\hat{P}(B) = P(\mathsf{X} \in B).$$

To get things finite, we have to assume $E\,|\mathsf{Y}| < \infty$; then

$$|\hat{Q}(B)| \le \int |\mathsf{Y}|\, P(d\omega, \mathsf{X} \in B) \le E\,|\mathsf{Y}|.$$

To show that $\hat{Q}$ is σ-additive, write it as

$$\hat{Q}(B) = \int_{\{\mathsf{X} \in B\}} \mathsf{Y}(\omega)\, dP.$$

Now $\{B_n\}$ disjoint implies $A_n = \{\mathsf{X} \in B_n\}$ disjoint, and

$$\hat{Q}(\cup B_n) = \int_{\cup A_n} \mathsf{Y}\, dP = \sum_n \int_{A_n} \mathsf{Y}\, dP = \sum_n \hat{Q}(B_n).$$

Also, $\hat{P}(B) = 0 \Rightarrow \hat{Q}(B) = 0$, thus $\hat{Q}$ is absolutely continuous with respect to $\hat{P}$. This allows the definition of $E(\mathsf{Y} \mid \mathsf{X} = x)$ as any version of $d\hat{Q}/d\hat{P}$.

Definition 4.12. *Let $E|\mathsf{Y}| < \infty$, then $E(\mathsf{Y} \mid \mathsf{X} = x)$ is any $\mathcal{B}_1$-measurable function satisfying*

$$\int_B E(\mathsf{Y} \mid \mathsf{X} = x)\, d\hat{P}(x) = \int_{\{\mathsf{X} \in B\}} \mathsf{Y}(\omega)\, dP(\omega), \quad \text{all } B \in \mathcal{B}_1.$$

Any two versions of $E(\mathsf{Y} \mid \mathsf{X} = x)$ are a.s. *equal under $\hat{P}$.*

Conditional expectations can also be looked at as random variables. Just as before, if $\varphi(x) = E(\mathsf{Y} \mid \mathsf{X} = x)$, $E(\mathsf{Y} \mid \mathsf{X})$ can be defined as $\varphi(\mathsf{X}(\omega))$. Again, we prefer the direct definition.

Definition 4.13. *Let $E|\mathsf{Y}| < \infty$; then $E(\mathsf{Y} \mid \mathsf{X})$ is any $\mathcal{F}(\mathsf{X})$-measurable function satisfying*

$$\int_A E(\mathsf{Y} \mid \mathsf{X})\, dP = \int_A \mathsf{Y}\, dP, \quad \text{all } A \in \mathcal{F}(\mathsf{X}). \tag{4.14}$$

The random variable Y trivially satisfies (4.14), but in general $\mathsf{Y} \neq E(\mathsf{Y} \mid \mathsf{X})$ because $\mathsf{Y}(\omega)$ is not necessarily $\mathcal{F}(\mathsf{X})$-measurable. This remark does discover the property that if $\mathcal{F}(\mathsf{Y}) \subset \mathcal{F}(\mathsf{X})$ then $E(\mathsf{Y} \mid \mathsf{X}) = \mathsf{Y}$ a.s. Another property in this direction is: Consider the space of $\mathcal{F}(\mathsf{X})$-measurable random variables. In this space, the random variable closest to Y is $E(\mathsf{Y} \mid \mathsf{X})$. (For a defined version of this statement see Problem 11.)

Curiously enough, conditional probabilities are a special case of conditional expectations. Because, by the definitions,

$$P(A \mid \mathsf{X} = x) = E(\chi_A(\omega) \mid \mathsf{X} = x) \quad \text{a.s. } \hat{P},$$
$$P(A \mid \mathsf{X}) = E(\chi_A(\omega) \mid \mathsf{X}) \quad \text{a.s. } P.$$

Therefore, the next section deals with the general definition of conditional expectation.

Definition 4.15. *Random variables* X, Y *on $(\Omega, \mathcal{F}, P)$ are said to have a joint density if the probability $\hat{P}(\cdot)$ defined on $\mathcal{B}_2$ by $\hat{P}(F) = P((\mathsf{Y}, \mathsf{X}) \in F)$ is absolutely continuous with respect to Lebesgue measure on $\mathcal{B}_2$, that is, if there exists $f(y, x)$ on $R^{(2)}$, measurable $\mathcal{B}_2$, such that*

$$P((\mathsf{Y}, \mathsf{X}) \in F) = \int_F f(y, x)\, dy\, dx, \quad \text{all } F \in \mathcal{B}_2.$$

Then, by Fubini's theorem (see Appendix A.37), defining $f(x) = \int f(y, x)\, dy$, *for all* $B \in \mathcal{B}_1$,

$$\hat{P}(B) = P(\mathsf{X} \in B) = \int_B f(x)\, dx.$$

(Actually, any σ-finite product measure on $\mathcal{B}_2$ could be used instead of $dy\, dx$.) If a joint density exists, then it can be used to compute the conditional probability and expectation. This is the point of Problems 2 and 3 below.

Problems

1. Let X take on only integer values. Show that $P(A \mid \mathsf{X} = x)$ as defined in 4.7 is any $\mathcal{B}_1$-measurable function $\varphi(x)$ satisfying

$$P(A, \mathsf{X} = j) = \varphi(j) P(\mathsf{X} = j), \quad \text{all } j.$$

Conclude that if $P(\mathsf{X} = j) > 0$, then any version of the above conditional probability satisfies

$$P(A \mid \mathsf{X} = j) = \frac{P(A, \mathsf{X} = j)}{P(\mathsf{X} = j)}.$$

2. Prove that if X, Y have a joint density, then for any $B \in \mathcal{B}_1$,

$$P(\mathsf{Y} \in B \mid \mathsf{X} = x) = \int_B \frac{f(y, x)}{f(x)}\, dy, \quad \text{a.s. } \hat{P}.$$

3. If (Y, X) have a joint density $f(y, x)$ and $E\,|\mathsf{Y}| < 0$, show that one version of $E(\mathsf{Y} \mid \mathsf{X} = x)$ is given by

$$\int y \frac{f(y, x)}{f(x)}\, dy.$$

4. Let $\mathsf{X}_1, \mathsf{X}_2$ take values in $\{1, 2, \ldots, N\}$. If $\mathcal{F}(\mathsf{X}_1) = \mathcal{F}(\mathsf{X}_2)$, then prove there is a permutation $\{i_1, i_2, \ldots, i_N\}$ of $\{1, 2, \ldots, N\}$ such that

$$A_j = \{\mathsf{X}_1 = j\} = \{\mathsf{X}_2 = i_j\}.$$

Let $P(A_j) > 0, j = 1, \ldots, N$. Prove that

$$P(A \mid \mathsf{X}_1) = P(A \mid \mathsf{X}_2) = \sum_1^N \chi_{A_j}(\omega) \cdot P(A \mid A_j) \quad \text{a.s.}$$

5. Let $\Omega = \{z \in [-1, +1]\}$, $\mathcal{F} = \mathcal{B}([-1, +1])$, $P = dz/2$, and $\mathsf{X}_1(z) = z^2$. Show that one version of $P(A \mid \mathsf{X}_1)$ is

$$P(A \mid \mathsf{X}_1) = \tfrac{1}{2}\chi_A(z) + \tfrac{1}{2}\chi_A(-z).$$

Let $X_2(z) = z^4$; find a version of $P(A \mid X_2)$. Find versions of $P(A \mid X_1 = x)$, $P(A \mid X_2 = x)$.

6. Given the situation of Problem 5, and with Y a random variable such that $E\,|Y(z)| < \infty$, show that

$$E(Y(z) \mid X_1(z)) = \tfrac{1}{2}Y(z) + \tfrac{1}{2}Y(-z), \quad \text{a.s. } dz.$$

Find a version of

$$E(Y(z) \mid X_1(z) = x).$$

7. If X and Y are independent, show that for any $B \in \mathcal{B}_1$, $P(Y \in B \mid X) = P(Y \in B)$ a.s. For $E\,|Y| < \infty$, show that $E(Y \mid X) = EY$ a.s.

8. Give an example to show that $E(Y \mid X) = EY$ a.s. does not imply that X and Y are independent.

9. (Borel paradox). Take Ω to be the unit sphere $S^{(2)}$ in $R^{(3)}$, $\mathcal{F}$ the Borel subsets of Ω, $P(\cdot)$ the extension of surface area. Choose two opposing points on $S^{(2)}$ as the poles and fix a reference half-plane passing through them. For any point $\mathbf{p}$, define its longtitude $\psi(\mathbf{p})$ as the angle between $-\pi$ and π that the half-plane of the great semi-circle through $\mathbf{p}$ makes with the reference half plane. Define its latitude $\theta(\mathbf{p})$ as the angle that the radius to $\mathbf{p}$ makes with the equatorial plane, $-\pi/2 < \theta(\mathbf{p}) \leq \pi/2$. Prove that the conditional probability of ψ given θ is uniformly distributed over $[-\pi, \pi)$ but that the conditional probability of θ given ψ is not uniformly distributed over $(-\pi/2, \pi/2]$. (See Kolmogorov [98, p. 50].)

2. A MORE GENERAL CONDITIONAL EXPECTATION

Section 1 pointed out that $E(Y \mid X)$ or $P(A \mid X)$ depended only on $\mathcal{F}(X)$. The point was that the relevant information contained in knowing $X(\omega)$ is the information regarding the location of ω. Let $(\Omega, \mathcal{F}, P)$ be a probability space, Y a random variable, $E\,|Y| < \infty$, $\mathfrak{D}$ any σ-field, $\mathfrak{D} \subset \mathcal{F}$.

Definition 4.16. *The conditional expectation $E(Y \mid \mathfrak{D})$ is any random variable measurable $(\Omega, \mathfrak{D})$ such that*

$$\textbf{(4.17)} \qquad \int_D E(Y \mid \mathfrak{D})\, dP = \int_D Y\, dP, \quad \text{all } D \in \mathfrak{D}.$$

As before, any two versions differ on a set of probability zero. *If $\mathfrak{D} = \mathcal{F}(X_1, X_2, \ldots)$, denote $E(Y \mid \mathfrak{D}) = E(Y \mid X_1, X_2, \ldots)$.*

If $\mathbf{X}$ is a random vector to $(R, \mathcal{B})$, then for $\mathbf{x} \in R$

Definition 4.18. *$E(Y \mid \mathbf{X} = \mathbf{x})$ is any random variable on $(R, \mathcal{B})$, where $\hat{P}(B) = P(\mathbf{X} \in B)$, satisfying*

$$\int_B E(Y \mid \mathbf{X} = \mathbf{x})\, d\hat{P} = \int_{\{\mathbf{X} \in B\}} Y\, dP, \quad \text{all } B \in \mathcal{B}.$$

The importance of this is mostly computational. By inspection verify that

$$\text{(4.19)} \qquad \varphi(\mathbf{x}) = E(\mathsf{Y} \mid \mathbf{X} = \mathbf{x}) \quad \Rightarrow \quad E(\mathsf{Y} \mid \mathbf{X}) = \varphi(\mathbf{X}).$$

Usually, $E(\mathsf{Y} \mid \mathbf{X} = \mathbf{x})$ is easier to compute, when densities exist, for example. Then (4.19) gets us $E(\mathsf{Y} \mid \mathbf{X})$.

Proposition 4.20. *A list of properties of* $E(\mathsf{Y} \mid \mathcal{D})$,

1) $E(\alpha\mathsf{Y}_1 + \beta\mathsf{Y}_2 \mid \mathcal{D}) = \alpha E(\mathsf{Y}_1 \mid \mathcal{D}) + \beta E(\mathsf{Y}_2 \mid \mathcal{D})$ a.s., *assuming* $E|\mathsf{Y}_1| < \infty$, $E|\mathsf{Y}_2| < \infty$.
2) $\mathsf{Y} \geq 0$, $E\mathsf{Y} < \infty$, $\Rightarrow E(\mathsf{Y} \mid \mathcal{D}) \geq 0$ a.s.
3) $E|\mathsf{Y}| < \infty$, $\mathcal{D} \subset \mathcal{E}$, *implies*

$$E\big(E(\mathsf{Y} \mid \mathcal{E}) \mid \mathcal{D}\big) = E(\mathsf{Y} \mid \mathcal{D}) \quad \text{a.s.}$$

4) $\mathcal{F}(\mathsf{Y})$ *independent of* $\mathcal{D}$, $E|\mathsf{Y}| < \infty$, $\Rightarrow E(\mathsf{Y} \mid \mathcal{D}) = E\mathsf{Y}$ a.s.

Proofs. These proofs follow pretty trivially from the definition 4.16. To improve technique I'll briefly go through them; the idea in all cases is to show that the integrals of both sides of (1), (2), (3), (4) over $\mathcal{D}$ sets are the same, (in 2, $\geq$). Let $D \in \mathcal{D}$. Then by 4.16 the integrals over D of the left hand sides of (1), (2), (3), (4) above are

$$\text{1)} \int_D (\alpha\mathsf{Y}_1 + \beta\mathsf{Y}_2)\, dP \qquad \text{2)} \int_D \mathsf{Y}\, dP$$

$$\text{3)} \int_D E(\mathsf{Y} \mid \mathcal{E})\, dP \qquad \text{4)} \int_D \mathsf{Y}\, dP.$$

The right-hand sides integrated over D are

$$\text{1')}\ \alpha\int_D \mathsf{Y}_1\, dP + \beta\int_D \mathsf{Y}_2\, dP \qquad \text{2')}\ 0$$

$$\text{3')} \int_D \mathsf{Y}\, dP \qquad \text{4')}\ (E\mathsf{Y}) \cdot P(D)$$

So (1) = (1′) is trivial, also (2) $\geq$ (2′). For (4), write

$$\int_D \mathsf{Y}\, dP = \int \chi_D \mathsf{Y}\, dP = E\chi_D E\mathsf{Y} = P(D) \cdot E\mathsf{Y}.$$

For (3), (3′) note that by 4.16,

$$\int_F E(\mathsf{Y} \mid \mathcal{E})\, dP = \int_F \mathsf{Y}\, dP, \quad \text{all } F \in \mathcal{E}.$$

But $\mathcal{D} \subset \mathcal{E}$, so (3) = (3′), all $D \in \mathcal{D}$.

An important property of the conditional expectation is, if $E|\mathsf{Y}| < \infty$,

$$E(E(\mathsf{Y} \mid \mathfrak{D})) = E\mathsf{Y}. \tag{4.21}$$

This follows quickly from the definitions.

Let $\mathsf{Y} = \chi_A(\omega)$; then the general definition of conditional probability is

Definition 4.22. *Let $\mathfrak{D}$ be a sub-σ-field of $\mathcal{F}$. The conditional probability of $A \in \mathcal{F}$ given $\mathfrak{D}$ is a random variable $P(A \mid \mathfrak{D})$ on $(\Omega, \mathfrak{D})$ satisfying*

$$\int_D P(A \mid \mathfrak{D})\, dP = P(A \cap D), \quad \text{all } D \in \mathfrak{D}.$$

By the properties in 4.20, a conditional probability acts almost like a probability, that is,

$$P(A \mid \mathfrak{D}) \geq 0 \quad \text{a.s.}, \qquad P(\Omega \mid \mathfrak{D}) = 1 \quad \text{a.s.} \tag{4.23}$$

$$A_1, \ldots, A_n \text{ disjoint} \Rightarrow P(\cup A_n \mid \mathfrak{D}) = \sum_1^n P(A_k \mid \mathfrak{D}) \quad \text{a.s.}$$

It is also σ-additive almost surely. This follows from

Proposition 4.24. *Let $\mathsf{Y}_n \geq 0$ be random variable such that $\mathsf{Y}_n \uparrow \mathsf{Y}$ a.s. and $E|\mathsf{Y}| < \infty$. Then $E(\mathsf{Y}_n \mid \mathfrak{D}) \to E(\mathsf{Y} \mid \mathfrak{D})$ a.s.*

Proof. Let $\mathsf{Z}_n = \mathsf{Y} - \mathsf{Y}_n$, so $\mathsf{Z}_n \downarrow 0$ a.s., and $E\mathsf{Z}_n \downarrow 0$. By 4.20(2) $\mathsf{Z}_n \geq \mathsf{Z}_{n+1} \Rightarrow E(\mathsf{Z}_n \mid \mathfrak{D}) \geq E(\mathsf{Z}_{n+1} \mid \mathfrak{D})$ a.s. Therefore the sequence $E(\mathsf{Z}_n \mid \mathfrak{D})$ converges monotonically downward a.s. to a random variable $\mathsf{U} \geq 0$. By the monotone convergence theorem,

$$\lim_n E(E(\mathsf{Z}_n \mid \mathfrak{D})) = E\mathsf{U}.$$

Equation (4.21) now gives $E\mathsf{U} = \lim_n E\mathsf{Z}_n = 0$. Thus $\mathsf{U} = 0$ a.s.

Let $A_k \in \mathcal{F}$, $\{A_k\}$ disjoint, and take

$$\mathsf{Y}_n = \sum_{k=1}^n \chi_{A_k}$$

in the above proposition to get from (4.23) to

$$P(\cup A_n \mid \mathfrak{D}) = \sum_n P(A_n \mid \mathfrak{D}) \quad \text{a.s.} \tag{4.25}$$

For A fixed, $P(A \mid \mathfrak{D})$ is an equivalence class of functions $f(A, \omega)$. It seems reasonable to hope from (4.25) that from each equivalence class a function $f^*(A, \omega)$ could be selected such that the resulting function $P^*(A \mid \mathfrak{D})$ on $\mathcal{F} \times \Omega$ would be a probability on $\mathcal{F}$ for every ω. If this can be done, then the entire business of defining $E(\mathsf{Y} \mid \mathfrak{D})$ would be unnecessary because it could be defined as

(4.26)
$$E(Y \mid \mathfrak{D}) = \int Y(\omega)P^*(d\omega \mid \mathfrak{D}).$$

Unfortunately, in general it is not possible to do this. What can be done is a question which is formulated and partially answered in the next section.

Problems

10. If $\mathfrak{D}$ has the property, $D \in \mathfrak{D} \Rightarrow P(D) = 0, 1$, then show $E(Y \mid \mathfrak{D}) = EY$ a.s. (if $E|Y| < \infty$).

11. Let Y be a random variable on $(\Omega, \mathcal{F})$, $EY^2 < \infty$. For any random variable X on $(\Omega, \mathcal{F})$, let $d^2(Y, X) = E|Y - X|^2$.

a) Prove that among all random variables X on $(\Omega, \mathfrak{D})$, $\mathfrak{D} \subset \mathcal{F}$, there is an a.s. unique random variable Y_0 which minimizes $d(Y, X)$. This random variable Y_0 is called the best predictor of Y based on $\mathfrak{D}$.

b) Prove that $Y_0 = E(Y \mid \mathfrak{D})$ a.s.

12. Let $X_0, X_1, X_2, \ldots, X_n$ be random variables having a joint normal distribution, $EX_k = 0$, $\Gamma_{ij} = E(X_iX_j)$. Show that $E(X_0 \mid X_1, X_2, \ldots, X_n) = \sum_1^n \lambda_j X_j$ a.s., and give the equations determining λ_j in terms of the Γ_{ij}. (See Chapter 9 for definitions.)

13. Let $\mathbf{X}$ be a random vector taking values in $(R, \mathcal{B})$, and $\varphi(\mathbf{x})$ a random variable on $(R, \mathcal{B})$. Let Y be a random variable on $(\Omega, \mathcal{F})$ and $E|Y| < \infty$, $E|\varphi(\mathbf{X})| < \infty$, $E|Y\varphi(\mathbf{X})| < \infty$. Prove that

a) $E(\varphi(\mathbf{X})Y \mid \mathbf{X}) = \varphi(\mathbf{X})E(Y \mid \mathbf{X})$ a.s. P.

b) $E(\varphi(\mathbf{X})Y \mid \mathbf{X} = \mathbf{x}) = \varphi(\mathbf{x})E(Y \mid \mathbf{X} = \mathbf{x})$ a.s. $\hat{P}$.

[This result concurs with the idea that if we know $\mathbf{X}$, then given $\mathbf{X}$, $\varphi(\mathbf{X})$ should be treated as a constant. To work Problem 13, a word to the wise: Start by assuming $\varphi(\mathbf{X}) \geq 0$, $Y \geq 0$, consider the class of random variables φ for which it is true, and apply 2.38.]

14. Let Y be a random variable on $(\Omega, \mathcal{F}, P)$ such that $E|Y| < \infty$ and $\mathbf{X}_1, \mathbf{X}_2$ random vectors such that $\mathcal{F}(Y, \mathbf{X}_1)$ is independent of $\mathcal{F}(\mathbf{X}_2)$, then prove that
$$E(Y \mid \mathbf{X}_1, \mathbf{X}_2) = E(Y \mid \mathbf{X}_1) \quad \text{a.s.}$$

15. Let $X_1, X_2, \ldots$ be independent, identically distributed random variables, $E|X_1| < \infty$, and denote $S_n = X_1 + \cdots + X_n$. Prove that
$$E(X_1 \mid S_n, S_{n+1}, \ldots) = S_n/n \quad \text{a.s.}$$

[Use symmetry in the final step.]

3. REGULAR CONDITIONAL PROBABILITIES AND DISTRIBUTIONS

Definition 4.27. *$P^*(A \mid \mathfrak{D})$ will be called a regular conditional probability on $\mathcal{F}_1 \subset \mathcal{F}$, given $\mathfrak{D}$ if*

a) *For $A \in \mathcal{F}_1$ fixed, $P^*(A \mid \mathfrak{D})$ is a version of $P(A \mid \mathfrak{D})$.*
b) *For any ω held fixed, $P^*(A \mid \mathfrak{D})$ is a probability on $\mathcal{F}_1$.*

If a regular conditional probability, given $\mathfrak{D}$, exists, all the conditional expectations can be defined through it.

Proposition 4.28. *If $P^*(A \mid \mathfrak{D})$ is a regular conditional probability on $\mathcal{F}_1$ and if Y is a random variable on $(\Omega, \mathcal{F}_1)$, $E|\mathsf{Y}| < \infty$, then*

$$E(\mathsf{Y} \mid \mathfrak{D}) = \int \mathsf{Y}(\omega) P^*(d\omega \mid \mathfrak{D}) \quad \text{a.s.}$$

Proof. Consider all nonnegative random variables on $(\Omega, \mathcal{F}_1)$ for which 4.28 holds. For the random variable χ_A, $A \in \mathcal{F}_1$, 4.28 becomes

$$P(A \mid \mathfrak{D}) = P^*(A \mid \mathfrak{D}) \quad \text{a.s.,}$$

which holds by 4.27. Hence 4.28 is true for simple functions. Now for $\mathsf{Y} \geq 0$, $E\mathsf{Y} < \infty$ take Y_n simple $\uparrow \mathsf{Y}$. Then by 4.24 and the monotone convergence theorem,

$$E(\mathsf{Y} \mid \mathfrak{D}) = \lim_n E(\mathsf{Y}_n \mid \mathfrak{D}) = \lim_n \int \mathsf{Y}_n \, dP^*(d\omega \mid \mathfrak{D}) = \int \mathsf{Y} P^*(d\omega \mid \mathfrak{D}) \quad \text{a.s.}$$

Unfortunately, a regular conditional probability on $\mathcal{F}_1$ given $\mathfrak{D}$ does *not* exist in general (see the Chapter notes). The difficulty is this: by (4.25), for $A_n \in \mathcal{F}_1$, disjoint, there is a set of probability zero such that

$$P(\cup A_n \mid \mathfrak{D}) \neq \sum_n P(A_n \mid \mathfrak{D}),$$

If $\mathcal{F}_1$ contains enough countable collections of disjoint sets, then the exceptional sets may pile up.

By doing something which is like passing over to representation space we can get rid of this difficulty. Let $\mathbf{Y}$ be a random vector taking values in a space $(R, \mathfrak{B})$.

Definition 4.29. *$\hat{P}(B \mid \mathfrak{D})$ defined for $B \in \mathfrak{B}$ and $\omega \in \Omega$ is called a regular conditional distribution for $\mathbf{Y}$ given $\mathfrak{D}$ if*

i) *for $B \in \mathfrak{B}$ fixed, $\hat{P}(B \mid \mathfrak{D})$ is a version of $P(\mathbf{Y} \in B \mid \mathfrak{D})$.*
ii) *for any $\omega \in \Omega$ fixed, $\hat{P}(B \mid \mathfrak{D})$ is a probability on $\mathfrak{B}$.*

If $\mathbf{Y}$ is a *random variable*, then by using the structure of $R^{(1)}$ in an essential way we prove the following theorem.

Theorem 4.30. *There always exists a regular conditional distribution for a random variable* Y *given* $\mathfrak{D}$.

Proof. The general proof we break up into steps.

Definition 4.31. $F(x \mid \mathfrak{D})$ *on* $R^{(1)} \times \Omega$ *is a conditional distribution function for* Y *given* $\mathfrak{D}$ *if*

i) $F(x \mid \mathfrak{D})$ *is a version of* $P(\mathsf{Y} < x \mid \mathfrak{D})$ *for every* x.
ii) *for every* ω, $F(x \mid \mathfrak{D})$ *is a distribution function.*

Proposition 4.32. *There always exists a conditional distribution function for* Y *given* $\mathfrak{D}$.

Proof. Let $R = \{r_j\}$ be the rationals. Select versions of $P(\mathsf{Y} < r_j \mid \mathfrak{D})$ and define

$$M_{ij} = \{\omega; P(\mathsf{Y} < r_j \mid \mathfrak{D}) < P(\mathsf{Y} < r_i \mid \mathfrak{D})\},$$
$$M = \bigcup_{r_j > r_i} M_{ij}.$$

So M is the set on which monotonicity is violated. By 4.20(2) $P(M) = 0$. Define

$$N_i = \left\{\omega; \lim_{r_j \uparrow r_i} P(\mathsf{Y} < r_j \mid \mathfrak{D}) \neq P(\mathsf{Y} < r_i \mid \mathfrak{D})\right\}, \quad N = \bigcup_i N_i.$$

Since $\chi_{(-\infty, r_j)}(\mathsf{Y}) \uparrow \chi_{(-\infty, r_i)}(\mathsf{Y})$ 4.24 implies $P(N_i) = 0$, or $P(N) = 0$. Finally take $r_j \uparrow \infty$ or $r_j \downarrow -\infty$ and observe that the set L on which $\lim_{r_j} P(\mathsf{Y} < r_j \mid \mathfrak{D})$ fails to equal one or zero, respectively, has probability zero. Thus, for ω in the complement of $M \cup N \cup L$, $P(\mathsf{Y} < x \mid \mathfrak{D})$ for $x \in R$ is monotone, left-continuous, zero and one at $-\infty$, $+\infty$, respectively. Take $G(x)$ an arbitrary distribution function and define:

$$F(x \mid \mathfrak{D}) = \begin{cases} G(x), & \omega \in M \cup N \cup L, \\ \lim_{r_j \uparrow x} P(\mathsf{Y} < r_j \mid \mathfrak{D}), & \text{otherwise.} \end{cases}$$

It is a routine job to verify that $F(x \mid \mathfrak{D})$ defined this way is a distribution function for every ω. Use 4.24 again and $\chi_{(-\infty, r_j)}(\mathsf{Y}) \uparrow \chi_{(-\infty, x)}(\mathsf{Y})$ to check that $F(x \mid \mathfrak{D})$ is a version of $P(\mathsf{Y} < x \mid \mathfrak{D})$.

Back to Theorem 4.30. For Y a random variable, define $\hat{P}(\cdot \mid \mathfrak{D})$, for each ω, as the probability on $(R^{(1)}, \mathcal{B}_1)$ extended from the distribution function $F(x \mid \mathfrak{D})$. Let $\mathcal{C}$ be the class of all sets $C \in \mathcal{B}_1$ such that $\hat{P}(C \mid \mathfrak{D})$ is a version of $P(\mathsf{Y} \in C \mid \mathfrak{D})$. By 4.23, $\mathcal{C}$ contains all finite disjoint unions of left-closed right-open intervals. By 4.24, $\mathcal{C}$ is closed under monotone limits. Hence $\mathcal{C} = \mathcal{B}_1$. Therefore, $\hat{P}(\cdot \mid \mathfrak{D})$ is the required conditional distribution.

For random vectors $\mathbf{Y}$ taking values in $(R, \mathcal{B})$, this result can be extended if $(R, \mathcal{B})$ looks enough like $(R^{(1)}, \mathcal{B}_1)$.

Definition 4.33. *Call $(R, \mathcal{B})$ a Borel space if there is a $E \in \mathcal{B}_1$ and a one-to-one mapping $\varphi: R \leftrightarrow E$ such that φ is $\mathcal{B}$-measurable and φ^{-1} is $\mathcal{B}_1$-measurable.*

Borel spaces include almost all the useful probability spaces. For example, see the proof in Appendix A that $(R^{(\infty)}, \mathcal{B}_\infty)$ is a Borel space. So, more easily, is $(R^{(n)}, \mathcal{B}_n)$, $n \geq 1$.

Theorem 4.34. *If $\mathbf{Y}$ takes values in a Borel space $(R, \mathcal{B})$, then there is a regular conditional distribution for $\mathbf{Y}$ given $\mathfrak{D}$.*

Proof. By definition, there is a one-to-one mapping $\varphi: R \leftrightarrow E \in \mathcal{B}_1$ with φ, φ^{-1} measurable $\mathcal{B}, \mathcal{B}_1$ respectively. Take $\mathsf{Y} = \varphi(\mathbf{Y})$; Y is a random variable so there is a regular conditional distribution $\hat{P}_0(A \mid \mathfrak{D}) = P(\mathsf{Y} \in A \mid \mathfrak{D})$ a.s. Define $\hat{P}(B \mid \mathfrak{D})$, for $B \in \mathcal{B}$, by

$$\hat{P}(B \mid \mathfrak{D}) = \hat{P}_0(\varphi(B) \mid \mathfrak{D}).$$

Because $\varphi(B)$ is the inverse set mapping of the measurable mapping $\varphi^{-1}(x)$, $\hat{P}(\cdot \mid \mathfrak{D})$ is a probability on $\mathcal{B}$ for each ω, and is also a version of $P(\mathbf{Y} \in B \mid \mathfrak{D})$ for every $B \in \mathcal{B}$.

Since the distribution of processes is determined on their range space, a regular conditional distribution will suit us just as well as a regular conditional probability. For instance,

Proposition 4.35. *Let $\mathbf{Y}$ be a random vector taking values in $(R, \mathcal{B})$, $\mathbf{y}$ a point in R, φ any $\mathcal{B}$-measurable function such that $E\,|\varphi(\mathbf{Y})| < \infty$. Then if $\hat{P}(\cdot \mid \mathfrak{D})$ is a regular conditional distribution for $\mathbf{Y}$ given $\mathfrak{D}$,*

$$E(\varphi(\mathbf{Y}) \mid \mathfrak{D}) = \int \varphi(\mathbf{y})\hat{P}(d\mathbf{y} \mid \mathfrak{D}) \quad \text{a.s.}$$

Proof. Same as 4.28 above.

If $\mathbf{Y}, \mathbf{X}$ are two random vectors taking values in $(R, \mathcal{B})$, $(S, \mathcal{E})$ respectively, then define a regular conditional distribution for $\mathbf{Y}$ given $\mathbf{X} = \mathbf{x}$, $\hat{P}(B \mid \mathbf{X} = \mathbf{x})$, in the analogous way: for each $\mathbf{x} \in S$, it is a probability on $\mathcal{B}$, and for B fixed it is a version of $\hat{P}(B \mid \mathbf{X} = \mathbf{x})$. Evidently the results 4.34 and 4.35 hold concerning $\hat{P}(B \mid \mathbf{X} = \mathbf{x})$. Some further useful results are:

Proposition 4.36. *Let $\varphi(\mathbf{x}, \mathbf{y})$ be a random variable on the product space $(R, \mathcal{B}) \times (S, \mathcal{E})$, $E\,|\varphi(\mathbf{X}, \mathbf{Y})| < \infty$. If $\hat{P}(\cdot \mid \mathbf{X} = \mathbf{x})$ is a regular conditional distribution for $\mathbf{Y}$ given $\mathbf{X} = \mathbf{x}$, then,*

$$E(\varphi(\mathbf{X}, \mathbf{Y}) \mid \mathbf{X} = \mathbf{x}) = \int \varphi(\mathbf{x}, \mathbf{y})\hat{P}(d\mathbf{y} \mid \mathbf{X} = \mathbf{x}) \quad \text{a.s.} \tag{4.37}$$

[*Note:* The point of 4.36 is that $\mathbf{x}$ is held constant in the integration occurring in the right-hand side of (4.37). Since $\varphi(\mathbf{x}, \mathbf{y})$ is jointly measurable in $(\mathbf{x}, \mathbf{y})$, for fixed $\mathbf{x}$ it is a measurable function of $\mathbf{y}$.]

Proof. Let $\varphi(\mathbf{x}, \mathbf{y}) = \chi_C(\mathbf{x})\chi_D(\mathbf{y})$, C, D measurable sets, then, by Problem 13(b). a.s. $\hat{P}$,

$$E(\chi_C(\mathbf{X})\chi_D(\mathbf{Y}) \mid \mathbf{X} = \mathbf{x}) = \chi_C(\mathbf{x})E(\chi_D(\mathbf{Y}) \mid \mathbf{X} = \mathbf{x})$$
$$= \chi_C(\mathbf{x})P(\mathbf{Y} \in D \mid \mathbf{X} = \mathbf{x}).$$

On the right in 4.37 is

$$\chi_C(\mathbf{x}) \int \chi_D(\mathbf{y})\hat{P}(d\mathbf{y} \mid \mathbf{X} = \mathbf{x}) = \chi_C(\mathbf{x})\hat{P}(D \mid \mathbf{X} = \mathbf{x})$$

which, by definition of $\hat{P}$, verifies (4.37) for this φ.

Now to finish the proof, just approximate in the usual way; that is, (4.37) is now true for all $\varphi(\mathbf{x}, \mathbf{y})$ of the form $\sum_{\alpha_i} \chi_{C_i}(\mathbf{x})\chi_{D_i}(\mathbf{y})$, C_i, D_i measurable sets, and apply now the usual monotonicity argument.

One can see here the usefulness of a regular conditional distribution. It is tempting to replace the right-hand side of (4.37) by $E(\phi(\mathbf{x}, \mathbf{Y}) \mid \mathbf{X} = \mathbf{x})$. But this object cannot be defined through the standard definition of conditional expectation (4.18).

A useful corollary is

Corollary 4.38. *Let* $\mathbf{X}$, $\mathbf{Y}$ *be independent random vectors. For* $\varphi(\mathbf{x}, \mathbf{y})$ *as in* 4.36,

$$E(\varphi(\mathbf{X}, \mathbf{Y}) \mid \mathbf{X} = \mathbf{x}) = E\varphi(\mathbf{x}, \mathbf{Y}) \quad \text{a.s.}$$

Proof. A regular conditional distribution for $\mathbf{Y}$ given $\mathbf{X} = \mathbf{x}$ is $P(\mathbf{Y} \in B)$. Apply this in 4.36.

Problems

16. Let I be any interval in $R^{(1)}$. A function $\varphi(x)$ on I measurable $\mathcal{B}_1(I)$ is called *convex* if for all $t \in [0, 1]$ and $x \in I$, $y \in I$

$$\varphi(tx + (1 - t)y) \leq t\varphi(x) + (1 - t)\varphi(y).$$

Prove that if Y is a random variable with range in I, and $E\,|\mathsf{Y}| < \infty$, then for $\varphi(x)$ convex on I, and $E\,|\varphi(\mathsf{Y})| < \infty$,

a) $\varphi(E\mathsf{Y}) \leq E[\varphi(\mathsf{Y})]$ (Jensen's inequality),

b) $\varphi(E(\mathsf{Y} \mid \mathcal{D})) \leq E(\varphi(\mathsf{Y}) \mid \mathcal{D})$ a.s.

[On (b) use the existence of a regular conditional distribution.]

17. Let **X**, **Y** be independent random vectors.

a) Show that a regular conditional probability for **Y** given $\mathbf{X} = \mathbf{x}$ is $P(\mathbf{Y} \in B)$.

b) If $\varphi(\mathbf{x}, \mathbf{y})$ is a random variable on the product of the range spaces, show that a regular conditional distribution for $Z = \varphi(\mathbf{X}, \mathbf{Y})$ given $\mathbf{X} = \mathbf{x}$ is $P(\varphi(\mathbf{x}, \mathbf{Y}) \in B)$.

NOTES

The modern definition of conditional probabilities and expectations, using the Radon-Nikodym theorem, was formulated by Kolmogorov in his monograph of 1933 [98].

The difficulty in getting a regular conditional probability on $\mathcal{F}(\mathbf{X})$, given any σ-field $\mathfrak{D}$, is similar to the extension problem. Once $\hat{P}(\cdot \mid \mathfrak{D})$ is gotten on the range space $(R, \mathcal{B})$ of **X**, if for any set $B \in \mathcal{B}$ contained in the complement of the range of **X**, $\hat{P}(B \mid \mathfrak{D}) \equiv 0$, then by 2.12 we can get a regular conditional probability $P^*(\cdot \mid \mathfrak{D})$ on $\mathcal{F}(\mathbf{X})$ given $\mathfrak{D}$. In this case it is sufficient that the range of **X** be a set in $\mathcal{B}$. Blackwell's article [8] also deals with this problem.

The counterexample referred to in the text is this: Let $\Omega = [0, 1]$, $\mathfrak{D} = \mathcal{B}([0, 1])$. Take C to be a nonmeasurable set of outer Lebesgue measure one and inner Lebesgue measure zero. The smallest σ-field $\mathcal{F}$ containing $\mathfrak{D}$ and C consists of all sets of the form $A = (C \cap B_1) \cup (C^c \cap B_2)$, where B_1, B_2 are in $\mathfrak{D}$. Define P on $\mathcal{F}$ by: If A has the above form,

$$P(A) = \tfrac{1}{2}l(B_1) + \tfrac{1}{2}l(B_2).$$

Because the B_1, B_2 in the definition of A are not unique, it is necessary to check that P is well defined, as well as a probability. There does not exist any regular conditional probability on $\mathcal{F}$ given $\mathfrak{D}$. This example can be found in Doob [39, p. 624].

CHAPTER 5

MARTINGALES

1. GAMBLING AND GAMBLING SYSTEMS

Since probability theory started from a desire to win at gambling, it is only sporting to discuss some examples from this area.

Example 1. Let $Z_1, Z_2, \ldots$ be independent, $Z_i = +1, -1$ with probability $p, 1 - p$. Suppose that at the nth toss, we bet the amount b. Then we receive the amount b if $Z_n = 1$, otherwise we lose the amount b. A gambling strategy is a rule which tells us how much to bet on the $(n + 1)$ toss. To be interesting, the rule will depend on the first n outcomes. In general, then, a gambling strategy is a sequence of functions $b_n:\{-1, +1\}^{(n)} \to [0, \infty)$ such that $b_n(Z_1, \ldots, Z_n)$ is the amount we bet on the $(n + 1)$ game. If we start with initial fortune S_0, then our fortune after n plays, S_n, is a random variable defined by

(5.1)
$$S_{n+1} = S_n + Z_{n+1}b_n(Z_1, \ldots, Z_n).$$

We may wish to further restrict the b_n by insisting that $b_n(Z_1, \ldots, Z_n) \le S_n$. Define the time of ruin n^* as the first n such that $S_n = 0$. One question that is certainly interesting is: What is $P(n^* < \infty)$; equivalently what is the probability of eventual ruin?

There is one property of the sequence of fortunes given by (5.1) that I want to focus on.

Proposition 5.2. *For* $p = \frac{1}{2}$,

$$E(S_{n+1} \mid S_n, S_{n-1}, \ldots, S_1) = S_n \quad \text{a.s.,} \quad n = 1, 2, \ldots$$

For $p \le \frac{1}{2}$,

$$E(S_{n+1} \mid S_n, S_{n-1}, \ldots, S_1) \le S_n \quad \text{a.s.,} \quad n = 1, 2, \ldots$$

Proof

$$\begin{aligned} E(S_{n+1} \mid Z_n, \ldots, Z_1) &= E(S_n + Z_{n+1}b_n(Z_1, \ldots, Z_n) \mid Z_n, \ldots, Z_1) \\ &= S_n + b_n(Z_1, \ldots, Z_n)E(Z_{n+1}). \end{aligned}$$

The last step follows because S_n is a function of $Z_1, \ldots, Z_n$. If $p = \frac{1}{2}$, then $EZ_{n+1} = 0$, otherwise $EZ_{n+1} < 0$, but b_n is nonnegative. Thus, for $p = \frac{1}{2}$, $E(S_{n+1} \mid Z_n, \ldots, Z_1) = S_n$ a.s. For $p \le \frac{1}{2}$, $E(S_{n+1} \mid Z_n, \ldots, Z_1) \le S_n$ a.s. Note that $\mathcal{F}(S_1, \ldots, S_n) \subset \mathcal{F}(Z_1, \ldots, Z_n)$. Taking conditional expectations of both sides above with respect to $S_1, \ldots, S_n$ concludes the proof.

Example 2. To generalize Example 1, consider any process $Z_1, Z_2, \ldots$ With any system of gambling on the successive outcomes of the $Z_1, Z_2, \ldots$ the player's fortune after n plays, S_n is a function of $Z_1, \ldots, Z_n$. We *assume* that corresponding to any gambling system is a sequence of real-valued functions φ_n measurable $\mathcal{B}_n$ such that the fortune S_n is given by

$$\varphi_n(Z_1, \ldots, Z_n), \quad \text{and} \quad E\,|S_n| < \infty.$$

Definition 5.3. *The sequence of games* $Z_1, Z_2, \ldots$, *under the given gambling system is called fair if*

$$E(S_{n+1} \mid S_n, \ldots, S_1) = S_n \quad \text{a.s.,} \quad n = 1, 2, \ldots,$$

unfavorable if

$$E(S_{n+1} \mid S_n, \ldots, S_1) \le S_n \quad \text{a.s.,} \quad n = 1, 2, \ldots$$

The first function of probability in gambling was involved in computation of odds. We wish to address ourselves to more difficult questions such as: Is there any system of gambling in fair or unfavorable games that will yield $S_n \to \infty$ a.s.? When can we assert that $P(n^* < \infty) = 1$ for a large class of gambling systems? This class of problems is a natural gateway to a general study of processes behaving like the sequences S_n.

2. DEFINITIONS OF MARTINGALES AND SUBMARTINGALES

Definition 5.4. *Given a process* $X_1, X_2, \ldots$ *It will be said to form a martingale* (MG) *if* $E\,|X_k| < \infty$, $k = 1, 2, \ldots$ *and*

$$E(X_{n+1} \mid X_n, \ldots, X_1) = X_n \quad \text{a.s.,} \quad n = 1, 2, \ldots \tag{5.5}$$

It forms a submartingale (SMG) *if* $E\,|X_k| < \infty$, $k = 1, 2, \ldots$, *and*

$$E(X_{n+1} \mid X_n, \ldots, X_1) \ge X_n \quad \text{a.s.,} \quad n = 1, 2, \ldots \tag{5.6}$$

Example 3. Sums of independent random variables. Let $Y_1, Y_2, \ldots$ be a sequence of independent random variables such that $E\,|Y_k| < \infty$, all k, $EY_k = 0$, $k = 1, 2, \ldots$ Define $X_n = Y_1 + \cdots + Y_n$; then $E\,|X_n| \le \sum_1^n E\,|Y_k| < \infty$. Furthermore, $\mathcal{F}(X_1, \ldots, X_n) = \mathcal{F}(Y_1, \ldots, Y_n)$, so

$$\begin{aligned} E(X_{n+1} \mid X_n, \ldots, X_1) &= E(X_{n+1} \mid Y_n, \ldots, Y_1) \\ &= E(Y_{n+1} \mid Y_n, \ldots, Y_1) + X_n = X_n. \end{aligned}$$

As mentioned in connection with fortunes, if the appropriate inequalities hold with respect to larger σ-fields, then the same inequalities hold for the process, that is,

Proposition 5.7. *Let* $X_1, X_2, \ldots$ *be a process with* $E|X_k| < \infty$, $k = 1, 2, \ldots$ *Let* $Y_1, \ldots,$ *be another process such that* $\mathcal{F}(X_1, \ldots, X_n) \subset \mathcal{F}(Y_1, \ldots Y_n)$, $n = 1, 2, \ldots$ *If*

$$E(X_{n+1} \mid Y_n, \ldots, Y_1) \underset{(=)}{\geq} X_n \quad \text{a.s.},$$

then $X_1, X_2, \ldots$ *is a* SMG(MG).

Proof. Follows from

$$E(E(X_{n+1} \mid Y_n, \ldots, Y_1) \mid X_n, \ldots, X_1) = E(X_{n+1} \mid X_n, \ldots, X_1).$$

[See 4.20 (3).]

For any $m \geq n$, if $X_1, \ldots$ is a SMG(MG), then

$$E(X_m \mid X_n, \ldots, X_1) \underset{(=)}{\geq} X_n \quad \text{a.s.},$$

because, for example,

$$\begin{aligned} E(X_{n+2} \mid X_n, \ldots, X_1) &= E(E(X_{n+2} \mid X_{n+1}, X_n, \ldots, X_1) \mid X_n, \ldots, X_1) \\ &\geq E(X_{n+1} \mid X_n, \ldots, X_1) \geq X_n \quad \text{a.s.} \end{aligned}$$

This remark gives us an equivalent way of defining SMG or MG.

Proposition 5.8. *Let* $X_1, X_2, \ldots$ *be a process with* $E|X_k| < \infty$, *all k. Then it is a* SMG(MG) *iff for every* $m \geq n$ *and* $A \in \mathcal{F}(X_1, \ldots, X_n)$,

$$\int_A X_m \, dP \underset{(=)}{\geq} \int_A X_n \, dP.$$

Proof. By definition, $\int_A X_m \, dP = \int_A E(X_m \mid X_n, \ldots, X_1) \, dP$.

Problem 1. Let $X_1, X_2, \ldots$ be a MG or SMG; let I be an interval containing the range of X_n, $n = 1, 2, \ldots$, and let $\varphi(x)$ be convex on I. Prove

a) If $E|\varphi(X_n)| < \infty$, $n = 1, 2, \ldots$ and $X_1, X_2, \ldots$ is a MG, then $X'_n = \varphi(X_n)$ is a SMG.
b) If $E|\varphi(X_n)| < \infty$, $n = 1, \ldots$, $\varphi(x)$ also nondecreasing on I, and $X_1, X_2, \ldots$ a SMG, then $X'_n = \varphi(X_n)$ is a SMG.

3. THE OPTIONAL SAMPLING THEOREM

One of the most powerful theorems concerning martingales is built around the idea of transforming a process by optional sampling. Roughly, the idea is this—starting with a given process $X_1, X_2, \ldots$, decide on the basis of observing $X_1, \ldots, X_n$ whether or not X_n will be the first value of the transformed process; then keep observing until on the basis of your observations a second value of the transformed process is chosen, and so forth. More precisely,

Definition 5.9. *Let $X_1, X_2, \ldots$ be an arbitrary process. A sequence $m_1, m_2, \ldots$ of integer-valued random variables will be called sampling variables if*

a) $1 \le m_1 \le m_2 \le \ldots,$
b) $\{m_k = j\} \in \mathcal{F}(X_1, \ldots, X_j).$

Then the sequence of random variables $\check{X}_n$ defined by $\check{X}_n = X_{m_n}$ is called the process derived by optional sampling from the original process.

Theorem 5.10. *Let $X_1, X_2, \ldots$ be a* SMG (MG), *$m_1, m_2, \ldots$ sampling variables, and $\check{X}_n$ the optional sampling process derived from $X_1, X_2, \ldots$ If*

a) $$E\,|\check{X}_k| < \infty, \quad \text{all } k,$$

b) $$\varliminf_N \int_{\{m_n > N\}} |X_N|\, dP = 0, \quad \text{all } n,$$

then the $\check{X}_1, \check{X}_2, \ldots$ process is a SMG(MG).

Proof. Let $A \in \mathcal{F}(\check{X}_1, \ldots, \check{X}_n)$. We must show that

$$\int_A \check{X}_{n+1}\, dP \underset{(=)}{\ge} \int_A \check{X}_n\, dP.$$

Let $D_j = A \cap \{m_n = j\}$. Then $A = \cup D_j$ and it suffices to show that

$$\int_{D_j} \check{X}_{n+1}\, dP \underset{(=)}{\ge} \int_{D_j} \check{X}_n\, dP, \quad \text{all } j.$$

We assert that $D_j \in \mathcal{F}(X_1, \ldots, X_j)$, since, letting $A = \{(\check{X}_1, \ldots, \check{X}_n) \in B\}$, $B \in \mathcal{B}_n$, then

$$\begin{aligned} D_j &= \{(\check{X}_1, \ldots, \check{X}_n) \in B,\ m_n = j\} \\ &= \bigcup_{\substack{j_1, \ldots, j_{n-1} \\ j_1 \le \cdots \le j_{n-1} \le j}} \{(\check{X}_1, \ldots, \check{X}_n) \in B,\ m_1 = j_1, \ldots,\ m_{n-1} = j_{n-1},\ m_n = j\} \\ &= \bigcup_{\substack{j_1 \cdots j_{n-1} \\ j_1 \le \cdots \le j_{n-1} \le j}} \{(X_{j_1}, \ldots, X_{j_{n-1}}, X_j) \in B,\ m_1 = j_1, \ldots,\ m_n = j\}. \end{aligned}$$

Evidently, each set in this union is in $\mathcal{F}(X_1, \ldots, X_j)$. Of course,

$$\int_{D_j} \check{X}_n\, dP = \int_{D_j} X_j\, dP.$$

Now, for arbitary $N > j$,

$$\int_{D_j} \breve{X}_{n+1}\, dP = \sum_{i=j}^{N} \int_{D_j \cap \{m_{n+1}=i\}} \breve{X}_{n+1}\, dP + \int_{D_j \cap \{m_{n+1}>N\}} \breve{X}_{n+1}\, dP$$

$$= \sum_{i=j}^{N} \int_{D_j \cap \{m_{n+1}=i\}} X_i\, dP + \int_{D_j \cap \{m_{n+1}>N\}} X_N\, dP$$

$$- \int_{D_j \cap \{m_{n+1}>N\}} (X_N - \breve{X}_{n+1})\, dP.$$

The first two terms on the right in the above telescope. Starting with the last term of the sum,

$$\int_{D_j \cap \{m_{n+1}=N\}} X_N\, dP + \int_{D_j \cap \{m_{n+1}>N\}} X_N\, dP = \int_{D_j \cap \{m_{n+1}\geq N\}} X_N\, dP.$$

But $\{m_{n+1} \geq N\} = \{m_{n+1} < N\}^c \in \mathcal{F}(X_1, \ldots, X_{N-1})$. By 5.8,

$$\int_{D_j \cap \{m_{n+1}\geq N\}} X_N\, dP \underset{(=)}{\geq} \int_{D_j \cap \{m_{n+1}\geq N\}} X_{N-1}\, dP = \int_{D_j \cap \{m_{n+1}> N-1\}} X_{N-1}\, dP.$$

Hence the first two terms reduce to being greater than or equal to

$$\int_{D_j \cap \{m_{n+1}\geq j\}} X_j\, dP.$$

But $D_j \subset \{m_{n+1} \geq j\}$, so we conclude that

$$\int_{D_j} \breve{X}_{n+1}\, dP \underset{(=)}{\geq} \int_{D_j} \breve{X}_n\, dP - \int_{D_j \cap \{m_{n+1}>N\}} (X_N - \breve{X}_{n+1})\, dP.$$

Letting $N \to \infty$ through an appropriate subsequence, by (b) we can force

$$\int_{D_j \cap \{m_{n+1}>N\}} X_N\, dP \to 0.$$

But $\{m_{n+1} > N\} \downarrow \emptyset$, and the bounded convergence theorem gives

$$\int_{D_j \cap \{m_{n+1}>N\}} \breve{X}_{n+1}\, dP \to 0,$$

since $E\,|\breve{X}_{n+1}| < \infty$ by (a). Q.E.D.

Corollary 5.11. *Assume the conditions of Theorem* 5.10 *are in force, and that in addition,* $\overline{\lim}\, E\,|X_n| < \infty$, *then*

1) $EX_1 \leq E\breve{X}_n \leq \overline{\lim}\, EX_n$,
2) $E\,|\breve{X}_n| \leq 2\,\overline{\lim}\, E\,|X_n| - EX_1$.

Proof. Without loss of generality, take $\mathsf{m}_1 \equiv 1$. By 5.10,

$$E(\check{\mathsf{X}}_n \mid \check{\mathsf{X}}_1) \geq \check{\mathsf{X}}_1 \Rightarrow E\check{\mathsf{X}}_n \geq E\check{\mathsf{X}}_1 = E\mathsf{X}_1.$$

Also,

$$E\check{\mathsf{X}}_n = \lim_N \sum_{j=1}^{N} \int_{\{\mathsf{m}_n=j\}} \mathsf{X}_j \, dP,$$

but

$$\sum_{j=1}^{N} \int_{\{\mathsf{m}_n=j\}} \mathsf{X}_j \, dP \leq \sum_{j=1}^{N} \int_{\{\mathsf{m}_n=j\}} \mathsf{X}_N \, dP = E\mathsf{X}_N - \int_{\{\mathsf{m}_n>N\}} \mathsf{X}_N \, dP.$$

By condition (b), we may take a subsequence of N such that

$$\int_{\{\mathsf{m}_n>N\}} \mathsf{X}_N \, dP \to 0.$$

Using this subsequence, the conclusion $E\check{\mathsf{X}}_n \leq \overline{\lim}\, E\mathsf{X}_n$ follows. For part (2), note that if X_n is a SMG, so is $\mathsf{X}_n^+ = \max(0, \mathsf{X}_n)$. Thus by 5.10 so is $\check{\mathsf{X}}_n^+$. Applying (1) we have

$$E\check{\mathsf{X}}_n^+ \leq \overline{\lim}\, E\mathsf{X}_n^+.$$

For the original process,

$$E\check{\mathsf{X}}_n^+ - E\check{\mathsf{X}}_n^- \geq E\mathsf{X}_1 \qquad \text{or} \qquad E\check{\mathsf{X}}_n^- \leq E\check{\mathsf{X}}_n^+ - E\mathsf{X}_1,$$

$$E\,|\check{\mathsf{X}}_n| = E\check{\mathsf{X}}_n^+ + E\check{\mathsf{X}}_n^- \leq 2\,\overline{\lim}\, E\mathsf{X}_n^+ - E\mathsf{X}_1 \leq 2\,\overline{\lim}\, E\,|\mathsf{X}_n| - E\mathsf{X}_1.$$

Proposition 5.12. *Under hypothesis (b) of Theorem* 5.10, *if* $\overline{\lim}\, E|\mathsf{X}_n| < \infty$, *then* $\sup_n E\,|\check{\mathsf{X}}_n| < \infty$, *so that (a) holds and the theorem is in force.*

Proof. We introduce truncated sampling variables: for $M > 0$,

$$\mathsf{m}_{n,M} = \begin{cases} \mathsf{m}_n & \text{if} \quad \mathsf{m}_n \leq M, \\ M & \text{if} \quad \mathsf{m}_n > M, \end{cases}$$

and $\check{\mathsf{X}}_{n,M} = \mathsf{X}_{\mathsf{m}_{n,M}}$. The reason for this is

$$E\,|\check{\mathsf{X}}_{n,M}| = \sum_{j=1}^{M} \int_{\{\mathsf{m}_{n,M}=j\}} |\mathsf{X}_j| \, dP \leq \sum_{j=1}^{M} E\,|\mathsf{X}_j| < \infty,$$

so that the conditions of Theorem 5.10 are in force for $\check{\mathsf{X}}_{n,M}$. By Corollary 5.11, $E\,|\check{\mathsf{X}}_{n,M}| \leq \alpha < \infty$, where α does not depend on n or M. Note now that $\lim_M \mathsf{m}_{n,M} = \mathsf{m}_n$; hence $\lim_M \check{\mathsf{X}}_{n,M} = \check{\mathsf{X}}_n$. By the Fatou lemma (Appendix A.27),

$$\int \underline{\lim}_M\, |\check{\mathsf{X}}_{n,M}| \, dP \leq \underline{\lim}_M\, E\,|\check{\mathsf{X}}_{n,M}| \leq \alpha$$

or

$$E\,|\check{\mathsf{X}}_n| \leq \alpha, \quad \text{all } n.$$

One nice application of optional sampling is

Proposition 5.13. *If* $\mathsf{X}_1, \mathsf{X}_2, \ldots$ *is a SMG then for any* $x > 0$

1) $$P\left(\max_{1\le j\le k} \mathsf{X}_j > x\right) \le \frac{1}{x} E\,|\mathsf{X}_k|,$$

2) $$P\left(\min_{1\le j\le k} \mathsf{X}_j < -x\right) \le \frac{1}{x} (E\,|\mathsf{X}_k| - E\mathsf{X}_1).$$

Proof. 1) Define sampling variables by

$$\mathsf{m}_1 = \begin{cases} \text{first } j \le k \quad \text{such that } \mathsf{X}_j > x, \quad \text{or} \\ k \quad \text{if no such } j \text{ exists.} \end{cases}$$

$$\mathsf{m}_n = k, \quad n \ge 2.$$

The conditions of 5.10 are satisfied, since $\{\mathsf{m}_n > N\} = \emptyset$, $N \ge k$, and $E\,|\check{\mathsf{X}}_n| \le \sum_1^k E\,|\mathsf{X}_j| < \infty$. Now

$$P\left(\max_{1\le j\le k} \mathsf{X}_j > x\right) = P(\mathsf{X}_{\mathsf{m}_1} > x) \le \frac{1}{x} \int_{\{\mathsf{X}_{\mathsf{m}_1} > x\}} \mathsf{X}_{\mathsf{m}_1}\, dP.$$

By the optional sampling theorem,

$$\int_{\{\mathsf{X}_{\mathsf{m}_1} > x\}} \mathsf{X}_{\mathsf{m}_1}\, dP \le \int_{\{\mathsf{X}_{\mathsf{m}_1} > x\}} \mathsf{X}_k\, dP \le E\,|\mathsf{X}_k|.$$

2) To go below, take $x > 0$ and define

$$\mathsf{m}_1 = 1,$$

$$\mathsf{m}_2 = \begin{cases} \text{first } j \le k \quad \text{such that } \mathsf{X}_j < -x, \text{ or} \\ k \quad \text{if no such } j \text{ exists,} \end{cases}$$

$$\mathsf{m}_n = k, \quad n \ge 3.$$

The conditions of 5.10 are again satisfied, so

$$E\mathsf{X}_{\mathsf{m}_2} \ge E\mathsf{X}_{\mathsf{m}_1} = E\mathsf{X}_1.$$

But

$$E\mathsf{X}_{\mathsf{m}_2} = \int_{\{\mathsf{X}_{\mathsf{m}_2} \ge -x\}} \mathsf{X}_{\mathsf{m}_2}\, dP + \int_{\{\mathsf{X}_{\mathsf{m}_2} < -x\}} \mathsf{X}_{\mathsf{m}_2}\, dP$$

$$\le E\,|\mathsf{X}_k| - xP(\mathsf{X}_{\mathsf{m}_2} < -x).$$

Therefore, since

$$P\left(\min_{1\le j\le k} \mathsf{X}_j < -x\right) = P(\mathsf{X}_{\mathsf{m}_2} < -x),$$

part (2) follows.

Problem 2. Use 5.13 and Problem 1 to prove Kolmogorov's extension of the Chebyshev inequality (see the notes to Chapter 3). That is, if $X_1, X_2, \ldots$ is a MG and $EX_n^2 < \infty$, then show

$$P\left(\max_{1 \le k \le n} |X_k| > \epsilon\right) \le \frac{1}{\epsilon^2} EX_n^2.$$

4. THE MARTINGALE CONVERGENCE THEOREM

One of the outstanding strong convergence theorems is

Theorem 5.14. *Let $X_1, X_2, \ldots$ be a SMG such that $\overline{\lim}\, E|X_n| < \infty$, then there exists a random variable X such that $X_n \to X$ a.s. and $E\,|X| < \infty$.*

Proof. To prove this, we are going to define sampling variables. Let $b > a$ be any two real numbers. Let

$$m_1^* = \begin{cases} +\infty & \text{if } X_n > a, \text{ all } n, \\ \text{first } n \text{ such that } X_n \le a, & \text{otherwise;} \end{cases}$$

$$m_2^* = \begin{cases} +\infty & \text{if } X_n < b, \text{ all } n \ge m_1^*, \\ \text{first } n \ge m_1^* \text{ such that } X_n \ge b, & \text{otherwise.} \end{cases}$$

In general,

$$m_{2n+1}^* = \begin{cases} +\infty & \text{if } X_n > a, \text{ all } n \ge m_{2n}^*, \\ \text{first } n \ge m_{2n}^* \text{ such that } X_n \le a & \text{otherwise;} \end{cases}$$

$$m_{2n+2}^* = \begin{cases} +\infty & \text{if } X_n < b, \text{ all } n \ge m_{2n+1}^*, \\ \text{first } n \ge m_{2n+1}^* \text{ such that } X_n \ge b, & \text{otherwise.} \end{cases}$$

That is, the m_n^* are the successive times that X_n drops below a or rises above b. Now, what we would like to conclude for any $b > a$ is that

$$P(X_n \le a \text{ i.o.}, X_n \ge b \text{ i.o.}) = 0.$$

This is equivalent to proving that if

$$S = \bigcap_1^\infty \{m_n^* < \infty\},$$

that $P(S) = 0$. Suppose we defined $\check{X}_n = X_{m_n^*}$, and

$$Z = (\check{X}_3 - \check{X}_2) + (\check{X}_5 - \check{X}_4) + \cdots$$

On S, $\check{X}_{2n+1} - \check{X}_{2n} \le -(b - a)$, so $Z = -\infty$, but if $\check{X}_n$ is a SMG, $E(\check{X}_{2n+1} - \check{X}_{2n}) \ge 0$, so $EZ \ge 0$, giving a contradiction which would be resolved only by $P(S) = 0$. To make this idea firm, define

$$\mathsf{m}_{n,M} = \min(M, \mathsf{m}_n^*), \quad \check{\mathsf{X}}_{n,M} = \mathsf{X}_{\mathsf{m}_{n,M}}.$$

Then the optional sampling theorem is in force, and $\check{\mathsf{X}}_{n,M}$ is a SMG. Define

$$\mathsf{Z}_M = (\check{\mathsf{X}}_{3,M} - \check{\mathsf{X}}_{2,M}) + (\check{\mathsf{X}}_{5,M} - \check{\mathsf{X}}_{4,M}) + \cdots,$$

noting that $\check{\mathsf{X}}_{2n+1,M} - \check{\mathsf{X}}_{2n,M} = 0$ for $m_{2n}^* \geq M$, which certainly holds for $2n \geq M$.

Let β_M be the largest n such that $\mathsf{m}_{2n}^* < M$. On the set $\{\beta_M = k\}$,

$$\begin{aligned}\mathsf{Z}_M &= (\check{\mathsf{X}}_{3,M} - \check{\mathsf{X}}_{2,M}) + \cdots + (\check{\mathsf{X}}_{2k+1,M} - \check{\mathsf{X}}_{2k,M}) \\ &= (\check{\mathsf{X}}_{3,M} - \check{\mathsf{X}}_{2,M}) + \cdots + (a - \check{\mathsf{X}}_{2k,M}) + (\check{\mathsf{X}}_{2k+1,M} - a).\end{aligned}$$

The last term above can be positive only if $\mathsf{m}_{2k+1}^* \geq M$, in which case it becomes $(\mathsf{X}_M - a)$. In any case, on $\{\beta_M = k\}$

$$\mathsf{Z}_M \leq -k(b - a) + (\mathsf{X}_M - a)^+,$$

or, in general,

$$\mathsf{Z}_M \leq -\beta_M(b - a) + (\mathsf{X}_M - a)^+. \tag{5.15}$$

On the other hand, since $\mathsf{X}_{n,M}$ is a SMG, $E\mathsf{Z}_M \geq 0$. Taking expectations of (5.15) gives

$$E\beta_M \leq \frac{E(\mathsf{X}_M - a)^+}{b - a}. \tag{5.16}$$

Now β_M is nondecreasing, and $\beta_M \uparrow \infty$ on S. This contradicts (5.16) since $\overline{\lim}\, E(\mathsf{X}_M - a)^+ \leq \overline{\lim}\, E\,|\mathsf{X}_M| + a$, unless $P(S) = 0$. This establishes

$$P(\cup \{\underline{\lim}\, \mathsf{X}_n \leq a < b \leq \overline{\lim}\, \mathsf{X}_n\}) = 0.$$

where the union is taken over all rational a, b. Then either a random variable X exists such that $\mathsf{X}_n \to \mathsf{X}$ a.s. or $|\mathsf{X}_n| \to \infty$ with positive probability. This latter case is eliminated by Fatou's lemma, that is,

$$\int \underline{\lim}\, |\mathsf{X}_n|\, dP \leq \underline{\lim} \int |\mathsf{X}_n|\, dP.$$

From this we also conclude that

$$\int |\mathsf{X}|\, dP \leq \underline{\lim} \int |\mathsf{X}_n|\, dP < \infty. \qquad \text{Q.E.D.}$$

In the body of the above proof, a result is obtained which will be useful in the future.

Lemma 5.17. *Let* $X_1, \ldots, X_M$ *be random variables such that* $E|X_n| < \infty$, $n = 1, \ldots, M$ *and* $E(X_{n+1} \mid X_n, \ldots, X_1) \geq X_n$, $n = 1, \ldots, M-1$. *Let* β_M *be the number of times that the sequence* $X_1, \ldots, X_M$ *crosses a finite interval* $[a, b]$ *from left to right. Then,*

$$E\beta_M \leq \frac{E(X_M - a)^+}{b - a}.$$

Problem 3. Now use the martingale convergence theorem to prove that for $X_1, X_2, \ldots$ independent random variables, $EX_k = 0$, $k = 1, 2, \ldots$,

$$\sum_1^\infty EX_k^2 < \infty \Rightarrow \sum_1^n X_k \xrightarrow{\text{a.s.}}.$$

5. FURTHER MARTINGALE THEOREMS

To go further and apply the basic convergence theorem, we need:

Definition 5.18. *Let* $X_1, X_2, \ldots$ *be a process. It is said to be uniformly integrable if* $E|X_k| < \infty$, *all k, and*

$$\lim_{x\uparrow\infty} \overline{\lim_n} \int_{\{|X_n|>x\}} |X_n| \, dP = 0.$$

Proposition 5.19. *Let* $X_1, X_2, \ldots$ *be uniformly integrable, then* $\overline{\lim}\, E|X_n| < \infty$. *If* $X_n \to X$ a.s, *then* $E|X - X_n| \to 0$, *and* $EX_n \to EX$.

Proof. First of all, for any $x > 0$,

$$E|X_n| \leq x + \int_{\{|X_n|>x\}} |X_n| \, dP.$$

Hence,

$$\overline{\lim}\, E|X_n| \leq x + \overline{\lim} \int_{\{|X_n|>x\}} |X_n| \, dP.$$

But the last term must be finite for some value of x sufficiently large. For the next item, use Fatou's lemma:

$$\int |X| \, dP = \int \underline{\lim}\, |X_n| \, dP \leq \underline{\lim} \int |X_n| \, dP.$$

Now, by Egoroff's theorem (Halmos [64, p. 88]), for any $\epsilon > 0$, we can take $A \in \mathcal{F}$ such that $P(A) < \epsilon$ and $V_n = |X_n - X| \to 0$ uniformly on A^c,

$$\overline{\lim}\, EV_n = \overline{\lim} \int_A |X_n - X| \, dP \leq \overline{\lim} \int_A |X_n| \, dP + \int_A |X| \, dP.$$

But bounds can be gotten by writing

$$\text{(5.20)}\qquad \int_A |X_n|\, dP = \int_{A\cap\{|X_n|\le x\}} |X_n|\, dP + \int_{A\cap\{|X_n|>x\}} |X_n|\, dP$$

$$\le x\epsilon + \int_{\{|X_n|>x\}} |X_n|\, dP.$$

Taking $\epsilon \downarrow 0$, we get

$$\overline{\lim}\, EV_n \le \overline{\lim_n} \int_{\{|X_n|>x\}} |X_n|\, dP + \int_{\{|X|>x\}} |X|\, dP,$$

since (5.20) holds for $|X|$ as well as $|X_n|$. Now letting $x \uparrow \infty$ gives the result. For the last result, use $|EX - EX_n| \le E|X - X_n|$.

We use this concept of uniform integrability in the proof of:

Theorem 5.21. *Let* $Z, Y_1, Y_2, \ldots,$ *be random variables on* $(\Omega, \mathcal{F}, P)$, such that $E|Z| < \infty$. Then

$$E(Z \mid Y_1, Y_2, \ldots, Y_n) \xrightarrow[1]{\text{a.s.}} E(Z \mid Y_1, Y_2, \ldots).$$

(Here $\xrightarrow[1]{\text{a.s.}}$ indicates both a.s. and first mean convergence.)

Proof. Let $X_n = E(Z \mid Y_1, \ldots, Y_n)$. Then $E|X_n| \le E|Z|$. By 4.20(3)

$$E(X_{n+1} \mid Y_1, \ldots, Y_n) = X_n \quad \text{a.s.}$$

Since

$$\mathcal{F}(X_1, \ldots, X_n) \subset \mathcal{F}(Y_1, \ldots, Y_n),$$

the sequence $X_1, X_2, \ldots$ is *a* MG. Since $\overline{\lim}\, E|X_n| < \infty$,

$$X_n \xrightarrow{\text{a.s.}} X, \qquad \text{and} \qquad E|X| \le \overline{\lim}\, E|X_n| \le E|Z|.$$

By convexity, $|X_n| \le E(|Z| \mid Y_1, \ldots, Y_n)$. Hence

$$\int_{\{|X_n|>x\}} |X_n|\, dP \le \int_{\{|X_n|>x\}} E(|Z| \mid Y_1, \ldots, Y_n)\, dP = \int_{\{|X_n|>x\}} |Z|\, dP$$

$$\le \int_{\{\sup_n |X_n|>x\}} |Z|\, dP.$$

Now,

$$U = \sup_n |X_n|$$

is a.s. finite; hence as $x \uparrow \infty$, the sets $\{U > x\}$ converge down to a set of probability zero. Thus the X_n sequence is uniformly integrable and $X_n \xrightarrow{1} X$. Let $A \in \mathcal{F}(Y_1, \ldots, Y_N)$; then

$$\lim_n \int_A X_n\, dP = \int_A X\, dP.$$

But for $n \geq N$,

$$\int_A \mathsf{X}_n \, dP = \int_A \mathsf{Z} \, dP.$$

This implies that $\mathsf{X} = \mathsf{Z}$ a.s.

Corollary 5.22. *If* $\mathcal{F}(\mathsf{Z}) \subset \mathcal{F}(\mathsf{Y}_1, \mathsf{Y}_2, \ldots)$, *and* $\mathsf{E}\,|\mathsf{Z}| < \infty$, *then*

$$E(\mathsf{Z} \mid \mathsf{Y}_1, \ldots, \mathsf{Y}_n) \xrightarrow{\text{a.s.}} \mathsf{Z}.$$

In particular, if $A \in \mathcal{F}(\mathsf{Y}_1, \ldots)$, *then*

$$P(A \mid \mathsf{Y}_1, \ldots, \mathsf{Y}_n) \xrightarrow{\text{a.s.}} \chi_A(\omega).$$

Proof. Immediate.

There is a converse of sorts, most useful, to 5.22.

Theorem 5.23. *Let* $\mathsf{X}_1, \mathsf{X}_2, \ldots$ *be a* SMG (MG), *and uniformly integrable. Then* $\mathsf{X}_n \xrightarrow{\text{a.s.}} \mathsf{X}$, *and*

$$\mathsf{X}_n \leq E(\mathsf{X} \mid \mathsf{X}_1, \ldots, \mathsf{X}_n)$$

(*with equality holding if* $\mathsf{X}_1, \mathsf{X}_2, \ldots$ *is MG*).

Proof. By definition, for every $m > n$,

$$E(\mathsf{X}_m \mid \mathsf{X}_1, \ldots, \mathsf{X}_n) \underset{(=)}{\geq} \mathsf{X}_n \text{ a.s.}$$

Thus, for every set $A \in \mathcal{F}(\mathsf{X}_1, \ldots, \mathsf{X}_n)$, and $m > n$,

$$\int_A \mathsf{X}_n \, dP \underset{(=)}{\leq} \int_A \mathsf{X}_m \, dP.$$

But by (5.19) $\overline{\lim}\, E\,|\mathsf{X}_n| < \infty$, hence $\mathsf{X}_n \xrightarrow{\text{a.s.}} \mathsf{X}$, implying $\mathsf{X}_n \xrightarrow{1} \mathsf{X}$, so

$$\int_A \mathsf{X}_n \, dP \underset{(=)}{\leq} \int_A \mathsf{X} \, dP, \Rightarrow \mathsf{X}_n \underset{(=)}{\leq} E(\mathsf{X} \mid \mathsf{X}_1, \ldots, \mathsf{X}_n) \text{ a.s.}$$

Therefore, every uniformly integrable MG sequence is a sequence of conditional expectations and every uniformly integrable SMG sequence is bounded above by a MG sequence.

In 5.21 we added conditions, that is, we took the conditional expectation of Z relative to increasing sequence σ-fields $\mathcal{F}(\mathsf{Y}_1, \ldots, \mathsf{Y}_n)$. We can also go the other way.

Theorem 5.24. *Let* $\mathsf{Z}, \mathsf{Y}_1, \ldots$ *be random variables,* $E\,|\mathsf{Z}| < \infty$, $\mathfrak{T}$ *the tail* σ-*field on the* $\mathsf{Y}_1, \mathsf{Y}_2, \ldots$ *process. Then*

$$E(\mathsf{Z} \mid \mathsf{Y}_n, \mathsf{Y}_{n+1}, \ldots) \underset{1}{\xrightarrow{\text{a.s.}}} E(\mathsf{Z} \mid \mathfrak{T}).$$

Proof. A bit of a sly dodge is used here. Define $X_{-n} = E(Z \mid Y_n, Y_{n+1}, \ldots)$, $n = 1, 2, \ldots$ Note that

$$E(X_{-n} \mid Y_{n+1}, \ldots) = X_{-n-1},$$

so Lemma 5.17 is applicable to the sequence $X_{-M}, X_{-M+1}, \ldots, X_{-1}$. For any interval $[a, b]$, if β_M is the number of times that this sequence crosses from below a to above b, then

$$E\beta_M \leq \frac{E(X_{-1} - a)^+}{b - a}.$$

By the same argument as in the main convergence theorem, X_{-n} either must converge to a random variable X a.s. or $|X_{-n}| \to \infty$ with positive probability. But $E|X_{-n}| \leq E|Z|$, so that Fatou lemma again gives $X_{-n} \to X$ a.s., and $E|X| \leq E|Z|$. Just as in the proof of 5.21,

$$\int_{\{|X_{-n}|>x\}} |X_{-n}| \, dP \leq \int_{\{|X_{-n}|>x\}} |Z| \, dP \leq \int_{\{\sup_{n\geq 1}|X_{-n}|>x\}} |Z| \, dP,$$

so the X_{-n} sequence is uniformly integrable; hence $E(Z \mid Y_n, Y_{n+1}, \ldots) \xrightarrow{1} X$. Since X is an a.s. limit of random variables measurable with respect to $\mathcal{F}(Y_n, \ldots)$, then X is measurable with respect to $\mathcal{F}(Y_n, \ldots)$ for every n. Hence X is a random variable on $(\Omega, \mathfrak{J})$. Let $A \in \mathfrak{J}$; then $A \in \mathcal{F}(Y_n, \ldots)$, and

$$\int_A X \, dP = \lim_n \int_A E(Z \mid Y_n, \ldots) \, dP = \int_A Z \, dP,$$

which proves that $X = E(Z \mid \mathfrak{J})$ a.s.

Problems

4. Show that $E|X_n| < \infty$, $E|X| < \infty$, $E|X_n - X| \to 0 \Rightarrow X_1, X_2, \ldots$ is a uniformly integrable sequence. Get the same conclusion if $X_n \xrightarrow{\text{a.s.}} X$, and $E|X_n| \to E|X|$.

5. Apply Theorem 5.21 to the analytic model for coin-tossing and the coin-tossing variables defined in Chapter 1, Section 5. Take $f(x)$ any Borel measurable function on $[0, 1)$ such that $\int |f(x)| \, dx < \infty$. Let $I_1, I_2, \ldots, I_N$ be the intervals

$$\left[0, \frac{1}{2^n}\right), \left[\frac{1}{2^n}, \frac{2}{2^n}\right), \ldots,$$

and define the step function $f_n(x)$ by

$$f_n(x) = \frac{1}{\|I_k\|} \int_{I_k} f(y) \, dy, \qquad x \in I_k, \qquad k = 1, \ldots, N.$$

Then prove that

$$f_n(x) \xrightarrow{\text{a.s.}} f(x).$$

6. Given a process $X_1, X_2, \ldots$, abbreviate $\mathcal{F}_n = \mathcal{F}(X_n, X_{n+1}, \ldots)$. Use 5.24 to show that the tail σ-field $\mathfrak{T}$ on the process has the zero-one property $(C \in \mathfrak{T} \Rightarrow P(C) = 0 \text{ or } 1)$ iff for every $A \in \mathcal{F}(\mathbf{X})$,

$$\lim_n \sup_{B \in \mathcal{F}_n} |P(A \cap B) - P(A)P(B)| = 0.$$

7. Use Problem 15, Chapter 4, to prove the strong law of large numbers (3.30) from the martingale results.

6. STOPPING TIMES

Definition 5.25. *Given a process $X_1, X_2, \ldots$, an extended stopping time is an extended integer-valued random variable $n^*(\omega) \geq 1$ such that $\{n^* = j\} \in \mathcal{F}(X_1, \ldots, X_j)$. The process $\check{X}_1, \check{X}_2, \ldots$ derived under stopping is defined by*

$$\check{X}_n = \begin{cases} X_n, & \text{if } n \leq n^*(\omega), \\ X_{n^*} & n > n^*(\omega). \end{cases}$$

If we define variables $m_n(\omega)$ as $\min(n, n^*(\omega))$, then obviously the $m_n(\omega)$ are optional sampling variables, and the $\check{X}_n$ defined in 5.25 above are given by $\check{X}_n = X_{m_n}$. Hence stopping is a special case of optional sampling. Furthermore,

Proposition 5.26. *Let $X_1, X_2, \ldots$ be a* SMG(MG), *and $\check{X}_1, \check{X}_2, \ldots$ be derived under stopping from $X_1, X_2, \ldots$ Then $\check{X}_1, \check{X}_2, \ldots$ is a* SMG(MG).

Proof. All that is necessary is to show that conditions (a) and (b) of the optional sampling Theorem 5.10 are met. Now

$$E|\check{X}_n| = \int_{\{n \leq n^*\}} |X_n| \, dP + \int_{\{n > n^*(\omega)\}} |X_{n^*}| \, dP$$

$$\leq E|X_n| + \sum_{j=1}^{n-1} \int_{\{n^*=j\}} |X_j| \, dP \leq \sum_{j=1}^{n} E|X_j|.$$

Noting that $m_n(\omega) = \min(n, n^*(\omega)) \leq n$, we know that $\{m_n > N\} = \emptyset$, for $N \geq n$, and

$$\int_{\{m_n > N\}} |X_N| \, dP = 0.$$

Not only does stopping appear as a transformation worthy of study, but it is also a useful tool in proving some strong theorems. For any set $B \in \mathcal{B}_1$, and a process $X_1, X_2, \ldots$, define an extended stopping time by

$$n^*(\omega) = \begin{cases} \text{first } n \text{ such that } X_n \in B, \\ \infty \quad \text{if no such } n \text{ exists.} \end{cases}$$

Then $\check{X}_1, \check{X}_2, \ldots$ is called *the process stopped on B*.

Proposition 5.27. *Let* $X_1, X_2, \ldots$ *be a* SMG, *B the set* $[a, \infty]$, $a > 0$. *If* $E[\sup_n (X_{n+1} - X_n)^+] < \infty$, *then for* $\check{X}_n$ *the process stopped on B,*

$$\overline{\lim}\, E\,|\check{X}_n| < \infty.$$

Proof. For any n, $\check{X}_n^+ \leq a + U$, where

$$U = \sup_n (X_{n+1} - X_n)^+.$$

But $\check{X}_1 = X_1$. By the fact that $\check{X}_n$ is a SMG, $E\check{X}_n \geq EX_1$, so $E\check{X}_n^- \leq E\check{X}_n^+ - EX_1$. Thus

$$E\,|\check{X}_n| \leq 2E\check{X}_n^+ - EX_1 \leq 2a + 2EU - EX_1.$$

Theorem 5.28. *Let* $X_1, X_2, \ldots$ *be a* MG *such that* $E(\sup_n |X_{n+1} - X_n|) < \infty$. *If the sets* A_1, A_2 *are defined by*

$$A_1 = \left\{\omega;\ \lim_n X_n(\omega) \text{ exists}\right\},$$

$$A_2 = \{\omega;\ \overline{\lim}\, X_n(\omega) = +\infty,\ \underline{\lim}\, X_n(\omega) = -\infty\},$$

then $A_1 \cup A_2 = \Omega$ a.s.

Proof. Consider the process $X_1, X_2, \ldots$ stopped on $[K, \infty]$. By 5.27 and the basic convergence theorem $\check{X}_n \xrightarrow{\text{a.s.}} X$. On the set $F_K = \{\sup_n X_n < K\}$, $X_n = \check{X}_n$, all n. Hence on F_K, $\lim_n X_n$ exists and is finite a.s. Thus this limit exists and is finite a.s. on the set $\bigcup_{K=1}^{\infty} F_K$, but this set is exactly the set $\{\overline{\lim}\, X_n < \infty\}$. By using now the MG sequence $-X_1, -X_2, \ldots$, conclude that $\lim_n X_n$ exists and is finite a.s. on the set $\{\underline{\lim}\, X_n > -\infty\}$. Hence $\lim X_n$ exists and is finite for almost all ω in the set $\{\overline{\lim}\, X_n < \infty\} \cup \{\underline{\lim}\, X_n > -\infty\}$, and the theorem is proved.

This theorem is something like a zero-one law. Forgetting about a set of probability zero, according to this theorem we find that for every ω, either $\lim X_n(\omega)$ exists finite, or the sequence $X_n(\omega)$ behaves badly in the sense that $\underline{\lim}\, X_n(\omega) = -\infty$, $\overline{\lim}\, X_n(\omega) = +\infty$. There are some interesting and useful applications. An elegant extension of the Borel-Cantelli lemma valid for arbitrary processes comes first.

Corollary 5.29 *(extended Borel-Cantelli lemma). Let* $Y_1, Y_2, \ldots$ *be any process, and* $A_n \in \mathcal{F}(Y_1, \ldots, Y_n)$. *Then almost surely*

$$\{\omega;\ \omega \in A_n \text{ i.o.}\} = \left\{\omega;\ \sum_1^{\infty} P(A_{n+1} \mid Y_n, \ldots, Y_1) = \infty\right\}.$$

Proof. Let

$$X_{n+1} = \sum_{k=1}^{n} [\chi_{A_{k+1}} - P(A_{k+1} \mid Y_k, \ldots, Y_1)], \qquad n \geq 1.$$

Note that

$$E(X_{n+1} \mid Y_n, \ldots, Y_1) = E([\chi_{A_{n+1}} - P(A_{n+1} \mid Y_n, \ldots, Y_1)] \mid Y_n, \ldots, Y_1) + X_n$$
$$= X_n.$$

Obviously, also $|X_{n+1}| \le n$, so $\{X_n\}$ is a MG sequence. To boot,

$$|X_{n+1} - X_n| \le 1,$$

so 5.28 is applicable. Now

$$\{\omega; \omega \in A_n \text{ i.o.}\} = \left\{\omega; \sum_1^\infty \chi_{A_n}(\omega) = \infty\right\}.$$

Let $D_1 = \{\lim_n X_n \text{ exists finite}\}$. Then on D_1,

$$\sum_1^\infty \chi_{A_n} = \infty \Leftrightarrow \sum_1^\infty P(A_{n+1} \mid Y_n, \ldots, Y_1) = \infty.$$

Let $D_2 = \{\underline{\lim}\, X_n = -\infty, \overline{\lim}\, X_n = +\infty\}$, then for all $\omega \in D_2$,

$$\sum_1^\infty \chi_{A_n} = \infty \quad \text{and} \quad \sum_1^\infty P(A_{n+1} \mid Y_n, \ldots, Y_1) = \infty.$$

Since $P(D_1 \cup D_2) = 1$, the corollary follows.

Some other applications of 5.28 are in the following problems.

Problems

8. A loose end, which is left over from the random signs problem, is that if $Y_1, Y_2, \ldots$ are independent random variables, $EY_k = 0$, by the zero-one law either $X_n = \sum_1^n Y_k$ converges to a finite limit a.s. or diverges a.s. The nature of the divergence can be gotten from 5.28 in an important special case. Show that if $Y_1, \ldots$ are independent, $|Y_k| \le \alpha < \infty$, all k, $EY_k = 0$, $S_n = Y_1 + \cdots + Y_n$, then either

a) $P\left(\lim_n S_n \text{ exists}\right) = 1$, or

b) $P(\overline{\lim}\, S_n = \infty, \underline{\lim}\, S_n = -\infty) = 1.$

9. For any process $X_1, X_2, \ldots$ and sets $A, B \in \mathcal{B}_1$, suppose that $P(X_m \in B$ for at least one $m > n \mid X_n, \ldots, X_1) \ge \delta > 0$, on $\{X_n \in A\}$. Then prove that

$$\{X_n \in A \text{ i.o.}\} \subset \{X_n \in B \text{ i.o.}\} \text{ a.s.}$$

[Let $F_N = \bigcup_N^\infty \{X_m \in B\}$, and use $P(F_N \mid X_n \cdots X_1) \to \chi_{F_N}$.]

10. Consider a process $X_1, X_2, \ldots$ taking values in $[0, \infty)$. Consider $\{0\}$ an absorbing state in the sense that $X_n = 0 \Rightarrow X_{n+m} = 0$, $m \geq 1$. Let D be the event that the process is eventually absorbed at zero, that is,

$$D = \{\exists n \text{ such that } X_n = 0\}.$$

If, for every x there exists a $\delta > 0$ such that

$$P(D \mid X_n, \ldots, X_1) \geq \delta, \quad \text{if} \quad X_n \leq x, n = 1, 2, \ldots,$$

prove that for almost every sample sequence, either $X_1, X_2, \ldots$ is eventually absorbed, or $X_n \to \infty$.

7. STOPPING RULES

Definition 5.30. *If an extended stopping time* n^* *is* a.s. *finite, call it a stopping time or a stopping rule. The stopped variable is* X_{n^*}.

It is clear that if we define $m_1(\omega) \equiv 1$, $m_n(\omega) = n^*(\omega)$, $n \geq 2$, then the m_n are optional sampling variables. From the optional sampling theorem we get the interesting

Corollary 5.31. *Let* n^* *be a stopping rule. If* $X_1, X_2, \ldots$ *is a* SMG(MG) *and if*

a) $$E\,|X_{n^*}| < \infty,$$

b) $$\varliminf_N \int_{\{n^* > N\}} |X_N|\, dP = 0,$$

then

(5.32) $$EX_{n^*} \underset{(=)}{\geq} EX_1.$$

Proof. Obvious.

The interest of 5.31 in gambling is as follows: If a sequence of games is unfavorable under a given gambling system, then the variables $-S_n$ form a SMG if $E\,|S_n| < \infty$. Suppose we use some stopping rule n^* which based on the outcome of the first j games tells us whether to quit or not after the jth game. Then our terminal fortune is S_{n^*}. But if (a) and (b) are in force, then

$$ES_{n^*} \leq ES_1,$$

with equality if the game is fair. Thus we cannot increase our expected fortune by using a stopping rule. Also, in the context of gambling, some illumination can be shed on the conditions (a) and (b) of the optional sampling theorem. Condition (b), which is pretty much the stickler, says roughly that the variables m_n cannot sample too far out in the sequence too fast. For example, if there are constants α_n such that $m_n \leq \alpha_n$, all n (even if $\alpha_n \to \infty$), then (b) is automatically satisfied. A counterexample where (b) is violated is in the honored rule "stop when you are ahead." Let $Y_1, Y_2, \ldots$ be independent random variables, $Y_i = \pm 1$ with probability $\frac{1}{2}$ each. Then $X_n = Y_1 + \cdots + Y_n$ is a MG sequence and represents the winnings after n plays in a coin-tossing game. From Problem 8, $\overline{\lim}\, X_n = +\infty$ a.s., hence we can define a stopping rule by

$$n^* = \{\text{first } n \quad \text{such that} \quad X_n = 1\},$$

that is, stop as soon as we win one dollar. If (5.32) were in force, then $EX_{n^*} = EX_1 = 0$, but $X_{n^*} \equiv 1$. Now (a) is satisfied because $E\,|X_{n^*}| = 1$, hence we must conclude that (b) is violated. This we can show directly. Note that $|X_n|$ is a SMG sequence, and

$$\{X_1 = -1, X_2 \neq 0, \ldots, X_{N-1} \neq 0\} \subset \{n^* > N\}.$$

Therefore

$$\int_{\{n^*>N\}} |X_N|\, dP \geq \int_{\{X_1=-1, X_2\neq 0, \ldots, X_{N-1}\neq 0\}} |X_N|\, dP$$
$$\geq \int_{\{X_1=-1, \ldots, X_{N-1}\neq 0\}} |X_{N-1}|\, dP$$
$$= \int_{\{X_1=-1, \ldots, X_{N-2}\neq 0\}} |X_{N-1}|\, dP.$$

Going down the ladder we find that

$$\int_{\{n^*>N\}} |X_N|\, dP \geq \int_{\{X_1=-1\}} |X_1|\, dP = \tfrac{1}{2}.$$

Here, as a matter of fact, n^* can be quite large. For example, note that $En^* = \infty$, because if $Y_1 = -1$, we have to wait for an equalization, in other words, wait for the first time that $Y_2 + \cdots + Y_n = 0$ before we can possibly get a dollar ahead. Actually, on the converse side, it is not difficult to prove.

Proposition 5.33. *Let $X_1, X_2, \ldots$ be a* SMG(MG) *and n^* a stopping rule. If $En^* < \infty$ and $E(|X_{n+1} - X_n| \mid X_n, \ldots, X_1) \leq \alpha < \infty$, $n \leq n^*$, then*

$$EX_{n^*} \underset{(=)}{\geq} EX_1.$$

Proof. All we need to do is verify the conditions of 5.31. Denote $Z_n = |X_n - X_{n-1}|$, $n > 1$, $Z_1 = |X_1|$, $Y = Z_1 + \cdots + Z_{n^*}$. Hence $|X_{n^*}| \leq Y$.

$$EY = \sum_{k=1}^{\infty} \int_{\{n^*=k\}} Y\, dP = \sum_{k=1}^{\infty} \sum_{j=1}^{k} \int_{\{n^*=k\}} Z_j\, dP.$$

Interchange the order of summation so that

$$EY = \sum_{j=1}^{\infty} \int_{\{n^* \geq j\}} Z_j \, dP.$$

The set $\{n^* \geq j\} = \{n^* < j\}^c$ is in $\mathcal{F}(X_1, \ldots, X_{j-1})$. Therefore

$$\int_{\{n^* \geq j\}} Z_j \, dP = \int_{\{n^* \geq j\}} E(Z_j \mid X_{j-1}, \ldots, X_1) \, dP \leq \alpha P(n^* \geq j),$$

and we get

$$EY \leq \alpha \sum_{1}^{\infty} P(n^* \geq j) = \alpha E n^*.$$

For 5.31(b), since $Z_1 + \cdots + Z_N \leq Y$ on the set $\{n^* > N\}$,

$$\int_{\{n^* > N\}} |X_N| \, dP \leq \int_{\{n^* > N\}} Y \, dP.$$

As $N \to \infty$, $\{n^* > N\} \downarrow \emptyset$ a.s. Apply the bounded convergence theorem to $\chi_{\{n^* > N\}} Y$ to get the result.

Proposition 5.33 has interesting applications to sums of independent random variables. Let $Y_1, Y_2, \ldots$ be independent and identically distributed, $S_n = Y_1 + \cdots + Y_n$, and assume that for some real $\lambda \neq 0$, $\varphi(\lambda) = Ee^{\lambda X_1}$ exists, and that $\varphi(\lambda) \geq 1$. Then

Proposition 5.34 *(Wald's identity). If n^* is a stopping time for the sums $S_1, S_2, \ldots$ such that $|S_n| < \gamma$, $n \leq n^*$, and $En^* < \infty$, then*

$$E\left(\frac{e^{\lambda S_{n^*}}}{\varphi(\lambda)^{n^*}}\right) = 1.$$

Proof. The random variables

$$X_n = \frac{e^{\lambda S_n}}{\varphi(\lambda)^n}$$

form a MG, since

$$E\left(\frac{e^{\lambda S_{n+1}}}{\varphi(\lambda)^{n+1}} \,\middle|\, S_n, \ldots, S_1\right) = \frac{e^{\lambda S_n}}{\varphi(\lambda)^{n+1}} E(e^{\lambda X_{n+1}}) = \frac{e^{\lambda S_n}}{\varphi(\lambda)^n}.$$

Obviously, $EX_1 = 1$, so if the second condition of 5.33 holds, then Wald's identity follows. The condition is

$$E\left(\left|\frac{e^{\lambda S_{n+1}}}{\varphi(\lambda)} - e^{\lambda S_n}\right| \,\middle|\, S_n, \ldots, S_1\right) \leq \alpha \varphi(\lambda)^n, \qquad n \leq n^*,$$

or

$$e^{\lambda S_n} E\left|\frac{e^{\lambda X_1}}{\varphi(\lambda)} - 1\right| \leq \alpha \varphi(\lambda)^n, \qquad n \leq n^*$$

which is clearly satisfied under 5.34.

Problems

11. For n^* a stopping time for sums $S_1, S_2, \ldots$ of independent, identically distributed random variables $Y_1, Y_2, \ldots, E|Y_1| < \infty$, prove that $En^* < \infty$ implies that

$$ES_{n^*} = EY_1 \cdot En^*.$$

Use this to give another derivation of Blackwell's equation (Problem 21, Chapter 3).

12. Let $Y_1, Y_2, \ldots$ be independent and equal ± 1 with probability $\frac{1}{2}$. Let r, s be positive integers. Define

$$n^* = \{\text{first } n \text{ such that } S_n = r \text{ or } -s\}.$$

Show

a) $$En^* < \infty,$$

b) $$P(S_{n^*} = r) = \frac{s}{r+s}.$$

For $r = s$, evaluate $Ee^{-\lambda n^*}$, $\lambda > 0$.

13. (See Doob, [39, p. 308].) For sums of independent, identically distributed random variables $Y_1, Y_2, \ldots$, define n^* as the time until the first positive sum, that is,

$$n^* = \min\{n;\ S_n > 0\}.$$

Prove that if $EY_1 = 0$, then $En^* = \infty$.

8. BACK TO GAMBLING

The reason that the strategy "quit when you are ahead" works in the fair coin-tossing game is that an infinite initial fortune is assumed. That is, there is no lower limit, say $-M$, such that if S_n becomes less than $-M$ play is ended.

A more realistic model of gambling would consider the sequence of fortunes $S_0, S_1, S_2, \ldots$ (that is, money in hand) as being nonnegative and finite. We now turn to an analysis of a sequence of fortunes under gambling satisfying

(5.35) i) $S_n \geq 0, \quad n = 0, 1, 2, \ldots, \quad S_0$ constant,

ii) $ES_n < \infty$ and $E(S_{n+1} \mid S_n, \ldots, S_0) \leq S_n$ a.s., $n \geq 0$.

In addition, in any reasonable gambling house, and by the structure of our monetary system, if we bet on the nth trial, there must be a lower bound to the amount we can win or lose. We formalize this by

Assumption 5.36. *There exists a $\delta > 0$ such that either*

$$S_{n+1} = S_n \quad \textit{or} \quad |S_{n+1} - S_n| \geq \delta.$$

Definition 5.37. *We will say that we bet on the nth game if* $|S_n - S_{n-1}| \geq \delta$. *Let* n^* *be the (possibly extended) time of the last bet, that is,*

$$n^* = \begin{cases} \text{largest } n \text{ such that } |S_n - S_{n-1}| \geq \delta, \\ +\infty \quad \text{if no such } n \text{ exists.} \end{cases}$$

where S_0 *is the starting fortune.*

Under (5.35), (i) and (ii), and 5.36, the martingale convergence theorem yields strong results. You can't win!

Theorem 5.38.

$$P(n^* < \infty) = 1, \quad ES_{n^*} \leq S_0.$$

Remark. The interesting thing about $P(n^* < \infty) = 1$ is the implication that in an unfavorable (or fair) sequence of games, one cannot keep betting indefinitely. There must be a last bet. Furthermore, $ES_{n^*} \leq S_0$ implies that the expected fortune after the last bet is smaller than the initial fortune.

Proof. Let $X_n = -S_n$; then the $X_0, X_1, \ldots$ sequence is a SMG. Furthermore, $E|X_n| = E|S_n| = ES_n$. Thus $E|X_n| \leq S_0$, all n. Hence there exists a random variable X such that $X_n \xrightarrow{\text{a.s.}} X$. Thus

$$P(|X_{n+1} - X_n| \geq \delta \text{ i.o.}) = 0,$$

or

$$P(|S_{n+1} - S_n| \geq \delta \text{ i.o.}) = 0 \qquad \text{or} \qquad P(n^* < \infty) = 1.$$

To prove the second part use the monotone convergence theorem

$$\int S_{n^*}\, dP = \lim_n \int_{\{n^* \leq n\}} S_{n^*}\, dP.$$

But on $\{n^* \leq n\}$, $S_{n^*} = S_n$; hence

$$ES_{n^*} = \lim_n \int_{\{n^* \leq n\}} S_n\, dP \leq \overline{\lim}\, ES_n \leq S_0.$$

Note that the theorem is actually a simple corollary of the martingale convergence theorem. Now suppose that the gambling house has a minimum bet of α dollars and we insist on betting as long as $S_n \geq \alpha$; then n^* becomes the time "of going broke," that is, $n^* = \{\text{first } n \text{ such that } S_n < \alpha\}$, and the obvious corollary of 5.38 is that the *persistent gambler goes broke with probability one.*

NOTES

Martingales were first fully explored by Doob, in 1940 [32] and systematically developed in his book [39] of 1953. Their widespread use in probability theory has mostly occurred since that time. However, many of the results had been scattered around for some time. In particular, some of them are due to Lévy, appearing in his 1937 book [103], and some to Ville [137, 1939]. Some of the convergence theorems in a measure-theoretic framework are due to Andersen and Jessen. See the Appendix to Doob's book [39] for a discussion of the connection with the Andersen-Jessen approach, and complete references. The important concepts of optional sampling, optional stopping, and the key lemma 5.17 are due to Doob.

David Freedman has pointed out to me that many of the convergence results can be gotten from the inequality 5.13. For example, here is a more elementary and illuminating proof of 5.21 for Z an $\mathcal{F}(\mathbf{X})$ measurable random variable. For any $\epsilon > 0$, take k, Z_k measurable $\mathcal{F}(\mathsf{X}_1, \ldots, \mathsf{X}_k)$ such that

$$E\,|\mathsf{Z} - \mathsf{Z}_k| \leq \epsilon.$$

Now,

$$E(\mathsf{Z}_k \mid \mathsf{X}_1, \ldots, \mathsf{X}_n) = \mathsf{Z}_k, \quad n \geq k.$$

Let

$$\mathsf{Y}_n = E(\mathsf{Z} - \mathsf{Z}_k \mid \mathsf{X}_1, \ldots, \mathsf{X}_n),$$

then the $\{\mathsf{Y}_n\}$ is a MG, $\{|\mathsf{Y}_n|\}$ is a SMG, and by 5.13,

$$P\left(\sup_{1 \leq n \leq N} |\mathsf{Y}_n| \geq x\right) \leq \frac{1}{x} E\,|\mathsf{Y}_N| \leq \frac{1}{x} E\,|\mathsf{Z} - \mathsf{Z}_k|\,.$$

Thus,

$$P\left(\overline{\lim_n}\, |E(\mathsf{Z} \mid \mathsf{X}_1, \ldots, \mathsf{X}_n) - \mathsf{Z}_k\,| \geq x\right) \leq \epsilon/x.$$

Take $\epsilon \downarrow 0$ fast enough so that $\mathsf{Z}_k \xrightarrow{\text{a.s.}} \mathsf{Z}$ to get the result

$$E(\mathsf{Z} \mid \mathsf{X}_1, \ldots, \mathsf{X}_n) \xrightarrow{\text{a.s.}} \mathsf{Z}.$$

For a fascinating modern approach to gambling strategies, see the book by Dubins and Savage [40].

CHAPTER 6

STATIONARY PROCESSES AND THE ERGODIC THEOREM

1. INTRODUCTION AND DEFINITIONS

The question here is: Given a process $X_1, X_2, \ldots$, find conditions for the almost sure convergence of $(X_1 + \cdots X_n)/n$. Certainly, if the $\{X_n\}$ are independent identically distributed random variables and $E|X_1| < \infty$, then,

$$\frac{X_1 + \cdots + X_n}{n} \xrightarrow{\text{a.s.}} EX_1.$$

A remarkable weakening of this result was proved by Birkhoff in 1931 [4]. Instead of having independent identically distributed random variables, think of requiring that the distribution of the process not depend on the placement of the time origin. In other words, assume that no matter when you start observing the sequence of random variables the resulting observations will have the same probabilistic structure.

Definition 6.1. *A process $X_1, X_2, \ldots$ is called stationary if for every k, the process $X_{k+1}, X_{k+2}, \ldots$ has the same distribution as $X_1, X_2, \ldots$, that is, for every $B \in \mathcal{B}_\infty$.*

(6.2) $$P((X_1, X_2, \ldots) \in B) = P((X_{k+1}, X_{k+2}, \ldots) \in B).$$

Since the distribution is determined by the distribution functions, (6.2) is equivalent to: For every $x_1, \ldots, x_n$, and integer $k > 0$,

(6.3) $$P(X_1 < x_1, \ldots, X_n < x_n) = P(X_{k+1} < x_1, \ldots, X_{k+n} < x_n).$$

In particular, if a process is stationary, then all the one-dimensional distribution functions are the same, that is,

$$P(X_1 < x) = P(X_k < x),\ k = 1, 2, \ldots$$

We can reduce (6.2) and (6.3) by noting

Proposition 6.4. *A process $X_1, X_2, \ldots$ is stationary if the process $X_2, X_3, \ldots$ has the same distribution as $X_1, X_2, \ldots$*

Proof. Let $X'_k = X_{k+1}$, $k = 1, 2, \ldots$ Then $X'_1, X'_2, \ldots$ has the same distribution as $X_1, X_2, \ldots$ Hence $X'_2, X'_3, \ldots$ has the same distribution as $X'_1, X'_2, \ldots$, and so forth.

Sometimes it is more convenient to look at stationary processes that consist of a double-ended sequence of random variables $\ldots, X_{-1}, X_0, X_1, \ldots$ In this context, what we have is an infinite sequence of readings, beginning in the infinitely remote past and continuing into the infinite future. Define such a process to be stationary if its distribution does not depend on choice of an origin, i.e., in terms of finite dimensional distributions:

$$P(X_1 < x_1, \ldots, X_n < x_n) = P(X_{k+1} < x_1, \ldots, X_{k+n} < x_n)$$

for all $x_1, \ldots, x_n$ and all k, both positive and *negative*.

The interesting point here is

Proposition 6.5. *Given any single-ended stationary process* $X_1, X_2, \ldots$, *there is a double-ended stationary process* $\ldots, \tilde{X}_{-1}, \tilde{X}_0, \tilde{X}_1, \ldots$ *such that* $\tilde{X}_1, \tilde{X}_2, \ldots$ *and* $X_1, X_2, \ldots$ *have the same distribution.*

Proof. From the Extension Theorem 2.26, all we need to define the $\tilde{X}_k$ process is a set of consistent distribution functions; i.e., we need to define

$$P(\tilde{X}_{-m} < x_{-m}, \ldots, \tilde{X}_0 < x_0, \ldots, \tilde{X}_n < x_n)$$

such that if either x_{-m} or $x_n \uparrow \infty$, then we drop down to the next highest distribution function. We do this by defining

$$P(\tilde{X}_{-m} < x_{-m}, \ldots, \tilde{X}_n < x_n) = P(X_1 < x_{-m}, \ldots, X_{n+m+1} < x_n),$$

that is, we slide the distribution functions of the $X_1, \ldots$ process to the left. Now $X_1, X_2, \ldots$ can be looked at as the continuation of a process that has already been going on an infinite length of time.

Starting with any stationary process, an infinity of stationary processes can be produced.

Proposition 6.6. *Let* $X_1, X_2, \ldots$ *be stationary,* $\varphi(\mathbf{x})$ *measurable* $\mathcal{B}_\infty$, *then the process* $Y_1, Y_2, \ldots$ *defined by*

$$Y_k = \varphi(X_k, X_{k+1}, \ldots)$$

is stationary.

Proof. On $R^{(\infty)}$ define $\varphi_k(\mathbf{x})$ as $\varphi(x_k, x_{k+1}, \ldots)$. The set

$$A = \{\mathbf{x};\ (\varphi_1(\mathbf{x}), \varphi_2(\mathbf{x}), \ldots) \in B\},$$

$B \in \mathcal{B}_\infty$, is in $\mathcal{B}_\infty$, because each $\varphi_k(\mathbf{x})$ is a random variable on $(R^{(\infty)}, \mathcal{B}_\infty)$. Note

$$\{\omega;\ (Y_1, Y_2, \ldots) \in B\} = \{\omega;\ (X_1, X_2, \ldots) \in A\}$$

and

$$\{\omega;\ (Y_2, Y_3, \ldots) \in B\} = \{\omega;\ (X_2, X_3, \ldots) \in A\},$$

which implies the stationarity of the Y_k sequence.

Corollary 6.7. *Let* $X_1, X_2, \ldots$ *be independent and identically distributed random variables,* $\varphi(\mathbf{x})$ *measurable* $\mathcal{B}_\infty$; *then*

$$Y_k = \varphi(X_k, X_{k+1}, \ldots)$$

is stationary.

Proof. The $X_1, X_2, \ldots$ sequence is stationary.

Problem 1. Look at the unit circle and define

$$f(x) = \begin{cases} 1, & 0 \le x < \pi, \\ 0, & \pi \le x < 2\pi. \end{cases}$$

Here Ω is the unit circle, $\mathcal{F}$ the Borel σ-field. Take P to be Lebesgue measure divided by 2π. Take θ to be an irrational angle. Define $x_1 = x$, $x_{k+1} = (x_k + \theta)[2\pi]$, and $X_k(x) = f(x_k)$. Hence the process is a sequence of zeros and ones, depending on whether x_k is in the last two quadrants or first two when x_1 is picked at random on the circumference. Prove that $X_1, X_2, \ldots$ is stationary, ($[\alpha]$ denotes modulo α).

2. MEASURE-PRESERVING TRANSFORMATIONS

Consider a probability space $(\Omega, \mathcal{F}, P)$ and a transformation T of Ω into itself. As usual, we will call T *measurable* if the inverse images under T of sets in $\mathcal{F}$ are again in $\mathcal{F}$; that is, if $T^{-1}A = \{\omega;\ T\omega \in A\} \in \mathcal{F}$, all $A \in \mathcal{F}$.

Definition 6.8. *A measurable transformation* T *on* $\Omega \to \Omega$ *will be called measure-preserving if* $P(T^{-1}A) = P(A)$, *all* $A \in \mathcal{F}$.

To check whether a given transformation T is measurable, we can easily generalize 2.28 and conclude that if $T^{-1}C \in \mathcal{F}$, for $C \in \mathcal{C}$, $\mathcal{F}(\mathcal{C}) = \mathcal{F}$, then T is measurable. Again, to check measure-preserving, both $P(T^{-1}A)$ and $P(A)$ are σ-additive probabilities, so we need check only their agreement on a class $\mathcal{C}$, closed under intersections, such that $\mathcal{F}(\mathcal{C}) = \mathcal{F}$.

Starting from measure-preserving transformations (henceforth assumed measurable) a large number of stationary processes can be generated. Let $X(\omega)$ be any random variable on $(\Omega, \mathcal{F}, P)$. Let T be measure-preserving and define a process $X_1, X_2, \ldots$ by $X_1(\omega) = X(\omega)$, $X_2(\omega) = X(T\omega)$, $X_3(\omega) = X(T^2\omega), \ldots$ Another way of looking at this is: If $X_1(\omega)$ is some measurement on the system at time one, then $X_n(\omega)$ is the same measurement after the system has evolved $n - 1$ steps so that $\omega \to T^{n-1}\omega$. It should be intuitively clear that the distribution of the $X_1, X_2, \ldots$ sequence does not depend on origin, since starting from any $X_n(\omega)$ we get $X_{n+1}(\omega)$ as $X_n(T\omega)$, and so forth. To make this firm, denote by T^0 the identity operator, and we prove

Proposition 6.9. *Let T be measure preserving on $(\Omega, \mathcal{F}, P)$, X a random variable on $(\Omega, \mathcal{F})$; then the sequence $\mathsf{X}_n(\omega) = \mathsf{X}(T^{n-1}\,\omega)$, $n = 1, 2, \ldots$ is a stationary sequence of random variables.*

Proof. First of all, $\mathsf{X}_n(\omega)$ is a random variable, because $\{\mathsf{X}_n(\omega) \in B\} = \{\mathsf{X}(T^{n-1}\omega) \in B\}$. Let $A = \{\mathsf{X} \in B\}$. Then

$$\{\mathsf{X}(T^{n-1}\omega) \in B\} = \{\omega;\ T^{n-1}\omega \in A\}.$$

Evidently, however, T measurable implies T^{n-1} measurable, or $T^{-n+1}A \in \mathcal{F}$. Now let $A = \{\omega;\ (\mathsf{X}_1, \mathsf{X}_2, \ldots) \in B\}$, $B \in \mathcal{B}_\infty$, thus

$$A = \{\omega;\ (\mathsf{X}(\omega), \mathsf{X}(T\omega), \ldots) \in B\}.$$

Look at $A_1 = \{\omega;\ (\mathsf{X}_2, \mathsf{X}_3, \ldots) \in B\}$. This similarly is the set $\{\omega;\ (\mathsf{X}(T\omega), \mathsf{X}(T^2\omega), \ldots) \in B\}$. Hence $\omega \in A_1 \Leftrightarrow T\omega \in A$ or $A_1 = T^{-1}A$. But, by hypothesis, $P(T^{-1}A) = P(A)$.

Can every stationary process be generated by a measure-preserving transformation? Almost! In terms of distribution, the answer is Yes. Starting from any stationary process $\mathsf{X}_1(\omega), \mathsf{X}_2(\omega), \ldots$ go to the coordinate representation process $\hat{\mathsf{X}}_1, \hat{\mathsf{X}}_2, \ldots$ on $(R^{(\infty)}, \mathcal{B}_\infty, \hat{P})$. By definition,

$$\hat{\mathsf{X}}_n(\mathbf{x}) = x_n.$$

Definition 6.10. *On $(R^{(\infty)}, \mathcal{B}_\infty)$ define the shift transformation $S: R^{(\infty)} \to R^{(\infty)}$ by $S(x_1, x_2, \ldots) = (x_2, x_3, \ldots)$.*

So, for example, $S(3, 2, 7, 1, \ldots) = (2, 7, 1, \ldots)$.

The point is that from the definitions, $\mathsf{X}_n(\mathbf{x}) = \mathsf{X}_1(S^{n-1}\mathbf{x})$. We prove below that S is measurable and measure-preserving, thus justifying the answer of "Almost" above.

Proposition 6.11. *The transformation S defined above is measurable, and if $\mathsf{X}_1, \mathsf{X}_2, \ldots$ is stationary, then S preserves $\hat{P}$ measure.*

Proof. To show S measurable, consider

$$B = S^{-1}\{\mathbf{x};\ x_1 \in I_1, \ldots, x_n \in I_n\} = S^{-1}C.$$

By definition, letting $(S\mathbf{x})_k$ be the kth coordinate of $S\mathbf{x}$, we find that

$$B = \{\mathbf{x};\ (S\mathbf{x})_1 \in I_1, \ldots, (S\mathbf{x})_n \in I_n\} = \{x;\ x_2 \in I_1, \ldots, x_{n+1} \in I_n\},$$

and that B is therefore obviously in $\mathcal{B}_\infty$. Furthermore, by the stationarity of $\hat{\mathsf{X}}_1, \hat{\mathsf{X}}_2, \ldots$

$$\hat{P}(B) = \hat{P}(S^{-1}C) = \hat{P}(\hat{\mathsf{X}}_2 \in I_1, \ldots, \hat{\mathsf{X}}_{n+1} \in I_n)$$
$$= \hat{P}(\hat{\mathsf{X}}_1 \in I_1, \ldots, \hat{\mathsf{X}}_n \in I_n) = \hat{P}(C),$$

So S is also measure-preserving.

Problems

2. Show that the following transformations are *measurable* and *measure-preserving*.

1) $\Omega = [0, 1)$, $\mathcal{F} = \mathcal{B}[0, 1)$, $P = dx$. Let λ be any number in $[0, 1)$ and define $Tx = (x + \lambda)[1]$.
2) $\Omega = [0, 1)$, $\mathcal{F} = \mathcal{B}[0, 1)$, $P = dx$, $Tx = (2x)[1]$.
3) Same as (2) above, but $Tx = (kx)[1]$, where $k > 2$ is integral.

3. Show that for the following transformations on $[0, 1)$, $\mathcal{B}[0, 1)$, there is no P such that $P(\text{single point}) = 0$, and the transformation preserves P.

1) $Tx = \lambda x, \quad 0 < \lambda < 1$.
2) $Tx = x^2$.

4. On $\Omega = [0, 1)$, define $T:\Omega \to \Omega$ by $Tx = (2x)[1]$. Use $\mathcal{F} = \mathcal{B}([0, 1))$, $P = dx$. Define

$$\mathsf{X}(x) = \begin{cases} 0, & 0 \le x < \frac{1}{2}, \\ 1, & \frac{1}{2} \le x < 1. \end{cases}$$

Show that the sequence $\mathsf{X}_n(x) = \mathsf{X}(T^{n-1}x)$ consists of independent zeros and ones with probability $\frac{1}{2}$ each.

Show that corresponding to every stationary sequence $\mathsf{X}_1(\omega), \mathsf{X}_2(\omega), \ldots$ such that $\mathsf{X}_n(\omega) \in \{0, 1\}$, there is a probability $Q(dx)$ on $\mathcal{B}[0, 1)$ such that $Tx = (2x)[1]$ preserves Q-measure, and such that the $\mathsf{X}_n(x)$ sequence defined above has the same distribution with respect to $\mathcal{B}[0, 1)$, $Q(dx)$ as $\mathsf{X}_1, \mathsf{X}_2, \ldots$

3. INVARIANT SETS AND ERGODICITY

Let T be a measure-preserving transformation on $(\Omega, \mathcal{F}, P)$.

Definition 6.12. *A set $A \in \mathcal{F}$ is invariant if $T^{-1}A = A$.*

If A is an invariant set, then the motion T of $\Omega \to \Omega$ carries A into A; that is, if $\omega \in A$, then $T\omega \in A$ (because $T^{-1}A^c = A^c$). A^c is also invariant, and for all n, T^n carries points of A into A and points of A^c into A^c. Because of the properties of inverse mappings we have

Proposition 6.13. *The class of invariant sets is a σ-field $\mathfrak{I}$.*

Proof. Just write down definitions.

In the study of dynamical systems, Ω is the phase space of the system, and if ω is the state of the system at $t = 0$, then its state at time t is given by $T_t\omega$, where $T_t:\Omega \to \Omega$ is the motion of the phase space into itself induced by the equations of motion. For a conservative system $T_t(T_\tau\omega) = T_{t+\tau}\omega$.

We discretize time and take $T = T_1$, so the state of the system at time n is given by $T^n\omega$. Suppose that $X(\omega)$ is some observable function of the state ω. Physically, in taking measurements, the observation time is quite long compared to some natural time scale of molecular interactions. So we measure, not $X(\omega)$, but the average of $X(\omega)$ over the different states into which ω passes with the evolution of the system. That is, we measure

$$\frac{1}{\tau}\int_0^\tau X(T_t\omega)\,dt,$$

for τ large, or in discrete time

$$\frac{1}{n}\sum_1^n X(T^{k-1}\omega),$$

for n large. The brilliant insight of Gibbs was the following argument: that in time, the point $\omega_t = T_t\omega$ wandered all over the phase space and that the density of the points ω_t in any neighborhood tended toward a limiting distribution. Intuitively, this limiting distribution of points had to be invariant under T. If there is such a limiting distribution, say a measure P, then we should be able to replace the limiting time average

$$\frac{1}{n}\sum_1^n X(T^{k-1}\omega)$$

by the phase average

$$\int X(\omega)P(d\omega).$$

Birkhoff's result was that this argument, *properly formulated*, was true! To put Gibb's conjecture in a natural setting, take Ω to be all points on a surface of constant energy. (This will be a subset of $R^{(6n)}$ where n is the number of particles.) Take $\mathcal{F}$ to be the intersection of $\mathcal{B}_{6n}$ with Ω and P the normalized surface area on Ω. By Liouville's theorem, $T:\Omega \to \Omega$ preserves P-measure. The point is now that $T^n\omega$ will never become distributed over Ω in accordance with P if there are invariant subsets A of Ω such that $P(A) > 0$, $P(A^c) > 0$, because in this case the points of A will remain in A; similarly, for A^c. Thus, the only hope for Gibb's conjecture is that every invariant set A has probability zero or one.

To properly formulate it, begin with

Definition 6.14. *Let T be measure-preserving on $(\Omega, \mathcal{F}, P)$. T is called ergodic if for every $A \in \mathfrak{J}$, $P(A) = 0$ or 1.*

One question that is relevant here is: Suppose one defined events A to be *a.s. invariant* if $P(A \,\triangle\, T^{-1}A) = 0$. Is the class of a.s. invariant events considerably different from the class of invariant events? Not so! It is exactly the completion of $\mathfrak{J}$ with respect to P and $\mathcal{F}$.

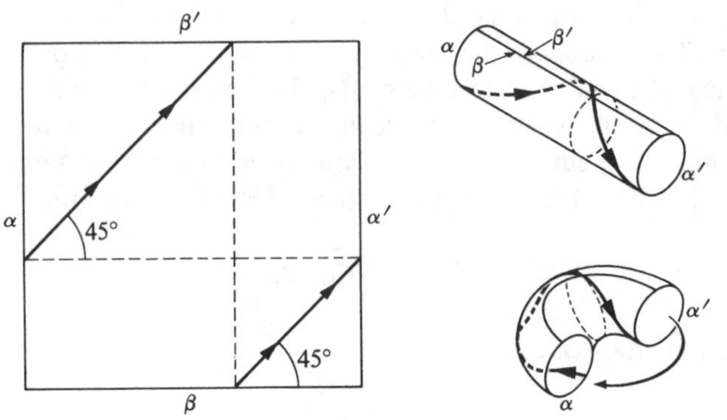

Figure 6.1

Proposition 6.15. *Let $A \in \mathcal{F}$ be* a.s. *invariant, then there is a set $A' \in \mathcal{F}$ which is invariant such that $P(A \,\triangle\, A') = 0$.*

Proof. Let

$$A'' = \bigcup_{0}^{\infty} T^{-n}A, \quad T^0A = A.$$

Then $A'' = A$ a.s., and $T^{-1}A'' \subset A''$. Let

$$A' = \bigcap_{0}^{\infty} T^{-n}A'',$$

noting that A'' is a.s. invariant gives $A' = A$ a.s., and $T^{-1}A'' \subset A''$ implies that A' is invariant.

The concept of ergodicity, as pointed out above, is a guarantee that the phase space does not split into parts of positive probability which are inaccessible to one another.

Example. Let Ω be the unit square, that is, $\Omega = \{(x, y);\ 0 \le x < 1, 0 \le y < 1\}$, $\mathcal{F}$ the Borel field. Let a be any positive number and define $T(x, y) = ((x + a)[1], (y + a)[1])$. Use $P = dx\,dy$; then it is easy to check that T is measure preserving. What we have done is to sew edges of Ω together (see Fig. 6.1) so that α and α' are together, β and β'. T moves points at a 45° angle along the sewn-together square. Just by looking at Fig. 6.1 you can see that T does not move around the points of Ω very much, and it is easy to construct invariant sets of any probability, for example, the shaded set as shown in Fig. 6.2.

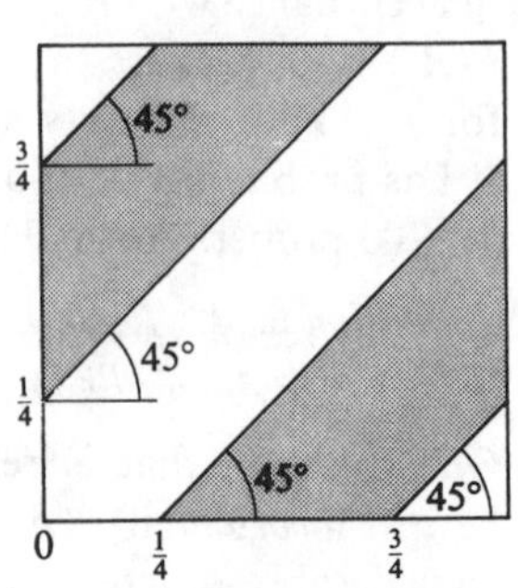

Figure 6.2

Problems

5. In Problem 2(1), show that if λ is rational, T is not ergodic.

6. We use this problem to illustrate more fully the dynamical aspect. Take $(\Omega, \mathcal{F})$ and let $T:\Omega \to \Omega$ be measurable. Start with any point $\omega \in \Omega$, then ω has the motion $\omega \to T\omega \to T^2\omega \to \cdots$ Let $\mathsf{N}_n(A, \omega)$ be the number of times that the moving point $T^k\omega$ enters the set $A \in \mathcal{F}$ during the first n motions; that is, $\mathsf{N}_n(A, \omega)$ is the number of times that $T^k\omega \in A$, $k = 0, 1, \ldots, n-1$. Keeping ω fixed, define probabilities $P_\omega^{(n)}(\cdot)$ on $\mathcal{F}$ by $P_\omega^{(n)}(\cdot) = \mathsf{N}_n(\cdot, \omega)/n$. That is, $P_\omega^{(n)}(A)$ is the proportion of times in the first n moves that the point is in the set A. Let X be any random variable on $(\Omega, \mathcal{F})$.

a) Show that

$$\frac{1}{n}\sum_{k=0}^{n-1} \mathsf{X}(T^k\omega) = \int \mathsf{X}(\cdot)\, dP_\omega^{(n)}(\cdot).$$

Assume that there is a probability $P_\omega(\cdot)$ on $\mathcal{F}$ such that for every $A \in \mathcal{F}$, $\lim_n P_\omega^{(n)} = P_\omega(A)$.

b) Show that T is $P_\omega(\cdot)$-preserving, that is,

$$P_\omega(T^{-1}A) = P_\omega(A), \quad \text{all } A \in \mathcal{F}.$$

c) Show that if $\mathsf{X}(\omega) \geq 0$ is bounded, then

$$\frac{1}{n}\sum_{k=0}^{n-1} \mathsf{X}(T^k\omega) \to \int \mathsf{X}(\cdot)\, dP_\omega(\cdot).$$

What is essential here is that the limit $\int \mathsf{X}\, dP_\omega$ not depend on where the system started at time zero. Otherwise, to determine the limiting time averages of (c) for the system, a detailed knowledge of the position ω in phase-space at $t = 0$ would be necessary. Hence, what we really need is the additional assumption that $P_\omega(\cdot)$ be the same for all ω, in other words, that the limiting proportion of time that is spent in the set A not depend on the starting position ω. Now substitute this stronger assumption, that is: There is a probability $P(\cdot)$ on $\mathcal{F}$ such that for every $A \in \mathcal{F}$ and $\omega \in \Omega$,

$$\lim P_\omega^{(n)}(A) = P(A).$$

d) Show that under the above assumption, A invariant $\Rightarrow A = \Omega$ or $\emptyset$.

This result shows not only that T is ergodic on $(\Omega, \mathcal{F}, P)$, but ergodic in a much stronger sense than that of Definition 6.14. The stronger assumption above is much too restrictive in the sense that most dynamical systems do not satisfy it. There are usually some starting states ω which are exceptional in that the motion under T does not mix them up very well. Take, for example, elastic two-dimensional molecules in a rectangular box. At $t = 0$ consider the state ω which is shown in Fig. 6.3, where all the molecules have the same x-coordinate and velocity. Obviously, there will be large chunks of phase-space that will never be entered if this is the starting state. What we want then, is some weaker version that says $P_\omega^{(n)}(\cdot) \to P(\cdot)$ for most starting states ω. With this weakening, the strong result of (d) above will no longer be true. We come back later to the appropriate weakening which, of course, will result in something like 6.14 instead of (d).

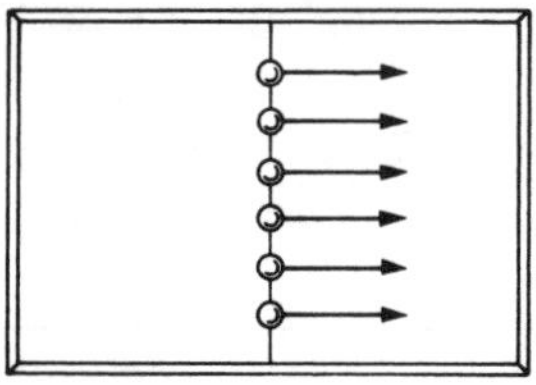

Figure 6.3

4. INVARIANT RANDOM VARIABLES

Along with invariant sets go invariant random variables.

Definition 6.16. *Let* $\mathsf{X}(\omega)$ *be a random variable on* $(\Omega, \mathcal{F}, P)$, *T measure-preserving; then* $\mathsf{X}(\omega)$ *is called an invariant random variable if* $\mathsf{X}(\omega) = \mathsf{X}(T\omega)$.

Note

Proposition 6.17. X *is invariant iff* X *is measurable* $\mathfrak{I}$.

Proof. If X is invariant, then for every x, $\{\mathsf{X} < x\} \in \mathfrak{I}$; hence X is measurable $\mathfrak{I}$. Conversely, if $\mathsf{X}(\omega) = \chi_A(\omega)$, $A \in \mathfrak{I}$, then

$$\mathsf{X}(T\omega) = \chi_A(T\omega) = \chi_{T^{-1}A}(\omega) = \chi_A(\omega).$$

Now consider the class $\mathfrak{L}$ of all random variables on $(\Omega, \mathfrak{I})$ which are invariant; clearly $\mathfrak{L}$ is closed under linear combinations, and $\mathsf{X}_n \in \mathfrak{L}$, $\mathsf{X}_n(\omega) \uparrow \mathsf{X}(\omega)$ implies

$$\mathsf{X}(T\omega) = \lim_n \mathsf{X}_n(T\omega) \doteq \lim_n \mathsf{X}_n(\omega) = \mathsf{X}(\omega).$$

Hence by 2.38, $\mathfrak{L}$ contains all nonnegative random variables on $(\Omega, \mathfrak{I})$. Thus clearly every random variable on $(\Omega, \mathfrak{I})$ is invariant.

The condition for ergodicity can be put very nicely in terms of invariant random variables.

Proposition 6.18. *Let T be measure-preserving on* $(\Omega, \mathcal{F}, P)$. *T is ergodic iff every invariant random variable* $\mathsf{X}(\omega)$ *is* a.s. *equal to a constant.*

Proof. One way is immediate; that is, for any invariant set A, let $\mathsf{X}(\omega) = \chi_A(\omega)$. Then $\mathsf{X}(\omega)$ constant a.s. implies $P(A) = 0, 1$. Conversely, suppose $P(\mathsf{X} < x) = 0, 1$ for all x. Since for $x \uparrow +\infty$, $P(\mathsf{X} < x) \to 1$, $P(\mathsf{X} < x) = 1$, for all x sufficiently large. Let $x_0 = \inf\{x;\ P(\mathsf{X} < x) = 1\}$. Then for every $\epsilon > 0$, $P(x_0 - \epsilon < \mathsf{X} < x_0 + \epsilon) = 1$, and taking $\epsilon \downarrow 0$ yields $P(\mathsf{X} = x_0) = 1$.

Obviously we can weaken 6.18 to read

Proposition 6.19. *T is ergodic iff every bounded invariant random variable is* a.s. *constant.*

In general, it is usually difficult to show that a given transformation is ergodic. Various tricks are used: For example, we can apply 6.19 to Problem 2(1).

Example. We show that if λ is irrational, then $Tx = (x + \lambda)$ [1] is ergodic. Let $f(x)$ be any Borel-measurable function on [0, 1). Assume it is in $L_2(dx)$, that is, $\int f^2\, dx < \infty$. Then we have

$$f(x) = \sum_{-\infty}^{+\infty} c_n e^{2\pi i n x}, \quad \text{a.s.}(dx),$$

where the sum exists as a limit in the second mean, and $\Sigma\, |c_n|^2 < \infty$. Therefore

$$f(Tx) = \sum_{-\infty}^{+\infty} c_n e^{2\pi i n \lambda} \cdot e^{2\pi i n x}.$$

For $f(x)$ to be invariant, $c_n(1 - e^{2\pi i n \lambda}) = 0$. This implies either $c_n = 0$ or $e^{2\pi i n \lambda} = 1$. The latter can never be satisfied for nonzero n and irrational λ. The conclusion is that $f(x) = c_0$ a.s.; by 6.19, T is ergodic.

Problems

7. Use the method of the above example to show that the transformation of Problem 2(2) is ergodic.

8. Using 2.38, show that if T is measure-preserving on $(\Omega, \mathcal{F}, P)$ and $\mathsf{X}(\omega)$ any random variable, that

(6.20) $$E[\mathsf{X}(\omega)] = E[\mathsf{X}(T\omega)].$$

5. THE ERGODIC THEOREM

One of the most remarkable of the strong limit theorems is the result usually referred to as the ergodic theorem.

Theorem 6.21. *Let T be measure-preserving on* $(\Omega, \mathcal{F}, P)$. *Then for* X *any random variable such that* $E\,|\mathsf{X}| < \infty$,

$$\lim \frac{1}{n} \sum_{k=0}^{n-1} \mathsf{X}(T^k\omega) = E(\mathsf{X} \mid \mathfrak{J}) \text{ a.s.}$$

To prove this result, we prove first an odd integration inequality:

Theorem 6.22 (*Maximal ergodic theorem*). *Let T be measure-preserving on $(\Omega, \mathcal{F}, P)$, and X a random variable such that $E\,|\mathsf{X}| < \infty$. Define*

$$\mathsf{S}_k(\omega) = \mathsf{X}(\omega) + \cdots + \mathsf{X}(T^{k-1}\,\omega), \quad and \quad \mathsf{M}_n(\omega) = \max\,(0, \mathsf{S}_1, \mathsf{S}_2, \ldots, \mathsf{S}_n).$$

Then

$$\int_{\{\mathsf{M}_n > 0\}} \mathsf{X}\, dP \geq 0.$$

Proof. We give a very simple recent proof of this due to Adriano Garsia [61]. For any $k \leq n$, $\mathsf{M}_n(T\omega) \geq \mathsf{S}_k(T\omega)$. Hence

$$\mathsf{X}(\omega) + \mathsf{M}_n(T\omega) \geq \mathsf{X}(\omega) + \mathsf{S}_k(T\omega) = \mathsf{S}_{k+1}(\omega).$$

Write this as

$$\mathsf{X}(\omega) \geq \mathsf{S}_{k+1}(\omega) - \mathsf{M}_n(T\omega), \quad k = 1, \ldots, n.$$

But trivially,

$$\mathsf{X}(\omega) \geq \mathsf{S}_1(\omega) - \mathsf{M}_n(T\omega),$$

since $\mathsf{S}_1(\omega) = \mathsf{X}(\omega)$ and $\mathsf{M}_n(\omega) \geq 0$. These two inequalities together give $\mathsf{X}(\omega) \geq \max\big(\mathsf{S}_1(\omega), \ldots, \mathsf{S}_n(\omega)\big) - \mathsf{M}_n(T\omega)$. Thus

$$\int_{\{\mathsf{M}_n > 0\}} \mathsf{X}\, dP \geq \int_{\{\mathsf{M}_n > 0\}} [\max\big(\mathsf{S}_1(\omega), \ldots, \mathsf{S}_n(\omega)\big) - \mathsf{M}_n(T\omega)]\, dP.$$

On the set $\{\mathsf{M}_n > 0\}$, $\max\,(\mathsf{S}_1, \ldots, \mathsf{S}_n) = \mathsf{M}_n$. Hence

$$\int_{\{\mathsf{M}_n > 0\}} \mathsf{X}\, dP \geq \int_{\{\mathsf{M}_n > 0\}} [\mathsf{M}_n(\omega) - \mathsf{M}_n(T\omega)]\, dP,$$

but

$$\int_{\{\mathsf{M}_n > 0\}} \mathsf{M}_n(\omega)\, dP - \int_{\{\mathsf{M}_n > 0\}} \mathsf{M}_n(T\omega)\, dP \geq \int \mathsf{M}_n(\omega)\, dP - \int \mathsf{M}_n(T\omega)\, dP = 0.$$

This last is by (6.20).

Completion of proof of 6.21. Assuming that $E(\mathsf{X} \mid \mathfrak{I}) = 0$, prove that the averages converge to zero a.s. Then apply this result to the random variable $\mathsf{X}(\omega) - E(\mathsf{X} \mid \mathfrak{I})$ to get the general case. Let $\overline{\mathsf{X}} = \overline{\lim}\, \mathsf{S}_n/n$, and for any $\epsilon > 0$, denote $D = \{\overline{\mathsf{X}} > \epsilon\}$. Note that $\overline{\mathsf{X}}(T\omega) = \overline{\mathsf{X}}(\omega)$, so $\overline{\mathsf{X}}$ and therefore D are invariant. Define the random variable

$$\mathsf{X}^*(\omega) = \big(\mathsf{X}(\omega) - \epsilon\big)\chi_D(\omega),$$

and using X^*, define S_k^*, M_n^* as above.

The maximal ergodic theorem gives

$$\int_{\{\mathsf{M}_n^* > 0\}} \mathsf{X}^*\, dP \geq 0.$$

The rest of the proof is easy sailing. The sets

$$F_n = \{\mathsf{M}_n^* > 0\} = \left\{\max_{1\le k\le n} \mathsf{S}_k^* > 0\right\}$$

converge upward to the set

$$F = \left\{\sup_{k\ge 1} \mathsf{S}_k^* > 0\right\} = \left\{\sup_{k\ge 1} \frac{\mathsf{S}_k^*}{k} > 0\right\}$$

$$= \left\{\sup_{k\ge 1} \frac{\mathsf{S}_k}{k} > \epsilon\right\} \cap D.$$

Since $\sup_{k\ge 1} \mathsf{S}_k/k \ge \overline{\mathsf{X}}$, $F = D$. The inequality $E\,|\mathsf{X}^*| \le E\,|\mathsf{X}| + \epsilon$ allows the use of the bounded convergence theorem, so we conclude that

$$\int_{F_n} \mathsf{X}^*\, dP \to \int_F \mathsf{X}^*\, dP.$$

Therefore,

$$\int_D \mathsf{X}^*\, dP \ge 0.$$

But

$$\int_D \mathsf{X}^*\, dP = \int_D \mathsf{X}\, dP - \epsilon P(D)$$

$$= \int_D E(\mathsf{X} \mid \mathfrak{J})\, dP - \epsilon P(D)$$

$$= -\epsilon P(D),$$

which implies $P(D) = 0$, and $\overline{\mathsf{X}} \le 0$ a.s. Apply the same argument to the random variable $-\mathsf{X}(\omega)$. Here the lim sup of the sums is

$$\overline{\lim}\left(-\frac{\mathsf{S}_n}{n}\right) = -\underline{\lim}\left(\frac{\mathsf{S}_n}{n}\right) = -\underline{\mathsf{X}}.$$

The conclusion above becomes $-\underline{\mathsf{X}} \le 0$ or $\underline{\mathsf{X}} \ge 0$ a.s. Putting these two together gives the theorem. Q.E.D.

A consequence of 6.21 is that if T is ergodic, time averages can be replaced by phase averages, in other words,

Corollary 6.23. *Let T be measure-preserving and ergodic on* $(\Omega, \mathcal{F}, P)$. *Then for* X *any random variable such that* $E\,|\mathsf{X}| < \infty$,

$$\lim_n \frac{1}{n} \sum_{k=0}^{n-1} \mathsf{X}(T^k\omega) = E\mathsf{X} \quad \text{a.s.}$$

Proof. Every set in $\mathfrak{J}$ has probability zero or one, hence

$$E(\mathsf{X} \mid \mathfrak{J}) = E\mathsf{X} \quad \text{a.s.}$$

6. CONVERSES AND COROLLARIES

It is natural to ask whether the conditions of the ergodic theorem are necessary and sufficient. Again the answer is—Almost. If X is a nonnegative random variable and $EX = \infty$, it is easy to show that for T measure-preserving and ergodic,

$$\frac{1}{n}\sum_{0}^{n-1} X(T^k\omega) \to \infty \quad \text{a.s.}$$

Because defining for $\alpha > 0$,

$$X_\alpha(\omega) = \begin{cases} X(\omega), & \text{if } X(\omega) \le \alpha, \\ 0, & \text{otherwise,} \end{cases}$$

of course, $E|X_\alpha| < \infty$. Thus the ergodic theorem can be used to get

$$\underline{\lim}\, \frac{1}{n}\sum_{0}^{n-1} X(T^k\omega) \ge \lim \frac{1}{n}\sum_{0}^{n-1} X_\alpha(T^k\omega) = EX_\alpha, \quad \text{a.s.}$$

Take $\alpha \uparrow \infty$ to get the conclusion.

But, in general, if $E|X| = \infty$, it does not follow that the averages diverge a.s. (see Halmos [65, p. 32]).

Come back now to the question of the asymptotic density of the points $\omega, T\omega, T^2\omega, \ldots$ In the ergodic theorem, for any $A \in \mathcal{F}$, take $X(\omega) = \chi_A(\omega)$. Then the conclusion reads, if T is ergodic,

$$\frac{1}{n}\sum_{k=0}^{n-1} \chi_A(T^k\omega) \xrightarrow{\text{a.s.}} P(A),$$

so that for almost every starting point ω, the asymptotic proportion of points in A is exactly $P(A)$. If Ω has a topology with a countable basis such that $P(N) > 0$ for every open neighborhood N, then this implies that for almost every ω, the set of points $\omega, T\omega, T^2\omega, \ldots$ is dense in Ω.

Another interesting and curious result is

Corollary 6.24. *Let $T: \Omega \to \Omega$ be measure-preserving and ergodic with respect to both $(\Omega, \mathcal{F}, P_1)$ and $(\Omega, \mathcal{F}, P_2)$. Then either $P_1 = P_2$ or P_1 and P_2 are orthogonal in the sense that there is a set $A \in \mathfrak{J}$ such that $P_1(A) = 1, P_2(A^c) = 1$.*

Proof. If $P_1 \ne P_2$, take $B \in \mathcal{F}$ such that $P_1(B) \ne P_2(B)$ and let $X(\omega) = \chi_B(\omega)$. Let A be the set of ω such that

$$\frac{1}{n}\sum_{0}^{n-1} X(T^k\omega) \to P_1(B).$$

By the ergodic theorem $P_1(A) = 1$. But A^c includes all ω such that

$$\frac{1}{n}\sum_{0}^{n-1} X(T^k\omega) \to P_2(B),$$

and we see that $P_2(A^c) = 1$.

Finally, we ask concerning convergence in the first mean of the averages to $E\mathsf{X}$. By the Lebesgue theorem we know that a.s. convergence plus just a little more gives first mean convergence. But here we have to work a bit to get the additional piece.

Corollary 6.25. *Under the conditions of the ergodic theorem*

$$\lim_n E\left|\frac{1}{n}\sum_{k=0}^{n-1}\mathsf{X}(T^k\omega) - E(\mathsf{X} \mid \mathfrak{J})\right| = 0.$$

Proof. We can assume that $E(\mathsf{X} \mid \mathfrak{J}) = 0$. Let

$$\mathsf{V}_n = \frac{1}{n}\sum_{k=0}^{n-1}\mathsf{X}(T^k\omega).$$

Since $\mathsf{V}_n \xrightarrow{\text{a.s.}} 0$, by Egoroff's theorem, for any $\epsilon > 0$, $\exists A \in \mathcal{F}$ such that $P(A) \leq \epsilon$ and $\mathsf{V}_n \to 0$ uniformly on A^c. Now,

$$\overline{\lim_n}\, E\,|\mathsf{V}_n| = \overline{\lim_n}\int_A |\mathsf{V}_n|\, dP \leq \overline{\lim_n}\,\frac{1}{n}\sum_0^{n-1}\int_A |\mathsf{X}_k|\, dP, \qquad \mathsf{X}_k = \mathsf{X}(T^k\omega).$$

These integrals can be estimated by

$$\begin{aligned}\int_A |\mathsf{X}_k|\, dP &= \int_{A\cap\{|\mathsf{X}_k|>N\}} |\mathsf{X}_k|\, dP + \int_{A\cap\{|\mathsf{X}_k|\leq N\}} |\mathsf{X}_k|\, dP \\ &\leq \int_{A\cap\{|\mathsf{X}_k|>N\}} |\mathsf{X}_k|\, dP + NP(A) \leq \int_{\{|\mathsf{X}|>N\}} |\mathsf{X}|\, dP + NP(A).\end{aligned}$$

Since ϵ is arbitrary, conclude that for any N,

$$\overline{\lim_n}\, E\,|\mathsf{V}_n| \leq \int_{\{|\mathsf{X}|>N\}} |\mathsf{X}|\, dP.$$

Let N go to infinity; then by the bounded convergence theorem, the right-hand side above goes to zero.

Problems

9. Another consequence of the ergodic theorem is a weak form of Weyl's equidistribution theorem. For any x in $[0, 1)$ and interval $I \subset [0, 1)$, let $\mathsf{R}_x^{(n)}(I)$ be the proportion of the points $\{(x + \lambda)[1], (x + 2\lambda)[1], \ldots, (x + n\lambda)[1]\}$ falling in the interval I. If $\lambda \geq 0$ is irrational, show that for x in a set of Lebesgue measure one, $\mathsf{R}_x^{(n)}(I) \to$ length I.

10. Let μ be a finite measure on $\mathcal{B}_1([0, 1))$ such that $Tx = 2x[1]$ preserves μ-measure. Show that μ is singular with respect to Lebesgue measure.

11. Let T be measurable on $(\Omega, \mathcal{F})$, and define $\mathcal{M}$ as the set of all probabilities P on $\mathcal{F}$ such that T is measure-preserving on $(\Omega, \mathcal{F}, P)$. Define real linear combinations by $(\alpha P_1 + \beta P_2)(B) = \alpha P_1(B) + \beta P_2(B)$, $B \in \mathcal{F}$. Show that

a) $\mathcal{M}$ is convex, that is, for $\alpha, \beta \geq 0$, $\alpha + \beta = 1$,

$$P_1, P_2 \in \mathcal{M} \Rightarrow \alpha P_1 + \beta P_2 \in \mathcal{M}.$$

An extreme point of $\mathcal{M}$ is a probability $P \in \mathcal{M}$ which is not a linear combination $\alpha P_1 + \beta P_2$, $\alpha, \beta > 0$, $\alpha + \beta = 1$, with $P_1, P_2 \in \mathcal{M}$. Show that

b) the extreme points of $\mathcal{M}$ are the probabilities P such that T is ergodic on $(\Omega, \mathcal{F}, P)$.

7. BACK TO STATIONARY PROCESSES

By the ergodic theorem and its corollary, if the shift-transformation S (see 6.10) is ergodic on $(R^{(\infty)}, \mathcal{B}_\infty, \hat{P})$, then

$$\frac{1}{n}\sum_1^n \hat{X}_k \rightarrow EX_1 \quad \text{a.s. and first mean.}$$

If S is ergodic, then, the same conclusions will hold for the original X_1, $X_2, \ldots$ process, because a.s. convergence and rth mean convergence depend only on the distribution of the process.

Almost all the material concerning invariance and ergodicity, can be formulated in terms of the original process $X_1, X_2, \ldots$ rather than going into representation space. If $B \in \mathcal{B}_\infty$ and $A = \{\mathbf{X} \in B\}$, then the inverse image under $\mathbf{X}$ of $S^{-1}B$, S the shift operator, is

$$\{\mathbf{X} \in S^{-1}B\} = \{\omega;\ (X_2, X_3, \ldots) \in B\}.$$

Hence, we reach

Definition 6.26. *An event $A \in \mathcal{F}$ is invariant if $\exists B \in \mathcal{B}_\infty$ such that for every $n \geq 1$*

$$A = \{(X_n, X_{n+1}, \ldots) \in B\}.$$

The class of invariant events is easily seen to be a σ-field. Similarly, we define a random variable Z to be invariant if there is a random variable φ on $(R^{(\infty)}, \mathcal{B}_\infty)$ such that

$$Z = \varphi(X_n, X_{n+1}, \ldots), \quad \text{all } n \geq 1. \tag{6.27}$$

The results of Section 4 hold again; Z is invariant iff it is $\mathfrak{J}$-measurable. The ergodic theorem translates as

Theorem 6.28. *If $X_1, X_2, \ldots$ is a stationary process, $\mathfrak{J}$ the σ-field of invariant events, and $E\,|X_1| < \infty$, then*

$$\frac{1}{n}\sum_1^n X_k \xrightarrow{\text{a.s.}} E(X_1 \mid \mathfrak{J}).$$

Proof. From the ergodic theorem, S_n/n converges a.s. to some random variable Y. It is not difficult to give an argument that the correct translation is $Y = E(X_1 \mid \mathfrak{I})$. But we can identify Y directly: take $Y = \overline{\lim}\, S_n/n$, then the sets $\{Y < y\}$ are invariant, hence Y is $\mathfrak{I}$-measurable. Take $A \in \mathfrak{I}$, then since we have first mean convergence,

$$\textbf{(6.29)} \qquad \frac{1}{n}\sum_1^n \int_A X_k\, dP \to \int_A Y\, dP.$$

Select $B \in \mathcal{B}_\infty$ so that $A = \{(X_k, \ldots) \in B\}$, for all $k \geq 1$. Now stationarity gives

$$\int_{\{(X_k,\ldots)\in B\}} X_k\, dP = \int_{\{(X_1,\ldots)\in B\}} X_1\, dP = \int_A X_1\, dP.$$

Use this in 6.29, to conclude

$$\int_A X_1\, dP = \int_A Y\, dP, \quad \text{all } A \in \mathfrak{I}.$$

By definition, $Y = E(X_1 \mid \mathfrak{I})$.

Definition 6.30. *A stationary process* $X_1, X_2, \ldots$ *is ergodic if every invariant event has probability zero or one.*

If $\mathfrak{I}$ has this zero-one property, of course the averages converge to EX_1, a.s.

Ergodicity, like stationarity, is preserved under taking functions of the process. More precisely,

Proposition 6.31. *Let* $X_1, X_2, \ldots$ *be a stationary and ergodic process,* $\varphi(\mathbf{x})$ *measurable* $\mathcal{B}_\infty$, *then the process* $Y_1, Y_2, \ldots$ *defined by*

$$Y_k = \varphi(X_k, X_{k+1}, \ldots)$$

is ergodic.

Proof. This is very easy to see. Use the same argument as in Proposition 6.6 to conclude that for any $B \in \mathcal{B}_\infty$, $\exists A \in \mathcal{B}_\infty$ such that

$$\{\omega;\ (Y_1, Y_2, \ldots) \in B\} = \{\omega;\ (X_1, X_2, \ldots) \in A\},$$
$$\{\omega;\ (Y_2, Y_3, \ldots) \in B\} = \{\omega;\ (X_2, X_3, \ldots) \in A\}.$$

Hence, every invariant event on the **Y**-process coincides with an invariant event on the **X**-process.

One result that is both interesting and useful in establishing ergodicity is

Proposition 6.32. *Let* $X_1, X_2, \ldots$ *be a stationary process. Then every invariant event A is a tail event.*

Proof. Take B so that $A = \{(X_n, X_{n+1}, \ldots) \in B\}$, $n \geq 1$. Hence $A \in \mathcal{F}(X_n, X_{n+1}, \ldots)$, all n.

Corollary 6.33. *Let* $X_1, X_2, \ldots$ *be independent and identically distributed; then the process is ergodic.*

Proof. Kolmogorov's zero-one law.

By this corollary we can include the strong law of large numbers as a consequence of the ergodic theorem, except for the converse.

Problems

12. Show that the event $\{X_n \in B \text{ i.o.}\}$, $B \in \mathcal{B}_1$, is invariant.

13. If $X_1, X_2, \ldots$ is stationary and if there $\exists B \in \mathcal{B}_\infty$ such that

$$A = \{(X_1, X_2, \ldots) \in B\} = \{(X_2, X_3, \ldots) \in B\} \quad \text{a.s.},$$

show that A is a.s. equal to an invariant event.

14. A process $X_1, X_2, \ldots$ is called a normal process if $X_1, \ldots, X_n$ have a joint normal distribution for every n. Let $EX_i = 0$, $\Gamma_{ij} = EX_iX_j$.

1) Prove that the process is stationary iff Γ_{ij} depends only on $|i - j|$.
2) Assume $\Gamma_{ij} = r(|i - j|)$. Then prove that $\lim_m r(m) = 0$ implies that the process is ergodic.

[Assume that for every n, the determinant of Γ_{ij}, $i, j = 1, \ldots, n$, is not zero. See Chapter 11 for the definition and properties of joint normal distributions.]

15. Show that $X_1, X_2, \ldots$ is ergodic iff for every $A \in \mathcal{B}_k$, $k = 1, 2, \ldots,$

$$\frac{1}{n}\sum_1^n \chi_A(X_n, \ldots, X_{n+k}) \xrightarrow{\text{a.s.}} P((X_1, \ldots, X_{k+1}) \in A).$$

16. Let $X_1, X_2, \ldots$ and $Y_1, Y_2, \ldots$ be two stationary, ergodic processes on $(\Omega, \mathcal{F}, P)$. Toss a coin with probability p of heads independently of $\mathbf{X}$ and $\mathbf{Y}$. If it comes up heads, observe $\mathbf{X}$, if tails, observe $\mathbf{Y}$. Show that the resultant process is stationary, but not ergodic.

8. AN APPLICATION

There is a very elegant application of the above ideas due to Spitzer, Kesten, and Whitman (see Spitzer, [130, pp. 35 ff]). Let $X_1, X_2, \ldots$ be a sequence of independent identically distributed random variables taking values in the integers. The range R_n of the first n sums is defined as the number of distinct points in the set $\{S_1, \ldots, S_n\}$. Heuristically, the more the tendency of the sums S_n to return to the origin, the smaller R_n will be, because if we are at a given point k at time n, the distribution of points around k henceforth looks like the distribution of points around the origin starting from $n = 1$. To pin this down, write

$$P(\text{no return}) = P(\mathsf{S}_n \neq 0, \quad n = 1, 2, \ldots),$$

then,

Proposition 6.34

$$\lim_n \frac{E\mathsf{R}_n}{n} = P(\text{no return}).$$

Proof. Write $\mathsf{R}_n = \sum_{k=0}^{n} \mathsf{W}_k$, where

$$\mathsf{W}_k = \begin{cases} 1 & \text{if } \ \mathsf{S}_j \neq \mathsf{S}_k, \quad j = 1, \ldots, k-1, \\ 0, & \text{otherwise.} \end{cases}$$

So now,

$$\begin{aligned} E\mathsf{W}_k &= P(\mathsf{S}_k - \mathsf{S}_{k-1} \neq 0, \mathsf{S}_k - \mathsf{S}_{k-2} \neq 0, \ldots, \mathsf{S}_k - \mathsf{S}_1 \neq 0) \\ &= P(\mathsf{X}_k \neq 0, \mathsf{X}_k + \mathsf{X}_{k+1} \neq 0, \ldots, \mathsf{X}_k + \cdots + \mathsf{X}_2 \neq 0) \\ &= P(\mathsf{S}_1 \neq 0, \ldots, \mathsf{S}_{k-1} \neq 0), \end{aligned}$$

the last equality holding because $(\mathsf{X}_k, \ldots, \mathsf{X}_2)$ has the same distribution as $(\mathsf{X}_1, \ldots, \mathsf{X}_{k-1})$. Therefore $\lim_k E\mathsf{W}_k = P(\text{no return})$.

The remarkable result is

Theorem 6.35

$$\lim \frac{\mathsf{R}_n}{n} = P(\text{no return}) \quad \text{a.s.}$$

Proof. Take N any positive integer, and let Z_k be the number of distinct points visited by the successive sums during the time $(k-1)N + 1$ to kN, that is, Z_k is the range of $\{\mathsf{S}_{(k-1)N+1}, \ldots, \mathsf{S}_{kN}\}$. Note that Z_k depends only on the X_n for n between $(k-1)N + 1$ and kN, so that the Z_k are independent, $|\mathsf{Z}_k| \leq N$, and are easily seen to be identically distributed. Use the obvious inequality $\mathsf{R}_{nN} \leq \mathsf{Z}_1 + \cdots + \mathsf{Z}_n$ and apply the law of large numbers:

$$\overline{\lim_n} \frac{\mathsf{R}_{nN}}{nN} \leq \overline{\lim_n} \frac{1}{Nn} (\mathsf{Z}_1 + \cdots + \mathsf{Z}_n) = \frac{1}{N} E\mathsf{Z}_1.$$

For n' not a multiple of N, $\mathsf{R}_{n'}$ differs by at most N from one of R_{nN}, so

$$\overline{\lim_n} \frac{\mathsf{R}_n}{n} \leq \frac{1}{N} E\mathsf{Z}_1 \quad \text{a.s.}$$

But $\mathsf{Z}_1 = \mathsf{R}_N$, hence letting $N \to \infty$, and using 6.34, we get

$$\overline{\lim_n} \frac{\mathsf{R}_n}{n} \leq P(\text{no return}) \quad \text{a.s.}$$

Going the other way is more interesting. Define

$$V_k = \begin{cases} 1 & \text{if } S_j \neq S_k, \; j > k, \\ 0, & \text{otherwise.} \end{cases}$$

That is, V_k is one if at time k, S_k is in a state which is never visited again, and zero otherwise. Now $V_1 + \cdots + V_n$ is the number of states visited in time n which are never revisited. R_n is the number of states visited in time n which are not revisited prior to time $n + 1$. Thus $R_n \geq V_1 + \cdots + V_n$. Now define

$$\varphi(X_1, X_2, \ldots) = \begin{cases} 1, & S_k \neq 0, \quad k = 1, \ldots, \\ 0, & \text{otherwise,} \end{cases}$$

and make the important observation that

$$V_k = \begin{cases} 1, & X_k \neq 0, X_k + X_{k+1} \neq 0, \ldots, \\ 0, & \text{otherwise,} \end{cases}$$

$$= \varphi(X_k, X_{k+1}, \ldots).$$

Use 6.31 and 6.33 to show that $V_1, V_2, \ldots$ is a stationary, ergodic sequence. Now the ergodic theorem can be used to conclude that

$$\underline{\lim} \frac{R_n}{n} \geq \underline{\lim} \frac{V_1 + \cdots + V_n}{n} = EV_1 \quad \text{a.s.}$$

Of course, $EV_1 = P(\text{no return})$, and this completes the proof.

9. RECURRENCE TIMES

For $X_0, X_1, \ldots$ stationary, and any set $A \in \mathcal{B}_1$, look at the times that the process enters A, that is, the n such that $X_n \in A$.

Definition 6.36. *For* $A \in \mathcal{B}_1$, $P(X_0 \in A) > 0$, *define*

$$R_1 = \min \{n; \; X_n \in A, n > 0\},$$

$$R_2 = \min \{n; \; X_n \in A, n > R_1\},$$

and so forth. These are the occurrence times of the set A. *The recurrence times* $T_1, T_2, \ldots$ *are given by*

(6.37) $$T_1 = R_1, \quad T_2 = R_2 - R_1, \quad T_k = R_k - R_{k-1}.$$

If $\{X_n\}$ is ergodic, then $P(X_n \in A \text{ i.o.}) = 1$; so the R_k are well defined a.s. But if $\{X_n\}$ is not ergodic a subsidiary condition has to be imposed to make the R_k well defined. At any rate, the smaller A is, the longer it takes to get back to it, and in fact we arrive at the following proposition:

Proposition 6.38. *Let* $X_0, X_1, \ldots$ *be a stationary process,* $A \in \mathcal{B}_1$ *such that*

$$P(X_n \in A \text{ at least once}) = 1, \tag{6.39}$$

then the R_k, $k = 1, \ldots$ *are finite* a.s. *On the sample space* $\Omega_A = \{\omega;\ X_0 \in A\}$ *the* $T_1, T_2, \ldots$ *form a stationary sequence under the probability* $P(\cdot \mid X_0 \in A)$, *and*

$$E(T_1 \mid X_0 \in A) = \frac{1}{P(X_0 \in A)}.$$

Remarks. This means that to get the $T_1, T_2, \ldots$ to be stationary, we have to start off on the set A at time zero. This seems too complicated because once we have landed in A then the returns should be stationary, that is, the $T_2, T_3, \ldots$ should be stationary under P. This is not so, and counterexamples are not difficult to construct (see Problem 20.)

Note that $P(X_0 \in A) > 0$, otherwise condition (6.39) is violated. Therefore, conditional probability given $\{X_0 \in A\}$ is well defined.

Proof. Extend $\{X_n\}$ to $0, \pm 1, \pm 2, \ldots$ By (6.39), $P(R_1 < \infty) = 1$. From the stationarity

$$P(T_k \leq n,\ R_{k-1} = j) = P(R_1 \leq n,\ C),$$

where $C \in \mathcal{F}(X_0, X_{-1}, \ldots)$. Let $n \to \infty$, so that we get

$$P(T_k < \infty,\ R_{k-1} = j) = P(R_{k-1} = j).$$

Go down the ladder to conclude that $P(R_1 < \infty) = 1$ implies $P(R_k < \infty) = 1$, $k \geq 1$.

To prove stationarity, we need to establish that

$$P(T_1 = n_1, \ldots, T_k = n_k \mid X_0 \in A)$$
$$= P(T_2 = n_1, \ldots, T_{k+1} = n_k \mid X_0 \in A).$$

This is not difficult to do, but to keep out of notational messes, I prove only that

$$P(T_1 = n \mid X_0 \in A) = P(T_2 = n \mid X_0 \in A).$$

The generalization is exactly the same argument. Define random variables $U_n = \chi_A(X_n)$, and sets C_k by

$$C_k = \{U_{-1} = 0, \ldots, U_{-k+1} = 0, U_{-k} = 1\}.$$

The $\{C_k\}$ are disjoint, and

$$\bigcup_1^\infty C_k = \{X_n \in A \quad \text{at least once, } n \leq -1\}.$$

I assert that $P\left(\bigcup_1^\infty C_k\right) = 1$, because

$$P\left(\bigcup_1^\infty C_k\right) = \lim_n P\left(\bigcup_{-n-1}^{-1} \{\mathsf{U}_k = 1\}\right).$$

By stationarity of the $\{\mathsf{U}_k\}$ process,

$$P\left(\bigcup_{-n-1}^{-1} \{\mathsf{U}_k = 1\}\right) = P\left(\bigcup_0^n \{\mathsf{U}_k = 1\}\right).$$

The limit of the right-hand side is $P(\mathsf{X}_n \in A$ at least once), which is one by (6.39). Now, using stationarity again, we find that

$$\begin{aligned} P(\mathsf{T}_2 = n, \mathsf{X}_0 \in A) &= \sum_{k=1}^\infty P(\mathsf{T}_2 = n, \mathsf{T}_1 = k, \mathsf{X}_0 \in A) \\ &= \sum_{k=1}^\infty P(\mathsf{T}_1 = n, \mathsf{X}_0 \in A, C_k) = P(\mathsf{T}_1 = n, \mathsf{X}_0 \in A). \end{aligned}$$

To compute $E(\mathsf{T}_1 \mid \mathsf{X}_0 \in A)$, note that

$$P(\mathsf{T}_1 \geq k, \mathsf{X}_0 \in A) = P(C_k).$$

Consequently,

$$\begin{aligned} E(\mathsf{T}_1 \mid \mathsf{X}_0 \in A) &= \sum_{k=1}^\infty P(\mathsf{T}_1 \geq k \mid \mathsf{X}_0 \in A) \\ &= \frac{1}{P(\mathsf{X}_0 \in A)} \sum_1^\infty P(C_k) = \frac{1}{P(\mathsf{X}_0 \in A)}. \end{aligned}$$

We can use the ergodic theorem to get a stronger result.

Theorem 6.40. *If the process $\{\mathsf{X}_n\}$ is ergodic, then the process $\{\mathsf{T}_k\}$ on $\{\mathsf{X}_0 \in A\}$ is ergodic under $P(\cdot \mid \mathsf{X}_0 \in A)$.*

Proof. By the ergodic theorem,

$$\frac{1}{n} \sum_{k=1}^n \chi_A(\mathsf{X}_k) \xrightarrow{\text{a.s.}} P(\mathsf{X}_0 \in A).$$

On every sequence such that the latter holds, $\mathsf{R}_n \to \infty$, and

$$\frac{1}{\mathsf{R}_n} \sum_{k=1}^{\mathsf{R}_n} \chi_A(\mathsf{X}_k) \to P(\mathsf{X}_0 \in A).$$

The sum $\sum_{k=1}^{\mathsf{R}_n} \chi_A(\mathsf{X}_k)$ is the number of visits X_k makes to A up to the time of the nth occurrence. Thus $\sum_1^{\mathsf{R}_n} \chi_A(\mathsf{X}_k) = n$, so

$$\frac{\mathsf{T}_1 + \cdots + \mathsf{T}_n}{n} \xrightarrow{\text{a.s.}} \frac{1}{P(\mathsf{X}_0 \in A)}.$$

Note that for every function f measurable $\mathcal{B}_\infty$, there is a function g measurable $\mathcal{B}_\infty(\{0, 1\})$ such that on the set $\{\mathsf{R}_{k-1} = j\}$,

$$f(\mathsf{T}_k, \mathsf{T}_{k+1}, \ldots) = g(\mathsf{U}_{j+1}, \mathsf{U}_{j+2}, \ldots),$$

where U_n denotes $\chi_A(\mathsf{X}_n)$ again. Therefore

$$\sum_{k=1}^{n} f(\mathsf{T}_k, \ldots) = \sum_{j=0}^{\mathsf{R}_{n-1}} \mathsf{U}_j g(\mathsf{U}_{j+1}, \ldots).$$

Since $\mathsf{R}_{n-1} \to \infty$ a.s., if $E|g| < \infty$ we can use the ergodic theorem as follows:

$$\text{(6.41)} \qquad \frac{1}{n} \sum_{k=1}^{n} f(\mathsf{T}_k, \ldots) = \frac{\mathsf{R}_{n-1}}{n} \cdot \frac{1}{\mathsf{R}_{n-1}} \sum_{j=0}^{\mathsf{R}_{n-1}} \mathsf{U}_j g(\mathsf{U}_{j+1}, \ldots)$$

$$\to \frac{1}{P(\mathsf{X}_0 \in A)} E(\mathsf{U}_0 g(\mathsf{U}_1, \ldots)) \quad \text{a.s.}$$

Because $\mathsf{U}_0 = \chi_A(\mathsf{X}_0)$, this limit is

$$E(g(\mathsf{U}_1, \ldots) \mid \mathsf{X}_0 \in A) \qquad \text{or} \qquad E(f(\mathsf{T}_1, \ldots) \mid \mathsf{X}_0 \in A).$$

On the space $\{\mathsf{X}_0 \in A\}$, take f to be any bounded invariant function of $\mathsf{T}_1, \ldots$, that is,

$$f(\mathsf{T}_1, \mathsf{T}_2, \ldots) = f(\mathsf{T}_k, \mathsf{T}_{k+1}, \ldots), \quad \text{all } k \geq 1.$$

Then (6.41) implies that $f(\mathsf{T}_1, \ldots)$ is a.s. constant on $\{\mathsf{X}_0 \in A\}$, implying in turn that the process $\mathsf{T}_1, \mathsf{T}_2, \ldots$ is ergodic.

10. STATIONARY POINT PROCESSES

Consider a class of processes gotten as follows: to every integer n, positive and negative, associate a random variable U_n which is either zero or one. A way to look at this process is as a sequence of points. If a point occurs at time n, then $\mathsf{U}_n = 1$, otherwise $\mathsf{U}_n = 0$. We impose a condition to ensure that these processes have a.s. no trivial sample points.

Condition 6.42. *There is probability zero that all* $\mathsf{U}_n = 0$.

For a point process to be time-homogeneous, the $\{\mathsf{U}_n\}$ have to form a stationary sequence.

Definition 6.43. *A stationary discrete point process is a stationary process* $\{\mathsf{U}_n\}$, $n = 0, \pm 1, \ldots$ *where* U_n *is either zero or one.*

Note: $P(\mathsf{U}_n = 1) > 0$, otherwise $P(\text{all } \mathsf{U}_n = 0) = 1$. Take $A = \{1\}$, then (6.38) implies that given $\{\mathsf{U}_0 = 1\}$ the times between points form a stationary sequence $T_1, T_2, \ldots$ of positive integer-valued random variables such that

$$ET_1 = \frac{1}{P(U_0 = 1)} \quad \text{or} \quad P(U_0 = 1) = \frac{1}{ET_1}.$$

The converse question is: Given a stationary sequence $T_1, T_2, \ldots$ of positive integer valued random variables, is there a stationary point process $\{U_n\}$ with interpoint distances having the same distribution as $T_1, T_2, \ldots$? The difficulty in extracting the interpoint distances from the point process was that an origin had to be pinned down; that is, $\{U_0 = 1\}$ had to be given. But here the problem is to start with the $T_1, T_2, \ldots$ and smear out the origin somehow. Define

$$V_k = \begin{cases} 1, & \exists n \text{ such that } T_1 + \cdots + T_n = k, \\ 0, & \text{otherwise.} \end{cases}$$

Suppose that there is a stationary process $\{U_k\}$ with interpoint distances having the same distribution as $T_1, \ldots$ Denote the probability on $\{U_n\}$ by Q. Then for $\mathbf{s}$ any sequence k-long of zeroes and ones

$$Q((U_k, \ldots, U_1) = \mathbf{s} \mid U_0 = 1) = P((V_k, \ldots, V_1) = \mathbf{s}).$$

This leads to

(6.44)

$$\begin{aligned} Q((U_k, \ldots, U_1) = \mathbf{s}) &= \sum_{j=0}^{\infty} Q((U_k, \ldots, U_1) = \mathbf{s}, U_0 = 0, \ldots, 0, U_{-j} = 1) \\ &= \sum_{j=0}^{\infty} Q((U_{k+j}, \ldots, U_{j+1}) = \mathbf{s}, U_j = 0, \ldots, U_0 = 1) \\ &= Q(U_0 = 1) \sum_{j=0}^{\infty} P((V_{k+j}, \ldots, V_{j+1}) = \mathbf{s}, T_1 > j) \\ &= \frac{1}{ET_1} \sum_{j=0}^{\infty} P((V_{k+j}, \ldots, V_{j+1}) = \mathbf{s}, T_1 > j). \end{aligned}$$

The right-hand side above depends only upon the probability P on the $T_1, T_2, \ldots$ process. So if a stationary point process exists with interpoint distances $T_1, T_2, \ldots$ it must be unique. Furthermore, if we define Q directly by means of (6.44), then it is not difficult to show that we get a consistent set of probabilities for a stationary point process having interpoint distances with distribution $T_1, T_2, \ldots$ Thus

Theorem 6.45. *Let $T_1, T_2, \ldots$ be a stationary sequence of positive integer valued random variables such that $ET_1 < \infty$. Then there is a unique stationary point process $\{U_n\}$ such that the interpoint distances given $\{U_0 = 1\}$ have the same distribution as $T_1, T_2, \ldots$*

Proof. The work is in showing that the Q defined in (6.44) is a probability for the desired process. To get stationarity one needs to verify that

$$Q((\mathsf{U}_{k+1}, \ldots, \mathsf{U}_2) = \mathbf{s}) = Q((\mathsf{U}_k, \ldots, \mathsf{U}_1) = \mathbf{s}).$$

Once this is done, then it is necessary to check that the interpoint distances given $\{\mathsf{U}_0 = 1\}$ have the same distribution as $\mathsf{T}_1, \mathsf{T}_2, \ldots$ For example, check that

$$Q(\mathsf{U}_j = 1, \mathsf{U}_{j-1} = 0, \ldots, \mathsf{U}_1 = 0 \mid \mathsf{U}_0 = 1) = P(\mathsf{T}_1 = j).$$

All this verification we leave to the reader.

Problems

17. Given the process $\mathsf{T}_1, \mathsf{T}_2, \ldots$ such that $\mathsf{T}_k = 2$, all k, describe the corresponding stationary point process.

18. If the interpoint distances for a stationary point process, given $\{\mathsf{U}_0 = 1\}$ are $\mathsf{T}_1, \mathsf{T}_2, \ldots,$ prove that the distribution of time n^* until the first point past the origin, that is,

$$\mathsf{n}^* = \min \{n;\ n \geq 1, \mathsf{U}_n = 1\}$$

is given by

$$P(\mathsf{n}^* = n) = \frac{1}{E\mathsf{T}_1} P(\mathsf{T}_1 > n).$$

19. If the interpoint distances $\mathsf{T}_1, \mathsf{T}_2, \ldots$ given $\{\mathsf{U}_0 = 1\}$ are independent random variables, then show that the unconditioned interpoint distances $\mathsf{T}_2, \mathsf{T}_3, \ldots$ are independent identically distributed random variables with the same distribution as the conditioned random variables and that $(\mathsf{T}_2, \mathsf{T}_3, \ldots)$ is independent of the time n^* of the first point past the origin.

20. Consider the stationary point process $\{\mathsf{U}_n\}$ having independent interpoint distances $\mathsf{T}_1, \mathsf{T}_2, \ldots$ with $P(\mathsf{T}_1 = 1) = 1 - \epsilon$, with ϵ very small. Now consider the stationary point process

$$\mathsf{V}_n = 1 - \mathsf{U}_n, \quad n = 0, \pm 1, \ldots$$

Show that for this latter process the random variables defined by the distance from the first point past the origin to the second point, from the second point to the third point, etc., do not form a stationary sequence.

NOTES

The ergodic theorem was proven by G. D. Birkhoff in 1931 [4]. Since that time significant improvements have been made on the original lengthy proof, and the proof given here is the most recent simplification in a sequence involving many simplifications and refinements.

The nicest texts around on ergodic theorems in a framework of measure-preserving transformations are Halmos [65], and Hopf [73] (the latter available only in German). From the point of view of dynamical systems, see Khintchine [90], and Birkhoff [5]. Recently, E. Hopf generalized the ergodic theorem to operators T acting on measurable functions defined on a measure space $(\Omega, \mathcal{F}, \mu)$ such that T does not increase L_1-norm or L_∞-norm. For this and other operator-theoretic aspects, see the Dunford-Schwartz book [41].

For $X_1, \ldots$ a process on $(\Omega, \mathcal{F}, P)$, we could try to define a set transformation S^{-1} on $\mathcal{F}(\mathbf{X})$ similar to the shift in sequence space as follows: If $A \in \mathcal{F}(\mathbf{X})$, then take $B \in \mathcal{B}_\infty$ such that $A = \{(X_1, X_2, \ldots) \in B\}$ and define $S^{-1}A = \{(X_2, X_3, \ldots) \in B\}$. The same difficulty comes up again; B is not uniquely determined, and if $A = \{\mathbf{X} \in B_1\}$, $A = \{\mathbf{X} \in B_2\}$, it is not true in general that

$$\{(X_2, X_3, \ldots) \in B_1\} = \{(X_2, X_3, \ldots) \in B_2\}.$$

But for stationary processes, it is easy to see that these latter two sets differ only by a set of probability zero. Therefore S^{-1} can be defined, not on $\mathcal{F}(\mathbf{X})$, but only on equivalence classes of sets in $\mathcal{F}(\mathbf{X})$, where sets $A_1, A_2 \in \mathcal{F}(\mathbf{X})$ are equivalent if $P(A_1 \triangle A_2) = 0$. For a deeper discussion of this and other topics relating to the translation into a probabilistic context of the ergodic theorem see Doob [39].

The fact that the expected recurrence time of A starting from A is $1/P(X_0 \in A)$ is due to Kac [81]. A good development of point processes and proofs of a more general version of 6.45 and the ergodic property of the recurrence times is in Ryll-Nardzewski [119].

CHAPTER 7

MARKOV CHAINS

1. DEFINITIONS

The basic property characterizing Markov chains is a probabilistic analogue of a familiar property of dynamical systems. If one has a system of particles and the position and velocities of all particles are given at time t, the equations of motion can be completely solved for the future development of the system. Therefore, any other information given concerning the past of the process up to time t is superfluous as far as future development is concerned. The present state of the system contains all relevant information concerning the future. Probabilistically, we formalize this by defining a Markov chain as

Definition 7.1. *A process $\mathsf{X}_0, \mathsf{X}_1, \ldots$ taking values in $F \in \mathcal{B}_1$ is called a Markov chain if*

$$P(\mathsf{X}_{n+1} \in A \mid \mathsf{X}_n, \ldots, \mathsf{X}_0) = P(\mathsf{X}_{n+1} \in A \mid \mathsf{X}_n) \text{ a.s.}$$

for all $n \geq 0$ and $A \in \mathcal{B}_1(F)$.

For each n, there is a version $p_n(A \mid x)$ of $P(\mathsf{X}_{n+1} \in A \mid \mathsf{X}_n = x)$ which is a regular conditional distribution. These are the *transition probabilities* of the process. We restrict ourselves to the study of a class of Markov chains which have a property similar to conservative dynamical systems. One way to state that a system is conservative is that if it goes from any state x at $t = 0$ to y in time τ, then starting from x at any time t it will be in y at time $t + \tau$. The corresponding property for Markov chains is that the transition probabilities do not depend on time.

Definition 7.2. *A Markov chain $\mathsf{X}_0, \mathsf{X}_1, \ldots$ on $F \in \mathcal{B}_1$ has stationary transition probabilities $p(A \mid x)$* if *$p(A \mid x)$ is a regular conditional distribution and if for each $A \in \mathcal{B}_1(F)$, $n \geq 0$, $p(A \mid x)$ is a version of $P(\mathsf{X}_{n+1} \in A \mid \mathsf{X}_n = x)$. The initial distribution is defined by*

$$\pi(A) = P(\mathsf{X}_0 \in A).$$

The transition probabilities and the initial distribution determine the distribution of the process.

Proposition 7.3. *For a Markov chain* $X_0, X_1, \ldots$ *on* $F \in \mathcal{B}_1$,

(7.4)

$$P(X_n \in A_n, X_{n-1} \in A_{n-1}, \ldots, X_0 \in A_0)$$

$$= \int_{A_{n-1}} \int_{A_{n-2}} \cdots \int_{A_0} p_{n-1}(A_n \mid x_{n-1}) p_{n-2}(dx_{n-1} \mid x_{n-2}) \cdots p_0(dx_1 \mid x_0) \pi(dx_0)$$

for all $A_n, \ldots, A_0 \in \mathcal{B}_1(F)$.

Proof. Use

$$P(X_n \in A_n, X_{n-1} \in A_{n-1}, \ldots, X_0 \in A_0) = \int_{\{X_{n-1} \in A_{n-1}, \ldots, X_0 \in A_0\}} P(X_n \in A_n \mid X_{n-1}, \ldots, X_0)\, dP$$

and proceed by induction.

The Markov property as defined in 7.1 simply states that the present state of the system determines the probability for one step into the future. This generalizes easily:

Proposition 7.5. *Let* $X_0, X_1, \ldots$ *be a Markov chain, and* C *any event in* $\mathcal{F}(X_{n+1}, X_{n+2}, \ldots)$. *Then*

$$P(C \mid X_n, \ldots, X_0) = P(C \mid X_n) \quad \text{a.s.}$$

Having stationary transition probabilities generalizes into

Proposition 7.6. *If the process* $X_0, X_1, \ldots$ *is Markov on* $F \in \mathcal{B}_1$ *with stationary transition probabilities, then for every* $B \in \mathcal{B}_\infty(F)$ *there are versions of*

$$P\big((X_{n+1}, X_{n+2}, \ldots) \in B \mid X_n = x\big)$$

and

$$P\big((X_1, X_2, \ldots) \in B \mid X_0 = x\big),$$

which are equal.

The proofs of both 7.5 and 7.6 are left to the reader.

Proposition 7.3 indicates how to do the following construction:

Proposition 7.7. *Let* $p(A \mid x)$ *be a regular conditional probability on* $\mathcal{B}_1(F)$ *for* $x \in F$, *and* $\pi(A)$ *a probability on* $\mathcal{B}_1(F)$. *Then there exists a Markov chain* $\hat{X}_0, \hat{X}_1, \hat{X}_2, \ldots$ *with stationary transition probabilities* $p(A \mid x)$ *and initial distribution* π.

Proof. Use (7.4) to define probabilities on rectangles. Then it is easy to check that all the conditions of 2.18 are met. Now extend and use the coordinate representation process. What remains is to show that any process satisfying (7.4) is a Markov chain with $p(A \mid x)$ as its transition probabilities. (That $P(X_0 \in A) = \pi(A)$ is obvious.) For this, use (7.4) for $n - 1$ to conclude that (7.4) can be written as

$$P(X_n \in A_n, \ldots) = \int_{A_{n-1}} p(A_n \mid x_{n-1}) P(X_{n-1} \in dx_{n-1}, X_{n-2} \in A_{n-2}, \ldots)$$

$$= \int_{\{x_{n-1} \in A_{n-1}, x_{n-2} \in A_{n-2}, \ldots\}} p(A_n \mid x_{n-1})\, P(X_{n-1} \in dx_{n-1}, X_{n-2} \in dx_{n-2}, \ldots).$$

But, by definition,

$$P(X_n \in A_n, \ldots) = \int_{\{X_{n-1} \in A_{n-1}, \ldots\}} P(X_n \in A_n \mid X_{n-1}, \ldots)\, dP.$$

Hence $p(A \mid x)$ is a version of $P(X_n \in A \mid X_{n-1} = x_{n-1}, \ldots)$ for all n.

Now we switch points of view. For the rest of this chapter forget about the original probability space $(\Omega, \mathcal{F}, P)$. Fix one version $p(A \mid x)$ of the stationary transition probabilities and consider $p(A \mid x)$ to be the given data. Each initial distribution π, together with $p(A \mid x)$, determines a probability on $(F^{(\infty)}, \mathcal{B}_\infty(F))$ we denote by P_π, and the corresponding expectation by E_π. Under P_π, the coordinate representation process $\hat{X}_0, \hat{X}_1, \ldots$ becomes a Markov chain with initial distribution π. Thus we are concerned with a family of processes having the same transition probability, but different initial distributions. If π is concentrated on the single point $\{x\}$, then denote the corresponding probability on $\mathcal{B}_\infty(F)$ by P_x, and the expectation by E_x. Under P_x, $\hat{X}_0, \hat{X}_1, \ldots$ is referred to as the "process starting from the point x." Now eliminate the $\hat{}$ over variables, with the understanding that henceforth all processes referred to are coordinate representation processes with P one of the family of probabilities $\{P_\pi\}$. For any π, and $B \in \mathcal{B}_\infty(F)$, always use the version of $P_\pi((X_0, \ldots) \in B \mid X_0 = x)$ given by $P_x((X_0, \ldots) \in B)$.

In exactly the same way as in Chapter 3, call a nonnegative integer-valued random variable n^* a *stopping time* or *Markov time* for a Markov chain $X_0, X_1, \ldots$ if

$$\{n^* = n\} \in \mathcal{F}(X_0, \ldots, X_n).$$

Define $\mathcal{F}(X_n, n \le n^*)$ to be the σ-field consisting of all events A such that $A \cap \{n^* \le n\} \in \mathcal{F}(X_0, \ldots, X_n)$. Then

Proposition 7.8. *If n^* is a Markov time for $X_0, X_1, \ldots$, then*

$$P((X_{n^*}, X_{n^*+1}, \ldots) \in B \mid X_n, n \le n^*) = P_{X_{n^*}}((X_0, X_1, \ldots) \in B) \quad \text{a.s. } P.$$

Note. The correct interpretation of the above is: Let

$$\varphi(x) = P_x((X_0, X_1, \ldots) \in B).$$

Then the right-hand side is $\varphi(X_{n^*})$.

Proposition 7.8 is called the *strong Markov property*.

Proof. Let $A \in \mathcal{F}(\mathsf{X}_n, n \le \mathsf{n}^*)$, then the integral of the left-hand side of 7.8 over A is

$$P((\mathsf{X}_{\mathsf{n}^*}, \mathsf{X}_{\mathsf{n}^*+1}, \ldots) \in B, A) = \sum_{n=0}^{\infty} P((\mathsf{X}_{\mathsf{n}^*}, \mathsf{X}_{\mathsf{n}^*+1}, \ldots) \in B, A, \mathsf{n}^* = n)$$

$$= \sum_{n=0}^{\infty} P((\mathsf{X}_n, \mathsf{X}_{n+1}, \ldots) \in B, A, \mathsf{n}^* = n).$$

The set $A \cap \{\mathsf{n}^* = n\} \in \mathcal{F}(\mathsf{X}_0, \ldots, \mathsf{X}_n)$. Hence

$$P((\mathsf{X}_n, \ldots) \in B, A, \mathsf{n}^* = n) = \int_{A \cap \{\mathsf{n}^* = n\}} P((\mathsf{X}_n, \ldots) \in B \mid \mathsf{X}_n, \ldots)\, dP$$

$$= \int_{A \cap \{\mathsf{n}^* = n\}} \varphi(\mathsf{X}_n)\, dP.$$

Putting this back in, we find that

$$P_x((\mathsf{X}_{\mathsf{n}^*}, \ldots) \in B, A) = \sum_{n=0}^{\infty} \int_{A \cap \{\mathsf{n}^* = n\}} \varphi(\mathsf{X}_n)\, dP$$

$$= \int_A \varphi(\mathsf{X}_{\mathsf{n}^*})\, dP.$$

A special case of a Markov chain are the successive sums $\mathsf{S}_0, \mathsf{S}_1, \mathsf{S}_2, \ldots$ of independent random variables $\mathsf{Y}_1, \mathsf{Y}_2, \ldots$, ($\mathsf{S}_0 = 0$ convention). This is true because independence gives

$$P(\mathsf{S}_{n+1} \in A \mid \mathsf{S}_n, \ldots, \mathsf{S}_0) = P(\mathsf{S}_{n+1} \in A \mid \mathsf{S}_n) \quad \text{a.s.}$$

If the summands are identically distributed, then the chain has stationary transition probabilities:

$$P(\mathsf{S}_{n+1} \in A \mid \mathsf{S}_n = x) = P(\mathsf{Y}_{n+1} \in A - x)$$
$$= F(A - x) \quad \text{a.s.,}$$

where $F(B)$ denotes the probability $P(\mathsf{Y}_1 \in B)$ on $\mathcal{B}_1$. In this case, take $F(A - x)$ as the fixed conditional probability distribution to be used. Now letting X_0 have any initial distribution π and using the transition probability $F(A - x)$ we get a Markov chain $\mathsf{X}_0, \mathsf{X}_1, \ldots$ having the same distribution as $\mathsf{Y}_0, \mathsf{Y}_0 + \mathsf{Y}_1, \mathsf{Y}_0 + \mathsf{Y}_1 + \mathsf{Y}_2, \ldots$ where $\mathsf{Y}_0, \mathsf{Y}_1, \ldots$ are independent, Y_0 has the distribution π, and $\mathsf{Y}_1, \mathsf{Y}_2, \ldots$ all have the distribution F. In particular, the process "starting from x" has the distribution of $\mathsf{S}_0 + x$, $\mathsf{S}_1 + x$, $\mathsf{S}_2 + x, \ldots$ Call any such Markov process a *random walk*.

Problems

1. Define the n-step transition probabilities $p^{(n)}(A \mid x)$ for all $A \in \mathcal{B}_1(F)$, $x \in F$ by

$$p^{(1)}(A \mid x) = p(A \mid x), \qquad p^{(n+1)}(A \mid x) = \int p^{(n)}(A \mid y) p(dy \mid x).$$

a) Show that $p^{(n)}(A \mid x)$ equals $P_x(X_n \in A)$, hence is a version of

$$P(X_n \in A \mid X_0 = x).$$

b) Show that $p^{(n)}(A \mid x)$ is a regular conditional probability on $\mathcal{B}_1(F)$ given $x \in F$, and that for all $A \in \mathcal{B}_1(F)$, $x \in F$, $n, m > 0$,

$$p^{(n+m)}(A \mid x) = \int p^{(n)}(A \mid y)p^{(m)}(dy \mid x).$$

2. Let $\varphi(\mathbf{x})$ be a bounded $\mathcal{B}_\infty$-measurable function. Demonstrate that $E_x\varphi(X_1, X_2, \ldots)$ is $\mathcal{B}_1$-measurable in x.

2. ASYMPTOTIC STATIONARITY

There is a class of limit theorems which state that certain processes are asymptotically stationary. Generally these theorems are formulated as: Given a process $X_1, X_2, \ldots$, a stationary process $X_1^*, X_2^*, \ldots$ exists such that for every $B \in \mathcal{B}_\infty$,

$$\lim_n P((X_n, X_{n+1}, \ldots) \in B) = P^*((X_1^*, X_2^*, \ldots) \in B).$$

The most well-known of these relate to Markov chains with stationary transition probabilities $p(A \mid x)$. Actually, what we would really like to show for Markov chains is that no matter what the initial distribution of X_0 is, convergence toward the same stationary limiting distribution sets in, that is, that for all $B \in \mathcal{B}_\infty(F)$, and initial distributions π,

$$\lim_n P_\pi((X_n, \ldots) \in B) = P^*((X_1^*, \ldots) \in B),$$

where $\{X_n^*\}$ is stationary.

Suppose, to begin, that there is a limiting distribution $\bar{\pi}(\cdot)$ on $\mathcal{B}_1(F)$ such that for all $x \in F$, $A \in \mathcal{B}_1(F)$,

$$\lim_n P_x(X_n \in A) = \bar{\pi}(A).$$

If this limiting distribution $\bar{\pi}(A)$ exists, then from

$$P_\pi(X_n \in A) = \int P(X_n \in A \mid X_0 = x)\pi(dx)$$
$$= \int P_x(X_n \in A)\pi(dx)$$

comes

$$P_\pi(X_n \in A) \to \bar{\pi}(A), \quad \text{all } \pi.$$

Also,

$$P_\pi(X_{n+1} \in A) = \int P(X_{n+1} \in A \mid X_n = x)P_\pi(X_n \in dx).$$

For A fixed, approximate $p(A \mid x)$ uniformly by simple functions of x. Taking limits implies that $\bar{\pi}(A)$ must satisfy

$$\text{(7.9)} \qquad \bar{\pi}(A) = \int p(A \mid x)\bar{\pi}(dx).$$

If $\bar{\pi}$ is the limiting distribution of X_n, what happens if we start the process off with the distribution $\bar{\pi}$? The idea is that if $\bar{\pi}$ is a limiting steady-state distribution, then, starting the system with this distribution, it should maintain a stable behavior. This is certainly true in the following sense—let us start the process with any initial distribution satisfying (7.9). Then this distribution maintains itself throughout time, that is,

$$P(X_n \in A) = P(X_{n-1} \in A) = \cdots = P(X_0 \in A).$$

This is established by iterating (7.9) to get $\bar{\pi}(A) = \int p^{(n)}(A \mid x)\bar{\pi}(dx)$. In this sense any solution of (7.9) gives stable initial conditions to the process.

Definition 7.10. *For transition probabilities $p(A \mid x)$, an initial distribution $\pi(A)$ will be called a stationary initial distribution if it satisfies* (7.9).

But if a stationary initial distribution is used for the process, much more is true.

Proposition 7.11. *Let $X_0, X_1, \ldots$ be a Markov process with stationary transition probabilities such that the initial distribution $\pi(A)$ satisfies* (7.9); *then the process is stationary.*

Proof. By 7.6 there are versions of $P(X_n \in A_n, \ldots, X_1 \in A_1 \mid X_0 = x)$, $P(X_{n+1} \in A_n, \ldots, X_2 \in A_1 \mid X_1 = x)$ which are equal. Since $P(X_1 \in A) = P(X_0 \in A) = \pi(A)$, integrating these versions over A_0 gives

$$P(X_n \in A_n, \ldots, X_0 \in A_0) = P(X_{n+1} \in A_n, \ldots, X_1 \in A_0),$$

which is sufficient to prove the process stationary.

Furthermore, Markov chains have the additional property that if $P_x(X_n \in A)$ converges to $\bar{\pi}(A)$, all $A \in \mathcal{B}_1(F)$, $x \in F$, then this one-dimensional convergence implies that the distribution of the entire process is converging to the distribution of the stationary process with initial distribution $\bar{\pi}$.

Proposition 7.12. *If $p^{(n)}(A \mid x) \to \bar{\pi}(A)$, all $A \in \mathcal{B}_1(F)$, $x \in F$, then for any $B \in \mathcal{B}_\infty(F)$, and all $x \in F$,*

$$P_x((X_{n+1}, \ldots) \in B) \to P_{\bar{\pi}}((X_1, X_2, \ldots) \in B).$$

Proof. Write

$$P_x((X_1, X_2, \ldots) \in B) = \varphi(x).$$

Then

$$P_x((X_{n+1}, \ldots) \in B \mid X_n = y) = \varphi(y), \quad \text{a.s.}$$

and, also,

$$P_{\bar{\pi}}((X_1, X_2, \ldots) \in B \mid X_0 = x) = \varphi(x), \quad \text{a.s.}$$

Now,

$$P((X_{n+1}, \ldots) \in B \mid X_0 = x)$$
$$= \int P((X_{n+1}, \ldots) \in B \mid X_n = y) P(X_n \in dy \mid X_0 = x)$$

or

$$P((X_{n+1}, \ldots) \in B \mid X_0 = x) = \int \varphi(y) p^{(n)}(dy \mid x).$$

Under the stated conditions, one can show, by taking simple functions that approximate $\varphi(x)$ uniformly, that

$$\int \varphi(y) p^{(n)}(dy \mid x) \to \int \varphi(y) \bar{\pi}(dy) = P_{\bar{\pi}}((X_1, \ldots) \in B).$$

Therefore, the asymptotic behavior problem becomes: How many stationary initial distributions does a given set of transition probabilities have, and does $p^{(n)}(A \mid x)$ converge to some stationary distribution as $n \to \infty$?

3. CLOSED SETS, INDECOMPOSABILITY, ERGODICITY

Definition 7.13. *A set $A \in \mathcal{B}_1(F)$ is closed if $p(A \mid x) = 1$ for all $x \in A$.*

The reason for this definition is obvious; if $X_0 \in A$, and A is closed, then $X_n \in A$ with probability one for all n. Hence if there are two closed disjoint sets A_1, A_2, then

$$p^{(n)}(A_1 \mid x) = \begin{cases} 1, & x \in A_1, \quad n \geq 0, \\ 0, & x \in A_2, \quad n \geq 0, \end{cases}$$

and there is no hope that $p^{(n)}(A_1 \mid x)$ converges to the same limit for all starting points x.

Definition 7.14. *A chain is called indecomposable if there are no two disjoint closed sets $A_1, A_2 \in \mathcal{B}_1(F)$.*

Use a stationary initial distribution $\bar{\pi}$, if one exists, to get the stationary process $X_0, X_1, \ldots$ If the process is in addition ergodic, then use the ergodic theorem to assert

$$\text{(7.15)} \qquad \frac{1}{n}\sum_{1}^{n} \chi_A(X_k) \to \bar{\pi}(A) \quad \text{a.s.}$$

Take conditional expectations of (7.15), given $X_0 = x$. Use the boundedness and proposition 4.24, to get

$$\frac{1}{n}\sum_{1}^{n} p^{(k)}(A \mid x) \to \bar{\pi}(A), \quad \text{a.s. } \bar{\pi}(dx).$$

Thus, from ergodicity, it is possible to get convergence of the averages of the $p^{(n)}(A \mid x)$ to $\bar{\pi}(dx)$ a.s. $\bar{\pi}(dx)$. The questions of ergodicity of the process uniqueness of stationary initial distributions, and indecomposability go together.

Theorem 7.16. *Let the chain be indecomposable. Then if a stationary initial distribution $\bar{\pi}$ exists, it is unique and the process gotten by using $\bar{\pi}$ as the initial distribution is ergodic.*

Proof. Let C be an invariant event under the shift transformation in $\mathcal{B}_\infty$, $\bar{\pi}$ a stationary initial distribution. Take $\varphi(x) = P_x(C)$. Now, using $\bar{\pi}$ as the initial distribution, write

$$P(C \mid X_0) = E\big(P(C \mid X_1, X_0) \mid X_0\big).$$

By the Markov property, since $C \in \mathcal{F}(X_n, X_{n+1}, \ldots)$ for all $n \geq 0$,

$$P(C \mid X_1, X_0) = P(C \mid X_1), \quad \text{a.s.}$$

By Proposition 7.6,

$$P(C \mid X_0 = x) = P(S^{-1}C \mid X_1 = x) \quad \text{a.s. } \bar{\pi},$$

and by the invariance of C, then

$$P(C \mid X_1 = x) = \varphi(x).$$

Therefore $\varphi(x)$ satisfies

$$\varphi(x) = \int \varphi(y) p(dy \mid x) \quad \text{a.s. } \bar{\pi}(dx). \tag{7.17}$$

By a similar argument, $P(C \mid X_n, \ldots, X_0) = P(C \mid X_n) = \varphi(X_n)$ a.s. and $E\big(\varphi(X_n) \mid X_{n-1}, \ldots, X_0\big) = \varphi(X_{n-1})$ a.s. Apply the martingale theorem to get $\varphi(X_n) \to \chi_C$ a.s. Thus for any $\epsilon > 0$,

$$\bar{\pi}(x;\ \epsilon \leq \varphi(x) \leq 1 - \epsilon) = 0$$

because the distribution of X_n is $\bar{\pi}$. So $\varphi(x)$ can assume only the two values 0, 1 a.s. $\bar{\pi}(dx)$. Define sets

$$A_1 = \{x;\ \varphi(x) = 1\} \qquad A_2 = \{x;\ \varphi(x) = 0\}.$$

Since $\varphi(x)$ is a solution of (7.17),

$$1 = \int \varphi(y) p(dy \mid x), \quad x \in A_1,$$

$$0 = \int \varphi(y) p(dy \mid x), \quad x \in A_2,$$

except for a set D, such that $\bar{\pi}(D) = 0$. Therefore

$$p(A_i \mid x) = 1, \quad x \in A_i - D, \quad i = 1, 2.$$

Let us define

$$A_i^{(0)} = A_i$$

$$A_i^{(1)} = A_i - D$$

$$A_i^{(n+1)} = A_i^{(n)} \cap \{x; p(A_i^{(n)} \mid x) = 1\}.$$

If $p(A_i^{(n)} \mid x) = 1$, $x \in A_i^{(n)}$, a.s. $\bar{\pi}$, then $\bar{\pi}(A_i^{(n)} - A_i^{(n+1)}) = 0$. Take $C_i = \bigcap_0^\infty A_i^{(n)}$. Then $\bar{\pi}(C_i) = \bar{\pi}(A_i)$, but the C_i are closed and disjoint. Hence one of C_1, C_2 is empty, $\varphi(x)$ is zero a.s. or one a.s., $P(C) = 0$ or 1, and the process is ergodic. Now, suppose there are two stationary initial distributions $\bar{\pi}_1$ and $\bar{\pi}_2$ leading to probabilities P_1 and P_2 on $(\Omega, \mathcal{F})$. By 6.24 there is an invariant C such that $P_1(C) = 1$, but $P_2(C) = 0$. Using the stationary initial distribution $\bar{\pi} = \frac{1}{2}\bar{\pi}_1 + \frac{1}{2}\bar{\pi}_2$ we get the probability

$$P = \tfrac{1}{2}P_1 + \tfrac{1}{2}P_2$$

which is again ergodic, by the above argument. But

$$P(C) = \tfrac{1}{2}P_1(C) + \tfrac{1}{2}P_2(C) = \tfrac{1}{2}. \qquad \text{Q.E.D.}$$

What has been left is the problem: When does there exist a $\bar{\pi}(A)$ such that $p^{(n)}(A \mid x) \to \bar{\pi}(A)$? If F is countable, this problem has a complete solution. In the case of general state spaces F, it is difficult to arrive at satisfactory conditions (see Doob, Chapter 6). But if a stationary initial distribution $\bar{\pi}(A)$ exists, then under the following conditions:

1) the state space F is indecomposable under $p(A \mid x)$;
2) the motion is *nonperiodic*; that is, F is indecomposable under the transition probabilities $p^{(n)}(A \mid x)$, $n = 2, 3, \ldots$,
3) for each $x \in F$, $p(A \mid x) \ll \bar{\pi}(A)$;

Doob [35] has shown that

Theorem 7.18. $\lim_n p^{(n)}(A \mid x) = \bar{\pi}(A)$ *for all* $A \in \mathcal{B}_1(F)$, $x \in F$.

The proof is essentially an application of the ergodic theorem and its refinements. [As shown in Doob's paper, (3) can be weakened somewhat.]

4. THE COUNTABLE CASE

The case where the state space F is countable is much easier to understand and analyze. It also gives some insight into the behavior of general Markov chains. Hence assume that F is a subset of the integers, that we have transition probabilities $p(j \mid k)$, satisfying

$$p(j \mid k) \geq 0, \qquad \sum_j p(j \mid k) = 1, \tag{7.19}$$

where the summation is over all states in F, and n-step transition probabilities $p^{(n)}(j \mid k)$ defined by

$$p^{(n+1)}(j \mid k) = \sum_i p^{(n)}(j \mid i)p(i \mid k).$$

This is exactly matrix multiplication: Denote the matrix $\{p(j \mid k)\}$ by $\mathbf{P}$; then the n-step transition probabilities are the elements of $\mathbf{P}^n$. Therefore, if F has a finite number of states the asymptotic stationarity problem can be studied in terms of what happens to the elements of matrix as it is raised to higher and higher powers. The theory in this case is complete and detailed. (See Feller [59, Vol. I, Chapter 16].) The idea that simplifies the theory in the countable case is the renewal concept. That is, if a Markov chain starts in state j, then every time it returns to state j, the whole process starts over again as from the beginning.

5. THE RENEWAL PROCESS OF A STATE

Let $X_0, X_1, \ldots$ be the Markov chain starting from state j. We ignore transitions to all other states and focus attention only on the returns of the process to the state j. Define random variables $U_1, U_2, \ldots$ by

$$U_n = \begin{cases} 1 & \text{if } X_n = j, \\ 0, & \text{otherwise.} \end{cases}$$

By the Markov property and stationarity of the transition probabilities, for any $B \in \mathcal{B}_\infty(\{0, 1\})$, $(s_1, \ldots, s_{n-1}) \in \{0, 1\}^{(n-1)}$,

$$\textbf{(7.20)} \quad P_j((U_{n+1}, \ldots) \in B \mid U_n = 1, U_{n-1} = s_{n-1}, \ldots) = P_j((U_1, \ldots) \in B).$$

This simple relationship partially summarizes the fact that once the process returns to j, the process starts anew. In general, any process taking values in $\{0, 1\}$ and satisfying (7.20) is called a *renewal process*. We study the behavior at a single state j by looking at the associated process $U_1, U_2, \ldots$ governing returns to j. Define the event G that a return to state j occurs at least once by $G = \bigcup_{n \ge 1} \{X_n = j\}$.

Theorem 7.21. *The following dichotomy is in force*

$$P_j(G) = 1 \Rightarrow \begin{cases} P_j(X_n = j \text{ i.o.}) = 1, \\ \sum_1^\infty P_j(X_n = j) = \infty. \end{cases}$$

$$P_j(G) < 1 \Rightarrow \begin{cases} P_j(X_n = j \text{ i.o.}) = 0, \\ \sum_1^\infty P_j(X_n = j) < \infty. \end{cases}$$

Proof. Let $F_n = \{U_n = 1, U_{n+k} = 0, \text{ all } k \geq 1\}$, $n \geq 1$; $F_0 = \{U_k = 0, \text{ all } k \geq 1\}$. Thus F_n is the event that the last return to j occurs at time n. Hence

$$\{X_n = j \text{ i.o.}\}^c = \bigcup_{n \geq 0} F_n.$$

The F_n are disjoint, so

$$1 - P_j(X_n = j \text{ i.o.}) = \sum_0^\infty P_j(F_n),$$

and by (7.20),

$$\textbf{(7.22)} \qquad P_j(F_n) = P_j(U_{n+k} = 0, \quad \text{all } k \geq 1 \mid U_n = 1)P_j(U_n = 1)$$
$$= P_j(F_0)P_j(U_n = 1).$$

According to the definitions, $F_0 = G^c$, hence

$$P_j(G) = 1 \Rightarrow P_j(F_0) = 0 \Rightarrow P_j(F_n) = 0,$$

for all $n \geq 0$; so $P_j(X_n = j \text{ i.o.}) = 1$. Then

$$\sum_1^\infty P_j(X_n = j) = \infty;$$

otherwise, the Borel-Cantelli lemma would imply $P_j(X_n = j \text{ i.o.}) = 0$. If $P_j(G) < 1$, then $P_j(F_0) > 0$, and we can use expression (7.22) to substitute for $P_j(F_n)$, getting

$$\textbf{(7.23)} \qquad 1 - P_j(X_n = j \text{ i.o.}) = P_j(F_0)\left(1 + \sum_1^\infty P_j(X_n = j)\right).$$

This implies $\sum_1^\infty P_j(X_n = j) < \infty$ and thus $P_j(X_n = j \text{ i.o.}) = 0$.

Definition 7.24. *Call the state j recurrent if $P_j(X_n = j \text{ i.o.}) = 1$, transient if $P_j(X_n = j \text{ i.o.}) = 0$.*

Note that 7.21 in terms of transition probabilities reads, "j is recurrent iff $\sum_1^\infty p^{(n)}(j \mid j) = \infty$."

Define R_k as the time of the nth return to j, and the times between returns as $T_1 = R_1$, $T_k = R_k - R_{k-1}$, $k > 1$. Then

Proposition 7.25. *If j is recurrent, then $T_1, T_2, \ldots$ are independent, identically distributed random variables under the probability P_j.*

Proof. T_1 is a Markov time for $X_0, X_1, \ldots$ By the strong Markov property 7.8,

$$P_j((X_{T_1+1}, X_{T_1+2}, \ldots) \in B \mid X_n, n \leq T_1) = P_j((X_1, X_2, \ldots) \in B).$$

Therefore, the process $(X_{T_1+1}, \ldots)$ has the same distribution as $(X_1, \ldots)$ and is independent of $\mathcal{F}(X_n, n \leq T_1)$, hence is independent of T_1.

The $\mathsf{T}_1, \mathsf{T}_2, \ldots$ are also called the *recurrence times* for the state j. The result of 7.25 obviously holds for any renewal process with $\mathsf{T}_1, \mathsf{T}_2, \ldots,$ the times between successive ones in the $\mathsf{U}_1, \mathsf{U}_2, \ldots$ sequence.

***Definition* 7.26.** *Call the recurrent state j*

$$\textit{positive-recurrent if } \; E_j\mathsf{T}_1 < \infty, \qquad \textit{null-recurrent if } \; E_j\mathsf{T}_1 = \infty.$$

***Definition* 7.27.** *The state j has period $d \geq 1$ if T_1 is distributed on the lattice L_d, $d \geq 1$ under P_j. If $d > 1$, call the state periodic; if $d = 1$, nonperiodic.* [*Recall from* (3.32) *that* $L_d = \{nd\}$, $n = 0, \pm 1, \ldots$]

For j recurrent, let the random vectors $\mathbf{Z}_k$ be defined by

$$\mathbf{Z}_k = (\mathsf{X}_{\mathsf{R}_k}, \ldots, \mathsf{X}_{\mathsf{R}_{k+1}-1}).$$

Then $\mathbf{Z}_k$ takes values in the space R of all finite sequences of integers with $\mathcal{B}$ the smallest σ-field containing all sets of the form

$$\{(x_1, \ldots, x_m) \in R;\; x_k = i\},\; k \leq m,$$

where i is any integer. Since the length of blocks is now variable, an interesting generalization of 7.25 is

***Theorem* 7.28.** *The $\mathbf{Z}_0, \mathbf{Z}_1, \ldots$ are independent and identically distributed random vectors under the probability P_j.*

Proof. This follows from 7.8 by seeing that T_1 is a Markov time. $\mathbf{Z}_0$ is measurable $\mathcal{F}(\mathsf{X}_n, n \leq \mathsf{T}_1)$ because for $B \in \mathcal{B}$,

$$\{\mathbf{Z}_0 \in B\} \cap \{\mathsf{T}_1 = k\} = \{(\mathsf{X}_0, \ldots, \mathsf{X}_{k-1}) \in B, \mathsf{T}_1 = k\} \in \mathcal{F}(\mathsf{X}_0, \ldots, \mathsf{X}_n), \quad k \leq n.$$

By 7.8, for $A \in \mathcal{B}_\infty$,

$$P_j((\mathsf{X}_{\mathsf{T}_1}, \mathsf{X}_{\mathsf{T}_1+1}, \ldots) \in A, \mathbf{Z}_0 \in B) = P_j((\mathsf{X}_0, \mathsf{X}_1, \ldots) \in A)P_j(\mathbf{Z}_0 \in B).$$

Now $\mathbf{Z}_1$ is the same function of the $\mathsf{X}_{\mathsf{T}_1}, \mathsf{X}_{\mathsf{T}_1+1}, \ldots$ process as $\mathbf{Z}_0$ is of the $\mathsf{X}_0, \mathsf{X}_1, \ldots$ process. Hence $\mathbf{Z}_1$ is independent of $\mathbf{Z}_0$ and has the same distribution.

Call events $\{A_n\}$ a *renewal event* $\mathcal{E}$ if the random variables χ_{A_n} form a renewal process. Problems 3 and 4 are concerned with these.

Problems

3. (Runs of length at least N). Consider coin-tossing (biased or fair) and define

$$A_n = \{\omega_n = T, \omega_{n-1} = H, \ldots, \omega_{n-N} = H\}.$$

Prove that $\{A_n\}$ form a renewal event. [Note that A_n is the event such that at time n a run of at least N heads has just finished.]

4. Given any sequence $\mathbf{t}_N$, N long of H, T, let $(\Omega, \mathcal{F}, P)$ be the coin-tossing game (fair or biased). Define $A_n = \{\omega;\; (\omega_n, \ldots, \omega_{n-N+1}) = \mathbf{t}_N\}$,

($A_n = \emptyset$ if $n < N$). Find necessary and sufficient conditions on $\mathbf{t}_N$ for the $\{A_n\}$ to form a renewal event.

5. If $P_j(X_n = j \text{ i.o.}) = 0$, then show that

$$P_j(X_n = j \text{ never occurs}) = \frac{1}{1 + \Sigma_1^\infty P_j(X_n = j)} \cdot$$

6. Use Problem 5 to show that for a biased coin

$$P(\text{no equalizations}) = |p - q|.$$

7. Use Theorem 7.21 to show that if $\{Z_n\}$ are the successive fortunes in a fair coin-tossing game,

$$P(Z_n = 0 \text{ i.o.}) = 1.$$

6. GROUP PROPERTIES OF STATES

Definition 7.29. *If there is an n such that $p^{(n)}(k \mid j) > 0$, j to k (denoted by $j \to k$) is a permissible transition. If $j \to k$ and $k \to j$, say that j and k communicate, and write $j \leftrightarrow k$.*

Communicating states share properties:

Theorem 7.30. *If $j \leftrightarrow k$, then j and k are simultaneously transient, null-recurrent, or positive-recurrent, and have the same period.*

Proof. Use the martingale result, Problem 9, Chapter 5, to deduce that under any initial distribution the sets $\{X_n = j \text{ i.o.}\}$ and $\{X_n = k \text{ i.o.}\}$ have the same probability. So both are recurrent or transient together. Let $T_1, T_2, \ldots$ be the recurrence times for j starting from j, and assume $E_jT_1 < \infty$. Let $V_n = 1$ or 0 as there was or was not a visit to state k between times R_{n-1} and R_n. The V_n are independent and identically distributed (7.28) with $P(V_n = 1) > 0$. Denote by n^* the first n such that $V_n = 1$, T the time of the first visit to state k. Then

$$E_j T \le \sum_{n=1}^{\infty} \int_{\{n^* = n\}} R_n \, dP_j = \sum_{n=1}^{\infty} \int_{\{n^* \ge n\}} T_n \, dP_j.$$

But $\{n^* \ge n\} = \{n^* < n\}^c$, and $\{n^* < n\} \in \mathcal{F}(X_k, k \le R_{n-1})$, thus is independent of T_n. Hence

$$E_j T \le E_j T_1 \cdot \sum_1^\infty P_j(n^* \ge n) = E_j T_1 \cdot E_j n^*.$$

Obviously, $E_j n^* < \infty$, so $E_j T < \infty$. Once k has occurred, the time until another occurrence of k is less than the time to get back to state j plus the time until another occurrence of k starting from j. This latter time has the same distribution as T. The former time must have finite expectation, otherwise $E_j T_1 = \infty$. Hence k has a recurrence time with finite expectation.

Take n_1, n_2 such that $p^{(n_1)}(k \mid j) > 0$, $p^{(n_2)}(j \mid k) > 0$. If $\mathsf{T}^{(k)}$ is the first return time to k, define $I_k = \{n;\ P_k(\mathsf{T}^{(k)} = n) > 0\}$, L_{d_2} is the smallest lattice containing I_k, L_{d_1} the smallest lattice containing I_j. Diagrammatically, we can go:

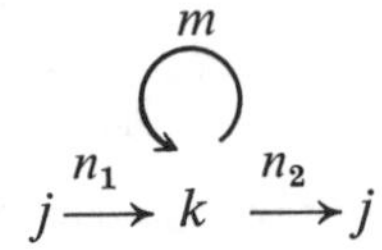

That is, if $m \in I_k$, then $n_1 + m + n_2 \in I_j$, or $I_k + n_1 + n_2 \subset I_j$. Hence $L_{d_2} \subset L_{d_1}$, and $d_2 \geq d_1$. The converse argument gives $d_1 \geq d_2$, so $d_1 = d_2$.

Q.E.D.

Let J be the set of recurrent states. Communication ($\leftrightarrow$) is clearly an equivalence relationship on J, so splits J into disjoint classes $C_1, C_2, \ldots$

Proposition 7.31. *For any C_l*

$$\sum_{k \in C_l} p(k \mid j) = 1, \quad \text{all } j \in C_l,$$

that is, each C_l is a closed set of states.

Proof. If j is recurrent and $j \to k$, then $k \to j$. Otherwise, $P_j(\mathsf{X}_n = j \text{ i.o.}) < 1$. Hence the set of states $\{k;\ j \to k\} = \{k;\ j \leftrightarrow k\}$, but this latter is exactly the equivalence class containing j. The sum of $p(k \mid j)$ over all k such that $j \to k$ is clearly one.

Take C to be a closed indecomposable set of recurrent states. They all have the same period d. If $d > 1$, define the relationship $\overset{d}{\leftrightarrow}$ as $j \overset{d}{\leftrightarrow} k$ if $p^{(n_1 d)}(k \mid j) > 0, p^{(n_2 d)}(j \mid k) > 0$ for some n_1, n_2. Since $j \overset{d}{\leftrightarrow} k$ is an equivalence relationship, C may be decomposed into disjoint sets $D_1, D_2, \ldots$ under $\overset{d}{\leftrightarrow}$.

Proposition 7.32. *There are d disjoint equivalence classes $D_1, D_2, \ldots, D_d$ under $\overset{d}{\leftrightarrow}$ and they can be numbered such that*

$$\sum_{k \in D_{l+1[d]}} p(k \mid j) = 1, \quad j \in D_l$$

or diagrammatically,

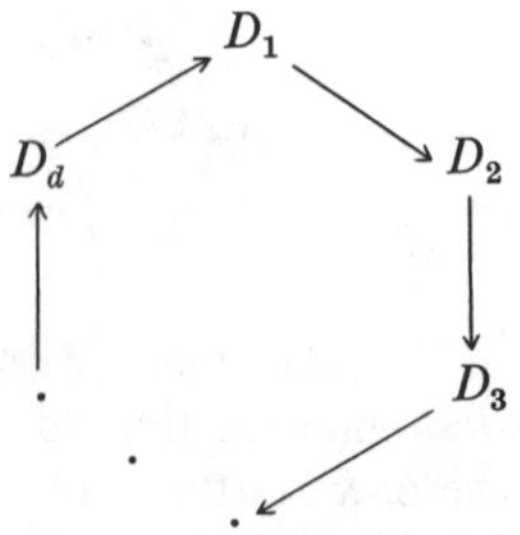

The $D_1, \ldots, D_d$ are called cyclically moving subsets.

Proof. Denote by $j \xrightarrow{l[d]} k$ the existence of an n such that $n[d] = l[d]$ and $p^{(n)}(k \mid j) > 0$. Fix j_1 and number D_1 as the equivalence class containing j_1. Take j_2 such that $p(j_2 \mid j_1) > 0$, and number D_2 the equivalence class containing j_2, and so on. See that $j \xrightarrow{0[d]} k$ implies that $k \xrightarrow{0[d]} j$. But

$$p^d(j_{d+1} \mid j_1) \geq p(j_{d+1} \mid j_d) \cdots p(j_2 \mid j_1) > 0,$$

so $j_1 \overset{d}{\leftrightarrow} j_{d+1} \Rightarrow D_{d+1} = D_1$. Consider any state k, such that $p(k \mid j) > 0$, $j \in D_1$ and $k \notin D_2$, say $k \in D_3$. Then $k \overset{d}{\leftrightarrow} j_3$. Look at the string

$$j_1 \xrightarrow{0[d]} j \xrightarrow{1[d]} k \xrightarrow{0[d]} j_3 \xrightarrow{1[d]} j_4 \rightarrow \cdots \xrightarrow{1[d]} j_{d+1} \xrightarrow{0[d]} j_1.$$

This string leads to $j_1 \xrightarrow{d-1[d]} j_1$. From this contradiction, conclude $k \in D_2$, and $\sum_{k \in D_2} p(k \mid j) = 1, j \in D_1$.

If C is a closed set of communicating nonperiodic states, one useful result is that for any two states j, k, all other states are common descendants. That is:

Proposition 7.33. *For j, k, l, any three states, there exists an n such that $p^{(n)}(l \mid j) > 0$, $p^{(n)}(l \mid k) > 0$.*

Proof. Take n_1, n_2 such that $p^{(n_1)}(l \mid j) > 0$, $p^{(n_2)}(l \mid k) > 0$. Consider the set $J = \{n;\ p^{(n)}(l \mid l) > 0\}$. By the nonperiodicity, the smallest lattice containing J is L_1. Under addition, J is closed, so every integer can be expressed as $s_1 m_1 - s_2 m_2$, $m_1, m_2 \in J$, s_1, s_2 nonnegative integers. Take $n_2 - n_1 = s_1 m_1 - s_2 m_2$, so $n_1 + s_1 m_1 = n_2 + s_2 m_2 = n$. Now check that $p^{(n)}(l \mid j) > 0$, $p^{(n)}(l \mid k) > 0$.

Problems

8. If the set of states is finite, prove that

a) there must be at least one recurrent state;

b) every recurrent state is positive recurrent;

c) there is a random variable n^* such that for $n \geq \mathsf{n}^*$, X_n is in the set of recurrent states.

9. Give an example of a chain with all states transient.

7. STATIONARY INITIAL DISTRIBUTIONS

Consider a chain $\mathsf{X}_0, \mathsf{X}_1, \ldots$ starting from an initial state i such that i is recurrent. Let $\mathsf{N}_n(j)$ be the number of visits of $\mathsf{X}_1, \ldots, \mathsf{X}_n$ to j, and $\pi(j)$ be the expected number of visits to j before return to i, $\pi(i) = 1$. The proof of Theorem 3.45 goes through, word for word. Use (3.46) again to conclude

$$\frac{\mathsf{N}_n(j)}{\mathsf{N}_n(l)} \xrightarrow{\text{a.s.}} \frac{\pi(j)}{\pi(l)}.$$

The relevance here is

Proposition 7.34

$$\pi(k) = \sum_j p(k \mid j)\pi(j)$$

for all k such that $i \to k$.

Proof. A visit to state j occurs on the nth trial before return to state i if $\{X_n = j, X_{n+1} \neq i, \ldots, X_1 \neq i\}$. Therefore

$$\pi(j) = \sum_{n=1}^{\infty} P_i(X_n = j, X_{n-1} \neq i, \ldots, X_1 \neq i),$$

so that

$$\sum_j p(k \mid j)\pi(j) = \sum_{n=1}^{\infty} \sum_{j \neq i} P(X_{n+1} = k \mid X_n = j) P_i(X_n = j, X_{n-1} \neq i, \ldots, X_1 \neq i) + p(k \mid i)$$

$$= \sum_{n=1}^{\infty} P_i(X_{n+1} = k, X_n \neq i, X_{n-1} \neq i, \ldots, X_1 \neq i) + P_i(X_1 = k)$$

$$= \pi(k), \quad k \neq i.$$

For $k = i$, $P_i(X_{n+1} = i, X_n \neq i, \ldots, X_1 \neq i) = P_i(T^{(i)} = n + 1)$, the right-hand side becomes $\sum_1^{\infty} P_i(T^{(i)} = n) = 1$, where $T^{(i)}$ is the time of first recurrence of state i.

The $\{\pi(j)\}$, therefore, form a stationary initial *measure* for the chain. By summing, we get

$$\sum_{j \neq i} \pi(j) = \sum_{n=1}^{\infty} P_i(X_n \neq i, \ldots, X_1 \neq i) = \sum_{n=1}^{\infty} P_i(T^{(i)} > n) = E_i T^{(i)} - 1$$

or

$$\sum_j \pi(j) = E_i T^{(i)}.$$

If i is positive-recurrent, then $\bar{\pi}(j) = \pi(j)/E_i T^{(i)}$ forms a stationary initial distribution for the process. By Proposition 6.38, if $T^{(j)}$ is the first recurrence time for state j, starting from j, then

$$\bar{\pi}(j) = \frac{1}{E_j T^{(j)}}.$$

Every equivalence class C of positive-recurrent states thus has the unique stationary initial distribution given by $\bar{\pi}(j)$. Note that $\bar{\pi}(j) > 0$ for all j in the class. Use the ergodic theorem to conclude

$$\frac{1}{n}\sum_{m=1}^{n} p^{(m)}(j \mid k) \to \bar{\pi}(j), \quad \text{all} \quad k \in C.$$

Restrict the state space to such a class C.

Proposition 7.35. *Let $\lambda(j)$ be a solution of*

$$\lambda(j) = \sum_k p(j \mid k)\lambda(k)$$

such that $\sum_j |\lambda(j)| < \infty$. Then there is a constant c such that $\lambda(j) = c\bar{\pi}(j)$.

Proof. By iteration

$$\lambda(j) = \sum_k p^m(j \mid k)\lambda(k), \quad m \geq 1.$$

Consequently,

$$\lambda(j) = \sum_k \left(\frac{1}{n}\sum_{m=1}^{n} p^{(m)}(j \mid k)\right)\lambda(k).$$

The inner term converges to $\bar{\pi}(j)$, and is always ≤ 1. Use the bounded convergence theorem to get

$$\lambda(j) = \bar{\pi}(j)\sum_k \lambda(k).$$

This proposition is useful in that it permits us to get the $\bar{\pi}(j)$ by solving the system $\bar{\pi}(j) = \sum_k p(j \mid k)\bar{\pi}(k)$.

8. SOME EXAMPLES

Example A. Simple symmetric random walks. Denote by I the integers, and by $I^{(m)}$ the space $\mathbf{j} = (j_1, \ldots, j_m)$ of all m-tuples of integers. If the particle is at $\mathbf{j}$, then it makes the transition to any one of the $2m$ nearest neighbors $(j_1 \pm 1, j_2, \ldots, j_m)$, $(j_1, j_2 \pm 1, \ldots, j_m), \ldots$ with probability $1/2m$. The distribution of this process starting from $\mathbf{j}$ is given by $\mathbf{j} + \mathbf{Y}_1 + \cdots + \mathbf{Y}_n$, where $\mathbf{Y}_1, \mathbf{Y}_2, \ldots$ are independent identically distributed random vectors taking values in $(\pm 1, 0, \ldots, 0)$, $(0, \pm 1, 0, \ldots, 0)$ with equal probabilities. All states communicate. The chain has period $d = 2$. Denote $\mathbf{0} = (0, 0, \ldots, 0)$. For $m = 1$, the process starting from 0 is the fair coin-tossing game. Thus for $m = 1$, all states are null-recurrent. Polya [116] discovered the interesting phenomenon that if $m = 2$, all states are again null-recurrent, but for $m \geq 3$ all states are transient. In fact, every random walk on $I^{(m)}$ that is genuinely m-dimensional is null-recurrent for $m \leq 2$, transient for $m \geq 3$. See Chung and Fuchs [18].

Example B. The renewal chain. Another way of looking at a renewal process which illuminates the use of the word *renewal* is the following: At time zero, a new light bulb is placed in a fixture. Let T_1 be the number of periods (integral) that it lasts. When it blows out at time n, it is replaced by another light bulb starting at time n that lasts T_2 periods; the kth light bulb lasts T_k periods. The light bulbs are of identical manufacture and a reasonable model is to assume that the $\mathsf{T}_1, \mathsf{T}_2, \ldots$ are independent and identically distributed random variables. Also assume each bulb lasts at least one period; that is, $P(\mathsf{T}_1 \geq 1) = 1$. Let A_n be the event that a light bulb blows out at time n. Intuitively, this starts the whole process over again. Mathematically, it is easy to show that $\{A_n\}$ form a renewal event. Formally, $A_n = \{\omega;\ \exists \text{ a } k \text{ such that } \mathsf{T}_1 + \cdots + \mathsf{T}_k = n\}$. Now the point is that 7.25 shows that the converse is true; given any renewal process $\mathsf{U}_1, \mathsf{U}_2, \ldots,$ and letting $\mathsf{T}_1, \mathsf{T}_2, \ldots$ be the times between occurrences of $\{\mathsf{U}_k = 1\}$, we find that if $P(\mathsf{T}_1 < \infty) = 1$, then

$$\{\mathsf{U}_n = 1\} = \{\omega; \exists \text{ a } k \text{ such that } \mathsf{T}_1 + \cdots + \mathsf{T}_k = n\},$$

where the $\mathsf{T}_1, \mathsf{T}_2, \ldots$ are independent and identically distributed.

For a Markov chain $\mathsf{X}_0, \mathsf{X}_1, \ldots$ starting from state j, the events $\{\mathsf{X}_n = j\}$ form a renewal event. Now we ask the converse question, given a renewal process $\mathsf{U}_1, \mathsf{U}_2, \ldots,$ is there a Markov process $\mathsf{X}_0, \mathsf{X}_1, \ldots$ starting from the origin such that the process $\hat{\mathsf{U}}_1, \hat{\mathsf{U}}_2, \ldots,$ defined by

$$\hat{\mathsf{U}}_n = \begin{cases} 1, & \mathsf{X}_n = 0, \\ 0, & \text{otherwise}, \end{cases}$$

has the same distribution as $\mathsf{U}_1, \mathsf{U}_2, \ldots$? Actually, we can define a Markov process $\mathsf{X}_0, \mathsf{X}_1, \ldots$ on the same sample space as the renewal process such that

$$\{\mathsf{U}_n = 1\} = \{\mathsf{X}_n = 0\}.$$

Definition 7.36. *For a renewal process* $\mathsf{U}_1, \mathsf{U}_2, \ldots,$ *add the convention* $\mathsf{R}_0 \equiv 0$, *and define the time of the last replacement prior to time n as*

$$\boldsymbol{\tau}_n = \mathsf{R}_k \text{ on the set } \{\omega;\ \mathsf{R}_k \leq n < \mathsf{R}_{k+1}\}.$$

The age of the current item is defined by

$$\mathsf{X}_n = n - \boldsymbol{\tau}_n.$$

Clearly,

$$\{\mathsf{X}_n = 0\} = \{\boldsymbol{\tau}_n = n\} = \{\exists \text{ a } k \text{ such that } \mathsf{R}_k = n\} = \{\mathsf{U}_n = 1\}.$$

The X_n process, as defined, takes values in the nonnegative integers, and $\mathsf{X}_0 \equiv 0$. What does it look like? If $\mathsf{X}_n = j$, then $\boldsymbol{\tau}_n = n - j$ and on this set there exists no k such that $n - j < \mathsf{R}_k \leq n$. Therefore, on the set $\boldsymbol{\tau}_n = n - j$, either $\boldsymbol{\tau}_{n+1} = n - j$ or $\boldsymbol{\tau}_{n+1} = n + 1$. So if $\mathsf{X}_n = j$, either

$X_{n+1} = j + 1$ or $X_{n+1} = 0$. Intuitively, either a renewal takes place at time $n + 1$ or the item ages one more time unit. Clearly, $X_0, X_1, \ldots$ is a Markov chain with the stationary transition probabilities

$$\begin{aligned} P(X_{n+1} = j + 1 \mid X_n = j) &= P(\tau_{n+1} = n - j \mid \tau_n = n - j) \\ &= P(T_1 > j + 1 \mid T_1 > j) \\ &= \frac{P(T_1 > j + 1)}{P(T_1 > j)}. \end{aligned}$$

All states communicate, the chain has period determined by the minimum lattice on which T_1 is distributed, and is null-recurrent or positive-recurrent as $ET_1 = \infty$ or $< \infty$.

If $ET_1 < \infty$, there is a stationary point process $U_1^*, U_2^*, \ldots$ having interpoint distances $T_1^*, T_2^*, \ldots$ with the same distribution as $T_1, T_2, \ldots$ For this process use R_k^* as the time of the kth point past the origin, and define τ_n^*, X_n^* as above. The transition probabilities of X_n^* are the same as for X_n, but X_n^* is easily seen to be stationary. At $n = 0$, the age of the current item is k on the set $\{U_0^* = 0, \ldots, U_{-k+1}^* = 0, U_{-k}^* = 1\}$. The probability of this set is, by stationarity,

$$P^*(U_k^* = 0, \ldots, U_1^* = 0, U_0^* = 1) = P(T_1 > k)/ET_1.$$

Hence $\bar{\pi}(k) = P(T_1 > k)/ET_1$ is a stationary initial distribution for the process. The question of asymptotic stationarity of $X_0, X_1, \ldots$ is equivalent to asking if the renewal process is asymptotically stationary in the sense

$$P((U_n, U_{n+1}, \ldots) \in B) \to P^*((U_1^*, \ldots) \in B)$$

for every $B \in \mathcal{B}_\infty(\{0, 1\})$.

Problem 10. Show that a sufficient condition for the $U_1, U_2, \ldots$ process to be asymptotically stationary in the above sense is

$$P(U_n = 1) \to \frac{1}{ET_1}.$$

Example C. Birth and death processes. These are a class of Markov chains in which the state space is the integers I or the nonnegative integers I^+ and where, if the particle is at j, it can move either to $j + 1$ with probability α_j, to $j - 1$ with probability β_j or remain at j with probability $1 - \alpha_j - \beta_j$. If the states are I, assume all states communicate. If the states are I^+, 0 is either an *absorbing state* (defined as any state i such that $p(i \mid i) = 1$) or *reflecting* $(p(1 \mid 0) > 0)$. Assume that all other states communicate between themselves, and can get to zero. Equivalently, $\alpha_j \neq 0$, $\beta_j \neq 0$, for $j > 0$. If 0 is absorbing, then all other states are transient, because $j \to 0$ but $0 \not\to j$, $j \neq 0$. Therefore, for almost every sample path, either $X_n \to \infty$ or $X_n \to 0$. If 0 is reflecting, the states can be transient or recurrent, either positive or null.

To get a criterion, let τ_j^* be the first passage time to state j starting from zero:

$$\tau_j^* = \min\{n; X_n = j\}, X_0 \equiv 0.$$

Let A_j be the event that a return to zero occurs between τ_j^* and τ_{j+1}^*,

$$A_j = \bigcup_n \{X_n = 0, \tau_j^* \le n \le \tau_{j+1}^*\}.$$

The τ_j^* are Markov times, and we use 7.8 to conclude that the A_j are independent events. By the Borel-Cantelli lemma, the process is transient or recurrent as $\sum_1^\infty P(A_j) < \infty$ or $\sum_1^\infty P(A_j) = \infty$. Let τ^* be the first time after τ_j^* that $X_n \ne j$. Then $P(A_j) = E(P(A_j \mid X_{\tau^*}))$. Now

$$P(A_j \mid X_{\tau^*} = j + 1) = 0.$$

On the set $X_{\tau^*} = j - 1$, A_j can occur if we return to zero before climbing to j or by returning to zero only after climbing to j but before climbing to $j + 1$. Since τ^* is a Markov time, by the strong Markov property

$$P(A_j|X_{\tau^*} = j-1) = P(A_{j-1}) + P(A^c_{j-1})P(A_j), \quad j > 1$$

Checking that $P(X_{\tau^*} = j-1) = \beta_j/(\alpha_j + \beta_j)$ gives the equation

$$P(A_j) = \frac{\beta_j}{\alpha_j + \beta_j}(P(A_{j-1}) + P(A^c_{j-1})P(A_j)),$$

or

$$P(A_j) = \frac{r_j P(A_{j-1})}{1 + r_j P(A_{j-1})}$$

where $r_j = \beta_j/\alpha_j$. Direct substitution verifies that

$$P(A_j) = \frac{\rho_j}{s_j}; \quad \text{where } \rho_j = \prod_1^j \frac{\beta_k}{\alpha_k}, \qquad s_j = 1 + \sum_1^j \rho_k.$$

Certainly, if $\sum_1^\infty \rho_j < \infty$ then $\sum_1^\infty P(A_j) < \infty$. To go the other way, note that since $s_j = s_{j-1}/(1 - \rho_j/s_j)$, then

$$\sum_1^\infty \frac{\rho_j}{s_j} < \infty \Rightarrow \sum_2^\infty \frac{\rho_j}{s_{j-1}} < \infty.$$

But

$$\sum_2^k \frac{\rho_j}{s_{j-1}} \ge \sum_2^k \log(1 + \frac{\rho_j}{s_{j-1}}) = \log(s_k) - \log(s_1).$$

We have proved

***Proposition* 7.37.** *A birth and death process on I^+ with the origin reflecting is transient iff*

$$\sum_{j=1}^\infty \prod_{k=1}^j \left(\frac{\beta_k}{\alpha_k}\right) < \infty.$$

(Due to Harris [68]. See Karlin [86, p. 204] for an alternative derivation.) To discriminate between null and positive recurrence is easier.

Problem 11. Use the condition that

$$\lambda(k) = \sum_j p(k \mid j)\lambda(j)$$

has no solutions such that $\Sigma |\lambda(k)| < \infty$ to find a necessary and sufficient condition that a recurrent birth and death process on I^+ be null-recurrent.

Example D. Branching processes. These processes are characterized as follows: If at time n there are k individuals present, then the jth one independently of the others gives birth to Y_j offspring by time $n+1, j = 1, \ldots, k$, where $P(\mathsf{Y}_j = l) = p_l$, $l = 0, 1, \ldots$ The $\{\mathsf{Y}_j = 0\}$ event corresponds to the death of the jth individual leaving no offspring. The state space is I^+, the transition probabilities for X_n, the population size at time n, are

$$p(m \mid k) = P(\mathsf{Y}_1 + \cdots + \mathsf{Y}_k = m),$$

where the $\mathsf{Y}_1, \ldots, \mathsf{Y}_k$ are independent and have the same distribution as Y_1. Zero is an absorbing state (unless the model is revised to allow the introduction of new individuals into the population). If $p_0 > 0$, then the same argument as for birth and death processes establishes the fact that every state except zero is transient. If $p_0 = 0$, then obviously the same result holds. For a complete and interesting treatment of these chains and their generalizations, see Harris [67].

Problem 12. In a branching process, suppose $E\mathsf{Y}_1 = m < \infty$. Use the martingale convergence theorem to show that X_n/m^n converges a.s.

Example E. The Ehrenfest urn scheme. Following the work of Gibbs and Boltzmann statistical mechanics was faced with this paradox. For a system of particles in a closed container, referring to the $6N$ position-velocity vector as the state of the system, then in the ergodic case every state is recurrent in the sense that the system returns infinitely often to every neighborhood of any initial state.

On the other hand, the observed macroscopic behavior is that a system seems to move irreversibly toward an equilibrium condition. Smoluchowski proposed the solution that states far removed from equilibrium have an enormously large recurrence time, thus the system over any reasonable observation time appears to move toward equilibrium. To illustrate this the Ehrenfests constructed a model as follows: consider two urns I and II, and a total of $2N$ molecules distributed within the two urns. At time n, a molecule is chosen at random from among the $2N$ and is transferred from whatever urn it happens to be in to the other urn. Let the state k of the chain be the number of molecules in urn I, $k = 0, \ldots, 2N$. The transition probabilities are given by

$$p(k+1 \mid k) = \frac{2N-k}{2N}, \qquad p(k-1 \mid k) = \frac{k}{2N}.$$

All states communicate, and since there are only a finite number, all are positive-recurrent. We can use the fact that the stationary distribution $\bar{\pi}(k) = 1/E_k\mathsf{T}^{(k)}$ to get the expected recurrence times.

Problem 13. Use the facts that

$$\bar{\pi}(k) = \sum_j p(k \mid j)\bar{\pi}(j), \qquad \sum_k \bar{\pi}(k) = 1$$

to show that

a)
$$\bar{\pi}(N+k) = \frac{(2N)!}{(N+k)!\,(N-k)!} \cdot 2^{-2N}.$$

Compare this with the derivation of the central limit theorem for coin-tossing, Chapter I, Section 3, and show that for N large, if T is the recurrence time for the states $\{k;\ |N-k| > x\sqrt{N/2}\}$

b)
$$E\mathsf{T} \simeq \frac{1}{\sqrt{2/\pi}\int_x^\infty e^{-t^2/2}\,dt}.$$

See Kac [80] for further discussion.

9. THE CONVERGENCE THEOREM

The fundamental convergence result for Markov chains on the integers is

Theorem 7.38. *Let C be a closed indecomposable set of nonperiodic recurrent states. If the states are null-recurrent, then for all $j, k \in C$*

$$p^{(n)}(j \mid k) \to 0.$$

If the states are positive-recurrent with stationary initial distribution $\bar{\pi}(j)$, then for all $j, k \in C$

$$p^{(n)}(j \mid k) \to \bar{\pi}(j).$$

There are many different proofs of this. Interestingly enough, the various proofs are very diverse in their origin and approach. One simple proof is based on

Theorem 7.39 *(The renewal theorem). For a nonperiodic renewal process,*

$$P(\mathsf{U}_n = 1) \to \begin{cases} \dfrac{1}{E\mathsf{T}_1}, & \text{if } E\mathsf{T}_1 < \infty, \\ 0, & \text{if } E\mathsf{T}_1 = \infty. \end{cases}$$

There is a nice elementary proof of this in Feller [59, Volume I], and we prove a much generalized version in Chapter 10.

The way this theorem is used in 7.38 is that for $\{\mathsf{U}_n\}$ the return process for state j, $P_j(\mathsf{U}_n = 1) = p^{(n)}(j \mid j)$; hence $p^{(n)}(j \mid j) \to \bar{\pi}(j)$ if $E_j\mathsf{T}^{(j)} < \infty$, or $p^{(n)}(j \mid j) \to 0$ if j is null-recurrent. No matter where the process is started, let $\mathsf{T}^{(j)}$ be the first time that j is entered. Then

(7.40)
$$p^{(n)}(j \mid k) = \sum_{m=1}^{n} p^{(n-m)}(j \mid j) P_k(\mathsf{T}^{(j)} = m)$$

by the Markov property. Argue that $P_k(\mathsf{T}^{(j)} < \infty) = 1$ (see the proof of 7.30). Now use the bounded convergence theorem to establish

$$\lim_n p^{(n)}(j \mid k) = \lim_n p^{(n)}(j \mid j) \sum_{m=1}^{\infty} P_k(\mathsf{T}^{(j)} = m).$$

We can also use some heavier machinery to get a stronger result due to Orey [114]. We give this proof, from Blackwell and Freedman [9], because it involves an interesting application of martingales and the Hewitt-Savage zero-one law.

Theorem 7.41. *Let C be a closed indecomposable set of nonperiodic recurrent states. Then for any states k, l in C*

$$\sum_{j\in C} |p^{(n)}(j \mid k) - p^{(n)}(j \mid l)| \to 0.$$

Remark. We can get 7.38 in the positive-recurrent case from 7.41 by noting that 7.41 implies

$$\sum_{l\in C} (p^{(n)}(j \mid k) - p^{(n)}(j \mid l))\bar{\pi}(l) \to 0.$$

In the null-recurrent case we use an additional fact: First, consider the event A_m that starting from j the last entry into j up to time n was at time $n - m$, $A_m = \{X_{n-m} = j, X_{n-m+1} \neq j, \ldots, X_n \neq j\}$. The A_m are disjoint and the union is the whole space. Furthermore, for the process starting from j,

$$\begin{aligned} P_j(A_m) &= P_j(X_{n-m} = j)P_j(X_{n-m+1} \neq j, \ldots, X_n \neq j \mid X_{n-m} = j) \\ &= p^{(n-m)}(j \mid j)P_j(T^{(j)} > m). \end{aligned}$$

Consequently,

$$1 = \sum_{m=0}^{n} p^{(n-m)}(j \mid j)P_j(T^{(j)} > m), \tag{7.42}$$

(where $p^{(0)}(j \mid j) = 1$).

Let $\overline{\lim}\, p^{(n)}(j \mid j) = \bar{p}$. Take a subsequence n' such that $p^{(n')}(j \mid j) \to \bar{p}$. By 7.41, for any other state, $p^{(n')}(j \mid k) \to \bar{p}$. Use this to get

$$p^{(n'+1)}(j \mid j) = \sum_k p^{(n')}(j \mid k)p(k \mid j) \to \bar{p}.$$

Then for any $r \geq 0$ fixed, $p^{(n'+r)}(j \mid j) \to \bar{p}$. Substitute $n = n' + r$ in (7.42), and chop off some terms to get

$$1 \geq \sum_{m=0}^{r} p^{(n'+r-m)}(j \mid j)P_j(T^{(j)} > m) \to \bar{p} \sum_{m=0}^{r} P_j(T^{(j)} > m).$$

Noting that

$$\sum_0^{\infty} P_j(T^{(j)} > m) = E_j T^{(j)} = \infty$$

implies $\bar{p} = 0$, and using 7.41 we can complete the proof that $p^{(n)}(j \mid k) \to 0$.

Proof of 7.41. Take π any initial distribution such that $\pi(l) > 0$. Let $\mathfrak{J}$ be the tail σ-field of the process $X_0, X_1, \ldots$ and suppose that $\mathfrak{J}$ has the zero-one property under P_π. Then an easy consequence of the martingale convergence theorem (see Problem 6, Chapter 5) is that for any $A \in \mathcal{F}(\mathbf{X})$,

$$\sup_{B \in \mathcal{F}(X_n, \ldots)} |P_\pi(A \cap B) - P_\pi(A)P_\pi(B)| \to 0.$$

Let $D \subset C$, $B = \{X_n \in D\}$, $A = \{X_0 = l\}$, so that

$$\sup_{D \subset C} |P_\pi(X_n \in D, X_0 = l) - P_\pi(X_n \in D)\pi(l)| \to 0$$

or

$$\sup_{D \subset C} |P_l(X_n \in D) - P_\pi(X_n \in D)| \to 0.$$

Write $C_n^+ = \{j;\ P_l(X_n = j) \geq P_\pi(X_n = j)\}$, C_n^- the complement of C_n^+ in C. Then by the above,

$$|P_l(X_n \in C_n^+) - P_\pi(X_n \in C_n^+)| \to 0, \qquad |P_l(X_n \in C_n^-) - P_\pi(X_n \in C_n^-)| \to 0,$$

implying

$$\sum_{j \in C} |P_l(X_n = j) - P_\pi(X_n = j)| \to 0.$$

Now use the initial distribution π which assigns mass $\frac{1}{2}$ each to the states l and k to get the stated result.

The completed proof is provided by

Theorem 7.43. *For any tail event A, either $P_j(A)$ is one for all $j \in C$, or zero for all $j \in C$ (under the conditions of* 7.41).

Proof. Consider the process starting from j. The random vectors $\mathbf{Z}_k = (X_{R_k}, \ldots, X_{R_{k+1}-1})$ are independent and identically distributed by 7.28. Clearly, $\mathcal{F}(\mathbf{X}) = \mathcal{F}(\mathbf{Z}_0, \mathbf{Z}_1, \ldots)$. Take W a tail random variable; that is, W is measurable $\mathfrak{J}$.

For every n, there is a random variable $\varphi_n(\mathbf{x})$ on $(R^{(\infty)}, \mathcal{B}_\infty)$ such that $W = \varphi_n(X_n, \ldots)$. So for every k,

$$W = \varphi_{R_k}(X_{R_k}, X_{R_k+1}, \ldots) = \theta(R_k, \mathbf{Z}_k, \mathbf{Z}_{k+1}, \ldots).$$

Now R_k is a symmetric function of $\mathbf{Z}_0, \ldots, \mathbf{Z}_{k-1}$. Hence W is a symmetric function of $\mathbf{Z}_0, \mathbf{Z}_1, \ldots$ The Hewitt-Savage zero-one law holds word by word for independent identically distributed random vectors, instead of random variables. Therefore W is a.s. constant, and $P_j(A)$ is zero or one for every tail event A.

For any two states j, k let l be a descendent, $p^{(n)}(l \mid j) > 0, p^{(n)}(l \mid k) > 0$. Write

$$P_j(A) = E_j(P_j(A \mid X_n, \ldots, X_0)).$$

Since A is measurable $\mathcal{F}(\mathsf{X}_{n+1}, \ldots)$,

$$P_j(A) = E_j(P_j(A \mid \mathsf{X}_n)) = \sum_i P_j(A \mid \mathsf{X}_n = i)p^{(n)}(i \mid j).$$

If $P_j(A) = 1$, then $P_j(A \mid \mathsf{X}_n = l) \neq 0$. But using k instead of j above, we get

$$P_k(A) \geq P_k(A \mid \mathsf{X}_n = l)p^{(n)}(l \mid k) > 0.$$

Hence $P_k(A) = 1$, and the theorem is proved.

From this fundamental theorem follows a complete description of the asymptotic behavior of the $p^{(n)}(j \mid k)$. If a closed communicating set of positive recurrent states has period d, then any one of the cyclically moving subclasses D_r, $r = 1, \ldots, d$ is nonperiodic and closed under the transition probability $p^{(d)}(j \mid k)$. Looking at this class at time steps d units apart, conclude that

$$p^{(nd)}(j \mid k) \rightarrow \frac{d}{E_j \mathsf{T}^{(j)}}, \quad j, k \in D_r.$$

If $k \xrightarrow{l[d]} j$, use (7.40) to get

$$p^{(nd+l)}(j \mid k) \rightarrow \frac{d}{E_j \mathsf{T}^{(j)}}.$$

If both transient states and positive-recurrent states are present then the asymptotic behavior of $p^{(n)}(j \mid k)$, j positive-recurrent and nonperiodic, k transient, depends on the probability $P(C \mid k)$ that starting from k the process will eventually enter the class of states C communicating with j. From (7.40), in fact,

$$p^{(n)}(j \mid k) \rightarrow \frac{P(C \mid k)}{E_j \mathsf{T}^{(j)}}.$$

When j is periodic, the behavior depends not only on $P(C \mid k)$ but also at what point in the cycle of motion in C the particle from k enters C.

10. THE BACKWARD METHOD

There is a simple device which turns out to be important, both theoretically and practically, in the study of Markov chains. Let $\mathsf{Z} = \varphi(\mathsf{X}_0, \mathsf{X}_1, \ldots)$ be any random variable on a Markov chain with stationary transition probabilities. Then the device is to get an equation for $f(x) = E(\mathsf{Z} \mid \mathsf{X}_0 = x)$ by using the fact that

$$E\big(E(\mathsf{Z} \mid \mathsf{X}_1 = y, \mathsf{X}_0 = x) \mid \mathsf{X}_0 = x\big) = E(\mathsf{Z} \mid \mathsf{X}_0 = x).$$

Of course, this will be useful only if $E(\mathsf{Z} \mid \mathsf{X}_1 = y, \mathsf{X}_0 = x)$ can be expressed in terms of f. The reason I call this the backward method is that it is the initial conditions of the process that are perturbed. Here are some examples.

Mostly for convenience, look at the countable cases. It is not difficult to see how the same method carries over to similar examples in the general case.

a) *Invariant random variable.* Let Z be a bounded invariant function, $\mathsf{Z} = \varphi(\mathsf{X}_n, \ldots)$, $n \geq 0$. Then if $E(\mathsf{Z} \mid \mathsf{X}_0 = j) = f(j)$,

$$E(\mathsf{Z} \mid \mathsf{X}_1 = k, \mathsf{X}_0 = j) = E(\varphi(\mathsf{X}_1, \ldots) \mid \mathsf{X}_1 = k) = f(k),$$

so that $f(j)$ is a bounded solution of

$$f(j) = \sum_k f(k)p(k \mid j). \tag{7.44}$$

There is an interesting converse. Let $f(j)$ be a bounded solution of (7.44). Then write (7.44)as

$$f(\mathsf{X}_n) = E(f(\mathsf{X}_{n+1}) \mid \mathsf{X}_n).$$

By the Markov property,

$$f(\mathsf{X}_n) = E(f(\mathsf{X}_{n+1}) \mid \mathsf{X}_n, \ldots, \mathsf{X}_0).$$

This says that $f(\mathsf{X}_n)$ is a martingale. Since it is bounded, the convergence theorem applies, and there is a random variable Y such that

$$f(\mathsf{X}_n) \xrightarrow{\text{a.s.}} \mathsf{Y}$$

If $\mathsf{Y} = \theta(\mathsf{X}_0, \mathsf{X}_1, \ldots)$, from

$$f(\mathsf{X}_{n+1}) \xrightarrow{\text{a.s.}} \mathsf{Y}$$

conclude that $\mathsf{Y} = \theta(\mathsf{X}_1, \mathsf{X}_2, \ldots)$ a.s. Thus Y is a.s. invariant, and

$$f(j) = E(\mathsf{Y} \mid \mathsf{X}_0 = j).$$

(Use here an initial distribution which assigns positive mass to all states.) Formally,

***Proposition** 7.45. Let $\pi(j) > 0$ for all $j \in F$, then there is a one-to-one correspondence between bounded* a.s. *invariant random variables and bounded solutions of* (7.44).

b) Absorption probabilities. Let $C_1, C_2, \ldots$ be closed sets of communicating recurrent states. Let A be the event that a particle starting from state k is eventually absorbed in C_l,

$$A = \{\mathsf{X}_n \in C_l, \text{ all } n \text{ sufficiently large}\}.$$

A is an invariant event, so $f(j) = P(A \mid \mathsf{X}_0 = j)$ satisfies (7.44). There are also the boundary conditions:

$$f(j) = \begin{cases} 1, & j \in C_l, \\ 0, & j \in C_h, \quad h \neq l. \end{cases}$$

If one solves (7.44) subject to these boundary conditions and boundedness, is the solution unique? No, in general, because if J is the set of all transient states, the event that the particle remains in J for all time is invariant, and

$$c(j) = P_j(\mathsf{X}_n \in J, n = 0, 1, \ldots)$$

is zero on all C_h, and any multiple of $c(j)$ may be added to any given solution satisfying the boundary conditions to give another solution. If the probability of remaining in transient states for all time is zero, then the solution is unique. For example, let g be such a solution, and start the process from state j. The process $g(\mathsf{X}_n)$, $n = 0, 1, \ldots$ is a martingale. Let n^* be the first time that one of the C_h is entered. This means that n^* is a stopping time for the $\mathsf{X}_0, \mathsf{X}_1, \ldots$ sequence. Furthermore $g(\mathsf{X}_n)$ and n^* satisfy the hypothesis of the optional stopping theorem. Therefore,

$$Eg(\mathsf{X}_{\mathsf{n}^*}) = Eg(\mathsf{X}_0) = g(j).$$

But

$$g(\mathsf{X}_{\mathsf{n}^*}) = \begin{cases} 1, & \mathsf{X}_{\mathsf{n}^*} \in C_l, \\ 0, & \mathsf{X}_{\mathsf{n}^*} \notin C_l. \end{cases}$$

Therefore $g(j) = P(A \mid \mathsf{X}_0 = j)$.

In fair coin-tossing, with initial fortune zero, what is the probability that we win M_1 dollars before losing M_2? This is the same problem as: For a simple symmetric random walk starting from zero, with absorbing states at M_1, $-M_2$, find the probability of being absorbed into M_1. Let $p^+(j)$ be the probability of being absorbed into M_1 starting from j, $-M_2 \leq j \leq M_1$. Then $p^+(j)$ must satisfy (7.44) which in this case is

$$p^+(j) = \tfrac{1}{2}p^+(j-1) + \tfrac{1}{2}p^+(j+1), \; -M_2 < j < M_1,$$

and the boundary conditions $p^+(-M_2) = 0$, $p^+(M_1) = 1$. This solution is easy to get:

$$p^+(j) = \frac{j + M_2}{M_1 + M_2}.$$

c) Two other examples. Among many others, not involving invariant sets, I pick two. Let n^* be the time until absorption into the class C of recurrent states, assuming $C \neq \emptyset$. Write $m(j) = E_j(\mathsf{n}^*)$. Check that

$$E(\mathsf{n}^* \mid \mathsf{X}_1 = k, \mathsf{X}_0 = j) = \begin{cases} 1, & k \in C, \quad j \notin C, \\ 1 + m(k), & k \notin C, \quad \notin C, \end{cases}$$

and apply the backward argument to give

(7.46) $$m(j) = \sum_{k \notin C} m(k)p(k \mid j) + 1, \quad j \notin C.$$

The boundary conditions are $m(j) = 0, j \in C$.

Now let N_i be the number of visits to state i before absorption into C. Denote $G(j, i) = E_j(\mathsf{N}_i)$. For $k \notin C, j \notin C$,

$$E_j(\mathsf{N}_i \mid \mathsf{X}_1 = k) = \begin{cases} G(k, i), & j \neq i, \\ G(k, i) + 1, & j = i. \end{cases}$$

So

(7.47)
$$G(j, i) = \delta(j, i) + \sum_{k \in C^c} G(k, i) p(k \mid j),$$

where $\delta(j, i) = 0$ or 1 as $j \neq i$ or $j = i$. The boundary conditions are $G(j, i) = 0, j \in C$. Of course, this makes no sense unless i is transient.

With these last two examples there is a more difficult uniqueness problem. For example, in (7.46) assume that

$$E_j(\mathsf{n}^*) < \infty, \quad \text{all } j.$$

Then any nonnegative solution $g(j)$ of (7.46) satisfying

$$\varliminf_N \int_{\{\mathsf{n}^* > N\}} g(\mathsf{X}_N)\, dP = 0$$

must be $E_j(\mathsf{n}^*)$. To prove this, check that

$$g(\mathsf{X}_n) - \sum_0^{n-1} \chi_J(\mathsf{X}_k)$$

is a martingale sequence, that $E_j\left(\sum_0^{\mathsf{n}^*-1} \chi_J(\mathsf{X}_k)\right) = E_j(\mathsf{n}^*)$, and apply optional stopping.

Problems

14. For simple symmetric random walk with absorbing states at $M_1, -M_2$, show that

a)
$$E_j(\mathsf{n}^*) = (M_2 + j)(M_1 - j),$$

b)
$$E_j(\mathsf{N}_i) = \begin{cases} \dfrac{(M_1 - j)(i + M_2)}{M_1 + M_2}, & j \geq i, \\[2ex] \dfrac{(M_2 + j)(M_1 - i)}{M_1 + M_2}, & j \leq i. \end{cases}$$

15. Let $\{\mathsf{X}_n\}$ be simple symmetric random walk. Derive the expressions for $p^+(j)$ and $E_j(\mathsf{n}^*)$ by showing that the sequences $\{\mathsf{X}_n\}$, $\{\mathsf{X}_n^2 - n\}$ are martingales and applying the stopping time results of Section 7, Chapter 5.

16. For simple symmetric random walk with absorbing states at $M_1, -M_2$, use the expression for $p^+(j)$ to evaluate $q(j) = P_j$ (at least one return to j). For $-M_2 < j < i < M_1$, $E_j(\mathsf{N}_i)$ is the probability that particle hits i

before $-M_2$ times the expected number of returns to i starting from i before absorption. Use $p^+(j)$, for absorbing states at $-M_2$, i, and $q(i)$ to evaluate $E_j(\mathsf{N}_i)$.

17. For any given set D of states, let A be the event that X_n stays in D for all n,

$$A = \{\mathsf{X}_n \in D, n = 0, 1, \ldots\}.$$

a) Show that $f(j) = P(A \mid j)$ satisfies

$$f(j) = \sum_{k \in D} f(k)p(k \mid j), \quad j \in D,$$

$$f(j) = 0, \quad j \in D^c.$$

b) Prove using (a) that a state h is transient iff there exists a bounded non-trivial solution to the equation

$$g(j) = \sum_{k \neq h} g(k)p(k \mid j), \quad j \neq h.$$

c) Can you use (b) to deduce 7.37?

NOTES

In 1906 A. A. Markov [110] proved the existence of stationary initial distributions for Markov chains with a finite number of states. His method is simple and clever, and the idea can be generalized. A good exposition is in Doob's book [38, pp. 170 ff]. The most fundamental work on general state spaces is due to W. Doeblin [25] in 1937 and [28] in 1940. Some of these latter results concerning the existence of invariant initial distributions are given in Doob's book. The basic restriction needed is a sort of compactness assumption to keep the motion from being transient or null-recurrent. But a good deal of Doeblin's basic work occurs before this restriction is imposed, and is concerned with the general decomposition of the state space. For an exposition of this, see K. L. Chung [15] or [17].

The difficulty in the general state space is that there is no way of classifying each state y by means of the process of returns to y. If, for example, $p(A \mid x)$ assigns zero mass to every one-point set, then the probability of a return to x is zero. You might hope to get around this by considering returns to a neighborhood of x, but then the important independence properties of the recurrence times no longer hold. It may be possible to generalize by taking smaller and smaller neighborhoods and getting limits, but this program looks difficult and has not been carried out successfully. Hence, in the general case, it is not yet clear what definition is most appropriate to use in classifying chains as recurrent or transient. For a fairly natural definition of recurrent chains Harris [66] generalized Doeblin's result by showing the existence of a possibly infinite, but always σ-finite measure $Q(dx)$ satisfying

$$Q(B) = \int p(B \mid x) Q(dx), \quad B \in \mathcal{B}_1(F).$$

His idea was very similar to the idea in the countable case: Select a set $A \in \mathcal{B}_1(F)$ so that an initial distribution $\pi_A(\cdot)$ exists concentrated on A such that every time the process returned to A, it had the distribution π_A. This could be done using Doeblin's technique. Then define $\pi(B)$, $B \in \mathcal{B}_1(F)$ as the expected number of visits to B between visits to A, using the initial distribution π_A.

The basic work when the state space is countable but not necessarily finite is due to Kolmogorov [95], 1936. The systematic application of the renewal theorem and concepts was done by Feller, see [55]. K. L. Chung's book [16] is an excellent source for a more complete treatment of the countable case.

The literature concerning applications of Markov chains is enormous. Karlin's book [86] has some nice examples; so does Feller's text [59, Vol. I]. A. T. Bharucha-Reid's book [3] is more comprehensive.

The proof of Proposition 7.37 given in the first edition of this book was incorrect. I am indebted to P. J. Thomson and K. M. Wilkinson for pointing out the error and supplying a correction.

CHAPTER 8

CONVERGENCE IN DISTRIBUTION AND THE TOOLS THEREOF

1. INTRODUCTION

Back in Chapter 1, we noted that if $\mathsf{Z}_n = \mathsf{Y}_1 + \cdots + \mathsf{Y}_n$, where the Y_i are independent and ± 1 with probability $\frac{1}{2}$, then

$$P\left(\frac{\mathsf{Z}_n}{\sqrt{n}} < x\right) \to \frac{1}{\sqrt{2\pi}} \int_{-\infty}^{x} e^{-t^2/2}\, dt.$$

Thus the random variables $\mathsf{Z}_n/\sqrt{n}$ have distribution functions $F_n(x)$ that converge for every value of x to $\Phi(x)$, but from Problem 23, Chapter 3, certainly the random variables $\mathsf{Z}_n/\sqrt{n}$ do not converge a.s. (or for that matter in L_1, or in any strong sense). What are convergent here are not the values of the random variables themselves, but the probabilities with which the random variables assume certain values. In general, we would like to say that the distribution of the random variable X_n converges to the distribution of X if $F_n(x) = P(\mathsf{X}_n < x) \to F(x) = P(\mathsf{X} < x)$ for every $x \in R^{(1)}$. But this is a bit too strong. For instance, suppose $\mathsf{X} \equiv 0$. Then we would want the values of X_k to be more and more concentrated about zero, that is for any $\epsilon > 0$ we would want

(8.1) $$P(\mathsf{X}_n < -\epsilon) \to 0, \qquad P(\mathsf{X}_n \geq \epsilon) \to 0.$$

Now $F(0) = 0$, but 8.1 could hold, even with $F_n(0) = 1$, for all n. Take $\mathsf{X}_n \equiv -1/n$, for example. What 8.1 says is that for all $x < 0$, $F_n(x) \to F(x)$, and for all $x > 0$, $F_n(x) \to F(x)$. Apparently, not much should be assumed about what happens for x a discontinuity point of $F(x)$. Hence we state the following:

Definition 8.2. *We say that* X_n *converges to* X *in distribution,* $\mathsf{X}_n \xrightarrow{\mathfrak{D}} \mathsf{X}$, *if* $F_n(x) \to F(x)$ *at every point* $x \in C(F)$, *the set of continuity points of* F. *That is,* $P(\mathsf{X} = x) = 0 \Rightarrow F_n(x) \to F(x)$. *We will also write in this case* $F_n \xrightarrow{\mathfrak{D}} F$.

Different terminology is sometimes used.

Definition 8.3. *By the law of* X, *written* $\mathfrak{L}(\mathsf{X})$, *is meant the distribution of* X. *Convergence in distribution is also called convergence in law and* $\mathfrak{L}(\mathsf{X}_n) \to \mathfrak{L}(\mathsf{X})$ *is equivalent notation for* $\mathsf{X}_n \xrightarrow{\mathfrak{D}} \mathsf{X}$. *If random variables* X *and* Y *have the same distribution, write either* $\mathfrak{L}(\mathsf{X}) = \mathfrak{L}(\mathsf{Y})$ *or* $\mathsf{X} \overset{\mathfrak{D}}{=} \mathsf{Y}$.

Recall from Chapter 2 that a function $F(x)$ on $R^{(1)}$ is the distribution function of a random variable iff

(8.4) i) $$x \leq y \Rightarrow F(x) \leq F(y)$$

ii) $$x \uparrow y \Rightarrow F(x) \uparrow F(y)$$

iii) $$F(-\infty) = 0, \quad F(+\infty) = 1.$$

Problems

1. Show that if $F(x) = P(\mathsf{X} < x)$, then $F(x^+) - F(x) = P(\mathsf{X} = x)$. Show that $C^c(F)$ is at most countable $\left(F(x^+) = \lim_{y \downarrow x} F(y)\right)$.

2. Let T be a dense set of points in $R^{(1)}$, $F_0(x)$ on T having properties (8.4 i, ii, and iii) with $x, y \in T$. Show that there is a unique distribution function $F(x)$ on $R^{(1)}$ such that $F(x) = F_0(x)$, $x \in T$.

3. Show that if, for each n, X_n takes values in the integers I, then $\mathsf{X}_n \xrightarrow{\mathfrak{D}} \mathsf{X}$ implies $P(\mathsf{X} \in I) = 1$ and $\mathsf{X}_n \xrightarrow{\mathfrak{D}} \mathsf{X} \Leftrightarrow P(\mathsf{X}_n = j) \to P(\mathsf{X} = j)$, all $j \in I$.

4. If F_n, F are distribution functions, $F_n \xrightarrow{\mathfrak{D}} F$, and $F(x)$ continuous, show that

$$\sup_x |F_n(x) - F(x)| \to 0.$$

5. If $\mathsf{X} \overset{\mathfrak{D}}{=} \mathsf{Y}$, and $\varphi(x)$ is $\mathcal{B}_1$-measurable, show that $\varphi(\mathsf{X}) \overset{\mathfrak{D}}{=} \varphi(\mathsf{Y})$. Give an example to show that if X, Y, Z are random variables defined on the same probability space, that $\mathsf{X} \overset{\mathfrak{D}}{=} \mathsf{Y}$ does not necessarily imply that $\mathsf{XZ} \overset{\mathfrak{D}}{=} \mathsf{YZ}$.

Define a random variable X to have a *degenerate distribution* if X is a.s. constant.

6. Show that if $\mathsf{X}_n \xrightarrow{\mathfrak{D}} \mathsf{X}$ and X has a degenerate distribution, then $\mathsf{X}_n \xrightarrow{P} \mathsf{X}$.

2. THE COMPACTNESS OF DISTRIBUTION FUNCTIONS

One of the most frequently used tools in $\mathfrak{D}$-convergence is a certain compactness property of distribution functions. They themselves are not compact, but we can look at a slightly larger set of functions.

Definition 8.5. *Let $\mathcal{M}$ be the class of all functions $G(x)$ satisfying* (8.4 i *and* ii), *with the addition of*

$$\text{(iii)} \qquad 0 \leq G(-\infty), \qquad G(+\infty) \leq 1.$$

As before $G, G_n \in \mathcal{M}$, $G_n \xrightarrow{\mathfrak{D}} G$ if $\lim_n G_n(x) = G(x)$ at all points of $C(G)$.

Theorem 8.6 *(Helly-Bray). $\mathcal{M}$ is sequentially compact under $\xrightarrow{\mathfrak{D}}$.*

Proof. Let $G_n \in \mathcal{M}$, take $T = \{x_k\}$, $k = 1, 2, \ldots$ dense in $R^{(1)}$. We apply Cantor's diagonalization method. That is, let $I_1 = \{n_1, n_2, \ldots\}$ be an ordered subset of the integers such that $G_n(x_1)$ converges as $n \to \infty$ through I_1. Let $I_2 \subset I_1$ be such that $G_n(x_2)$ converges as $n \to \infty$ through I_2. Continue this way getting decreasing ordered subsets $I_1, I_2, \ldots$ of the integers. Let n_m be the mth member of I_m. For $m \geq k$, $n_m \in I_k$, so for every $x_k \in T$, $G_{n_m}(x_k)$ converges. Define $G_0(x)$ on T by $G_0(x_k) = \lim_m G_{n_m}(x_k)$. Define $G(x)$ on $R^{(1)}$ by

$$G(x) = \lim_{x_k \uparrow x} G_0(x_k).$$

It is easy to check that $G \in \mathcal{M}$. Let $x \in C(G)$. Then

$$\lim_{x_k \uparrow x} G_0(x_k) = G(x),$$

by definition, but also check that

$$\lim_{x_k \downarrow x} G_0(x_k) = G(x).$$

Take $x_k' < x < x_k''$, x_k', $x_k'' \in T$. Then

$$G_{n_m}(x_k') \leq G_{n_m}(x) \leq G_{n_m}(x_k''),$$

implying

$$G_0(x_k') \leq \underline{\lim}\, G_{n_m}(x) \leq \overline{\lim}\, G_{n_m}(x) \leq G_0(x_k'').$$

Letting $x_k' \uparrow x$, $x_k'' \downarrow x$ gives the result that G_{n_m} converges to G at every $x \in C(G)$.

A useful way of looking at the Helly-Bray theorem is

Corollary 8.7. *Let $G_n \in \mathcal{M}$. If there is a $G \in \mathcal{M}$ such that for every $\mathfrak{D}$-convergent subsequence G_{n_m}, $G_{n_m} \xrightarrow{\mathfrak{D}} G$, then the full sequence $G_n \xrightarrow{\mathfrak{D}} G$.*

Proof. If $G_n \overset{\mathfrak{D}}{\not\to} G$, there exists an $x_0 \in C(G)$ such that $G_n(x_0) \not\to G(x_0)$. But every subsequence of the G_n contains a convergent subsequence G_{n_m} and $G_{n_m}(x_0) \to G(x_0)$.

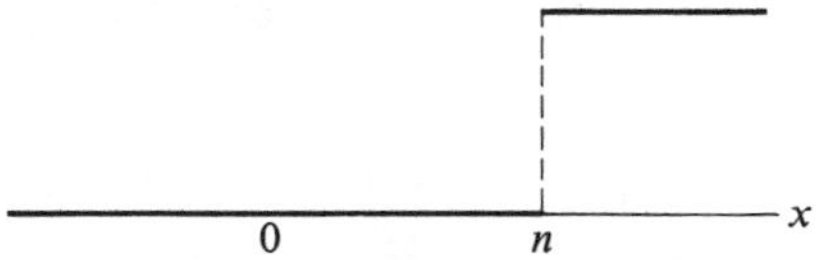

Fig. 8.1 $F_n(x)$.

Unfortunately, the class of distribution functions itself is not compact. For instance, take $\mathsf{X}_n \equiv n$ (see Fig. 8.1). Obviously $\lim_n F_n(x) = 0$ identically. The difficulty here is that mass floats out to infinity, disappearing in the limit. We want to use the Helly-Bray theorem to get some compactness properties for distribution functions. But to do this we are going to have to impose additional restrictions to keep the mass from moving out to infinity. We take some liberties with the notation.

Definition 8.8. *$F(B)$, $B \in \mathcal{B}_1$, will denote the extension of $F(x)$ to a probability measure on $\mathcal{B}_1$, that is, if $F(x) = P(X < x)$, $F(B) = P(X \in B)$.*

Definition 8.9. *Let $\mathcal{N}$ denote the set of all distribution functions. A subset $\mathcal{L} \subset \mathcal{N}$ will be said to be mass-preserving if for any $\epsilon > 0$, there is a finite interval I such that $F(I^c) < \epsilon$, all $F \in \mathcal{L}$.*

Proposition 8.10. *Let $\mathcal{L} \subset \mathcal{N}$. Then $\mathcal{L}$ is conditionally compact in $\mathcal{N}$ if and only if $\mathcal{L}$ is mass-preserving (that is, $F_n \in \mathcal{L} \Rightarrow \exists F_{n_m}$ such that $F_{n_m} \xrightarrow{\mathcal{D}} F \in \mathcal{N}$).*

Proof. Assume $\mathcal{L}$ mass-preserving, $F_n \in \mathcal{L}$. There exists $G \in \mathcal{M}$ such that $F_{n_m} \xrightarrow{\mathcal{D}} G$. For any ϵ, take a, b such that $F_n([a, b)) \geq 1 - \epsilon$. Take $a' < a$, $b' > b$ so that $a', b' \in C(G)$. Then

$$F_{n_m}([a', b')) = F_{n_m}(b') - F_{n_m}(a') \to G(b') - G(a'),$$

with the conclusion that $G(b') - G(a') \geq 1 - \epsilon$, or $G(+\infty) = 1$, $G(-\infty) = 0$, hence $G \in \mathcal{N}$. On the other hand, let $\mathcal{L}$ be conditionally compact in $\mathcal{N}$. If $\mathcal{L}$ is not mass-preserving, then there is an $\epsilon > 0$ such that for every finite I,

$$\inf_{F \in \mathcal{L}} F(I) < 1 - \epsilon.$$

Take $F_n \in \mathcal{L}$ such that for every n, $F_n([-n, +n)) < 1 - \epsilon$. Now take a subsequence $F_{n_m} \xrightarrow{\mathcal{D}} F \in \mathcal{N}$. Let $a, b \in C(F)$; then $F_{n_m}([a, b)) \to F([a, b))$, but for n_m sufficiently large $[a, b) \subset [-n_m, +n_m)$. Thus $F([a, b)) \leq 1 - \epsilon$ for any $a, b \in C(F)$ which implies $F \notin \mathcal{N}$.

One obvious corollary of 8.10 is

Corollary 8.11. *If $F_n \xrightarrow{\mathcal{D}} F$, $F \in \mathcal{N}$, then $\{F_n\}$ is mass-preserving.*

Problems

7. For $-\infty < a < b < +\infty$, consider the class of all distribution functions such that $F(a) = 0$, $F(b) = 1$. Show that this class is sequentially compact.

8. Let $F_n \xrightarrow{\mathcal{D}} F$, and F_n, F be distribution functions. Show that for B, any Borel set, it is not necessarily true that $F_n(B) = 1$, for all $n \Rightarrow F(B) = 1$. Show that if B is closed, however, then $F_n(B) = 1$, for all $n \Rightarrow F(B) = 1$.

9. Let $g(x)$ be $\mathcal{B}_1$-measurable, such that $|g(x)| \to \infty$ as $x \to \pm\infty$. If $\mathcal{L} \subset \mathcal{N}$ is such that $\sup_{F \in \mathcal{L}} \int |g| \, dF < \infty$, then $\mathcal{L}$ is mass-preserving.

10. Show that if there is an $r > 0$ such that $\overline{\lim}\, E|X_n|^r < \infty$, then $\{F_n\}$ is mass-preserving.

11. The support of F is the smallest closed set C such that $F(C) = 1$. Show that such a minimal closed set exists. A point of increase of F is defined as a point x such that for every neighborhood N of x, $F(N) > 0$. Show that the set of all points of increase is exactly the support of F.

12. Define a Markov chain with stationary transition probabilities $p(\cdot \mid x)$ to be *stable* if for any sequence of initial distributions π_n $\mathfrak{D}$-converging to an initial distribution π, the probabilities $\int p(\cdot \mid x)\pi_n(dx)$ $\mathfrak{D}$-converge to the probability $\int p(\cdot \mid x)\pi(dx)$.

If the state space of a stable Markov chain is a compact interval, show that there is at least one invariant initial distribution. [Use Problem 7 applied to the probabilities $1/n \sum_1^n p^{(k)}(\cdot \mid x)$ for x fixed.]

3. INTEGRALS AND $\mathfrak{D}$-CONVERGENCE

Suppose $F_n \xrightarrow{\mathfrak{D}} F$, $F_n, F \in \mathcal{N}$, does it then follow that for any reasonable measurable function $f(x)$, that $\int f(x)\,dF_n \to \int f(x)\,dF$? The answer is No! For example, let

$$\mathsf{X}_n \equiv 1/n, \mathsf{X} \equiv 0, \quad \text{so } F_n \xrightarrow{\mathfrak{D}} F.$$

Now take $f(x) = 0$, $x \le 0$, and $f(x) = 1$, $x > 0$. Then $\int f\,dF_n = 1$, but $\int f\,dF = 0$. But it is easy to see that it works for f bounded and continuous. Actually, a little more can be said.

Proposition 8.12. *Let $F_n, F \in \mathcal{N}$ and $F_n \xrightarrow{\mathfrak{D}} F$. If $f(x)$ is bounded on $R^{(1)}$, measurable $\mathcal{B}_1$ and the discontinuity points of f are a set S with $F(S) = 0$, then*

$$\int f\,dF_n \to \int f\,dF.$$

Remark. The set of discontinuity points of a $\mathcal{B}_1$-measurable function is in $\mathcal{B}_1$, so $F(S)$ is well-defined. (See Hobson [72, p. 313].)

Proof. Take $a, b \in C(F)$, $I_1, \ldots, I_k$ a partition $\mathcal{I}_k$ of $I = [a, b)$, where $I_i = [a_i, b_i)$ and $a_i, b_i \in C(F)$. Define on I

$$f_k^+(x) = \sum_1^k \sup_{x \in I_i} (f(x))\chi_{I_i}(x),$$

$$f_k^-(x) = \sum_1^k \inf_{x \in I_i} (f(x))\chi_{I_i}(x).$$

Then

$$\int_I f_k^-\,dF_n \le \int_I f\,dF_n \le \int_I f_k^+\,dF_n.$$

Clearly, the right- and left-hand sides above converge, and

$$\int_I f_k^-\,dF \le \varliminf \int_I f\,dF_n \le \varlimsup \int_I f\,dF_n \le \int_I f_k^+\,dF.$$

Let $\|\mathcal{T}_k\| \to 0$. At every point x which is a continuity point of $f(x)$,

$$f_k^-(x) \uparrow f(x), \qquad f_k^+(x) \downarrow f(x).$$

By the Lebesgue bounded convergence theorem,

$$\lim_k \int_I f_k^- \, dF = \lim_k \int_I f_k^+ \, dF = \int_I f \, dF.$$

Let $M = \sup |f(x)|$, then since $\{F_n\}$ is mass-preserving, for any $\epsilon > 0$ we can take I such that $F_n(I^c) \le \epsilon/2M$ and $F(I^c) \le \epsilon/2M$. Now

$$\overline{\lim} \left| \int f \, dF_n - \int f \, dF \right| \le \overline{\lim} \left| \int_I f \, dF_n - \int_I f \, dF \right| + \epsilon = \epsilon.$$

Corollary 8.13. *In the above proposition, eliminate the condition that f be bounded on $R^{(1)}$, then*

$$\int |f| \, dF \le \underline{\lim} \int |f| \, dF_n.$$

Proof. Define

$$g_\alpha(x) = \begin{cases} |f(x)|, & \text{if} \quad |f(x)| \le \alpha, \\ \alpha, & \text{otherwise.} \end{cases}$$

Every continuity point of f is a continuity point of g_α. Apply Proposition 8.12 to g_α to conclude

$$\int g_\alpha \, dF = \lim \int g_\alpha \, dF_n \le \underline{\lim} \int |f| \, dF_n.$$

Let $\alpha \uparrow \infty$, then $g_\alpha \uparrow |f|$. By the monotone convergence theorem $\int g_\alpha \, dF \to \int |f| \, dF$.

Problems

13. Let F_n, $F \in \mathcal{N}$, $F_n \xrightarrow{\mathfrak{D}} F$. For any set $E \subset R^{(1)}$, define the boundary of E as $bd(E) = \bar{E} \cap \overline{E^c}$, ($\bar{E}$ = closure of E). Prove that for any $B \in \mathcal{B}_1$ such that $F(bd(B)) = 0$, $F_n(B) \to F(B)$.

14. Let $F_n \xrightarrow{\mathfrak{D}} F$, and $h(x)$, $g(x)$ be continuous functions such that

$$|g(x)| \to +\infty \quad \text{as} \quad x \to \pm\infty,$$

$$\left| \frac{h(x)}{g(x)} \right| \to 0 \quad \text{as} \quad x \to \pm\infty.$$

Show that $\overline{\lim} \int |g(x)| \, dF_n < \infty$ implies

$$\int h(x) \, dF_n \to \int h(x) \, dF.$$

4. CLASSES OF FUNCTIONS THAT SEPARATE

Definition 8.14. *A set $\mathcal{E}$ of bounded continuous functions on $R^{(1)}$ will be called $\mathcal{N}$-separating if for any $F, G \in \mathcal{N}$,*

$$\int f\,dF = \int f\,dG, \quad \text{all } f \in \mathcal{E}$$

implies $F = G$.

We make this a bit more general (also ultimately, more convenient) by allowing the functions of $\mathcal{E}$ to be complex-valued. That is, we consider functions $f(x)$ of the form $f(x) = f_1(x) + if_2(x), f_1, f_2$ real-valued, continuous, and bounded. Now, of course, $|f(x)|$ has the meaning of the absolute value of a complex number. As usual, then, $\int f\,dF = \int f_1\,dF + i\int f_2\,dF$.

The nice thing about such a class $\mathcal{E}$ of functions is that we can check whether $F_n \xrightarrow{\mathcal{D}}$ by looking at the integrals of these functions. More specifically:

Proposition 8.15. *Let $\mathcal{E}$ be $\mathcal{N}$-separating, and $\{F_n\}$ mass-preserving. Then there exists an $F \in \mathcal{N}$ such that $F_n \xrightarrow{\mathcal{D}} F$ if and only if*

$$\lim_n \int f\,dF_n \text{ exists}, \quad \text{all } f \in \mathcal{E}.$$

If this holds, then $\lim_n \int f\,dF_n = \int f\,dF$, all $f \in \mathcal{E}$.

Proof. One way is clear. If $F_n \xrightarrow{\mathcal{D}} F$, then $\lim_n \int f\,dF_n = \int f\,dF$ by 8.12. To go the other way, take any $\mathcal{D}$-convergent subsequence F_{n_k} of F_n. By mass-preservation $F_{n_k} \xrightarrow{\mathcal{D}} F \in \mathcal{N}$. Take any other convergent subsequence $F_{n'_k} \xrightarrow{\mathcal{D}} G$. Then for $f \in \mathcal{E}$, by 8.12,

$$\lim_k \int f\,dF_{n_k} = \int f\,dF, \lim_k \int f\,dF_{n'_k} = \int f\,dG,$$

so $\int f\,dF = \int f\,dG$, all $f \in \mathcal{E} \Rightarrow F = G$. All $\mathcal{D}$-convergent subsequences of F_n have the same limit F, implying $F_n \xrightarrow{\mathcal{D}} F$.

Corollary 8.16. *Let $\mathcal{E}$ be $\mathcal{N}$-separating and $\{F_n\}$ mass-preserving. If $F \in \mathcal{N}$ is such that $\int f\,dF_n \to \int f\,dF$, all $f \in \mathcal{E}$, then $F_n \xrightarrow{\mathcal{D}} F$.*

The relevance of looking at integrals of functions to $\mathcal{D}$-convergence can be clarified by the simple observation that $F_n \xrightarrow{\mathcal{D}} F$ is equivalent to $\int f\,dF_n \to \int f\,dF$ for all functions f of the form

$$f(x) = \begin{cases} 1, & x < x_0, \\ 0, & x \geq x_0, \end{cases}$$

for any $x_0 \in C(F)$.

What classes of functions are $\mathcal{N}$-separating? Take $\mathcal{E}_0$ to be all functions f of the form below (see Fig. 8.2) with any a, b finite and any $\epsilon > 0$.

Proposition 8.17. *$\mathcal{E}_0$ is $\mathcal{N}$-separating.*

Proof. For any $F, G \in \mathcal{N}$, take $a, b \in C(F) \cap C(G)$. Assume that for any f as described,

$$\int f\,dF = \int f\,dG.$$

Then

$$F([a, b)) \le \int f\,dF = \int f\,dG \le G([a - \epsilon, b + \epsilon)),$$

and conversely,

$$G([a, b)) \le F([a - \epsilon, b + \epsilon)).$$

Let $\epsilon \downarrow 0$, to get $F([a, b)) = G([a, b))$. The foregoing being true for all $a, b \in C(F) \cap C(G)$ implies $F = G$.

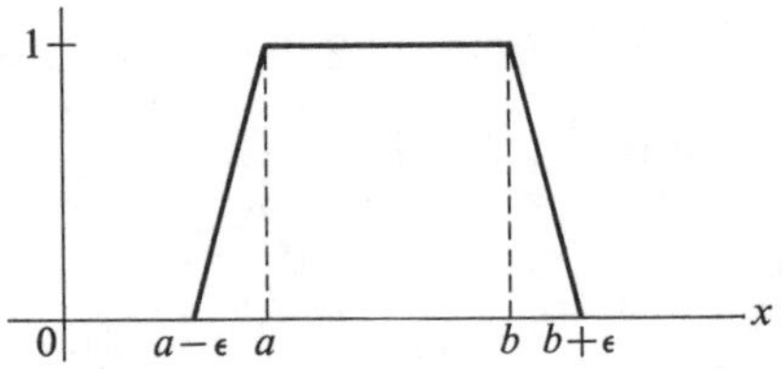

Figure 8.2

However $\mathcal{E}_0$ is an awkward set of functions to work with. What is really more important is

Proposition 8.18. *Let $\mathcal{E}$ be a class of continuous bounded functions on $R^{(1)}$ with the property that for any $f_0 \in \mathcal{E}_0$, there exist $f_n \in \mathcal{E}$ such that* $\sup |f_n(x)| \le M$, *all n, and* $\lim_n f_n(x) = f_0(x)$ *for every $x \in R^{(1)}$. Then $\mathcal{E}$ is $\mathcal{N}$-separating.*

Proof. Let $\int f\,dF = \int f\,dG$, all $f \in \mathcal{E}$. For any $f_0 \in \mathcal{E}_0$, take $f_n \in \mathcal{E}$ converging to f_0 as in the statement 8.18 above. By the Lebesgue bounded convergence theorem

$$\lim_n \int f_n\,dF = \int f_0\,dF, \qquad \lim_n \int f_n\,dG = \int f_0\,dG,$$

so $\int f_0\,dF = \int f_0\,dG$, all $f_0 \in \mathcal{E}_0$ implying $F = G$.

5. TRANSLATION INTO RANDOM-VARIABLE TERMS

The foregoing is all translatable into random-variable terms. For example:

i) If X_n are random variables, their distribution functions are mass-preserving iff for any $\epsilon > 0$, there is a finite interval I such that

$$P(X_n \in I^c) < \epsilon, \quad \text{all } n.$$

ii) If $|g(x)| \to \infty$ as $x \to \pm\infty$, then the distribution functions of X_n are mass-preserving if $\sup_n E|g(X_n)| < \infty$ (Problem 9).

iii) If X_n have mass-preserving distribution functions and $\mathcal{E}$ is an $\mathcal{N}$-separating set of functions, then there exists a random variable X such that $X_n \xrightarrow{\mathfrak{D}} X$ if and only if $\lim_n Ef(X_n)$ exists, all $f \in \mathcal{E}$.

We will switch freely between discussion in terms of distribution functions and in terms of random variables, depending on which set of terms is more illuminating.

Proposition 8.19. *Let* $X_n \xrightarrow{\mathfrak{D}} X$, *and let* $\varphi(x)$ *be measurable* $\mathcal{B}_1$, *with its set* S *of discontinuities such that* $P(X \in S) = 0$. *Then*

$$\varphi(X_n) \xrightarrow{\mathfrak{D}} \varphi(X).$$

Proof. Let $Z_n = \varphi(X_n)$, $Z = \varphi(X)$. If $Ef(Z_n) \to Ef(Z)$, for all $f \in \mathcal{E}_0$, then $Z_n \xrightarrow{\mathfrak{D}} Z$. Let $g(x) = f(\varphi(x))$. This function g is bounded, measurable $\mathcal{B}_1$, and continuous wherever φ is continuous. By 8.12, $Eg(X_n) \to Eg(X)$.

We can't do any better with a.s. convergence. This is illustrated by the following problem.

Problem 15. If $\varphi(x)$ is as in 8.19, and $X_n \xrightarrow{\text{a.s.}} X$, then show $\varphi(X_n) \xrightarrow{\text{a.s.}} \varphi(X)$. Give an example to show that in general this is not true if $\varphi(x)$ is only assumed measurable.

6. AN APPLICATION OF THE FOREGOING

With only this scanty background we are already in a position to prove a more general version of the central limit theorem. To do this we work with the class of functions defined by

$\mathcal{E}_1$ *consists of all continuous bounded* f *on* $R^{(1)}$ *such that* $f''(x)$ *exists for all* x, $\sup_x |f''(x)| < \infty$, *and* $f''(x)$ *is uniformly continuous on* $R^{(1)}$.

It is fairly obvious that $\mathcal{E}_1$ satisfies the requirements of 8.18 and hence is $\mathcal{N}$-separating. We use $\mathcal{E}_1$ to establish a simple example of what has become known as the "invariance principle."

Theorem 8.20. *If there is one sequence* $X_1^*, X_2^*, \ldots$ *of independent, identically distributed random variables,* $EX_1^* = 0$, $E(X_1^*)^2 = \sigma^{*2} < \infty$, *such that*

$$\frac{X_1^* + \cdots + X_n^*}{\sigma^*\sqrt{n}} \xrightarrow{\mathfrak{D}} X,$$

then for all sequences $X_1, X_2, \ldots$ *of independent, identically distributed random variables such that* $EX_1 = 0$, $EX_1^2 = \sigma^2 < \infty$,

$$\frac{X_1 + \cdots + X_n}{\sigma\sqrt{n}} \xrightarrow{\mathcal{D}} X.$$

Proof. Let $f \in \mathcal{E}_1$, and define

$$\delta(h) = \sup_{|x-y| \le h} |f''(x) - f''(y)|.$$

By definition $\lim \delta(h) = 0$ as $h \downarrow 0$. We may as well assume $\sigma^* = \sigma = 1$, otherwise we use X_k^*/σ^*, X_k/σ. Let

$$Z_n = \frac{X_1 + \cdots + X_n}{\sqrt{n}}, \qquad Z_n^* = \frac{X_1^* + \cdots + X_n^*}{\sqrt{n}}.$$

Since $EZ_n^2 = 1$, $E(Z_n^*)^2 = 1$, both sequences are mass-preserving. By 8.15

$$Ef(Z_n^*) \to Ef(X), \quad \text{all } f \in \mathcal{E}_1,$$

and by 8.16 it suffices to show that

$$Ef(Z_n) \to Ef(X), \quad \text{all } f \in \mathcal{E}_1.$$

Since only the distributions are relevant here, we can assume that $\mathbf{X}^*$, $\mathbf{X}$ are defined on a common sample space and are independent of each other. Now write

$$f(Z_n) - f(Z_n^*) = f\left(\frac{X_1 + \cdots + X_{n-1} + X_n}{\sqrt{n}}\right) - f\left(\frac{X_1 + \cdots + X_{n-1} + X_n^*}{\sqrt{n}}\right)$$

$$+ f\left(\frac{X_1 + \cdots + X_{n-1} + X_n^*}{\sqrt{n}}\right) - f\left(\frac{X_1 + \cdots + X_{n-2} + X_{n-1}^* + X_n^*}{\sqrt{n}}\right)$$

$$+ \cdots + f\left(\frac{X_1 + X_2^* + \cdots + X_n^*}{\sqrt{n}}\right) - f\left(\frac{X_1^* + \cdots + X_n^*}{\sqrt{n}}\right).$$

Define random variables V_k, U_k by

$$V_k = f\left(\frac{X_1 + \cdots + X_k + X_{k+1}^* + \cdots + X_n^*}{\sqrt{n}}\right)$$

$$- f\left(\frac{X_1 + \cdots + X_{k-1} + X_k^* + \cdots + X_n^*}{\sqrt{n}}\right),$$

$$U_k = \frac{X_1 + \cdots + X_{k-1} + X_{k+1}^* + \cdots + X_n^*}{\sqrt{n}}.$$

Use Taylor's expansion around U_k to get

$$\textbf{(8.21)}\quad \mathsf{V}_k = f\left(\mathsf{U}_k + \frac{\mathsf{X}_k}{\sqrt{n}}\right) - f\left(\mathsf{U}_k + \frac{\mathsf{X}_k^*}{\sqrt{n}}\right)$$

$$= \frac{\mathsf{X}_k - \mathsf{X}_k^*}{\sqrt{n}} f'(\mathsf{U}_k) + \frac{\mathsf{X}_k^2}{2n} f''\left(\mathsf{U}_k + \frac{\boldsymbol{\theta}\mathsf{X}_k}{\sqrt{n}}\right)$$

$$- \frac{\mathsf{X}_k^{*^2}}{2n} f''\left(\mathsf{U}_k + \frac{\boldsymbol{\theta}^*\mathsf{X}_k^*}{\sqrt{n}}\right),$$

where $\boldsymbol{\theta}, \boldsymbol{\theta}^*$ are random variables such that $0 \le \boldsymbol{\theta}, \boldsymbol{\theta}^* \le 1$. Both X_k, X_k^* are independent of U_k, so

$$E[(\mathsf{X}_k - \mathsf{X}_k^*)f'(\mathsf{U}_k)] = 0, \qquad (|f'(x)| \le c_0 + c_1|x| \Rightarrow E|f'(\mathsf{U}_k)| < \infty).$$

$$E\left(\mathsf{X}_k^2 f''\left(\mathsf{U}_k + \frac{\boldsymbol{\theta}\mathsf{X}_k}{\sqrt{n}}\right)\right) = (E\mathsf{X}_k^2)(Ef''(\mathsf{U}_k)) + \alpha E\left(\mathsf{X}_k^2\delta\left(\frac{|\mathsf{X}_k|}{\sqrt{n}}\right)\right), \qquad |\alpha| \le 1.$$

Let $h_n(x) = x^2\,\delta(|x|/\sqrt{n})$. Take the expectation of (8.21) and use $E\mathsf{X}_k^2 = E(\mathsf{X}_k^*)^2$ to get

$$|E\mathsf{V}_k| \le (1/2n)[Eh_n(\mathsf{X}_k) + Eh_n(\mathsf{X}_k^*)] = (1/2n)[Eh_n(\mathsf{X}_1) + Eh_n(\mathsf{X}_1^*)],$$

this latter by the identical distribution. Note that

$$f(\mathsf{Z}_n) - f(\mathsf{Z}_n^*) = \mathsf{V}_n + \cdots + \mathsf{V}_1,$$

so

$$|Ef(\mathsf{Z}_n) - Ef(\mathsf{Z}_n^*)| \le \tfrac{1}{2}Eh_n(\mathsf{X}_1) + \tfrac{1}{2}Eh_n(\mathsf{X}_1^*)$$

Let $M = \sup_x |f''(x)|$; then $\delta(h) \le 2M$, all h, so $h_n(\mathsf{X}_1) \le 2M\mathsf{X}_1^2$. But $h_n(\mathsf{X}_1) \to 0$ a.s. Since X_1^2 is integrable, the bounded convergence theorem yields $Eh_n(\mathsf{X}_1) \to 0$. Similarly for $Eh_n(\mathsf{X}_1^*)$. Thus, it has been established that

$$|Ef(\mathsf{Z}_n) - Ef(\mathsf{Z}_n^*)| \to 0$$

implying

$$Ef(\mathsf{Z}_n) \to Ef(\mathsf{X}). \qquad \text{Q.E.D.}$$

This proof is anachronistic in the sense that there are much simpler methods of proving the central limit theorem if one knows some more probability theory. But it is an interesting proof. We know that if we take $\mathsf{X}_1^*, \mathsf{X}_2^*, \ldots$ to be fair coin-tossing variables, that

$$\frac{\mathsf{X}_1^* + \cdots + \mathsf{X}_n^*}{\sqrt{n}} \to \mathcal{N}(0, 1),$$

where the notation $\mathcal{N}(0,1)$ is clarified by

Definition 8.22. *The normal distribution with mean μ and variance σ^2, denoted $\mathcal{N}(\mu, \sigma)$, is the distribution of a random variable $\sigma \mathsf{X} + \mu$, where*

$$P(\mathsf{X} < x) = \frac{1}{\sqrt{2\pi}} \int_{-\infty}^{x} e^{-t^2/2}\, dt.$$

So we have proved

Corollary 8.23. *Let $\mathsf{X}_1, \mathsf{X}_2, \ldots$ be independent, identically distributed random variables, $E\mathsf{X}_1 = 0$, $E\mathsf{X}_1^2 = \sigma^2 < \infty$. Then*

$$\frac{\mathsf{X}_1 + \cdots + \mathsf{X}_n}{\sigma\sqrt{n}} \xrightarrow{\mathcal{D}} \mathcal{N}(0, 1).$$

7. CHARACTERISTIC FUNCTIONS AND THE CONTINUITY THEOREM

The class of functions of the form $\{e^{iux}\}$, $u \in R^{(1)}$, is particularly important and useful in studying convergence in distribution. To begin with

Theorem 8.24. *The set of all complex exponentials $\{e^{iux}\}$, $u \in R^{(1)}$, is $\mathcal{N}$-separating.*

Proof. Let $\int e^{iux}\, dF = \int e^{iux}\, dG$, all u. Then for α_k, $k = 1, \ldots, m$ any complex numbers, and $u_1, \ldots, u_m$ real,

$$\int \Big(\sum_k \alpha_k e^{iu_k x}\Big)\, dF = \int \Big(\sum_k \alpha_k e^{iu_k x}\Big)\, dG. \tag{8.25}$$

Let f_0 be in $\mathcal{E}_0$, let $\epsilon_n \downarrow 0$, $\epsilon_n \le 1$, and consider the interval $[-n, +n]$. Any continuous function on $[-n, +n]$ equal at endpoints can be uniformly approximated by a trigonometric polynomial; that is, there exists a finite sum

$$f_n = \sum_k a_k e^{i\pi kx/n}$$

such that $|f_0(x) - f_n(x)| \le \epsilon_n$, $x \in [-n, +n]$. Since f_n is periodic, and $\epsilon_n \le 1$, then for all n, $\sup_x |f_n(x)| \le 2$. By (8.25) above $\int f_n\, dF = \int f_n\, dG$. This gives $\int f_0\, dF = \int f_0\, dG$ or $F = G$.

Definition 8.26. *Given a distribution function $F(x)$, its characteristic function $f(u)$ is a complex-valued function defined on $R^{(1)}$ by*

$$f(u) = \int e^{iux} F(dx).$$

If F is the distribution function of the random variable X, then equivalently,

$$f(u) = E\, e^{iu\mathsf{X}}.$$

Note quickly that

Proposition 8.27. *Any characteristic function $f(u)$ has the properties*

i) $f(0) = 1$,
ii) $|f(u)| \leq 1$,
iii) *$f(u)$ is uniformly continuous on $R^{(1)}$,*
iv) $f(-u) = \bar{f}(u)$.

Proof

i) Obvious;

ii) $$\left|\int e^{iux} dF\right| \leq \int |e^{iux}|\, dF = 1;$$

iii) $$|f(u + h) - f(u)| = \left|\int e^{iux}(e^{ixh} - 1)\, dF\right| \leq \int |e^{ixh} - 1 \,|\, dF = \delta(h),$$

by the bounded convergence theorem $\delta(h) \to 0$ as $h \to 0$;

iv) $f(-u) = \bar{f}(u)$ is obvious.

Theorem 8.24 may be stated as: No two distinct distribution functions have the same characteristic function. However, examples are known (see Loève [108, p. 218]) of distribution functions $F_1 \neq F_2$ such that $f_1(u) = f_2(u)$ for all u in the interval $[-1, +1]$. Consequently, the set of functions $\{e^{iux}\}$, $-1 \leq u \leq +1$, is not $\mathcal{N}$-separating.

The condition that $F_n \xrightarrow{\mathfrak{D}} F$ can be elegantly stated in terms of the associated characteristic functions.

Theorem 8.28 **(*The continuity theorem*).** *If F_n are distribution functions with characteristic functions $f_n(u)$ such that*

a) *$\lim_n f_n(u)$ exists for every u, and*

b) *$\lim_n f_n(u) = h(u)$ is continuous at $u = 0$,*

then there is a distribution function F such that $F_n \xrightarrow{\mathfrak{D}} F$ and $h(u)$ is the characteristic function of F.

Proof. Since $\{e^{iux}\}$ is $\mathcal{N}$-separating and $\lim_n \int e^{iux}\, dF_n$ exists for every member of $\{e^{iux}\}$, by 8.15, all we need to do is show that $\{F_n\}$ is mass-preserving. To do this, we need

Proposition 8.29. *There exists a constant α, $0 < \alpha < \infty$, such that for any distribution function F with characteristic function f, and any $u > 0$,*

$$F\left(\left[-\frac{1}{u}, +\frac{1}{u}\right]^c\right) \leq \frac{\alpha}{u}\int_0^u (1 - Rlf(v))\, dv.$$

Proof. $Rlf(u) = \int \cos ux\, F(dx)$, so

$$\frac{1}{u}\int_0^u [1 - Rlf(v)]\, dv = \frac{1}{u}\int_0^u \int (1 - \cos vx) F(dx)\, dv$$
$$= \int \left[\frac{1}{u}\int_0^u (1 - \cos vx)\, dv\right] F(dx)$$
$$= \int \left(1 - \frac{\sin ux}{ux}\right) F(dx)$$
$$\geq \int_{\{|xu| \geq 1\}} \left(1 - \frac{\sin ux}{ux}\right) F(dx)$$
$$\geq \inf_{|t|\geq 1} \left(1 - \frac{\sin t}{t}\right) F\left(\left[-\frac{1}{u}, +\frac{1}{u}\right]^c\right).$$

Letting

$$\alpha = \left[\inf_{|t|\geq 1}\left(1 - \frac{\sin t}{t}\right)\right]^{-1}$$

does it.

Now back to the main theorem. By the above inequality,

$$\overline{\lim}\, F_n\left(\left[-\frac{1}{u}, +\frac{1}{u}\right]^c\right) \leq \frac{\alpha}{u}\, \overline{\lim} \int_0^u [1 - Rlf_n(v)]\, dv.$$

The bounded convergence theorem gives

$$\overline{\lim}\, F_n\left(\left[-\frac{1}{u} + \frac{1}{u}\right]^c\right) \leq \frac{\alpha}{u}\int_0^u (1 - Rlh(v))\, dv.$$

Now $f_n(0) = 1 \Rightarrow h(0) = 1$. By continuity of h at zero, $\lim_{v\to 0} Rlh(v) = 1$. Therefore,

$$\lim_{u\to 0} \overline{\lim}\, F_n\left(\left[-\frac{1}{u}, +\frac{1}{u}\right]^c\right) = 0.$$

By this, for any $\epsilon > 0$ we may take a so that $\overline{\lim}\, F_n([-a, +a]^c) < \epsilon/2$. So there is an n_0 such that for $n > n_0$, $F_n([-a, +a]^c) < \epsilon$. Take $b > a$ such that $F_k([-b, +b]^c) < \epsilon$ for $k = 1, 2, \ldots, n_0$. From these together $\sup_n F_n([-b, +b]^c) < \epsilon$. Q.E.D.

Corollary 8.30. *Let F_n be distribution functions, f_n their characteristic functions If there is a distribution function F with characteristic function f such that $\lim_n f_n(u) = f(u)$ for every u, then $F_n \xrightarrow{\mathfrak{D}} F$.*

Proof. Obvious from 8.28.

Clearly, if $F_n \xrightarrow{\mathfrak{D}} F$, then the characteristic functions $f_n(u)$ converge at every point u to $f(u)$. We strengthen this to

Proposition 8.31. *If $F_n \xrightarrow{\mathfrak{D}} F$, then the corresponding characteristic functions $f_n(u)$ converge uniformly to $f(u)$ on every finite interval I. (Denote this kind of convergence by $\xrightarrow{uc}$).*

Proof. This result follows from the fact that the f_n, f form an equicontinuous family. That is, if we fix a finite interval I, then for any n, u, and h,

$$|f_n(u + h) - f_n(u)| \leq \left| \int_I (e^{i(u+h)x} - e^{iux})\, dF_n(x) \right| + 2F_n(I^c)$$
$$\leq \sup_{x \in I} |e^{ihx} - 1| + 2F_n(I^c).$$

Thus, since the $\{F_n\}$ are mass-preserving,

$$\sup_n |f_n(u + h) - f_n(u)| \leq \delta(h),$$

where $\delta(h) \downarrow 0$ as $h \downarrow 0$. Now the usual argument works: Divide I up into points $u_1, u_2, \ldots, u_m$ such that $|u_{k+1} - u_k| \leq h$. For $u \in I$,

$$|f_n(u) - f(u)| \leq |f_n(u) - f_n(u_k)| + |f(u) - f(u_k)| + |f_n(u_k) - f(u_k)|,$$

where u_k is the point of the partition nearest u. Therefore

$$\overline{\lim_n} \sup_{u \in I} |f_n(u) - f(u)| \leq 2\delta(h),$$

because $f(u)$ also satisfies $|f(u + h) - f(u)| \leq \delta(h)$. Taking $h \to 0$ now completes the proof.

The continuity theorem gives us a strong basic tool. Now we start reaping limit theorems from it by using some additional technical details.

Problems

16. A random variable X has a symmetric distribution if $P(\mathsf{X} \in B) = P(\mathsf{X} \in -B)$, where $-B = \{-x;\ x \in B\}$. Prove that the characteristic function of X is real for all u iff X has a symmetric distribution.

17. A natural question is, what continuous complex-valued functions $f(u)$ on $R^{(1)}$ are characteristic functions? Say that such a function is nonnegative definite if for any complex numbers $\lambda_1, \ldots, \lambda_n$, and points $u_1, \ldots, u_n \in R^{(1)}$,

$$\sum_{i,j=1}^{n} f(u_i - u_j)\lambda_i\bar{\lambda}_j \geq 0.$$

A complete answer to the question is given by the following theorem.

Bochner's Theorem. *Let $f(u)$ be continuous on $R^{(1)}$, $f(0) = 1$. Then f is a characteristic function if and only if it is nonnegative definite.*

Prove that if f is a characteristic function, then it is nonnegative definite. (See Loève [108, pp. 207 ff.] for a proof of the other direction.)

18. Find the characteristic function for a Poisson distribution with parameter λ.

19. Find the characteristic function of S_n for coin-tossing.

20. If $Y = aX + b$, show that

$$f_Y(u) = e^{iub} f_X(au).$$

21. A random variable X is called a displaced lattice random variable if there are numbers a, d such that

$$\sum_{n=-\infty}^{+\infty} P(X = nd + a) = 1.$$

Show that X is a displaced lattice if and only if there is a $u \neq 0$ such that $|f_X(u)| = 1$. If u_1, u_2 are irrational with respect to each other, and $|f_X(u_1)| = |f_X(u_2)| = 1$, show that X is a.s. constant, hence $|f_X(u)| \equiv 1$. Show that X is distributed on a lattice L_d, $d > 0$ iff there is a $u \neq 0$ such that $f_X(u) = 1$.

8. THE CONVERGENCE OF TYPES THEOREM

Look at the question: Suppose that $X_n \xrightarrow{\mathfrak{D}} X$, and X is nondegenerate. Can we find constants a_n, b_n such that $a_nX_n + b_n \xrightarrow{\mathfrak{D}} X'$ where X' has a law not connected with that of X in any reasonable way? For example, if $X_1, X_2, \ldots$ are independent and identically distributed, $EX_1 = 0$, $EX_1^2 < \infty$, can we find constants λ_n such that S_n/λ_n $\mathfrak{D}$-converges to something not $\mathcal{N}(\mu, \sigma)$? And if S_n/λ_n $\mathfrak{D}$-converges, what can be said about the size of λ_n compared with $\sqrt{n}$, the normalizing factor we have been using? Clearly, we cannot get the result that $X_n \xrightarrow{\mathfrak{D}} X$, $a_nX_n + b_n \xrightarrow{\mathfrak{D}} X'$ implies $\lim a_n = a$ exists, because if X_n has a symmetric distribution, then $X_n \xrightarrow{\mathfrak{D}} X \Rightarrow (-1)^n X_n \xrightarrow{\mathfrak{D}} X$, since $-X_n$ and X_n have the same law. But if we rule this out by requiring $a_n > 0$, then the kind of result we want holds.

Theorem 8.32 (Convergence of types theorem). *Let $X_n \xrightarrow{\mathfrak{D}} X$, and suppose there are constants $a_n > 0$, b_n such that $a_nX_n + b_n \xrightarrow{\mathfrak{D}} X'$, where X and X' are nondegenerate. Then there are constants a, b such that $\mathfrak{L}(X') = \mathfrak{L}(aX + b)$ and $b_n \to b$, $a_n \to a$.*

Proof. Use characteristic functions and let $f_n = f_{X_n}$ so that

$$f_{a_nX_n+b_n}(u) = e^{iub_n} f_n(a_nu).$$

By 8.31, if f', f are the characteristic functions of X', X respectively, then

$$e^{iub_n} f_n(a_n u) \xrightarrow{uc} f'(u), \qquad f_n(u) \xrightarrow{uc} f(u).$$

Take n_m such that $a_{n_m} \to a$, where a may be infinite. Since

$$|f_n(a_n u)| \xrightarrow{uc} |f'(u)|, \ |f_n(u)| \xrightarrow{uc} |f(u)|,$$

if $a_n \to \infty$, substitute $v_n = u/a_n$, $u \in I$, to get

$$|f_n(u)| = |f_n(a_n v_n)| \to |f'(0)|.$$

Thus $|f(u)| \equiv 1$, implying X degenerate by Problem 21. Hence a is finite. Using uc-convergence $|f_n(a_n u)| \to |f(au)|$; thus $|f'(u)| = |f(au)|$. Suppose $a_{n_m} \to a$, $a_{n'_m} \to a'$ and $a \neq a'$. Use $|f(au)| = |f(a'u)|$, assume $a' < a$, so $|f(u)| = |f((a'/a)u)| = \cdots = |f((a'/a)^N u)|$ by iterating N times. Let $N \to \infty$ to get the contradiction $|f(u)| \equiv 1$. Thus there is a unique $a \geq 0$ such that $a_n \to a$. So $f_n(a_n u) \to f(au)$. Hence e^{iub_n} must converge for every u such that $f(au) \neq 0$, thus in some interval $|u| < \delta$. Obviously then, $\overline{\lim}\, |b_n| < \infty$, and if b, b' are two limit-points of b_n, then $e^{iub} = e^{iub'}$ for all $|u| < \delta$, which implies $b = b'$. Thus $b_n \to b$, $e^{iub_n} \to e^{iub}$, and $f'(u) = e^{iub} f(au)$.

9. CHARACTERISTIC FUNCTIONS AND INDEPENDENCE

The part that is really important and makes the use of characteristic functions so natural is the multiplicative property of the complex exponentials and the way that this property fits in with the independence of random variables.

Proposition 8.33. *Let $X_1, X_2, \ldots, X_n$ be random variables with characteristic functions $f_1(u), \ldots, f_n(u)$. The random variables are independent iff for all $u_1, \ldots, u_n$,*

$$E\left(\exp\left[i \sum_1^n u_k X_k\right]\right) = \prod_1^n f_k(u_k).$$

Proof. Suppose X, Y are independent random variables and f, g are complex-valued measurable functions, $f = f_1 + if_2$, $g = g_1 + ig_2$, and f_1, f_2, g_1, g_2 are $\mathcal{B}_1$-measurable. Then I assert that if $E|f(X)| < \infty$, $E|g(Y)| < \infty$,

$$E(f(X)g(Y)) = (Ef(X))(Eg(Y)), \tag{8.34}$$

so splitting into products does carry over to complex-valued functions. To show this, just verify

$$\begin{aligned} &E[(f_1(X) + if_2(X))(g_1(Y) + ig_2(Y))] \\ &\qquad = Ef_1(X)g_1(Y) + iEf_2(X)g_1(Y) + iEf_1(X)g_2(Y) - Ef_2(X)g_2(Y). \end{aligned}$$

All the expectations are those of real-valued functions. We apply the ordinary result to each one and get (8.34). Thus, inducing up to n variables, conclude that if $X_1, X_2, \ldots, X_n$ are independent, then

$$E\left(\prod_1^n e^{iu_kX_k}\right) = \prod_1^n Ee^{iu_kX_k}.$$

To go the other way, we make use of a result which will be proved in Chapter 11. If we consider the set of functions on $R^{(n)}$,

$$\{e^{i(u_1x_1+\cdots+u_nx_n)}\}, \quad \text{all } (u_1, \ldots, u_n),$$

then these separate the n-dimensional distribution functions. Let $F_n(\mathbf{x})$ be the distribution function of $X_1, \ldots, X_n$; then the left-hand side of the equation in proposition 8.33 is simply

$$\int \exp\,[i(u_1x_1 + \cdots + u_nx_n)]F_n(\mathbf{dx}).$$

But the right-hand side is the integral of

$$e^{i(u_1x_1+\cdots+u_nx_n)}$$

with respect to the distribution function $\prod_1^n F_k(x_k)$. Hence $F(x_1, \ldots, x_n) = \prod_1^n F_k(x_k)$, thus establishing independence.

Notation. *To keep various variables and characteristic functions clear, we denote by* $f_X(u)$ *the characteristic function of the random variable* X.

Corollary 8.35. *If* $X_1, \ldots, X_n$ *are independent, then the characteristic function of* $S_n = X_1 + \cdots + X_n$ is given by

$$f_{S_n}(u) = \prod_1^n f_{X_k}(u).$$

The proof is obvious.

See that X_1, X_2 independent implies that $Ee^{iu(X_1+X_2)} = f_{X_1}(u)f_{X_2}(u)$. But having this hold for all u is not sufficient to guarantee that X_1, X_2 are independent. (See Loève [108, p. 263, Example 1].)

Recall that in Chapter 3, we got the result that if $X_1, X_2, \ldots$ are independent, $\sum_1^n X_k$ converges a.s. iff $\sum_1^n X_k$ converges in probability, hence iff $\sum_m^n X_k \xrightarrow{P} 0$. The one obvious time that $\xrightarrow{\mathfrak{D}}$ and $\xrightarrow{P}$ coincide is when $Y_n \xrightarrow{P} c \Leftrightarrow Y_n \xrightarrow{\mathfrak{D}} \mathfrak{L}(\text{degenerate at } c)$. This observation will lead to

Proposition 8.36. *For* $X_1, X_2, \ldots$ *independent,* $\sum_1^n X_k \xrightarrow{\text{a.s.}}$ *iff* $\sum_1^n X_k \xrightarrow{\mathfrak{D}}$,

because for degenerate convergence, we can prove the following proposition.

Proposition 8.37. *If* Y_n *are random variables with characteristic functions* $f_n(u)$, *then* $Y_n \xrightarrow{\mathfrak{D}} 0$ *iff* $f_n(u) \to 1$ *in some neighborhood of the origin.*

Proof. One way is obvious: $Y_n \xrightarrow{\mathfrak{D}} 0$ implies $f_n(u) \to 1$ for all u. Now let $f_n(u) \to 1$ in $[-\delta, +\delta]$. Proposition 8.29 gives

$$F_n\left(\left[-\frac{1}{\delta}, +\frac{1}{\delta}\right]^c\right) \le \frac{\alpha}{\delta}\int_0^\delta (1 - Rlf_n(v))\, dv.$$

The right-hand side goes to zero as $n \to \infty$, so the F_n are mass-preserving. Let n' be any subsequence such that $F_{n'} \xrightarrow{\mathfrak{D}} F$. Then the characteristic function of F is identically one in $[-\delta, +\delta]$, hence F is degenerate at zero. By 8.7, the full sequence F_n converges to the law degenerate at zero.

This gives a criterion for convergence based on characteristic functions. Use the notation $f_k(u) = f_{X_k}(u)$.

Theorem 8.38. $\sum_1^n X_k \xrightarrow{\text{a.s.}}$ *iff* $\prod_1^\infty f_k(u)$ *converges to* $h(u)$ *in some neighborhood* N *of the origin, and* $|h(u)| > 0$ *on* N.

Proof. Certainly $\sum_1^n X_k \xrightarrow{\text{a.s.}}$ implies $\prod_1^\infty f_k(u)$ converges everywhere to a characteristic function. To go the other way, the characteristic function of $\sum_m^n X_k$ is $\prod_{k=m}^n f_k(u)$. Because $\prod_1^n f_k(u) \to h(u) \neq 0$ on N, $\prod_m^n f_k(u) \to 1$ on N. Use 8.37 to complete the proof, and note that 8.36 is a corollary.

Problems

22. For $Y_1, Y_2, \ldots$ independent and ± 1 with probability $\frac{1}{2}$, use 8.38 to show that $\Sigma\, c_k Y_k$ converges a.s. $\Leftrightarrow \Sigma\, c_k^2 < \infty$.

23. Show that the condition on $f_k(u)$ in 8.38 can be partly replaced by—if $\sum_1^\infty |1 - f_k(u)|$ converges in some neighborhood N of the origin, then $\sum_1^n X_k \xrightarrow{\text{a.s.}}$.

10. FOURIER INVERSION FORMULAS

To every characteristic function corresponds one and only one distribution function. Sometimes it is useful to know how, given a characteristic function, to find the corresponding distribution function, although by far the most important facts regarding characteristic functions do not depend on knowing how to perform this inversion. The basic inversion formula is the Fourier transform inversion formula. There are a lot of different versions of this; we give one particularly useful version.

Theorem 8.39. *Let $f(u)$ be the characteristic function of a distribution function $F(dx)$ such that*

$$\int |f(u)|\, du < \infty.$$

Then $F(dx)$ has a bounded continuous density $h(x)$ with respect to Lebesgue measure given by

$$h(x) = \frac{1}{2\pi}\int e^{-iux} f(u)\, du. \tag{8.40}$$

Proof. Assume that (8.40) holds true for one distribution function $G(dx)$ with density $g(x)$ and characteristic function $\varphi(u)$. Then we show that it holds true in general. Write

$$\frac{1}{2\pi}\int f(u)\varphi(u)e^{-iuy}\, du = \frac{1}{2\pi}\int e^{iux}e^{-iuy}\varphi(u)\, dF(x)\, du.$$

Then, interchanging order of integration on the right:

$$\frac{1}{2\pi}\int f(u)\varphi(u)e^{-iuy}\, du = \int g(y-x)\, dF(x) \tag{8.41}$$

If X has distribution $F(dx)$, and Y has distribution $G(dx)$, then the integral on the right is the density for the distribution of $\mathsf{X} + \mathsf{Y}$ where they are taken to be independent. Instead of Y, now use $\epsilon\mathsf{Y}$, in (8.41), because if the distribution of Y satisfies (8.40), you can easily verify that so does that of $\epsilon\mathsf{Y}$, for ϵ any real number. As $\epsilon \to 0$ the characteristic function $\varphi_\epsilon(u)$ of $\epsilon\mathsf{Y}$ converges to one everywhere. Use the bounded convergence theorem to conclude that the left-hand side of (8.41) converges to

$$h(y) = \frac{1}{2\pi}\int f(u)e^{-iuy}\, du.$$

The left-hand side is bounded by $\int |f(u)|\, du$ for all y, so the integral of the left-hand side over any finite interval I converges to

$$\int_I h(y)\, dy.$$

If the endpoints of I are continuity points of $F(x)$, then since $\mathcal{L}(\mathsf{X} + \epsilon\mathsf{Y}) \to \mathcal{L}(\mathsf{X})$, the right-hand side of (8.41) converges to $F(I)$. Thus the two measures $F(B)$ and $\int_B h(y)\, dy$ on $\mathcal{B}_1$ agree on all intervals, therefore are identical. The continuity and boundedness of $h(x)$ follows directly from the expression (8.40). To conclude, all I have to do is produce one $G(x)$, $\varphi(u)$ for which (8.40) holds. A convenient pair is

$$g(x) = \tfrac{1}{2}e^{-|x|}, \qquad \varphi(u) = \frac{1}{1+u^2}. \tag{8.42}$$

To verify (8.42) do a straightforward contour integration.

We can use the same method to prove

Proposition 8.43. *Let $\varphi_n(u)$ be any sequence of characteristic functions converging to one for all u such that for each n,*

$$\int |\varphi_n(u)|\, du < \infty.$$

If b and a are continuity points of any distribution function $F(x)$, with characteristic function $f(u)$, then

$$F(b) - F(a) = \lim_n \frac{1}{2\pi} \int \left(\frac{e^{-iub} - e^{-iua}}{-iu} \right) \varphi_n(u) f(u)\, du$$

Proof. Whether or not F has a density or $f(u)$ is integrable, (8.41) above still holds, where now the right-hand side is the density of the distribution of $X + Y_n$, X, Y_n independent, $\varphi_n(u)$ the characteristic function of Y_n. Since $\varphi_n(u) \to 1$, $Y_n \xrightarrow{P} 0$, $\mathcal{L}(X + Y_n) \to \mathcal{L}(X)$. The integral of the right-hand side over $[a, b)$ thus converges to $F(b) - F(a)$. The integral of the left-hand side is

$$\frac{1}{2\pi} \int \left(\frac{e^{-iub} - e^{-iua}}{-iu} \right) \varphi_n(u) f(u)\, du.$$

This all becomes much simpler if X is distributed on the lattice L_d, $d > 0$. Then

$$f(u) = \sum_{n=-\infty}^{+\infty} e^{iund} P(X = nd),$$

so that $f(u)$ has period $2\pi/d$. The inversion formula is simply

$$P(X = nd) = \frac{d}{2\pi} \int_{-\pi/d}^{+\pi/d} e^{-iund} f(u)\, du$$

Problem 24. Let $X_1, X_2, \ldots$ be independent, identically distributed integer-valued random variables. Show that their sums are recurrent iff

$$\lim_{r \uparrow 1} \int_{-\pi}^{+\pi} \frac{1}{1 - rf(u)}\, du = \infty,$$

where $f(u)$ is the common characteristic function of $X_1, X_2, \ldots$

11. MORE ON CHARACTERISTIC FUNCTIONS

There are some technical results concerning characteristic functions which we will need later. These revolve around expansions, approximation, and similar results.

Proposition 8.44. *If $E\,|\mathsf{X}|^k < \infty$, then the characteristic function of* X *has the expansion*

$$f(u) = \sum_{j=0}^{k-1} \frac{(iu)^j}{j!} E\mathsf{X}^j + \frac{(iu)^k}{k!}(E\mathsf{X}^k + \delta(u)),$$

where $\delta(u)$ denotes a function of u, such that $\lim_{u\to 0} \delta(u) = 0$, *and* satisfying $|\delta(u)| \le 3E\,|\mathsf{X}|^k$ *for all u.*

Proof. Use the Taylor expansion with remainder on $\sin y$, $\cos y$ for y real to get

$$\cos y + i \sin y = \sum_0^{k-1} \frac{(iy)^j}{j!} + \frac{(iy)^k}{k!}(\cos \theta_1 y + i \sin \theta_2 y),$$

where θ_1, θ_2 are real numbers such that $|\theta_1| \le 1$, $|\theta_2| \le 1$. Thus

$$e^{iu\mathsf{X}} = \sum_0^k \frac{(iu\mathsf{X})^j}{j!} + \frac{(iu\mathsf{X})^k}{k!}[\cos(\boldsymbol{\theta}_1 u\mathsf{X}) + i \sin(\boldsymbol{\theta}_2 u\mathsf{X}) - 1].$$

Now $\boldsymbol{\theta}_1, \boldsymbol{\theta}_2$ are random, but still $|\boldsymbol{\theta}_1| \le 1$, $|\boldsymbol{\theta}_2| \le 1$. Now

$$|\mathsf{X}^k[\cos(\boldsymbol{\theta}_1 u\mathsf{X}) - 1 + i \sin(\boldsymbol{\theta}_2 u\mathsf{X})]| \le 3\,|\mathsf{X}|^k,$$

which establishes $|\delta(u)| \le 3E\,|\mathsf{X}|^k$. Use the bounded convergence theorem to get

$$\lim_{u\to 0} \delta(u) = \lim_{u\to 0} E\,|\mathsf{X}^k[\cos(\boldsymbol{\theta}_1 u\mathsf{X}) - 1 + i \sin(\boldsymbol{\theta}_2 u\mathsf{X})]| = 0.$$

Another point that needs discussion is the logarithm of a complex number. For z complex, $\log z$ is a many-valued function defined by

(8.45) $$z = e^{\log z}.$$

For any determination of $\log z$, $\log z + 2n\pi i$, $n = 0, \pm 1, \ldots$ is another solution of (8.45). Write $z = re^{i\theta}$; then $\log z = \log r + i\theta$. We always will pick that determination of θ which satisfies $-\pi < \theta \le \pi$, unless we state otherwise. With this convention, $\log z$ is uniquely determined.

Proposition 8.46. *For z complex,*

$$\log(1 + z) = z(1 + \epsilon(z)),$$

where $|\epsilon(z)| \le |z|$ for $|z| \le \frac{1}{2}$.

Proof. For $|z| < 1$, the power series expansion is

$$\log(1 + z) = z - \frac{z^2}{2} + \frac{z^3}{3} - \frac{z^4}{4} + \cdots = z\left(1 - \frac{z}{2} + \frac{z^2}{3} - \cdots\right).$$

For $|z| \le \frac{1}{2}$,

$$\left|-\frac{z}{2} + \frac{z^2}{3} \cdots\right| \le \frac{|z|}{2}(1 + |z| + |z|^2 + \cdots) = \frac{|z|}{2}\frac{1}{1 - |z|} \le |z|.$$

One remark: Given a sequence $f_n(u)$ of characteristic function, frequently we will take $l_n(u) = \log f_n(u)$, and show that $l_n(u) \to \varphi(u)$ for some evaluation of the log function. Now $l_n(u)$ is not uniquely determined.

$$l'_n(u) = l_n(u) + 2\pi i N_n(u),$$

$N_n(u)$ integer-valued, is just as good a version of $\log f_n(u)$. However, if $l_n(u) \to \varphi(u)$, and $\varphi(u)$ is continuous at the origin for one evaluation of $l_n(u)$, then because $f_n(u) = e^{l_n(u)} \to e^{\varphi(u)}$ the continuity theorem is in force.

12. METHOD OF MOMENTS

Suppose that all moments of a sequence of distribution functions F_n exists and for every integer $k \geq 0$, the limit of

$$\int x^k F_n(dx)$$

exists. Does it follow that there is a distribution F such that $F_n \xrightarrow{\mathfrak{D}} F$? Not necessarily! The reason that the answer may be "No" is that the functions x^k do not separate. There are examples [123] of distinct distribution functions F and G such that $\int |x|^k \, dF < \infty$, $\int |x|^k \, dG < \infty$ for all $k \geq 0$, and

$$\int x^k \, dF = \int x^k \, dG, \qquad k = 0, 1, \ldots$$

Start to argue this way: If $\overline{\lim} \int x^2 \, dF_n < \infty$, then (Problem 10) the $\{F_n\}$ are mass-preserving. Take a subsequence $F_{n'} \xrightarrow{\mathfrak{D}} F$. Then (Problem 14)

$$\lim_{n'} \int x^k \, dF_{n'} = \int x^k \, dF,$$

so for the full sequence

(8.47) $$\lim_n \int x^k \, dF_n = \int x^k \, dF.$$

If there is only one F such that (8.47) holds, then every convergent subsequence of F_n converges to F, hence $F_n \xrightarrow{\mathfrak{D}} F$. Thus

Theorem 8.48. *If there is at most one distribution function F such that*

$$\lim_n \int x^k \, dF_n = \int x^k \, dF,$$

then $F_n \xrightarrow{\mathfrak{D}} F$.

The question is now one of uniqueness. Let

$$\mu_k = \lim_n \int x^k \, dF_n.$$

If F is uniquely determined by (8.47), then the moment problem given by the μ_k is said to be determined. In general, if the μ_k do not grow too fast, then uniqueness holds. A useful sufficient condition is

Proposition 8.49. *If*

$$\overline{\lim_k} \frac{|\mu_k|^{1/k}}{k} < \infty,$$

then there is at most one distribution function F satisfying

$$\mu_k = \int x^k F(dx).$$

Proof. Let

$$r = \overline{\lim} \frac{|\mu_k|^{1/k}}{k};$$

then for any $\epsilon > 0$ and $k \geq k_0$, using the even moments to get bounds for the odd moments,

$$\int |x|^k \, dF \leq k^k (r + \epsilon)^k.$$

Hence, by the monotone convergence theorem,

$$\int e^{|x\xi|} \, dF = \lim_n \sum_0^n \frac{|\xi|^k}{k!} \int |x|^k \, dF < \infty,$$

for $|\xi| < 1/re$. Consider

$$\varphi(z) = \int e^{xz} F(dx).$$

By the above, $\varphi(z)$ is analytic in the strip $|Rlz| < 1/re$. For $|z| < 1/re$,

(8.50) $$\varphi(z) = \sum_0^\infty \frac{z^k}{k!} \mu_k$$

This holds for any distribution function F having moments μ_k. Since $\varphi(z)$ in the strip is the analytic continuation of $\varphi(z)$ given by (8.50), then $\varphi(z)$ is completely determined by μ_k. But for $Rlz = 0$, $\varphi(z)$ is the characteristic function and thus uniquely determines F.

13. OTHER SEPARATING FUNCTION CLASSES

For restricted classes of distribution functions, there are separating classes of functions which are sometimes more useful than the complex exponentials. For example, consider only nonnegative random variables; their distribution functions assign zero mass to $(-\infty, 0)$. Call this class of distribution functions $\mathcal{M}^+$.

Proposition 8.51. *The exponentials* $\{e^{-\lambda x}\}$, λ *real and nonnegative, separate in* $\mathcal{M}^+$.

Proof. Suppose F and G are in $\mathcal{M}^+$ and for all $\lambda \geq 0$,

$$\int e^{-\lambda x} F(dx) = \int e^{-\lambda x} G(dx).$$

Then substitute $e^{-x} = y$, so

$$\int_0^1 y^\lambda \, dF(-\log y) = \int_0^1 y^\lambda \, dG(-\log y). \tag{8.52}$$

In particular (8.52) holds for λ ranging through the nonnegative integers. Thus for any polynomial $P(y)$,

$$\int_0^1 P(y) \, dF(-\log y) = \int_0^1 P(y) \, dG(-\log y),$$

hence equality holds for any continuous function on [0, 1]. Use an approximation argument to conclude now that $F \equiv G$.

As before, if $F_n \in \mathcal{M}^+$ and $\int e^{-\lambda x} F_n(dx)$ converges for all $\lambda \geq 0$, then there is at most one distribution function F such that $F_n \xrightarrow{\mathfrak{D}} F$. Let the limit of $\int e^{-\lambda x} \, dF_n(x)$ be $h(\lambda)$. Then by the bounded convergence theorem,

$$\lim_n \int \left(\frac{1 - e^{-\lambda x}}{\lambda x} \right) dF_n(x) = \frac{1}{\lambda} \int_0^\lambda h(\lambda) \, d\lambda.$$

So conclude, just as in the continuity theorem, that if

$$\lim_{\lambda \downarrow 0} h(\lambda) = 1,$$

then the sequence $\{F_n\}$ is mass-preserving. Hence there is a unique distribution function F such that $F_n \xrightarrow{\mathfrak{D}} F$.

For X taking on nonnegative integer values, the *moment-generating* function is defined as

$$\varphi(z) = Ez^{\mathsf{X}},$$

for z complex, $|z| \leq 1$.

Problem 25. Prove that the functions z^x, $|z| \leq 1$ are separating in the class of distribution functions of nonnegative integer-valued random variables. If $\{\mathsf{X}_n\}$ are a set of such random variables and

$$\varphi_n(z) = Ez^{\mathsf{X}_n}$$

converges for all $|z| \leq 1$ to a function continuous at $z = 1$, then show there is a random variable X such that $\mathsf{X}_n \xrightarrow{\mathfrak{D}} \mathsf{X}$.

NOTES

More detailed background on distribution functions, etc., can be found in Loève's book [108]. For material on the moment problem, consult Shohat and Tamarkin [123]. For Laplace transforms of distributions $\int e^{-\lambda x}\, dF(x)$ see Widder [140]. Although the central limit theorem for coin-tossing was proved early in the nineteenth century, a more general version was not formulated and proved until 1901 by Lyapunov [109]. The interesting proof we give in Section 6 is due to Lindeberg [106].

An important estimate for the rate of convergence in the central limit theorem is due to Berry and Eseen (see Loève, [108, pp. 282 ff.]). They prove that there is a universal constant c such that if $S_n = X_1 + \cdots + X_n$ is a sum of independent, identically distributed random variables with $EX_1 = 0$, $EX_1^2 = \sigma^2 < \infty$, $E\,|X_1|^3 < \infty$, and if $\Phi(x)$ is the distribution function of the $\mathcal{N}(0, 1)$ law, then

$$\sup_x \left| P\left(\frac{S_n}{\sigma\sqrt{n}} < x\right) - \Phi(x) \right| \le c\,\frac{E\,|X_1|^3}{\sigma^3\sqrt{n}}.$$

It is known that $c \le 4$, (Le Cam [99]) and unpublished calculations give bounds as low as 2.05. By considering coin-tossing, note that the $1/\sqrt{n}$ rate of convergence cannot be improved upon.

CHAPTER 9

THE ONE-DIMENSIONAL CENTRAL LIMIT PROBLEM

1. INTRODUCTION

We know already that if $X_1, X_2, \ldots$ are independent and identically distributed, $EX_1 = 0$, $EX_1^2 = \sigma^2 < \infty$, then

$$\frac{X_1 + \cdots + X_n}{\sigma\sqrt{n}} \xrightarrow{\mathfrak{D}} \mathcal{N}(0, 1).$$

Furthermore, by the convergence of types theorem, no matter how S_n is normalized, if $S_n/A_n \xrightarrow{\mathfrak{D}}$ then the limit is a normal law or degenerate. So this problem is pretty well solved, with the exception of the question: Why is the normal law honored above all other laws? From here there are a number of directions available; the identically distributed requirement can be dropped. This leads again to a normal limit if some nice conditions on moments are satisfied. So the condition on moments can be dropped; take $X_1, X_2, \ldots$ independent, identically distributed but $EX_1^2 = \infty$. Now a new class of laws enters as the limits of S_n/A_n for suitable A_n, the so-called stable laws.

In a completely different direction is the law of rare events, convergence to a Poisson distribution. But this result is allied to the central limit problem and there is an elegant unification via the infinitely divisible laws. Throughout this chapter, unless explicitly stated otherwise, equations involving logs of characteristic functions are supposed to hold modulo additive multiples of $2\pi i$.

2. WHY NORMAL?

There is really no completely satisfying answer to this question. But consider, if $X_1, X_2, \ldots,$ are independent, identically distributed, and if

$$Z_n = \frac{X_1 + \cdots + X_n}{\sqrt{n}} \xrightarrow{\mathfrak{D}} X$$

what are the properties that X must have? Look at

$$Z_{2n} = \frac{X_1 + \cdots + X_n + X_{n+1} + \cdots + X_{2n}}{\sqrt{2n}}.$$

Now $Z_{2n} \xrightarrow{\mathcal{D}} X$. But

$$Z_{2n} = \frac{X_1 + \cdots + X_n}{\sqrt{2}\sqrt{n}} + \frac{X_{n+1} + \cdots + X_{2n}}{\sqrt{2}\sqrt{n}} = \frac{1}{\sqrt{2}}(Z'_n + Z''_n).$$

The variables Z'_n, Z''_n are independent, and $Z'_n \xrightarrow{\mathcal{D}} X$, $Z''_n \xrightarrow{\mathcal{D}} X$, since they have the same distributions as Z_n. This (we would like to believe) implies that X has the same distribution as $1/\sqrt{2}\,(X' + X'')$, where X', X'' are independent and have the same distribution as X. To verify this, note that

$$Ee^{iuZ_{2n}} = (Ee^{iuZ'_n/\sqrt{2}})(Ee^{iuZ''_n/\sqrt{2}})$$

or

$$f_{2n}(u) = f_n\left(\frac{u}{\sqrt{2}}\right)f_n\left(\frac{u}{\sqrt{2}}\right).$$

Since $f_n(u) \xrightarrow{uc} f(u)$, where $f(u)$ is the characteristic function of X, it follows that $f(u) = f(u/\sqrt{2})^2$. But the right-hand side of this is the characteristic function of $(X' + X'')/\sqrt{2}$. So our expectation is fulfilled. Now the point is:

Proposition 9.1. *If a random variable* X *satisfies* $EX^2 < \infty$, *and*

$$X \overset{\mathcal{D}}{=} \frac{X' + X''}{\sqrt{2}},$$

where X', X'' *are independent and* $\mathcal{L}(X) = \mathcal{L}(X') = \mathcal{L}(X'')$, *then* X *has a* $\mathcal{N}(0, \sigma^2)$ *distribution.*

Proof. The proof is simple. Let $X_1, X_2, \ldots$ be independent, $\mathcal{L}(X_k) = \mathcal{L}(X)$. EX must be zero, since $EX = (EX' + EX'')/\sqrt{2}$ implies $EX = \sqrt{2}EX$. By iteration,

$$X \overset{\mathcal{D}}{=} \frac{X_1 + \cdots + X_n}{\sqrt{n}}, \quad n = 2^m.$$

But the right-hand sums, divided by σ, converge in distribution to $\mathcal{N}(0, 1)$.

Actually, this result holds without the restriction that $EX^2 < \infty$. A direct proof of this is not difficult, but it also comes out of later work we will do with stable laws, so we defer it.

3. THE NONIDENTICALLY DISTRIBUTED CASE

Let $X_1, X_2, \ldots$ be independent. Then

Theorem 9.2. *Let* $EX_k = 0$, $EX_k^2 = \sigma_k^2 < \infty$, $E\,|X_k^3| < \infty$, *and* $s_n^2 = \sum_1^n \sigma_k^2$. *If*

$$\text{(9.3)} \qquad \overline{\lim_n}\, \frac{1}{s_n^3} \sum_1^n E\,|X_k|^3 = 0,$$

then

$$\frac{X_1 + \cdots + X_n}{s_n} \xrightarrow{\mathfrak{D}} \mathcal{N}(0, 1).$$

Proof. Very straightforward and humdrum, using characteristic functions. Let f_k be f_{X_k}, g_n the characteristic function of S_n/s_n. Then

$$g_n(u) = \prod_1^n f_k\left(\frac{u}{s_n}\right) = \prod_1^n \left[1 + \left(f_k\left(\frac{u}{s_n}\right) - 1\right)\right].$$

Using the Taylor expansion, we get from (8.44)

$$f_k\left(\frac{u}{s_n}\right) = 1 - \frac{u^2}{2s_n^2}\left(\sigma_k^2 + \delta_k\left(\frac{u}{s_n}\right)\right),$$

$$\left|f_k\left(\frac{u}{s_n}\right) - 1\right| \leq 2u^2 \frac{\sigma_k^2}{s_n^2}.$$

Now $(E|X_k^2|)^{3/2} < E|X_k|^3$, or $\sigma_k^3 \leq E|X_k|^3$. Then condition 9.3 implies that $\sup_{k\leq n} \sigma_k/s_n \to 0$. So $\sup_{k\leq n} |f_k(u/s_n) - 1| \to 0$ as n goes to infinity. Therefore use the log expansion

$$\log(1 + z) = z(1 + \theta z),$$

where $|\theta| \leq 1$ for $|z| \leq \frac{1}{2}$, to get

$$\log g_n(u) = \sum_1^n \left(f_k\left(\frac{u}{s_n}\right) - 1\right) + \theta \sum_1^n \left(f_k\left(\frac{u}{s_n}\right) - 1\right)^2,$$

where the equality holds modulo $2\pi i$. Consider the second term above,

$$\begin{aligned}\left|\theta \sum_1^n \left(f_k\left(\frac{u}{s_n}\right) - 1\right)^2\right| &\leq \sup_{k\leq n}\left|f_k\left(\frac{u}{s_n}\right) - 1\right| \cdot \sum_1^n \left|f_k\left(\frac{u}{s_n}\right) - 1\right| \\ &\leq \sup_{k\leq n}\left|f_k\left(\frac{u}{s_n}\right) - 1\right| \cdot \sum_1^n 2u^2 \frac{\sigma_k^2}{s_n^2} \\ &\leq 2u^2 \sup_{k\leq n}\left|f_k\left(\frac{u}{s_n}\right) - 1\right|.\end{aligned}$$

This bound goes to zero as $n \to \infty$. Apply the Taylor expansion,

$$f_k\left(\frac{u}{s_n}\right) - 1 = -\frac{u^2}{2}\frac{\sigma_k^2}{s_n^2} + \theta_k \frac{u^3}{s_n^3} E|X_k|^3, \quad |\theta_k| \leq 1,$$

to the first term above to get

$$\sum_1^n \left(f_k\left(\frac{u}{s_n}\right) - 1\right) = -\frac{u^2}{2} + \frac{u^3\theta}{s_n^3}\sum_1^n E|X_k|^3, \quad |\theta| \leq 1,$$

which converges to $-u^2/2$.

We conclude that

$$g_n(n) \to e^{-u^2/2}$$

for every u. Since the theorem holds for identically distributed random variables, it follows that $e^{-u^2/2}$ must be the characteristic function of the $\mathcal{N}(0, 1)$ distribution. Apply the continuity theorem to complete the proof.

Note that we got, in this proof, the additional dividend that if X is $\mathcal{N}(0, 1)$, then

$$Ee^{iuX} = e^{-u^2/2}.$$

4. THE POISSON CONVERGENCE

For $X_1, X_2, \ldots$ independent and identically distributed, $EX_1 = 0$, $EX_1^2 = \sigma^2$, let

$$M_n = \max\left(\frac{|X_1|}{\sqrt{n}}, \ldots, \frac{|X_n|}{\sqrt{n}}\right).$$

Write, for $x > 0$,

$$\begin{aligned} P(M_n < x) &= P(|X_1| < \sqrt{n}x, \ldots, |X_n| < \sqrt{n}x) \\ &= [P(|X_1| < \sqrt{n}x)]^n. \end{aligned}$$

Now

$$P(|X_1| \geq \sqrt{n}\,x) \leq \frac{1}{nx^2}\int_{|y| \geq \sqrt{n}x} y^2\, dF(y) = \frac{1}{n}\,\theta_n(x),$$

where $\lim_n \theta_n(x) = 0$. This leads to

$$P(M_n < x) \geq \left(1 - \frac{1}{n}\,\theta_n(x)\right)^n \approx e^{-\theta_n(x)},$$

the point being that

$$\lim_n P(M_n < x) = 1, \quad \text{all } x,$$

or $M_n \xrightarrow{\mathfrak{D}} 0$. In this case, therefore, we are dealing with sums

$$Z_n = \left(\frac{X_1}{\sqrt{n}}\right) + \cdots + \left(\frac{X_n}{\sqrt{n}}\right) = X_1^{(n)} + \cdots + X_n^{(n)}$$

of independent random variables such that the maximum of the individual summands converges to zero. I have gone through this to contrast it to the situation in which we have a sequence of coins, 1, 2, ..., with probabilities of heads $p_1, p_2, \ldots$, where $p_n \to 0$, and the nth coin is tossed n times. For the nth coin, let $X_k^{(n)}$ be one if heads comes on the kth trial, zero otherwise. So

$$S_n = X_1^{(n)} + \cdots + X_n^{(n)}$$

is the number of heads gotten using the nth coin. Think!—the probability of heads on each individual trial is p_n and that is going to zero. However, the total number of trials is getting larger and larger. Is it possible that S_n converges in distribution? Compute

$$P(S_n = 0) = (1 - p_n)^n.$$

This will converge if and only if $np_n \to \lambda$, $0 \le \lambda < \infty$. If $\lambda = 0$, then $P(S_n = 0) \to 1$, and henceforth rule this case out.

For $\lambda > 0$, $P(M_n = 0) = P(S_n = 0) \to e^{-\lambda}$, so $M_n \overset{\mathcal{D}}{\not\to} 0$. Take characteristic functions, noting that

$$Ee^{iuX_k^{(n)}} = p_n e^{iu} + q_n.$$

For n sufficiently large, these are close to one, for u fixed, and we can write

$$f_{S_n}(u) = \big(1 + p_n(e^{iu} - 1)\big)^n,$$

$$\log f_{S_n}(u) = np_n(e^{iu} - 1) + \theta np_n^2(e^{iu} - 1)^2, \quad |\theta| \le 1.$$

Since $p_n \to 0$, $np_n \to \lambda$, $np_n^2 \to 0$, this gives

$$\log f_{S_n}(u) \to \lambda(e^{iu} - 1),$$

so

Theorem 9.4. $S_n \xrightarrow{\mathcal{D}}$ *if and only if* $np_n \to \lambda$, *then the limit has characteristic function* $\exp[\lambda(e^{iu} - 1)]$. *The limit random variable* X *takes values in* $\{0,1,2,\ldots\}$, *so*

$$Ee^{iuX} = \sum_0^\infty e^{iuk} P(X = k).$$

Expanding,

$$e^{\lambda(e^{iu}-1)} = \sum_0^\infty e^{-\lambda} \cdot \frac{\lambda^k}{k!} e^{iuk},$$

so

$$P(X = k) = \frac{\lambda^k}{k!} e^{-\lambda}.$$

Definition 9.5. *A random variable* X *taking values in* $\{0, a, 2a, 3a, \ldots\}$ *will be said to have Poisson distribution* $\mathcal{P}(\lambda)$, $\lambda \ge 0$ *with jump size* a, *if*

$$P(X = ak) = \frac{\lambda^k}{k!} e^{-\lambda}, \qquad \text{or} \qquad f_X(u) = \exp[\lambda(e^{iua} - 1)].$$

Look now at the $X_1^{(n)}, \ldots, X_n^{(n)}$. Since $P(X_k^{(n)} = 0) = q_n \to 1$, usually the $X_k^{(n)}$ are zero, but once in a while along comes a blip. Again, take $M_n = \max(X_1^{(n)}, \ldots, X_n^{(n)})$. Now M_n can only take the values 0 or 1, and $M_n \not\to 0$ unless $\lambda = 0$. Here the contrast obviously is that M_n must equal 1 with positive probability, or $S_n \xrightarrow{\mathcal{D}} 0$. It is the difference between the sum of uniformly small smears, versus the sum of occasionally large blips. That this is pretty characteristic is emphasized by

Proposition 9.6. *Let* $S_n = X_1^{(n)} + \cdots + X_n^{(n)}$, *where* $X_1^{(n)}, \ldots, X_n^{(n)}$ *are independent and identically distributed. If* $S_n \xrightarrow{\mathcal{D}} X$ *then* $M_n \xrightarrow{\mathcal{D}} 0$ *if and only if* X *is normal.*

Proof. Deferred until Section 7.

The Poisson convergence can be generalized enormously. For example, suppose $S_n = X_1^{(n)} + \cdots + X_n^{(n)}$, the $X_k^{(n)}$ independent and identically distributed with

$$X_k^{(n)} = \begin{cases} 0, & \text{with probability} \quad q_n \\ x_i, & \text{with probability} \quad p_n^{(i)}, \quad i = 1, \ldots, j. \end{cases}$$

and $S_n \xrightarrow{\mathcal{D}} X$. We could again show that this is possible only if $np_n^{(i)} \to \lambda_i$, $0 \le \lambda_i < \infty$, and if so, then

$$f_X(u) = \exp\left(\sum_{i=1}^{j} \lambda_i(e^{iux_i} - 1)\right).$$

Two interesting points are revealed in this result.

First: The expected number of times that $X_k^{(n)} = x_i$ is $np_n^{(i)}$. So λ_i is roughly the expected number of times that one of the summands is x_i.

Second: Since

$$f_X(u) = \prod_{i=1}^{j} \exp\left(\lambda_i(e^{iux_i} - 1)\right),$$

X is distributed as $X_1 + \cdots + X_j$, where the X_i are independent random variables and X_i has Poisson distribution $\mathcal{P}(\lambda_i)$ with jump size x_i. So the jumps do not interact; each jump size x_i contributes an independent Poisson component.

5. THE INFINITELY DIVISIBLE LAWS

To include both Poisson and $\mathcal{N}(0, 1)$ convergence, ask the following question: Let

$$S_n = X_1^{(n)} + \cdots + X_n^{(n)}, \tag{9.7}$$

where the $X_k^{(n)}$, $k = 1, \ldots, n$, *are independent and identically distributed. If* $S_n \xrightarrow{\mathcal{D}} X$, *what are the possible distributions of* X?

S_n is the sum of many independent components; heuristically X must have this same property.

Definition 9.8. *X will be said to have an infinitely divisible distribution if for every n, there are independent and identically distributed random variables $X_1^{(n)}, \ldots, X_n^{(n)}$ such that $\mathcal{L}(X) = \mathcal{L}(X_1^{(n)} + \cdots + X_n^{(n)})$.*

Proposition 9.9. *A random variable X is a limit in distribution of sums of the type (9.7) if and only if it has an infinitely divisible distribution.*

Proof. If X has an infinitely divisible distribution, then by definition there are sums S_n of type (9.7) with distribution exactly equal to X. The other way: Consider

$$S_{2n} = (X_1^{(2n)} + \cdots + X_n^{(2n)}) + (X_{n+1}^{(2n)} + \cdots + X_{2n}^{(2n)}) = Y_n + Y_n'.$$

The random variables Y_n and Y_n' are independent with the same distribution. If $S_n \xrightarrow{\mathcal{D}} X$, the distributions of Y_n are mass-preserving, because

$$(P(Y_n > y))^2 = P(Y_n > y, Y_n' > y) \leq P(S_{2n} > 2y),$$

and similarly,

$$(P(Y_n < -y))^2 \leq P(S_{2n} < -2y).$$

Take a subsequence $\{n'\}$ such that $Y_{n'} \xrightarrow{\mathcal{D}} Y$. Obviously, $f_X(u) = [f_Y(u)]^2$; so $\mathcal{L}(X) = \mathcal{L}(Y + Y')$, Y, Y' independent. This can be repeated to get X equal in distribution to the sum of $Y_1^{(n)} + \cdots + Y_n^{(n)}$ by considering S_{nm}.

If $S_n \xrightarrow{\mathcal{D}} X$, do the components $X_1^{(n)}, \ldots, X_n^{(n)}$ have to get smaller and smaller in any reasonably formulated way? Note that in both the Poisson and $\mathcal{N}(0, 1)$ convergence, for any $\epsilon > 0$, $P(|X_1^{(n)}| \geq \epsilon) \to 0$; that is, $X_1^{(n)} \xrightarrow{\mathcal{D}} 0$ [so, of course, $\sup_k P(|X_k^{(n)}| \geq \epsilon) \to 0$, since these probabilities are the same for all $k = 1, \ldots, n$]. This holds in general.

Proposition 9.10. *If $S_n \xrightarrow{\mathcal{D}} X$, then $X_1^{(n)} \xrightarrow{\mathcal{D}} 0$.*

Proof. Since $f_{S_n}(u) \xrightarrow{uc} f_X(u)$, there is a neighborhood N of the origin such that $Rlf_{S_n}(u) \geq \delta > 0$, $u \in N$, all $n \geq 1$. On N, $|\text{Arg } f_{S_n}(u)|$ is bounded away from π. Let $f_n(u) = f_{X_1^{(n)}}(u)$; so

$$f_{S_n}(u) = [f_n(u)]^n.$$

On N, then

$$|f_{S_n}(u)| = |f_n(u)|^n,$$

$$\text{Arg } f_{S_n}(u) = n \text{ Arg } f_n(u).$$

So $f_n(u) \to 1$, for $u \in N$, and now apply 8.37.

Now I turn to the problem of characterizing the infinitely divisible distributions. Let $f(u)$ be the characteristic function of X. Therefore, since

$\mathcal{L}(\mathsf{X}) = \mathcal{L}(\mathsf{X}_1^{(n)} + \cdots + \mathsf{X}_n^{(n)})$, there is a characteristic function $f_n(u)$ such that $f(u) = [f_n(u)]^n$, and by 9.10, $f_n(u) \xrightarrow{uc} 1$. Then,

$$\log f(u) = n \log [1 - (1 - f_n(u))]$$

(9.11)
$$= n(f_n(u) - 1)(1 + \epsilon_n(u)).$$

Since $f_n(u) \xrightarrow{uc} 1$, it follows that $\epsilon_n(u) \xrightarrow{uc} 0$. Also, $|f(u)| > 0$, all u, otherwise $f(u_0) = 0$ implies $f_n(u_0) = 0$, all n, contradicting $f_n(u) \xrightarrow{uc} 1$. Denote by F_n the distribution function of $\mathsf{X}_1^{(n)}$; then

(9.12)
$$\log f(u) = (1 + \epsilon_n(u)) \int (e^{iux} - 1) n \, dF_n.$$

If we set up approximating sums of the integral in (9.12), we get

$$f(u) = \exp \left[\sum \lambda_i (e^{iux_i} - 1)\right],$$

exactly like the general Poisson case looked at before. Note also that if we put $\mu_n(B) = nF_n(B)$, μ_n a nonnegative measure on $\mathcal{B}_1$, then

$$\mu_n(B) = E\,(\text{number of } \mathsf{X}_k^{(n)} \in B).$$

Since $\log f(u) = (1 + \epsilon_n) \int (e^{iux} - 1) \mu_n\,(dx)$, if μ_n converges to a measure μ such that for continuous bounded functions $\varphi(x)$, $\int \varphi(x) \mu_n\,(dx) \to \int \varphi(x) \mu(dx)$, then we could conclude

$$f(u) = e^{\int (e^{iux} - 1)\mu(dx)}.$$

This is the basic idea, but there are two related problems. First, the total mass of μ_n is $\mu_n(R^{(1)}) = n$, hence $\mu_n(R^{(1)}) \to \infty$. Certainly, then, for $\varphi(x) = 1$ there is no finite μ such that $\int \varphi\, d\mu_n \to \int \varphi\, d\mu$. Second, how can the $\mathcal{N}(0, 1)$ characteristic function $e^{-u^2/2}$ be represented as above? Now, for any neighborhood N of the origin, we would expect more and more of the $\mathsf{X}_k^{(n)}$ to be in N; that is, $\mu_n(N) \to \infty$. But in analogy with Poisson convergence, the number of times that $\mathsf{X}_k^{(n)}$ is sizeable enough to take values outside of N should be bounded; that is, $\overline{\lim}\, \mu_n(N^c) < \infty$. We can prove even more than this.

Proposition 9.13

$$\overline{\lim}\, \mu_n[-a, +a]^c \le \alpha a \int_0^{1/a} |Rl \log f(v)| \, dv.$$

Proof. By inequality 8.29,

$$\mu_n[-a, +a]^c \le \alpha a n \int_0^{1/a} (1 - Rl f_n(v)) \, dv.$$

Take the real part of Eq. (9.11), and pass to the limit to get

$$\lim_n n(1 - Rlf_n(u)) = -Rl \log f(u).$$

Use $|f(u)| > 0$, all u, and the bounded convergence theorem for the rest.

Since

$$Rl \log f(0) = 0, \quad \overline{\lim_{a \uparrow \infty}}\ \overline{\lim_{n}}\ \mu_n[-a, +a]^c = 0,$$

so the μ_n sequence is in this sense mass-preserving. What is happening is that the mass of μ_n is accumulating near the origin, and behaving nicely away from the origin as $n \to \infty$. But if then $\varphi(0) = 0$, there is some hope that $\int \varphi(x)\mu_n(dx)$ may converge. This is true to some extent, more exactly,

Proposition 9.14

$$\overline{\lim} \int_{[-1,+1]} x^2 \mu_n(dx) < \infty.$$

Proof.

$$n(1 - Rlf_n(1)) = \int (1 - \cos x)\mu_n(dx)$$
$$\geq \int_{[-1,+1]} (1 - \cos x)\mu_n(dx).$$

By Taylor's expansion, $\cos x = 1 - x^2/2 \cos x\alpha$, $|\alpha| \leq 1$, so there is a β, $0 < \beta < \infty$, such that $\cos x \leq 1 - \beta x^2$, for $|x| \leq 1$. Thus

$$n(1 - Rlf_n(1)) \geq \beta \int_{[-1,+1]} x^2 \mu_n(dx).$$

However, $n(1 - Rlf_n(1)) \to Rl \log f(1)$, giving the result, since $|f(1)| \neq 0$.

By 9.13 and 9.14, if we define $\nu_n(B) = \int_B \varphi(x)\mu_n(dx)$, where $\varphi(x)$ is bounded and behaves like x^2 near the origin, the $\nu_n(B)$ is a bounded sequence of measures and we can think of trying to apply the Helly-Bray theorem. The choice of $\varphi(x)$ is arbitrary, subject only to boundedness and the right behavior near zero. The time honored custom is to take $\varphi(x)$ to be $x^2/(1 + x^2)$. Thus let $\alpha_n = \int (y^2/(1 + y^2))\mu_n(dy)$,

$$G_n(x) = \frac{1}{\alpha_n} \int_{(-\infty,x)} \left(\frac{y^2}{1 + y^2}\right)\mu_n(dy),$$

making $G_n(x)$ a distribution function. By 9.13 and 9.14 $\overline{\lim}\ \alpha_n < \infty$. We can write

$$\log f(u) = (1 + \epsilon_n)\alpha_n \int (e^{iux} - 1) \frac{x^2 + 1}{x^2} G_n(dx),$$

but the integrand blows up as $x \to 0$. So we first subtract the infinity by writing,

$$\log f(u) = (1 + \epsilon_n)\int \left(e^{iux} - 1 - \frac{iux}{1 + x^2}\right) d\mu_n + (1 + \epsilon_n)\int \frac{iux}{1 + x^2}\, d\mu_n.$$

Then we write $\beta_n = \int x/(1 + x^2)\, d\mu_n$, so that

$$\text{(9.15)}\quad \log f(u) = (1 + \epsilon_n)\alpha_n \int \left(e^{iux} - 1 - \frac{iux}{1 + x^2}\right)\frac{1 + x^2}{x^2}\, G_n(dx) + i(1 + \epsilon_n)\beta_n u.$$

If the integral term converges we see that $\{\beta_n\}$ can contain no subsequence going to infinity. If it did, then

$$\exp\,[i\beta_n u(1 + \epsilon_n(u))] \xrightarrow{uc} h(u)$$

would imply that on substituting $u = v/\beta_n$ and going along the subsequence, we would get $e^{iv} = h(0)$ for all v. If $\{\beta_n\}$ has two limit points β, β', then $e^{i\beta u} = e^{i\beta' u}$, hence $\beta = \beta'$. Thus, *uc* convergence of the first term entails convergence of the second term to $i\beta u$.

The integrand in (9.15)

$$\varphi(x, u) = \left(e^{iux} - 1 - \frac{iux}{1 + x^2}\right)\frac{1 + x^2}{x^2}$$

is a continuous bounded function of x for $x \neq 0$. As $x \to 0$, it has the limit $-u^2/2$. By defining $\varphi(0, u) = -u^2/2$, $\varphi(x, u)$ is jointly continuous everywhere. By 9.13, $\{G_n\}$ is mass-preserving. If $\underline{\lim}\ \alpha_n = 0$, take n' such that $\alpha_{n'} \to 0$ and conclude from (9.15) that X is degenerate. Otherwise, take n' such that $\alpha_{n'} \to \alpha > 0$ and $G_{n'} \xrightarrow{\mathfrak{D}} G$. Then G is a distribution function. Go along the n' sequence in 9.15. The $\beta_{n'}$ sequence must converge to some limit β since the integral term converges *uc*. Therefore

$$\log f(u) = \alpha \int \left(e^{iux} - 1 - \frac{iux}{1 + x^2}\right)\frac{1 + x^2}{x^2}\, G(dx) + i\beta u.$$

Suppose $G(\{0\}) > 0$, then

$$\text{(9.16)}\qquad \log f(u) = i\beta u - \frac{\sigma^2 u^2}{2} + \alpha \int_{\{0\}^c} \varphi(x, u) G(dx).$$

We have now shown part of

Theorem 9.17. *X has infinitely divisible distribution if and only if its characteristic function $f(u)$ is given by*

$$\text{(9.18)}\quad \log f(u) = i\beta u - \frac{\sigma^2 u^2}{2} + \int \left(e^{iux} - 1 - \frac{iux}{1 + x^2}\right)\frac{1 + x^2}{x^2}\, \nu(dx),$$

where ν is a finite measure that assigns zero mass to the origin.

To complete the proof: It has to be shown that any random variable whose characteristic function is of the form (9.18) has infinitely divisible distribution. To begin, assume that any function $f(u)$ whose log is of the form (9.18) is a characteristic function. Then it is trivial because if $f_n(u)$ is defined by

$$\log f_n(u) = \frac{1}{n} \log f(u),$$

then $\log f_n(u)$ is again of the form (9.18); so $f_n(u)$ is a characteristic function. Since $f(u)$ now is given by $[f_n(u)]^n$ for any n, the corresponding distribution is infinitely divisible. The last point is to show that (9.18) always gives a characteristic function. Take partitions $\mathcal{I}_n$ of $R^{(1)}$ into finite numbers of intervals such that the Riemann sums in (9.18) converge to the integral, that is,

$$\sum_{i=1}^{j_n} \left(e^{iux_i} - 1 - \frac{iux_i}{1 + x_i^2}\right)\frac{1 + x_i^2}{x_i^2}\, \nu(I_i) \to \int \varphi(x, u)\nu(dx).$$

Put $\beta_n = \beta - \sum_i (\nu(I_i)/x_i)$, denote

$$\lambda_i = \nu(I_i) \cdot \frac{1 + x_i^2}{x_i^2},$$

and write

$$g_n(u) = (\exp [i\beta_n u - (\sigma^2/2)u^2]) \cdot (\exp [\sum_i \lambda_i(e^{iux_i} - 1)]).$$

See that $g_n(u)$ is the product of a characteristic function of a $\mathcal{N}(\beta_n, \sigma)$ distribution and characteristic functions of Poisson distributions $\mathcal{P}(\lambda_i)$ with jump x_i. Therefore by Corollary 8.35, $g_n(u)$ is a characteristic function. This does it, because

$$\lim_n g_n(u) = f(u)$$

for every u. Check that anything of the form (9.18) is continuous at $u = 0$. Certainly the first two terms are. As to the integral, note that $\sup_x |\varphi(x, u)| \le M$ for all $|u| \le 1$. Also, $\lim_{u\to 0} \varphi(x, u) = 0$ for every x, and apply the bounded convergence theorem to get

$$\lim_{u\to 0} \int \varphi(x, u)\nu(dx) = 0, \qquad \text{so} \quad \lim_{u\to 0} f(u) = 1.$$

By the continuity theorem, $f(u)$ is a characteristic function. Q.E.D.

6. THE GENERALIZED LIMIT PROBLEM

Just as before, it becomes reasonable to ask what are the possible limit laws of

$$\mathsf{S}_n = \mathsf{X}_1^{(n)} + \cdots + \mathsf{X}_n^{(n)}$$

if the restriction that the $\mathsf{X}_1^{(n)}, \ldots, \mathsf{X}_n^{(n)}$ be identically distributed is lifted.

Some restriction is needed; otherwise, take

$$\mathsf{X}_2^{(n)} = \cdots = \mathsf{X}_n^{(n)} = 0, \qquad \mathsf{X}_1^{(n)} = \mathsf{X},$$

to get any limit distribution desired. What is violated in the spirit of the previous work is the idea of a sum of a large number of components, each one *small on the average*. That is, we had, in the identically distributed case, that for every $\epsilon > 0$,

$$\text{(A)} \qquad \sup_k P(|\mathsf{X}_k^{(n)}| > \epsilon) \to 0, \qquad \text{as} \quad n \to \infty.$$

This is the restriction that we retain in lieu of identical distribution. It is just about the weakest condition that can be imposed on the summands in order to prevent one of the components from exerting a dominant influence on the sum. With condition (A) a surprising result comes up.

Theorem 9.19. *If the sums* $\mathsf{S}_n \xrightarrow{\mathfrak{D}} \mathsf{X}$, *then* X *has infinitely divisible distribution.*

So in a strong sense the infinitely divisible laws are the limit laws of large sums of independent components, each one small on the average. The proof of 9.19 proceeds in exactly the same way as that of Theorem 9.17, the only difference being that $\mu_n(B) = \sum_1^n F_k^{(n)}(B)$ instead of $nF_n(B)$, but the same inequalities are used. It is the same proof except that one more subscript is floating around.

Problem 1. Let X have infinitely divisible distribution,

$$\log f(u) = i\beta u + \int \varphi(x, u)\nu(dx)$$

if $\nu(\{0\}) = 0$, and if ν assigns all its mass to a countable set of points, prove that the distribution of X is of pure type. [Use the law of pure types.]

7. UNIQUENESS OF REPRESENTATION AND CONVERGENCE

Let X have an infinitely divisible distribution with characteristic function $f(u)$. Then by (9.18), there is a finite measure $\gamma(dx)$ (possibly with mass at the origin) and a constant β such that

$$\textbf{(9.20)} \qquad \log f(u) = i\beta u + \int \varphi(x, u)\gamma(dx) \qquad [2\pi i],$$

and $\varphi(x, u)$ is continuous in both x and u and bounded for $x \in R^{(1)}$, $u \in [-U, +U]$, $U < \infty$. Log $f(u)$ is defined up to additive multiples of $2\pi i$. Because $|f(u)| \neq 0$, there is a unique version of $\log f(u)$ which is zero when u is zero and is a continuous function of u on $R^{(1)}$. Now (9.20) states that this version is given by the right-hand side above.

Proposition 9.21. *$\gamma(dx)$ and β are uniquely determined by* (9.20).

Proof. Let $\psi(u) = \log f(u)$; this is the continuous version of $\log f(u)$. Then (following Loève [108]), take

$$\theta(u) = \psi(u) - \int_0^1 \left[\frac{\psi(u+h) + \psi(u-h)}{2}\right] dh$$

so that $\theta(u)$ is determined by ψ, hence by f. Note that

$$\int_0^1 \left[\frac{\varphi(x, u+h) + \varphi(x, u-h)}{2}\right] dh$$

$$= \varphi(x, u) - e^{iux}\int_0^1 (1 - \cos(hx))\, dh \cdot \frac{1+x^2}{x^2}$$

$$= \varphi(x, u) - e^{iux}\left(1 - \frac{\sin x}{x}\right) \cdot \frac{1+x^2}{x^2}.$$

Hence, using (9.20),

$$\theta(u) = \int e^{iux} g(x)\gamma(dx),$$

where

$$g(x) = \left(1 - \frac{\sin x}{x}\right)\frac{1+x^2}{x^2}.$$

It is easy to check that $0 < \inf g(x) < \sup g(x) < \infty$. But $\theta(u)$ uniquely determines the measure $\nu(B) = \int_B g(x)\gamma(dx)$, and thus γ is determined as $\gamma(B) = \int_B [g(x)]^{-1}\nu(dx)$. If, therefore,

$$\psi(u) = i\beta' u + \int \varphi(x, u)\gamma'(dx),$$

then $\gamma \equiv \gamma'$ implying $\beta = \beta'$.

The fact that $\gamma(dx)$ is unique gives us a handhold on conditions for $\mathsf{S}_n \xrightarrow{\mathfrak{D}} \mathsf{X}$. Let $\gamma(dx) = \alpha G(dx)$ where G is a distribution function. Recall that α, $G(x)$ were determined by taking any convergent subsequence $\alpha_{n'}$ of α_n, $\alpha = \lim_{n'} \alpha_{n'}$, and taking $G(x)$ as any limit distribution of the $G_{n'}(x)$ sequence. Since α, $G(x)$ are unique, then $\alpha_n \to \alpha$, $G_n \xrightarrow{\mathfrak{D}} G$. Consequently β_n, defined by

$$\beta_n = \int \frac{x}{1+x^2}\, d\mu_n,$$

converges to β. Thus, letting $\gamma(x) = \gamma(-\infty, x)$, and

$$\gamma_n(x) = \int_{(-\infty, x)} \frac{y^2}{1+y^2}\mu_n(dy),$$

then $\mathsf{S}_n \xrightarrow{\mathfrak{D}} \mathsf{X}$ implies $\gamma_n \xrightarrow{\mathfrak{D}} \gamma$, and $\{\gamma_n\}$ is mass-preserving in the sense that for any $\epsilon > 0$, there exists a finite interval I such that $\sup_n \gamma_n(I^c) \le \epsilon$. These conditions are also sufficient:

Theorem 9.22. $\mathsf{S}_n \xrightarrow{\mathfrak{D}} \mathsf{X}$ *where* X *has characteristic function given by*

$$\log f(u) = i\mu\beta + \int \varphi(x, u)\gamma(dx)$$

for $\gamma(dx)$ *a finite measure if and only if the measures* $\gamma_n(dx)$ *are mass-preserving in the above sense and*

i) $\gamma_n \xrightarrow{\mathfrak{D}} \gamma$, ii) $\beta_n \to \beta$.

Proof. All that is left to do is show sufficiency. This is easy. Since

$$\int n \frac{y^2}{1 + y^2} F_n(dy) = \gamma_n(+\infty) \le M < \infty,$$

it follows that

$$\int \frac{y^2}{1 + y^2} F_n(dy) \to 0;$$

hence F_n converges to the law degenerate at zero. Thus for all u in a finite interval, we can write

$$\log f_{\mathsf{S}_n}(u) = (1 + \epsilon_n(u))[i\beta_n u + \int \varphi(x, u)\gamma_n(dx)],$$

where $\epsilon_n(u) \xrightarrow{uc} 0$. Thus

$$\log f_{\mathsf{S}_n}(u) \to i\beta\mu + \int \varphi(x, u)\gamma(dx).$$

Now we can go back and get the proof of 9.6. If $\mathsf{S}_n = \mathsf{X}_1^{(n)} + \cdots + \mathsf{X}_n^{(n)}$, and $\mathsf{S}_n \xrightarrow{\mathfrak{D}} \mathsf{X}$, then clearly X is normal if and only if γ_n converges to a measure γ concentrated on the origin. Equivalently, for every $x > 0$, $\gamma_n((-x, +x)^c) \to 0$. Since

$$\gamma_n((-x, +x)^c) = \int_{|y| \ge x} \frac{y^2}{1 + y^2} nF_n(dy),$$

this is equivalent to

$$nF_n((-x, +x)^c) \to 0, \quad x > 0 \qquad \text{or} \qquad nP(|\mathsf{X}_1^{(n)}| \ge x) \to 0.$$

But

$$P(\mathsf{M}_n < x) = P(|\mathsf{X}_1^{(n)}| < x, \ldots, |\mathsf{X}_n^{(n)}| < x) = [1 - P(|\mathsf{X}_1^{(n)}| \ge x)]^n.$$

Because $\mathsf{X}_1^{(n)} \xrightarrow{\mathfrak{D}} 0$,

$$\log P(\mathsf{M}_n < x) = -n(1 + \delta_n(x))P(|\mathsf{X}_1^{(n)}| \ge x),$$

where $\delta_n(x) \to 0$ for $x > 0$. Therefore,

$$\mathsf{M}_n \xrightarrow{\mathfrak{D}} 0 \Leftrightarrow nP(|\mathsf{X}_1^{(n)}| \geq x) \to 0, \quad x > 0,$$

which completes the proof.

8. THE STABLE LAWS

Let $\mathsf{X}_1, \mathsf{X}_2, \ldots$ be identically distributed, nondegenerate, independent random variables. What is the class of all possible limit laws of normed sums

$$\text{(9.23)} \qquad \mathsf{S}_n = \frac{\mathsf{X}_1 + \cdots + \mathsf{X}_n}{A_n} - B_n, \quad A_n > 0?$$

Since S_n may be written as

$$\mathsf{S}_n = \mathsf{X}_1^{(n)} + \cdots + \mathsf{X}_n^{(n)},$$

where

$$\mathsf{X}_k^{(n)} = \frac{\mathsf{X}_k}{A_n} - \frac{B_n}{n},$$

the requirement $\mathsf{S}_n \xrightarrow{\mathfrak{D}} \mathsf{X}$ implies that X is infinitely divisible. The condition $\mathsf{X}_1^{(n)} \xrightarrow{\mathfrak{D}} 0$ implies $A_n \to \infty$, $B_n/n \to 0$. This class of limit laws is the most interesting set of distributions following the normal and Poisson. Of course, if $E\mathsf{X}_1^2 < \infty$, then $A_n \sim \sqrt{n}$ and X must be normal. So the only interesting case is $E\mathsf{X}_1^2 = \infty$. Two important questions arise.

First: What is the form of the class of all limit distributions X such that $\mathsf{S}_n \xrightarrow{\mathfrak{D}} \mathsf{X}$?

Second: Find necessary and sufficient conditions on the common distribution function of $\mathsf{X}_1, \mathsf{X}_2, \ldots$ so that $\mathsf{S}_n \xrightarrow{\mathfrak{D}} \mathsf{X}$.

These two questions lead to the stable laws and the domains of attraction of the stable laws.

Definition 9.24. *A random variable X is said to have a stable law if for every integer $k > 0$, and $\mathsf{X}_1, \ldots, \mathsf{X}_k$ independent with the same distribution as X, there are constants $a_k > 0$, b_k such that*

$$\mathfrak{L}(\mathsf{X}_1 + \cdots + \mathsf{X}_k) = \mathfrak{L}(a_k\mathsf{X} + b_k).$$

This approach is similar to the way we intrinsically characterized the normal law; by breaking S_{nk} up into k blocks, we concluded that the limit of $\mathsf{S}_n/\sqrt{n}$ must satisfy $\mathfrak{L}(\mathsf{X}_1 + \cdots + \mathsf{X}_k) = \mathfrak{L}(\sqrt{k}\mathsf{X})$.

Proposition 9.25. *X is the limit in distribution of normed sums* (9.23) *if and only if X has a stable law.*

Proof. One way is quick: If X is stable, then $\mathcal{L}(\mathsf{X}_1 + \cdots + \mathsf{X}_n) = \mathcal{L}(a_n\mathsf{X} + b_n)$. Then (check this by characteristic functions),

$$\mathcal{L}(\mathsf{X}) = \mathcal{L}\left(\frac{\mathsf{X}_1 + \cdots + \mathsf{X}_n}{a_n} - \frac{b_n}{a_n}\right),$$

and we can take $A_n = a_n$, $B_n = b_n/a_n$ to get

$$\frac{\mathsf{X}_1 + \cdots + \mathsf{X}_n}{A_n} - B_n \xrightarrow{\mathfrak{D}} \mathsf{X}$$

(actually $\overset{\mathfrak{D}}{=} \mathsf{X}$ is true here).

To go the other way, suppose

$$\mathsf{Z}_n = \frac{\mathsf{X}_1 + \cdots + \mathsf{X}_n}{A_n} - B_n \xrightarrow{\mathfrak{D}} \mathsf{X}.$$

Then $\mathsf{Z}_{nk} \xrightarrow{\mathfrak{D}} \mathsf{X}$ as $n \to \infty$ for all k. Repeat the trick we used for the normal law:

$$\mathsf{Z}_{nk} = \frac{1}{A_{nk}}(\mathsf{S}_n^{(1)} + \cdots + \mathsf{S}_n^{(k)}) - B_{nk},$$

where $\mathsf{S}_n^{(1)} = \mathsf{X}_1 + \cdots + \mathsf{X}_n$, $\mathsf{S}_n^{(2)} = \mathsf{X}_{n+1} + \cdots + \mathsf{X}_{2n}, \ldots$ Thus

$$\left(\frac{\mathsf{S}_n^{(1)}}{A_n} - B_n\right) + \cdots + \left(\frac{\mathsf{S}_n^{(k)}}{A_n} - B_n\right) = \frac{A_{nk}}{A_n}\mathsf{Z}_{nk} + C_{n,k},$$

where $C_{n,k} = (A_{nk}/A_n)B_{nk} - kB_n$. By the law of convergence of types,

$$\frac{A_{nk}}{A_n} \to a_k, \qquad C_{n,k} \to b_k.$$

Therefore,

$$\mathcal{L}(\mathsf{X}_1 + \cdots + \mathsf{X}_k) = \mathcal{L}(a_k\mathsf{X} + b_k).$$

This not only proves 9.25 but contains the additional information that if $\mathsf{S}_n \xrightarrow{\mathfrak{D}} \mathsf{X}$, then $A_{nk}/A_n \to a_k$ for all k. By considering the limit of A_{nmk}/A_n as $n \to \infty$, we conclude that the constants a_k must satisfy

(9.26) $$a_{mk} = a_m a_k, \quad \text{all } k, m \geq 1.$$

9. THE FORM OF THE STABLE LAWS

Theorem 9.27. *Let X have a stable law. Then either X has a normal distribution or there is a number α, $0 < \alpha < 2$, called the exponent of the law and constants $m_1 \geq 0$, $m_2 \geq 0$, β such that*

$$\log f_{\mathsf{X}}(u) = iu\beta + m_1\int_0^{\infty}\left(e^{iux} - 1 - \frac{iux}{1 + x^2}\right)\frac{dx}{x^{1+\alpha}}$$
$$+ m_2\int_{-\infty}^{0}\left(e^{iux} - 1 - \frac{iux}{1 + x^2}\right)\frac{dx}{|x|^{1+\alpha}}.$$

Proof. Since X is stable, it is infinitely divisible,

$$\psi(u) = \log f_X(u) = iu\beta + \int \varphi(x, u)\gamma(dx).$$

In terms of characteristic function, the definition of stability becomes

$$[e^{\psi(u)}]^k = e^{ib_k u}e^{\psi(a_k u)},$$

or

$$\textbf{(9.28)} \qquad k\psi(u) = ib_k u + \psi(a_k u).$$

Separate the situation into two cases.

$$\text{I.}\quad \gamma(\{0\}) = 0, \qquad \text{II.}\quad \gamma(\{0\}) > 0.$$

CASE I. Define a measure μ:

$$\mu(B) = \int_B \frac{1 + x^2}{x^2}\,\gamma(dx).$$

Then μ is σ-finite, $\mu[-a, +a]^c < \infty$, for any $a > 0$, $\int_{[-a,+a]} x^2\, d\mu < \infty$, and

$$\psi(u) = iu\beta + \int \left(e^{iux} - 1 - \frac{iux}{1 + x^2}\right)\mu(dx),$$

$$\psi(a_k u) = ia_k\beta u + \int \left(e^{iua_kx} - 1 - \frac{iua_kx}{1 + x^2}\right)\mu(dx)$$

$$= id_k u + \int \left(e^{iua_kx} - 1 - \frac{iua_kx}{1 + a_k^2x^2}\right)\mu(dx),$$

where

$$d_k = a_k\beta + a_k\int \left[\frac{x}{1 + a_k^2x^2} - \frac{x}{1 + x^2}\right]\mu(dx).$$

This last integrand behaves like x^3 near the origin, and is bounded away from the origin, so the integral exists. Define a change of variable measure μ_k by

$$\mu_k(B) = \mu(z;\ a_k z \in B),$$

to get

$$\psi(a_k u) = id_k u + \int \left(e^{iux} - 1 - \frac{iux}{1 + x^2}\right)\mu_k(dx).$$

Therefore, (9.28) becomes

$$iu\beta k + \int \left(e^{iux} - 1 - \frac{iux}{1 + x^2}\right)k\mu(dx) = i(b_k + d_k)u$$

$$+ \int \left(e^{iux} - 1 - \frac{iux}{1 + x^2}\right)\mu_k(dx).$$

By the uniqueness of representation of infinitely divisible characteristic function we get the central result

$$k\mu \equiv \mu_k. \tag{9.29}$$

Let $M^+(x) = \mu[x, \infty)$, $x > 0$. From (9.29),

$$kM^+(x) = M^+(x/a_k), \quad x > 0, k > 0. \tag{9.30}$$

Similarly, let $M^-(x) = \mu(-\infty, x)$, $x < 0$. Again, $kM^-(x) = M^-(x/a_k)$, $x < 0, k > 0$.

Proposition 9.31. *$a_k = k^\lambda$, $\lambda > 0$, and*

$$M^+(x) = x^{-1/\lambda}M^+(1),$$
$$M^-(x) = |x|^{-1/\lambda}\, M^-(-1).$$

Proof. $M^+(x)$ is nonincreasing. The relation $kM^+(1) = M^+(1/a_k)$ implies a_k increasing in k, and we know $a_k \to \infty$. For any k, $a_n a_k = a_{nk}$ gives

$$a_{k^j} = (a_k)^j.$$

For $n > k$, take j such that $k^j \le n \le k^{j+1}$. Then

$$a_{k^j} \le a_n \le a_{k^{j+1}}$$

or

$$\log(a_{k^j}) \le \log a_n \le \log(a_{k^{j+1}}),$$
$$j \log a_k \le \log a_n \le (j+1)\log a_k.$$

Dividing by $j \log k$, we get

$$\frac{\log a_k}{\log k} \le \frac{\log a_n}{\log n}\left[\frac{\log n}{j \log k}\right] \le \frac{j+1}{j}\frac{\log a_k}{\log k}.$$

Now let $n \to \infty$; consequently $j \to \infty$, and $(\log n)/(j \log k) \to 1$, implying

$$\frac{\log a_k}{\log k} = \lim \frac{\log a_n}{\log n} = \lambda > 0.$$

To do the other part, set $x = (k/n)^\lambda$; then in (9.30),

$$kM^+((k/n)^\lambda) = M^+(1/n^\lambda), \quad \text{all } k, n \ge 1.$$

For $k = n$, this is $nM^+(1) = M^+(1/n^\lambda)$. Substituting this above gives

$$kM^+((k/n)^\lambda) = nM^+(1)$$

or

$$M^+((k/n)^\lambda) = M^+(1)/(k/n).$$

For all x in the dense set $\{(k/n)^\lambda\}$ we have shown $M^+(x) = x^{-1/\lambda}M^+(1)$. The fact that $M^+(x)$ is nonincreasing makes this hold true for all x. Similarly for $M^-(x)$.

The condition $\int_{-1}^{1} x^2\, d\mu < \infty$ implies $\int_{-1}^{+1} x^2\, |x|^{-(1/\lambda)-1}\, dx < \infty$ so that $-1/\lambda + 1 > -1$, or finally $\lambda > \frac{1}{2}$. For $\psi(u)$ the expression becomes

$$\psi(u) = iu\beta + m_1\int_0^\infty \left(e^{iux} - 1 - \frac{iux}{1 + x^2}\right)\frac{dx}{x^{1+\alpha}}$$

$$+ m_2\int_{-\infty}^{0} \left(e^{iux} - 1 - \frac{iux}{1 + x^2}\right)\frac{dx}{|x|^{1+\alpha}},$$

where $m_1 = M^+(1)\cdot\dfrac{1}{\lambda}$, $m_2 = M^-(-1)\cdot\dfrac{1}{\lambda}$, and $\alpha = \dfrac{1}{\lambda}$; so $0 < \alpha < 2$.

CASE II. If $\gamma(\{0\}) = \sigma^2 > 0$, then

$$\psi(u) = iu\beta - \frac{\sigma^2}{2}u^2 + \int_{\{0\}^c} \varphi(x, u)\gamma(dx).$$

The coefficient σ^2 is uniquely determined by $\lim_{u\to\infty}\psi(u)/u^2 = -\sigma^2/2$, because $\sup_{x,u} |\varphi(x, u)/u^2| < \infty$ and $\varphi(x, u)/u^2 \to 0$ for $x \neq 0$, as $u \to \infty$. Apply the bounded convergence theorem to $\int_{\{0\}^c} (\varphi(x, u)/u^2)\gamma(dx)$ to get the result. Therefore, dividing (9.28) by u^2 and letting $u \to \infty$ gives

$$k = a_k^2 \qquad \text{which implies} \quad \lambda = \tfrac{1}{2}.$$

So (9.28) becomes

$$k\psi(u) = ib_k u + \psi(\sqrt{k}u)$$

$$\psi(u) = i\frac{b_k}{k}u + u^2\frac{\psi(\sqrt{k}u)}{ku^2}.$$

As $k \to \infty$, $\psi(\sqrt{k}u)/ku^2 \to -\sigma^2/2$. This entails $b_k/k \to \beta$ and

$$\psi(u) = iu\beta - (\sigma^2/2)u^2. \qquad \text{Q.E.D.}$$

It is not difficult to check that every characteristic function of the form given in (9.27) is the characteristic function of a stable law. This additional fact completes our description of the form of the characteristic function for stable laws.

Problems

2. Use the methods of this section to show that Proposition 9.1 holds without the restriction $EX^2 < \infty$.

3. Use Problem 2 to prove that if $X_1, X_2, \ldots$ are independent, identically distributed random variables and

$$\frac{X_1 + \cdots + X_n}{\sqrt{n}} \xrightarrow{\mathfrak{D}} X,$$

then X is normal or degenerate.

4. Show that if $\alpha < 1$, then $\psi(u)$ can be written as

$$\psi(u) = iu\beta + m_1 \int_0^\infty (e^{iux} - 1) \frac{dx}{x^{1+\alpha}} + m_2 \int_{-\infty}^0 (e^{iux} - 1) \frac{dx}{|x|^{1+\alpha}}.$$

Then prove that $\beta = 0$, $m_2 = 0$, implies $X \geq 0$ a.s.

10. THE COMPUTATION OF THE STABLE CHARACTERISTIC FUNCTIONS

When the exponent α is less than 2, the form of the stable characteristic function is given by 9.27. By doing some computations, we can evaluate these integrals in explicit form.

Theorem 9.32. *$f(u) = e^{\psi(u)}$ is the characteristic function of a stable law of exponent α, $0 < \alpha < 1$, and $1 < \alpha < 2$ if and only if it has the form*

$$\psi(u) = iuc - d\,|u|^\alpha \left(1 + i\theta \frac{u}{|u|} \tan \frac{\pi}{2} \alpha\right),$$

where c is real, d real and positive, and θ real such that $|\theta| \leq 1$. For $\alpha = 1$, the form of the characteristic function is given by

$$\psi(u) = iuc - d\,|u| \left(1 + i\theta \frac{u}{|u|} \frac{2}{\pi} \log |u|\right),$$

with c, d, θ as above.

Proof. Let

$$I_1(u) = \int_0^\infty \left(e^{iux} - 1 - \frac{iux}{1 + x^2}\right) \frac{dx}{x^{1+\alpha}}.$$

Since $I_1(-u) = \overline{I_1(u)}$, we evaluate $I_1(u)$ only for $u > 0$. Also,

$$I_2(u) = \int_{-\infty}^0 \left(e^{iux} - 1 - \frac{iux}{1 + x^2}\right) \frac{dx}{|x|^{1+\alpha}} = I_1(-u).$$

Consider first $0 < \alpha < 1$; then

$$I_1(u) = \int_0^\infty \left(e^{iux} - 1\right) \frac{dx}{x^{1+\alpha}} - iu \int_0^\infty \frac{1}{1 + x^2} \frac{dx}{x^\alpha}.$$

Substitute $ux = y$ in the first term to get, for $u \geq 0$,

$$I_1(u) = iuc + u^\alpha H(\alpha),$$

where

$$H(\xi) = \int_0^\infty (e^{iy} - 1)\frac{dy}{y^{\xi+1}}.$$

For $1 < \alpha < 2$, integrate by parts, getting

$$\begin{aligned} I_1(u) &= \frac{iu}{\alpha}\int_0^\infty \left(e^{iux} - \frac{d}{dx}\frac{x}{1+x^2}\right)\frac{dx}{x^\alpha} \\ &= iuc + \frac{iu}{\alpha}\int_0^\infty (e^{iux} - 1)\frac{dx}{x^\alpha}, \end{aligned}$$

where

$$c = \frac{1}{\alpha}\int_0^\infty \left(1 - \frac{d}{dx}\frac{x}{1+x^2}\right)\frac{dx}{x^\alpha}.$$

Substitute $ux = y$ again, so

$$I_1(u) = iuc + u^\alpha \cdot \frac{i}{\alpha} H(\alpha - 1), \quad u \geq 0.$$

If $\alpha = 1$, the integration by parts gives

$$I_1(u) = iu \lim_{T\to\infty} \int_0^T \left(e^{iux} - \frac{d}{dx}\frac{x}{1+x^2}\right)\frac{dx}{x}.$$

Let

$$J(T, u) = \int_0^T \left(e^{iux} - \frac{d}{dx}\frac{x}{1+x^2}\right)\frac{dx}{x}.$$

Then for $u_2 \geq u_1 > 0$,

$$\begin{aligned} J(T, u_2) - J(T, u_1) &= \int_0^T (e^{iu_2x} - e^{iu_1x})\frac{dx}{x} \\ &= i\int_0^T dx\left[\int_{u_1}^{u_2} e^{ivx}\, dv\right] = \int_{u_1}^{u_2}\left(\frac{e^{ivT} - 1}{v}\right) dv. \end{aligned}$$

Now, by the Riemann-Lebesgue lemma (see Chapter 10, Section 2),

$$\lim_T \int_{u_1}^{u_2} \frac{e^{ivT}}{v}\, dv = \lim_T \int e^{ivT}\chi_{[u_1,u_2]}(v) \cdot \frac{dv}{v} = 0$$

since $\chi_{[u_1,u_2]}(v)/v$ is a bounded measurable function vanishing outside a finite interval. This gives

$$\lim_{T\to\infty} [J(T, u_2) - J(T, u_1)] = \log(u_1/u_2).$$

Consequently, checking that $\lim_{T\to\infty} J(T, 1)$ exists,

$$\lim_{T\to\infty} J(T, u) = -\log u + c,$$

and

$$I_1(u) = iuc - iu \log u, \quad u \geq 0.$$

For the first time, the constant c appearing in the linear term is complex:

$$\begin{aligned} c &= \int_0^\infty \left(e^{ix} - \frac{d}{dx}\frac{x}{1+x^2}\right)\frac{dx}{x} \\ &= i\int_0^\infty \frac{\sin x}{x}\, dx + c_1, \end{aligned}$$

where c_1 is real. The integral $\int_0^\infty (\sin x/x)\, dx$ is well known and equals $\pi/2$. Finally, then,

$$I_1(u) = iuc_1 - (\pi/2)u - iu \log u, \quad u \geq 0.$$

The remaining piece of work is to evaluate

$$H(\xi) = \int_0^\infty (e^{iy} - 1)\frac{dy}{y^{\xi+1}}, \quad 0 < \xi < 1.$$

This can be done by contour integration (see Gnedenko and Kolmogorov, [62, p. 169]), with the result that

$$H(\xi) = e^{-i(\pi/2)\xi}L(\xi),$$

where $L(\xi)$ is real and negative. Putting everything together, we get

$$m_1 I_1(u) + m_2 I_1(-u) = m_1 I_1(u) + m_2 \overline{I_1(u)}.$$

For $0 < \alpha < 1$, and $n \geq 0$,

$$\begin{aligned} \psi(u) &= icu + u^\alpha\left(m_1 H(\alpha) + m_2 \overline{H(\alpha)}\right) \\ &= icu - du^\alpha(1 + i\theta \tan(\pi/2)\alpha), \end{aligned}$$

where $d = -(m_1 + m_2)RlH(\alpha)$ is real and positive, $\theta = (m_1 - m_2)/(m_1 + m_2)$ is real with range $[-1, +1]$, and c is real. For $1 < \alpha < 2$,

$$\psi(u) = icu + \frac{iu^\alpha}{\alpha}\left(m_1 H(\alpha - 1) - m_2 \overline{H(\alpha - 1)}\right).$$

Now

$$H(\alpha - 1) = e^{-i(\pi/2)(\alpha-1)}L(\alpha - 1) = ie^{-i(\pi/2)\alpha}L(\alpha - 1),$$

so

$$\psi(u) = icu - du^\alpha(1 + i\theta \tan(\pi/2)\alpha), \quad u \geq 0.$$

Here $d = ((m_1 + m_2)/\alpha)Rl(e^{-i(\pi/2)\alpha}L(\alpha - 1))$ is real and positive, and θ is again $(m_1 - m_2)/(m_1 + m_2)$ with range $[-1, +1]$. If $\alpha = 1$, then

$$\begin{aligned}\psi(u) &= icu - (\pi/2)u(m_1 + m_2) + iu(m_1 - m_2)\log u \\ &= icu - du(1 + i\theta \cdot (2/\pi)\log u), \quad u \geq 0,\end{aligned}$$

θ as above, d real and positive, c real. Q.E.D.

Problem 5. Let $\mathsf{S}_n = \mathsf{X}_1 + \cdots + \mathsf{X}_n$ be consecutive sums of independent random variables each having the same symmetric stable law of exponent α. Use the Fourier inversion theorem and the technique of Problem 24, Chapter 8, to show that the sums are transient for $0 < \alpha < 1$, recurrent for $\alpha \geq 1$. The case $\alpha = 1$ provides an example where $E|\mathsf{X}_1| = \infty$, but the sums are recurrent. Show that for all $\alpha > 0$,

$$P(\overline{\lim}\, \mathsf{S}_n = \infty) = P(\overline{\lim}\, \mathsf{S}_n = -\infty) = 1.$$

Conclude that the sums change sign infinitely often for any $\alpha > 0$.

11. THE DOMAIN OF ATTRACTION OF A STABLE LAW

Let $\mathsf{X}_1, \mathsf{X}_2, \ldots$ be independent and identically distributed. What are necessary and sufficient conditions on their distribution function $F(x)$ such that S_n suitably normalized converges in distribution to X where X is nondegenerate? Of course, the limit random variable X must be stable.

Definition 9.33. *The distribution $F(x)$ is said to be in the domain of attraction of a stable law with exponent $\alpha < 2$ if there are constants A_n, B_n such that*

$$\frac{\mathsf{S}_n}{A_n} - B_n \xrightarrow{\mathcal{D}} \mathsf{X}$$

and X has exponent α. Denote this by $F \in D(\alpha)$.

Complete conditions on $F(x)$ are given by

Theorem 9.34. *$F(x)$ is in the domain of attraction of a stable law with exponent $\alpha < 2$ if and only if there are constants M^+, $M^- \geq 0$, $M^+ + M^- > 0$, such that as $y \to \infty$:*

i)
$$\lim \frac{F(-y)}{1 - F(y)} = \frac{M^-}{M^+};$$

ii) *For every $\xi > 0$,*

$$M^+ > 0 \Rightarrow \lim \frac{1 - F(\xi y)}{1 - F(y)} = \frac{1}{\xi^\alpha},$$

$$M^- > 0 \Rightarrow \lim \frac{F(-\xi y)}{F(-y)} = \frac{1}{\xi^\alpha}.$$

Proof. We show first *necessity*. By Theorem 9.22, $\gamma_n \xrightarrow{\mathcal{D}} \gamma$. Now take $b_n = B_n/n$. Then

$$\gamma_n(B) = n\int_B \frac{y^2}{1+y^2} F_n(dy),$$

where

$$F_n(x) = P\left(\frac{X_1}{A_n} - b_n < x\right) = F(A_n(x+b_n)),$$

and

$$\gamma(B) = \int_B \frac{y^2}{1+y^2}\mu(dy).$$

Thus, for any $x_1 < 0$, $x_2 > 0$, since the γ_n must be mass-preserving,

$$\int_{[x_1,x_2)^c} \frac{1+z^2}{z^2}\gamma_n(dz) \to \int_{[x_1,x_2)^c} \frac{1+z^2}{z^2}\gamma(dz).$$

This is

$$n\int_{[x_1,x_2)^c} F_n(dz) \to \int_{[x_1,x_2)^c} \mu(dz).$$

By taking one or the other of x_1, x_2 to be infinite, we find that the condition becomes

$$n[1 - F_n(x)] \to M^+(x), \quad x > 0,$$
$$nF_n(x) \to M^-(x), \quad x < 0.$$

Then

$$n[1 - F_n(A_n(x+b_n))] \to M^+(x), \quad x > 0,$$
$$nF_n(A_n(x+b_n)) \to M^-(x), \quad x < 0.$$

Since $b_n \to 0$, for any $\epsilon > 0$ and n sufficiently large,

$$F(A_n(x - \epsilon + b_n)) \le F(A_n x) \le F(A_n(x + \epsilon + b_n)).$$

We can use this to get

$$M^+(x+\epsilon) \le \underline{\lim}\, n[1 - F(A_n x)]$$
$$\le \overline{\lim}\, n[1 - F(A_n x)] \le M^+(x - \epsilon), \quad x > 0,$$
$$M^-(x - \epsilon) \le \underline{\lim}\, nF(A_n x)$$
$$\le \overline{\lim}\, nF(A_n x) \le M^-(x+\epsilon), \quad x < 0.$$

We know that $M^+(x)$, $M^-(x)$ are continuous, so we can conclude that

$$n[1 - F(A_n x)] \to M^+(x), \quad x > 0,$$
$$nF(A_n x) \to M^-(x), \quad x < 0.$$

Now if we fix $x > 0$, for any $y > 0$ sufficiently large, there is an n such that

$$A_n x \le y \le A_{n+1} x.$$

Then

$$F(A_n x) \le F(y) \le F(A_{n+1}x),$$
$$F(-A_{n+1}x) \le F(-y) \le F(-A_n x).$$

So for any $\xi > 0$,

$$\frac{1 - F(A_{n+1}\xi x)}{1 - F(A_n x)} \le \frac{1 - F(\xi y)}{1 - F(y)} \le \frac{1 - F(A_n \xi x)}{1 - F(A_{n+1}x)},$$

$$\frac{M^+(\xi x)}{M^+(x)} \le \varliminf_{y} \frac{1 - F(\xi y)}{1 - F(y)} \le \varlimsup_{y} \frac{1 - F(\xi y)}{1 - F(y)} \le \frac{M^+(\xi x)}{M^+(x)}$$

$$\Rightarrow \lim_{y \to \infty} \frac{1 - F(\xi y)}{1 - F(y)} = \frac{1}{\xi^\alpha}.$$

In exactly the same way, conclude

$$\lim_{y \to \infty} \frac{F(-\xi y)}{F(-y)} = \frac{1}{\xi^\alpha}.$$

Also,

$$\frac{F(-A_{n+1}x)}{1 - F(A_n x)} \le \frac{F(-y)}{1 - F(y)} \le \frac{F(-A_n x)}{1 - F(A_{n+1}x)},$$

which leads to

$$\lim \frac{F(-y)}{1 - F(y)} = \frac{M^-(-x)}{M^+(x)} = \frac{M^-(-1)}{M^+(1)}.$$

To get *sufficiency*, assume, for example, that $M^+ > 0$. I assert we can define constants $A_n > 0$ such that

$$n(1 - F(A_n)) \to M^+.$$

Condition (ii) implies $1 - F(x) > 0$, all $x > 0$. Take A_n such that for any $\epsilon > 0$, $n(1 - F(A_n)) \ge M^+$, but $n(1 - F(A_n + \epsilon)) \le M^+$. Then if $\varlimsup n(1 - F(A_n)) = M^+(1 + \delta)$, $\delta > 0$, there is a subsequence n' such that for every $\epsilon > 0$,

$$\varlimsup \frac{1 - F(A_{n'} + \epsilon)}{1 - F(A_{n'})} \le \frac{1}{1 + \delta}.$$

This is ruled out by condition (ii). So

$$n(1 - F(A_n x)) = n(1 - F(A_n)) \cdot \frac{1 - F(A_n x)}{1 - F(A_n)} \to M^+ \cdot \frac{1}{x^\alpha}.$$

Similarly,

$$nF(-A_n x) \to M^- \cdot \frac{1}{x^\alpha}.$$

Take F to be the distribution function of a random variable X_1, and write

$$\mu_n(B) = nP\left(\frac{X_1}{A_n} \in B\right), \quad B \in \mathcal{B}_1.$$

Then

$$\mu_n([x, \infty)) \to \frac{M^+}{x^\alpha}, \quad x > 0,$$

$$\mu_n((-\infty, x)) \to \frac{M^-}{|x|^\alpha}, \quad x < 0.$$

Therefore, the $\{\mu_n\}$ sequence is mass-preserving, and so is the $\{\gamma_n\}$ sequence defined, as before, by

$$\gamma_n(B) = \int_B \frac{y^2}{1 + y^2} \mu_n(dy), \quad B \in \mathcal{B}_1.$$

Take $f(x)$ any bounded continuous function, and $\epsilon > 0$:

$$\int f(x)\gamma_n(dx) = \int_{(-\epsilon,+\epsilon)^c} f(x)\gamma_n(dx) + \int_{(-\epsilon,+\epsilon)} f(x)\gamma_n(dx).$$

Let $g(x) = f(x)x^2/(1 + x^2)$. The first integral converges to

$$\alpha M^+ \int_\epsilon^\infty g(x) \frac{dx}{x^{\alpha+1}} + \alpha M^- \int_{-\infty}^{\epsilon} g(x) \frac{dx}{|x|^{\alpha+1}}$$
$$= \alpha M^+ \int_0^\infty g(x) \frac{dx}{x^{\alpha+1}} + \alpha M^- \int_{-\infty}^0 g(x) \frac{dx}{|x|^{\alpha+1}} + \delta_1(\epsilon),$$

where $\delta_1(\epsilon) \to 0$ as $\epsilon \to 0$. Thus, defining $\gamma(x)$ in the obvious way,

$$\overline{\lim} \left| \int f(x)\gamma_n(dx) - \int f(x)\gamma(dx) \right| \le \delta_1(\epsilon) + \overline{\lim} \int_{|x|<\epsilon} |f(x)|\, \gamma_n(dx).$$

To complete the proof that $\gamma_n \xrightarrow{\mathcal{D}} \gamma$, we need

Proposition 9.35

$$\overline{\lim} \int_{|x|<\epsilon} x^2 \mu_n(dx) = \delta_2(\epsilon),$$

where $\delta_2(\epsilon) \to 0$ *as* $\epsilon \to 0$.

In order not to interrupt the main flow of the proof, I defer the proof of 9.35 until the end. Since $A_n \to \infty$, the characteristic function $g_n(u)$ of $X_1 + \cdots + X_n/A_n$ is given, as before, by

$$\log g_n(u) = (1 + \epsilon_n(u)) \int \varphi(x, u)\gamma_n(dx) + (1 + \epsilon_n(u))iu\beta_n,$$

where

$$\beta_n = \int \frac{x}{1 + x^2} \mu_n(dx).$$

and $\epsilon_n(u) \xrightarrow{uc} 0$. Since $\gamma_n \xrightarrow{\mathfrak{D}} \gamma$, the first term tends toward $\int \varphi(x, u)\gamma(dx)$. The characteristic function of $S_n/A_n - \beta_n$ is $e^{-iu\beta_n}g_n(u)$. So, if $\epsilon_n(u)\beta_n \to 0$, then the theorem (except for 9.35) follows with $B_n = \beta_n$. For n sufficiently large,

$$|\epsilon_n(u)| \leq |1 - f(u/A_n)|,$$

where f is the characteristic function of X_1. But

$$f\left(\frac{u}{A_n}\right) - 1 = \frac{1}{n}\int (e^{iux} - 1)\mu_n(dx) = \frac{1}{n}\left[\int \varphi(x, u)\gamma_n(dx) + iu\beta_n\right],$$

so it is sufficient to show that $\beta_n^2/n \to 0$. For any $\epsilon > 0$, use the Schwarz inequality to get

$$\begin{aligned} \overline{\lim}\, \frac{\beta_n^2}{n} &\leq \overline{\lim}\, \frac{1}{n}\left[\int_{|x|<\epsilon} |x|\, \mu_n(dx)\right]^2 \\ &\leq \overline{\lim}\left[\frac{1}{n}\mu_n((-\epsilon, +\epsilon))\int_{|x|<\epsilon} x^2\mu_n(dx)\right] \\ &\leq \overline{\lim} \int_{|x|<\epsilon} x^2\mu_n(dx). \end{aligned}$$

Apply 9.35 to reach the conclusion.

Proof of 9.35. We adapt a proof due to Feller [59, Vol. II]. Write

$$I(x) = \int_0^x y(1 - F(y))\, dy.$$

We begin by showing that there is a constant c such that

$$I(x) \leq cx^2(1 - F(x)).$$

For $t > \tau$,

$$I(tx) = I(\tau x) + x^2\int_\tau^t \xi(1 - F(x\xi))\, d\xi. \tag{9.36}$$

Fix $x > 1$. For any $\epsilon > 0$, take τ so large that for $\xi > \tau$,

$$\frac{1 - F(x\xi)}{1 - F(\xi)} \geq (1 - \epsilon)x^{-\alpha}.$$

From (9.36)

$$I(tx) \geq I(\tau x) + x^{2-\alpha}(1 - \epsilon)(I(t) - I(\tau)).$$

This inequality implies $I(t) \to \infty$ as $t \to \infty$. Then, dividing by $I(t)$ and letting $t \to \infty$ yields

$$\underline{\lim}\, \frac{I(tx)}{I(t)} \geq x^{2-\alpha}.$$

Now in 9.36 take $\tau = 1$, so

$$\frac{I(tx)}{I(x)} = 1 + \frac{x^2(1 - F(x))}{I(x)} \int_1^t \frac{1 - F(x\xi)}{1 - F(x)} \cdot \xi \, d\xi.$$

The integrand is bounded by ξ, so

$$\frac{I(tx)}{I(x)} \le 1 + t^2 \frac{x^2(1 - F(x))}{I(x)}, \quad t > 1.$$

Thus, taking lim inf on x,

$$t^{2-\alpha} \le 1 + t^2 \,\underline{\lim}\, \frac{x^2(1 - F(x))}{I(x)}.$$

This does it.

To relate this to 9.35 integrate by parts on $\int_{[0,\epsilon)} x^2\mu_n(dx)$, assuming ϵ is a continuity point of μ_n for all n. This gives

$$\int_{[0,\epsilon)} x^2\mu_n(dx) \le 2\int_{[0,\epsilon)} x\mu_n([x, \infty))\, dx.$$

From $\mu_n([x, \infty)) = n(1 - F(A_n x))$, we get

(9.37) $$\int_{[0,\epsilon)} x^2\mu_n(dx) \le \frac{2n}{A_n^2}\int_0^{\epsilon A_n} y(1 - F(y))\, dy = \frac{2n}{A_n^2} I(\epsilon A_n).$$

Because of the inequality proved above,

$$\int_{[0,\epsilon)} x^2\mu_n(dx) \le 2nc\epsilon^2(1 - F(A_n\epsilon)).$$

Therefore,

$$\overline{\lim} \int_{[0,\epsilon)} x^2\mu_n(dx) \le 2cM^+\epsilon^{2-\alpha}.$$

The integral over the range $(-\epsilon, 0)$ is treated in the same way, with $F(-x)$ in place of $1 - F(x)$. Q.E.D.

Problems

6. Show that $n/A_n^2 \to 0$. [For any $\delta > 0$, take x_0 such that for $x \ge x_0$,

$$\frac{1 - F(2x)}{1 - F(x)} \ge 2^{-(\alpha+\delta)}.$$

Define k by $2^{k-1} < A_n \le 2^k$; then

$$1 - F(A_n) \ge 1 - F(2^k) = \prod_{j=m}^{k} \frac{1 - F(2^j)}{1 - F(2^{j-1})}(1 - F(2^{m-1})).$$

Now select m appropriately.]

7. Show, using the same methods used to get $n/A_n^2 \to 0$, that for $F(x) \in D(\alpha)$,

$$\lim_{x\to\infty} x^\lambda[1 - F(x)] = \begin{cases} \infty, & \lambda > \alpha, \\ 0, & \lambda < \alpha. \end{cases}$$

Conclude that

$$\int |x|^\lambda\, F(dx) \begin{cases} < \infty, & \lambda < \alpha, \\ = \infty, & \lambda > \alpha. \end{cases}$$

8. For $F(x) \in D(\alpha)$, $\alpha < 1$, by using the same methods as in the proof of 9.35, show that

$$\overline{\lim} \int_{|x|\le\epsilon} |x|\, \mu_n(dx) = \delta(\epsilon), \quad \delta(\epsilon) \to 0 \quad \text{as} \quad \epsilon \to 0.$$

Conclude that β_n converges to a finite constant β; hence B_n can be defined to be zero, all n.

9. For $F(x) \in D(\alpha)$, $1 < \alpha < 2$, show that

$$\int \frac{x}{1 + x^2}\, \mu_n(dx) - \int x\mu_n(dx)$$

converges, so B_n/n can be taken as $-E\mathsf{X}_1/A_n$; that is, $\mathsf{X}_1, \mathsf{X}_2, \ldots$ are replaced by $\mathsf{X}_1 - E\mathsf{X}_1, \mathsf{X}_2 - E\mathsf{X}_2, \ldots$

10. For $F(x) \in D(\alpha)$, $\alpha < 1$, if $F(x) = 0$, $x \le 0$, then

$$\log f_{\mathsf{S}_n/A_n}(u) = \int_0^\infty (e^{iux} - 1)\mu_n(dx) \cdot \big(1 + \epsilon_n(u)\big).$$

Prove, using Problem 8 and the computations of the various integrals defining the stable laws, that

$$\frac{\mathsf{S}_n}{A_n} \xrightarrow{\mathfrak{D}} \mathsf{X}, \qquad \text{where} \qquad \log f_{\mathsf{X}}(u) = M^+ L(\alpha) e^{-i(\pi/2)\alpha} u^\alpha, \quad u \ge 0.$$

12. A COIN-TOSSING EXAMPLE

The way to recognize laws in $D(\alpha)$ is

Definition 9.38. *A function $H(x)$ on $(0, \infty)$ is said to be slowly changing if for all $\xi \in (0, \infty)$,*

$$\lim_{x\to\infty} \frac{H(\xi x)}{H(x)} = 1.$$

For example, $\log x$ is slowly changing, so is $\log \log x$, but not x^α, $\alpha \ne 0$.

Proposition 9.39. *$F(x) \in D(\alpha)$ if and only if, defining $H^+(x)$, $H^-(x)$ on $(0, \infty)$ by*

$$1 - F(x) = \frac{1}{x^\alpha} H^+(x), \qquad F(-x) = \frac{1}{x^\alpha} H^-(x),$$

there are constants $M^+ \geq 0$, $M^- \geq 0$, $M^+ + M^- > 0$ *such that*

$M^+ > 0 \Rightarrow H^+(x)$ *slowly changing,* $M^- > 0 \Rightarrow H^-(x)$ *slowly changing, and, as* $x \to \infty$,

$$\frac{H^+(x)}{H^-(x)} \to \frac{M^+}{M^-}.$$

Proof. Obvious.

Now, in fair coin-tossing, if T_1 is the time until the first return to equilibrium, then by Problem 19, Chapter 3,

$$P(\mathsf{T}_1 > n) = \sqrt{\frac{2}{\pi}} \cdot \frac{1}{n^{1/2}} (1 + \delta_n),$$

where $\delta_n \to 0$, so $1 - F(n) \sim cn^{-1/2}$. Thus for the range $n \leq x \leq n + 1$, $1 - F(x) \sim cn^{-1/2}$ or finally, $1 - F(x) = cx^{-1/2}(1 + \delta(x))$, $\delta(x) \to 0$. Therefore, by 9.39, $F(x) \in D(\frac{1}{2})$. To get the normalizing A_n, the condition is $n(1 - F(A_n))$ converges to M^+, that is,

$$cn \cdot \frac{1}{\sqrt{A_n}} (1 + \delta(A_n)) \to M^+.$$

Take $A_n = n^2$; then $M^+ = c = \sqrt{2}/\sqrt{\pi}$. We conclude, if R_n is the time of the nth return to equilibrium, then

Theorem 9.40. $\mathsf{R}_n/n^2 \xrightarrow{\mathcal{D}} \mathsf{X}$, *where*

$$\log f_{\mathsf{X}}(u) = \frac{\sqrt{2}}{\sqrt{\pi}} u^{\alpha} e^{-i(\pi/2)\alpha} L(\alpha), \quad \alpha = \tfrac{1}{2}, \quad u \geq 0.$$

Proof. See Problem 10.

13. THE DOMAIN OF ATTRACTION OF THE NORMAL LAW

If we look at the various possible distribution functions $F(x)$ and ask when can we normalize S_n so that $\mathsf{S}_n/A_n - B_n$ converges to a nondegenerate limit, we now have some pretty good answers. For $F(x)$ to be in $D(\alpha)$, the tails of the distribution must behave in a very smooth way—actually the mass of the distribution out toward ∞ must mimic the behavior of the tails of the limiting stable law. So for $0 < \alpha < 2$, only a few distribution functions $F(x)$ are in $D(\alpha)$. But the normal law is the limit for a wide class of distributions, including all $F(x)$ such that $\int x^2 F(dx) < \infty$. The obvious unanswered question is: What else does this class contain? In other words, for what distributions $F(x)$ does $\mathsf{S}_n/A_n - B_n \xrightarrow{\mathcal{D}} \mathcal{N}(0, 1)$ for an appropriate choice of A_n, B_n?

We state the result only—see Gnedenko and Kolmogorov, [62, pp. 172 ff.], for the proof.

Theorem 9.41. *There exist* A_n, B_n *such that* $\mathsf{S}_n/A_n - B_n \xrightarrow{\mathfrak{D}} \mathcal{N}(0, 1)$ *if and only if*

$$\textbf{(9.42)} \qquad \lim_{x\to\infty} \frac{x^2 \int_{|y|>x} F(dy)}{\int_{|y|\le x} y^2 F(dy)} = 0.$$

Problem 11. Show that $\int y^2F(dy) < \infty$ implies that the limit in (9.42) is zero. Find a distribution such that $\int y^2F(dy) = \infty$, but (9.42) is satisfied.

NOTES

The central limit theorem 9.2 dates back to Lyapunov [109, 1901]. The general setting of the problem into the context of infinitely divisible laws starts with Kolmogorov [94, 1932], who found all infinitely divisible laws with finite second moment, and Paul Lévy [102, 1934], who derived the general expression while investigating processes depending on a continuous-time parameter (see Chapter 14). The present framework dates to Feller [51] and Khintchine [89] in 1937. Stable distributions go back to Paul Lévy [100, 1924], and also [103, 1937]. In 1939 and 1940 Doeblin, [27] and [29], analyzed the problem of domains of attraction. One fascinating discovery in his later paper was the existence of universal laws. These are laws $\mathfrak{L}(\mathsf{X})$ such that for $\mathsf{S}_n = \mathsf{X}_1 + \cdots + \mathsf{X}_n$ sums of independent random variables each having the law $\mathfrak{L}(\mathsf{X})$, there are normalizing constants A_n, B_n such that for Y having *any* infinitely divisible distribution, there is a a subsequence $\{n_k\}$ with

$$\frac{\mathsf{S}_{n_k}}{A_{n_k}} - B_{n_k} \xrightarrow{\mathfrak{D}} \mathsf{Y}.$$

For more discussion of stable laws see Feller's book [59, Vol. II]. A much deeper investigation into the area of this chapter is given by Gnedenko and Kolmogorov [62].

CHAPTER 10

THE RENEWAL THEOREM AND LOCAL LIMIT THEOREM

1. INTRODUCTION

By sharpening our analytic tools, we can prove two more important weak limit theorems regarding the distribution of sums of independent, identically distributed random variables. We group these together because the methods are very similar, involving a more delicate use of characteristic functions and Fourier analytical methods. In the last sections, we apply the local limit theorem to get occupation time laws. This we do partly because of their own interest, and also because they illustrate the use of Tauberian arguments and the method of moments.

2. THE TOOLS

A basic necessity is the

Riemann-Lebesgue lemma. *Let $f(x)$ be $\mathcal{B}_1$-measurable and $\int |f(x)|\, dx < \infty$; then*

(10.1)
$$\lim_{u \to \pm\infty} \int e^{iux} f(x)\, dx = 0.$$

Proof. For any $\epsilon > 0$, take I a finite interval such that

$$\int_{I^c} |f(x)| \le \epsilon$$

and take M such that

$$\int_{\{|f(x)| > M\}} |f(x)|\, dx \le \epsilon.$$

Therefore,

$$\left| \int e^{iux} f(x)\, dx - \int_{I \cap \{|f(x)| \le M\}} e^{iux} f(x)\, dx \right| \le 2\epsilon,$$

so it is sufficient to prove this lemma for bounded $f(x)$ vanishing off of finite intervals I. Then (see Problem 5, Chapter 5) for any $\epsilon > 0$, there are disjoint intervals $I_1, \ldots, I_n$ such that

$$\int \left| f(x) - \sum_k \alpha_k \chi_{I_k}(x) \right| dx \le \epsilon.$$

By direct computation,

$$\int e^{iux}\chi_{I_k}(x)\,dx \to 0$$

as u goes to $\pm\infty$.

Next, we need to broaden the concept of convergence in distribution. Consider the class $\mathcal{Q}$ of σ-additive measures μ on $\mathcal{B}_1$ such that $\mu(I)$ is finite for every finite interval I.

Definition 10.2. *Let $\mu_n, \mu \in \mathcal{Q}$. Say that μ_n converges weakly to μ, $\mu_n \xrightarrow{w} \mu$, if for every continuous function $\varphi(x)$ vanishing off of a finite interval*

$$\int \varphi(x)\mu_n(dx) \to \int \varphi(x)\mu(dx). \tag{10.3}$$

If the μ_n, μ have total mass one, then weak convergence coincides with convergence in distribution. Some of the basic results concerning convergence in distribution have easy extensions.

Proposition 10.4. *$\mu_n \xrightarrow{w} \mu$ iff for every Borel set A contained in a finite interval such that $\mu(bd(A)) = 0$,*

$$\mu_n(A) \to \mu(A).$$

The proof is exactly the same as in Chapter 8. The Helly-Bray theorem extends to

Theorem 10.5. *If $\mu_n \in \mathcal{Q}$ and for every finite interval I, $\overline{\lim}\ \mu_n(I) < \infty$, then there is a subsequence μ_{n_k} converging weakly to $\mu \in \mathcal{Q}$.*

Proof. For $I_k = [-k, +k]$, $k = 1, 2, \ldots$, use the Helly-Bray theorem to get an ordered subset N_k of the positive integers such that on the interval I_k, $\mu_n \xrightarrow{w} \mu^{(k)}$ as n runs through N_k, and $N_{k+1} \subset N_k$. Here we use the obvious fact that the Helly-Bray theorem holds for measures whose total mass is bounded by the same constant. Let μ be the measure that agrees with $\mu^{(k)}$ on Borel subsets of I_k. Since $N_{k+1} \subset N_k$, μ is well defined. Let n_k be the kth member of N_k. Then clearly, $\mu_{n_k} \xrightarrow{w} \mu$.

There are also obvious generalizations of the ideas of separating classes of functions. But the key result we need is this:

Definition 10.6. *Let $\mathcal{H}$ be the class of $\mathcal{B}_1$-measurable complex-valued functions $h(x)$ such that*

$$\int |h(x)|\,dx < \infty$$

and

$$h(x) = \int e^{iux}\hat{h}(u)\,du,$$

where $\hat{h}(u)$ is real and vanishes outside of a finite interval I.

Note that if $h \in \mathcal{H}$, then for any real v, the function $e^{ivx}h(x)$ is again in $\mathcal{H}$.

Theorem 10.7. *Let μ_n, $\mu \in \mathcal{Q}$. Suppose that there is an everywhere positive function $h_0 \in \mathcal{H}$ such that $\int h_0\, d\mu$ is finite, and*

$$\int h\, d\mu_n \to \int h\, d\mu$$

for all functions $h \in \mathcal{H}$ of the form $e^{ivx} h_0(x)$, v real. Then $\mu_n \xrightarrow{w} \mu$.

Proof. Let $\alpha_n = \int h_0\, d\mu_n$, $\alpha = \int h_0\, d\mu$. Note that h_0, or for that matter any $h \in \mathcal{H}$, is continuous. If $\alpha = 0$, then $\mu \equiv 0$, and since $\alpha_n \to \alpha$, for any finite interval I, $\mu_n(I) \to 0$. Hence, assume $\alpha > 0$. Define probability measures ν_n, ν on $\mathcal{B}_1$ by

$$\nu_n(B) = \frac{1}{\alpha_n}\int_B h_0(x)\mu_n(dx), \qquad \nu(B) = \frac{1}{\alpha}\int_B h_0(x)\mu(dx).$$

By the hypothesis of the theorem, for all real v,

$$\int e^{ivx}\nu_n(dx) \to \int e^{ivx}\nu(dx),$$

so that $\nu_n \xrightarrow{w} \nu$. Thus, by 8.12, for any bounded continuous $g(x)$,

$$\int g(x)h_0(x)\mu_n(dx) \to \int g(x)h_0(x)\mu(dx).$$

For any continuous $\varphi(x)$ vanishing off of a finite interval I, take $g(x) = \varphi(x)/h_0(x)$ to conclude

$$\int \varphi(x)\mu_n(dx) \to \int \varphi(x)\mu(dx).$$

A question remains as to whether $\mathcal{H}$ contains any everywhere positive functions. To see that it does, check that

$$h_\lambda(x) = \left(\frac{\sin \lambda x}{\lambda x}\right)^2, \quad \lambda > 0$$

in $\mathcal{H}$, with

$$\hat{h}_\lambda(u) = \begin{cases} (2\lambda - |u|)/4\lambda^2, & |u| \le 2\lambda, \\ 0, & |u| > 2\lambda. \end{cases}$$

Now, for λ_1, λ_2 not rational multiples of each other, the function $h_0(x) = h_{\lambda_1}(x) + h_{\lambda_2}(x)$ is everywhere positive and in $\mathcal{H}$.

3. THE RENEWAL THEOREM

For independent, identically distributed random variables $X_1, X_2, \ldots$, define, as usual, $S_n = X_1 + \cdots + X_n$, $S_0 = 0$. In the case that the sums are transient, so that $P(S_n \in I \text{ i.o.}) = 0$, all finite I, there is one major result concerning the interval distribution of the sums. It interests us chiefly in the case of finite nonzero first moment, say $0 < E\mathsf{X}_1 < \infty$. The law of large numbers guarantees $\mathsf{S}_n \to \infty$ a.s. But there is still the possibility of a more or less regular progression of the sums out to ∞. Think of $\mathsf{X}_1, \mathsf{X}_2, \ldots$ as the successive life spans of light bulbs in a given socket. After a considerable number of years of operation, the distributions of this replacement process should be invariant under shifts of time axis. For instance, the expected number of failures in any interval of time I should depend only on the length of I. This is essentially the renewal theorem: Let l_d assign mass d to every point of L_d, l_0 is Lebesgue measure on L_0.

Theorem 10.8. *Suppose* X_1 *is distributed on the lattice* L_d, $d \geq 0$. *Let* $\mathsf{N}(I)$ *be the number of members of the sequence* $\mathsf{S}_0, \mathsf{S}_1, \mathsf{S}_2, \ldots$ *landing in the finite interval I. Then as* $y \to \infty$ *through* L_d,

$$\lim_y E\mathsf{N}(I + y) = \frac{l_d(I)}{E\mathsf{X}_1}.$$

Remarks. The puzzle is why the particular limit of 10.8 occurs. To get some feel for this, look at the nonlattice case. Let $\mathsf{N}(B)$ denote the number of landings of the sequence $\mathsf{S}_0, \mathsf{S}_1, \ldots$ in any Borel set B. Suppose that $\lim_y E\mathsf{N}(B + y)$ existed for all $B \in \mathcal{B}_1$. Denote this limit by $Q(B)$. Note that B_1, B_2 disjoint implies $\mathsf{N}(B_1 \cup B_2) = \mathsf{N}(B_1) + \mathsf{N}(B_2)$. Hence $Q(B)$ is a nonnegative, additive set function. With a little more work, its σ-additivity and σ-finiteness can be established. The important fact now is that for any $x \in R^{(1)}, B \in \mathcal{B}_1, Q(B + x) = Q(B)$; hence Q is invariant under translations. Therefore Q must be some constant multiple of Lebesgue measure, $Q(dx) = \alpha\, dx$. To get α, by adding up disjoint intervals, see that

$$E\mathsf{N}([0, x]) \sim \alpha x.$$

Let $\mathsf{N}(x) = \mathsf{N}([0, x])$, and assume in addition that $\mathsf{X}_1 \geq 0$ a.s., then

$$\mathsf{N}(x) = \max\{n;\ \mathsf{S}_n \leq x\}.$$

By the law of large numbers,

$$\frac{\mathsf{S}_n}{n} \xrightarrow{\text{a.s.}} E\mathsf{X}_1. \tag{10.9}$$

Along every sample sequence such that (10.9) holds $\mathsf{N}(x) \to \infty$, and

$$\lim_{x\to\infty} \frac{\mathsf{S}_{\mathsf{N}(x)}}{\mathsf{N}(x)} = E\mathsf{X}_1, \qquad \lim_{x\to\infty} \frac{\mathsf{S}_{\mathsf{N}(x)+1}}{\mathsf{N}(x) + 1} = E\mathsf{X}_1.$$

Since

$$\frac{\mathsf{S}_{\mathsf{N}(x)}}{\mathsf{N}(x)} \leq \frac{x}{\mathsf{N}(x)} < \frac{\mathsf{S}_{\mathsf{N}(x)+1}}{\mathsf{N}(x)},$$

then for the sequences such that (10.9) holds,

$$\frac{\mathsf{N}(x)}{x} \to \frac{1}{E\mathsf{X}_1}.$$

If we could take expectations of this then $\alpha = 1/E\mathsf{X}_1$. (This argument is due to Doob [35].)

Proof of 10.8. Define the renewal measure on $\mathcal{B}_1$ by

$$H(B) = E\mathsf{N}(B), \quad B \in \mathcal{B}_1.$$

The theorem states that $H(\cdot + y) \xrightarrow{w} l_d(\cdot)/E\mathsf{X}_1$ as y goes to plus infinity through L_d. For technical reasons, it is easier to define

$$\mu_y(B) = H(B + y) + H(-B - y), \quad B \in \mathcal{B}_1.$$

The second measure converges weakly to zero as y goes to plus infinity through L_d. Because (from Chapter 3, Section 7) if I is any interval of length a, then

$$E\mathsf{N}(I) \leq P(\mathsf{S}_n \in I \text{ at least once})\, E\mathsf{N}([-a, +a]).$$

So for $I = [b - a, b]$

$$E\mathsf{N}(I - y) \leq P\left(\inf_n \mathsf{S}_n \leq b - y\right) E\mathsf{N}([-a, +a]).$$

The fact that $E\mathsf{X}_1 > 0$, implies that the sums are transient, and $E\mathsf{N}(I) < \infty$ for all finite intervals I. Since $\mathsf{S}_n \to \infty$ a.s., $\inf_n \mathsf{S}_n$ is a.s. finite. Hence the right-hand side above goes to zero as $y \to +\infty$. So it is enough to show that $\mu_y \xrightarrow{w} l_d/E\mathsf{X}_1$. Let $h \in \mathcal{H}$. The first order of business is to evaluate $\int h(x)\mu_y(dx)$. Since

$$\mathsf{N}(B) = \sum_0^\infty \chi_B(\mathsf{S}_n), \quad B \in \mathcal{B}_1,$$

then

$$H(B) = \sum_0^\infty P(\mathsf{S}_n \in B).$$

Using

$$h(x) = \int e^{iux}\hat{h}(u)\, du$$

gives

$$\int h(x)H(dx + y) = \sum_0^\infty \iint \hat{h}(u)e^{iux}P(\mathsf{S}_n \in dx + y)\, du,$$

where we assume either that $h(x)$ is nonnegative, or $\int |h(x)|\, \mu_y(dx) < \infty$. Note that

$$\int e^{iux}P(\mathsf{S}_n \in dx + y) = e^{-iuy}\int e^{iux}P(\mathsf{S}_n \in dx) = e^{-iuy}[f(u)]^n,$$

so

$$\int h(x)H(dx + y) = \sum_0^\infty \int e^{-iuy}\hat{h}(u)[f(u)]^n \, du.$$

We would like to take the sum inside the integral sign, but the divergence at $u = 0$ of $\sum_0^\infty [f(u)]^n$ is troublesome. Alternatively, let

$$H_r(B) = \sum_0^\infty r^n P(\mathsf{S}_n \in B), \quad 0 \le r < 1.$$

Compute as above, that

$$\int h(x)H_r(dx + y) = \sum_0^\infty r^n \int e^{-iuy}\hat{h}(u)[f(u)]^n \, du.$$

Now there is no trouble in interchanging to get

$$\int h(x)H_r(dx + y) = \int e^{-iuy} \frac{\hat{h}(u)}{1 - rf(u)} \, du.$$

In the same way,

$$\int h(x)H_r(-dx - y) = \int e^{-iuy} \frac{\hat{h}(u)}{1 - r\bar{f}(u)} \, du,$$

where $\bar{f}(u)$ denotes the complex conjugate of $f(u)$. Then

$$\textbf{(10.10)} \qquad \int h(x)\mu_y(dx) = 2 \lim_{r\uparrow 1} \int e^{-iuy}\hat{h}(u) Rl\left(\frac{1}{1 - rf(u)}\right) du.$$

The basic result used now is a lemma due to Feller and Orey [60].

Lemma 10.11. *Let $f(u)$ be a characteristic function such that $f(u) \neq 1$ for $0 < |u| \le b$. Then on $|u| \le b$, the measure with density $Rl(1/1 - rf(u))$ converges weakly as $r \uparrow 1$ to the measure with density $Rl(1/1 - f(u))$, plus a point mass of amount $\pi/\mathsf{E}\mathsf{X}_1$ at $u = 0$. Also, the integral of $|Rl(1/1 - f(u))|$ is finite on $|u| \le b$.*

This lemma is really the core of the proof. Accept it for the moment. In the nonlattice case, $f(u) \neq 1$ for every $u \neq 0$. Thus, the limit on the right-hand side of (10.10) exists and

$$\int h(x)\mu_y(dx) = 2\pi \frac{\hat{h}(0)}{\mathsf{E}\mathsf{X}_1} + 2\int e^{-iuy}\hat{h}(u) Rl\left(\frac{1}{1 - f(u)}\right) du.$$

Apply the Riemann-Lebesgue lemma to the integral on the right, getting

$$\lim_{y\to\infty} \int h(x)\mu_y(dx) = \frac{2\pi}{\mathsf{E}\mathsf{X}_1}\hat{h}(0).$$

The inversion formula (8.40) yields

$$\lim_{y\to\infty} \int h(x)\mu_y(dx) = \frac{1}{EX_1}\int h(x)\,dx,$$

which proves the theorem in the nonlattice case. In the lattice case, say $d = 1$, put

$$g(u) = \sum_{-\infty}^{+\infty} \hat{h}(u + 2\pi k).$$

Since $y \in L_1$, use the notation μ_n instead of μ_y. Then (10.10) becomes, since $f(u)$ is periodic,

$$\int h(x)\mu_n(dx) = 2\lim_{r\uparrow 1}\int_{-\pi}^{+\pi} e^{-iun}g(u)Rl\left(\frac{1}{1 - rf(u)}\right)du.$$

Furthermore, since L_1 is the minimal lattice for $X_1, f(u) \neq 1$ on $0 < |u| \leq \pi$. Apply lemma 10.11 again:

$$\int h(x)\mu_n(dx) = 2\pi\frac{g(0)}{EX_1} + 2\int_{-\pi}^{+\pi} e^{-iun}g(u)Rl\left(\frac{1}{1 - f(u)}\right)du.$$

The Riemann-Lebesgue lemma gives

$$\lim_n \int h(x)\mu_n(dx) = 2\pi\frac{g(0)}{EX_1}.$$

Now look:

$$h(m) = \int e^{ium}\hat{h}(u)\,du = \int_{-\pi}^{+\pi} e^{ium}g(u)\,du.$$

By the inversion formula for Fourier series, taking $\hat{h}(u)$ so that $\sum_{-\infty}^{+\infty}|h(m)| < \infty$,

$$g(u) = \frac{1}{2\pi}\sum_{-\infty}^{+\infty} e^{-ium}h(m),$$

$$g(0) = \frac{1}{2\pi}\int h(x)l_1(dx),$$

finishing the proof in the lattice case.

Now for the proof of the lemma. Let $\varphi(u)$ be any real-valued continuous function vanishing for $|u| \geq b$, such that $\sup|\varphi| \leq 1$. Consider

$$\int \varphi(u)Rl\left[\frac{1}{1 - rf(u)} - \frac{1}{1 - f(u)}\right]du$$
$$= -\int \frac{(1 - r)\varphi(u)}{|1 - rf(u)|^2}Rl\left[\frac{f(u)(1 - r\bar{f}(u))}{1 - f(u)}\right]du.$$

For any $\epsilon > 0$, the integrand converges boundedly to zero on $(-\epsilon, +\epsilon)^c$ as $r \uparrow 1$. Consider the integral over $(-\epsilon, +\epsilon)$. The term in brackets equals.

(10.12)
$$\frac{f(u)(1 - \bar{f}(u))}{1 - f(u)} + (1 - r)\frac{|f(u)|^2}{1 - f(u)}.$$

Using $|1 - rf| \geq 1 - r$, we find that the integral above containing the second term of (10.12) over $(-\epsilon, +\epsilon)$ is dominated by

$$\int_{|u|<\epsilon} |Rl(1/1 - f)|\, du.$$

Assuming that $|Rl(1/1 - f)|$ has a finite integral, we can set this term arbitrarily small by selection of ϵ. The function $(1 - \bar{f})/(1 - f)$ is continuous for $0 < |u| \leq b$. Use the Taylor expansion,

$$f(u) = 1 + iu(EX + \delta(u)),$$

where $\delta(u) \to 0$ as $u \to 0$, to conclude that the limit of $(1 - \bar{f})/(1 - f)$ exists equal to -1 as $u \to 0$. Define its value at zero to be -1 making $(1 - \bar{f})/(1 - f)$ a continuous function on $|u| \leq b$. Then the integral containing the first term of (10.12) is given by

(10.13)
$$\lim_{r\uparrow 1} \int_{|u|<\epsilon} \frac{1 - r}{|1 - rf(u)|^2}\, g(u)\, du,$$

where $g(u)$ is a continuous function such that $g(0) = \varphi(0)$. Denote $m = EX$. Use the Taylor expansion again to see that for any $\Delta > 0$, ϵ can be selected so that for $|u| < \epsilon$

$$(1 - \Delta)[(1 - r)^2 + u^2m^2] \leq |1 - rf(u)|^2 \leq (1 + \Delta)[(1 - r)^2 + u^2m^2].$$

Combine this with

$$\lim_{r\uparrow 1} \int_{|u|<\epsilon} \frac{1 - r}{(1 - r)^2 + u^2m^2}\, du = \lim_{r\uparrow 1} \int_{|v|<\epsilon/1-r} \frac{1}{1 + v^2m^2}\, dv$$
$$= \int_{-\infty}^{+\infty} \frac{1}{1 + v^2m^2}\, dv = \frac{\pi}{m}$$

to conclude that the limit of (10.13) is $\pi\varphi(0)/m$. The last fact we need now is the integrability of

$$Rl\left(\frac{1}{1 - f(u)}\right) = \frac{Rl(1 - f(u))}{|1 - f(u)|^2}$$

on $|u| \leq b$. Since $|1 - f(u)|^2 \geq m^2u^2/2$ in some neighborhood of the origin, it is sufficient to show that $Rl(1 - f(u))/u^2$ has a finite integral. Write

$$\frac{1}{u^2} Rl(f(u) - 1) = \frac{1}{u^2} \int Rl(e^{iux} - 1 - i \sin ux)\, dF(x),$$

so that

$$\int \frac{1}{u^2} Rl(1 - f(u))\, du \le \iint \frac{1}{u^2} |e^{iux} - 1 - i \sin ux|\, dF(x)\, du$$

$$\le J \int |x|\, dF(x),$$

where the orders of integration have been interchanged, and

$$J = \int \frac{1}{u^2 x^2} |e^{iux} - 1 - i \sin ux|\, |x|\, du$$

$$= \int \frac{1}{v^2} |e^{iv} - 1 - i \sin v|\, dv < \infty.$$

4. A LOCAL CENTRAL LIMIT THEOREM

For fair coin-tossing, a combinatorial argument gives $P(S_{2n} = 0) \sim 1/\sqrt{\pi n}$, a result that has been useful. What if the $X_1, X_2, \ldots$ are integer-valued, or distributed on the lattice L_d? More generally, what about estimating $P(S_n \in I)$? Look at this optimistic argument:

(10.14) $$P(S_n < \sigma\sqrt{n}x) \to \Phi(x)$$

if $EX_1 = 0$, $EX_1^2 = \sigma^2 < \infty$. By Problem 4, Chapter 8, the supremum of the difference in (10.14) goes to zero. Thus by substituting $y = \sigma\sqrt{n}\, x$,

$$P(y_1 \le S_n \le y_2) - \frac{1}{\sqrt{2\pi}} \int_{y_1/\sigma\sqrt{n}}^{y_2/\sigma\sqrt{n}} e^{-x^2/2}\, dx \to 0.$$

Substitute in the integral, and rewrite as

(10.15) $$P(y_1 \le S_n < y_2) - \frac{1}{\sigma\sqrt{2\pi n}} \int_{y_1}^{y_2} e^{-\xi^2/2\sigma^2 n}\, d\xi \to 0.$$

If the convergence is so rapid that the difference in (10.15) is $o(1/\sqrt{n})$, then multiplying by $\sigma\sqrt{2\pi n}$ gets us

$$\sigma\sqrt{2\pi n} P(y_1 \le S_n < y_2) - \int_{y_1}^{y_2} e^{-\xi^2/2\sigma^2 n}\, d\xi \to 0.$$

The integrand goes uniformly to one, so rewrite this as

$$\sigma\sqrt{2\pi n} P(y_1 \le S_n < y_2) \to y_2 - y_1.$$

This gives the surprising conclusion that estimates like $P(S_{2n} = 0) \sim 1/\sqrt{\pi n}$ may not be peculiar to fair coin-tossing, but may hold for all sums of independent, identically distributed random variables with zero means and finite variance. The fact that this delicate result is true for most distributions is a special case of the local central limit theorem. It does not hold generally—look at coin-tossing again: $P(S_n = 0) = 0$ for n odd. The next definition restricts attention to those $X_1, X_2, \ldots$ such that the sums do not have an unpleasant periodicity.

Definition 10.16. *A random variable X is called a centered lattice random variable if there exists $d > 0$ such that $P(X \in L_d) = 1$, and there is no $d' > d$ and α such that $P(X \in \alpha + L_{d'}) = 1$. X is called centered nonlattice if there are no numbers α and $d > 0$ such that $P(X \in \alpha + L_d) = 1$.*

For example, a random variable X with $P(X = 1) = P(X = 3) = \frac{1}{2}$ is not centered lattice, because L_1 is the minimal lattice with $P(X \in L) = 1$, but $P(X \in 1 + L_2) = 1$.

As before, let l_d assign mass d to every point of L_d, and l_0 denotes Lebesgue measure on L_0.

Theorem 10.17. *Let $X_1, X_2, \ldots$ be independent, identically distributed random variables, either centered lattice on L_d or centered nonlattice on L_0, with $EX_1^2 = \sigma^2 < \infty$. Then for any finite interval I,*

$$\sigma\sqrt{2\pi n}P(S_n \in I) \to l_d(I). \tag{10.18}$$

Proof. In stages—first of all, if X is centered nonlattice, then by Problem 21, Chapter 8, $|f(u)| < 1$, $u \neq 0$. If X is centered lattice on L_d, then $f(u)$ has period $2\pi/d$ and the only points at which $|f(u)| = 1$ are $\{2\pi k/d\}$, $k = 0, \pm 1, \ldots$ Eq. (10.18) is equivalent to the assertion that the measures μ_n defined on $\mathcal{B}_1$ by

$$\mu_n(B) = \sigma\sqrt{2\pi n}P(S_n \in B),$$

converge weakly to the measure l_d. The plan of this proof is to show that for every $h(x) \in \mathcal{H}$

$$\sigma\sqrt{2\pi n}Eh(S_n) \to \int h(x)l_d(dx), \tag{10.19}$$

and then to apply Theorem 10.7.

Now to prove (10.19): Suppose first that $|f(u)| \neq 1$ on $J - \{0\}$, J some finite closed interval, and that $\hat{h}(u)$ vanishes on J^c. Write

$$Eh(S_n) = \iint e^{iux}\hat{h}(u)\,du\,dF_n(x),$$

where F_n is the distribution function of S_n. Then

$$Eh(S_n) = \int [f(u)]^n\hat{h}(u)\,du.$$

From 8.44, $f(u) = 1 - (\sigma^2u^2/2)(1 + \delta(u))$, where $\delta(u) \to 0$ as $u \to 0$. Take $N = (-b, +b)$, so small that on N, $|\delta(u)| \le \frac{1}{2}$, $\sigma^2u^2 \le 1$. On $J - N$, $|f(u)| \le 1 - \beta$, $0 < \beta < 1$. Letting $\|\hat{h}\| = \sup |\hat{h}(u)|$, we get

$$\textbf{(10.20)}\quad Eh(\mathsf{S}_n) = \int_N [f(u)]^n \hat{h}(u)\, du + \theta_n \|\hat{h}\| (1 - \beta)^n, \qquad |\theta_n| \le \|J\|.$$

On N,

$$\textbf{(10.21)}\qquad |f(u)| \le \left|1 - \frac{\sigma^2u^2}{2}\right| + \frac{\sigma^2u^2}{2}|\delta(u)|$$

$$\le 1 - \frac{\sigma^2u^2}{2} + \frac{\sigma^2u^2}{4} \le 1 - \frac{\sigma^2u^2}{4} \le e^{-\sigma^2u^2/4}.$$

By the substitution $u = v/\sqrt{n}$,

$$\int_{\{|u|<b\}} [f(u)]^n \hat{h}(u)\, du = \int_{\{|v|<b\sqrt{n}\}} \left[f\left(\frac{v}{\sqrt{n}}\right)\right]^n \hat{h}\left(\frac{v}{\sqrt{n}}\right)\frac{dv}{\sqrt{n}}.$$

By (10.21) the integrand on the right is dominated for all v by the integrable function

$$\|\hat{h}\|\, e^{-\sigma^2v^2/4}.$$

But just as in the central limit theorem, $[f(v/\sqrt{n})]^n \to e^{-\sigma^2v^2/2}$. Since $\hat{h}(v/\sqrt{n}) \to \hat{h}(0)$, use the dominated convergence theorem for

$$\sqrt{n}\int_N [f(u)]^n \hat{h}(u)\, du \to \hat{h}(0)\int e^{-\sigma^2v^2/2}\, dv = \hat{h}(0)\frac{\sqrt{2\pi}}{\sigma}.$$

Use (10.20) to get

$$\textbf{(10.22)}\qquad \sqrt{n}Eh(\mathsf{S}_n) \to \frac{\sqrt{2\pi}}{\sigma}\hat{h}(0).$$

When X is centered nonlattice, this holds for all finite J. Furthermore, the Fourier inversion theorem gives

$$\hat{h}(u) = \frac{1}{2\pi}\int e^{-iux}h(x)\, dx.$$

By putting $u = 0$ we can prove the assertion. In the lattice case, assume $d = 1$ for convenience, so $f(u)$ has period 2π. Let $g(u) = \sum_{-\infty}^{+\infty} \hat{h}(u + 2\pi k)$ so that

$$Eh(\mathsf{S}_n) = \frac{1}{2\pi}\int_{-\pi}^{+\pi} [f(u)]^n g(u)\, du.$$

The purpose of this is that now $|f(u)| \ne 1$ on $[-\pi, +\pi] - \{0\}$, so (10.22) holds in the form

$$\sqrt{n}Eh(\mathsf{S}_n) \to \frac{\sqrt{2\pi}}{\sigma} g(0).$$

Just as in the proof of the renewal theorem,

$$g(0) = \frac{1}{2\pi}\int h(x)l_1(dx)$$

which proves the theorem in the lattice case. Q.E.D.

Problem 1. Under the conditions of this section, show that

$$\sigma\sqrt{2\pi n}P(\mathsf{S}_n \in I + x) \to l_d(I)$$

uniformly for x in bounded subsets of L_d. [See Problem 4, Chapter 8.]

5. APPLYING A TAUBERIAN THEOREM

Let $\mathsf{X}_1, \mathsf{X}_2, \ldots$ be centered lattice. Suppose we want to get information concerning the distribution of T, the time of the first zero of the sums $\mathsf{S}_1, \mathsf{S}_2, \ldots$ From (7.42), for all $n \geq 0$,

$$1 = \sum_{m=0}^{n} P(\mathsf{S}_{n-m} = 0)P(\mathsf{T} > m)$$

with the convention $\mathsf{S}_0 \equiv 0$. Multiply this equation by r^n, $0 \leq r < 1$, and sum from $n = 0$ to ∞,

$$\frac{1}{1-r} = \left(\sum_0^\infty r^n P(\mathsf{S}_n = 0)\right)\left(\sum_0^\infty r^n P(\mathsf{T} > n)\right).$$

The local limit theorem gives

$$P(\mathsf{S}_n = 0) \sim \frac{d}{\sigma\sqrt{2\pi n}},$$

so as $r \uparrow 1$,

$$P(r) = \sum_0^\infty r^n P(\mathsf{S}_n = 0)$$

blows up. Suppose we can use the asymptotic expression for $P(\mathsf{S}_n = 0)$ to get the rate at which $P(r)$ blows up. Since

$$T(r) = \sum_0^\infty r^n P(\mathsf{T} > n)$$

is given by

$$T(r) = \frac{1}{(1-r)P(r)},$$

we have information regarding the rate at which $T(r) \to \infty$ when $r \to 1$. Now we would like to reverse direction and get information about $P(\mathsf{T} > n)$ from the rate at which $T(r)$ blows up. The first direction is called an Abelian argument; the reverse and much more difficult direction is called a Tauberian argument.

To get some feeling for what is going on, consider

$$\varphi(r) = \sum_1^\infty n^\alpha r^n, \quad \alpha > -1.$$

Put $r = e^{-s}$. For s small, a good approximation to $\varphi(r)$ is given by the integral

$$\int_0^\infty e^{-sx} x^\alpha \, dx = \frac{1}{s^{\alpha+1}} \int_0^\infty e^{-x} x^\alpha \, dx.$$

The integral is the gamma function $\Gamma(\alpha + 1)$. Since $s \sim 1 - r$ as $r \uparrow 1$, this can be made to provide a rigorous proof of

(10.23) $$\varphi(r) \sim \frac{\Gamma(\alpha + 1)}{(1 - r)^{\alpha+1}}, \qquad \text{as } r \uparrow 1.$$

Now suppose (10.23) holds for $\varphi(r) = \sum_0^\infty a_n r^n$. Can we reverse and conclude that $a_n \sim n^\alpha$? Not quite; what is true is the well-known theorem:

Theorem 10.24. *Let $\varphi(r) = \sum_0^\infty a_n r^n$, $a_n \geq 0$, $n = 0, \ldots$ Then as $n \to \infty$*

$$a_0 + \cdots + a_n \sim cn^\alpha, \quad \alpha > 0,$$

if and only if, as $r \uparrow 1$,

$$\varphi(r) \sim c(1 - r)^{-\alpha} \Gamma(\alpha + 1).$$

For a nice proof of this theorem, see Feller [59, Vol. II, p. 423]. The implication from the behavior of $\varphi(r)$ to that of $a_1 + \cdots + a_n$ is the hard part and is a special case of Karamata's Tauberian theorem [85].

We use the easy part of this theorem to show that

$$P(r) \sim (1 - r)^{-1/2} \frac{d}{\sigma\sqrt{2}} \cdot$$

This follows immediately from the local limit theorem, the fact that $\Gamma(\frac{1}{2}) = \sqrt{\pi}$, and

$$\sum_1^n \frac{1}{\sqrt{k}} \sim 2\sqrt{n}.$$

[Approximate the sums above and below by integrals of $1/\sqrt{x}$ and $1/\sqrt{x + 1}$ respectively.]

Of course, this gives

$$T(r) \sim \frac{\sigma\sqrt{2}}{d} (1 - r)^{-1/2}, \qquad \text{as } r \uparrow 1.$$

Theorem 10.25

$$P(\mathsf{T} > n) \sim \frac{\sigma}{d}\frac{\sqrt{2}}{\sqrt{n\pi}}.$$

Proof. By Karamata's Tauberian theorem, putting $p_n = P(\mathsf{T} > n)$, $c = 2^{3/2}\sigma/d\sqrt{\pi}$, we get

$$p_0 + \cdots + p_n \sim c\sqrt{n}.$$

Since the p_n are nonincreasing, for any $m \le n$ write

$$(n - m)p_n \le p_{m+1} + \cdots + p_n \le (n - m)p_{m+1}.$$

Divide by $\sqrt{n}$, put $m = [\lambda n]$, $0 < \lambda < 1$, and let $n \to \infty$ to get

$$(1 - \lambda)\,\overline{\lim}\,\sqrt{n}p_n \le c(1 - \sqrt{\lambda}) \le \frac{(1 - \lambda)}{\sqrt{\lambda}}\,\underline{\lim}\,\sqrt{m}p_m.$$

Let $\lambda \to 1$; then $(1 - \sqrt{\lambda})/1 - \lambda \to \frac{1}{2}$, so $\sqrt{n}p_n \to c/2$.

Problem 2. Let $\mathsf{N}_n(0)$ be the number of times that the sums $\mathsf{S}_1, \ldots, \mathsf{S}_n$ visit the state zero. Use the theory of stable laws and 10.25 to find constants $A_n \uparrow \infty$ such that $\mathsf{N}_n(0)/A_n \xrightarrow{\mathfrak{D}} \mathsf{X}$ where X is nondegenerate.

6. OCCUPATION TIMES

One neat application of the local central limit theorem is to the problem: Given a finite interval I, $\mathsf{S}_1, \mathsf{S}_2, \ldots$ sums of independent, identically distributed random variables $\mathsf{X}_1, \mathsf{X}_2, \ldots$ such that $E\mathsf{X}_1 = 0$, $E\mathsf{X}_1^2 = \sigma^2 < \infty$, and take $\mathsf{S}_0 \equiv 0$. Let N_n be the number of visits of $\mathsf{S}_0, \mathsf{S}_1, \ldots, \mathsf{S}_n$ to the interval I. Is there a normalization such that $\mathsf{N}_n/A_n \xrightarrow{\mathfrak{D}} \mathsf{X}$, X nondegenerate? N_n is the amount of time in the first n trials that the sums spend in the interval I, hence the name "occupation-time problem." Let $\mathsf{X}_1, \mathsf{X}_2, \ldots$ be centered lattice or nonlattice, so the local limit theorem is in force. To get the size of A_n, compute

$$E\mathsf{N}_n = \sum_0^n E\chi_I(\mathsf{S}_k) \sim \sum_0^n \frac{l_d(I)}{\sigma\sqrt{2\pi n}} \sim c\sqrt{n}.$$

Hence we take $A_n = \sqrt{n}$. The way from here is the method of moments. Let

$$\mu_k(n) = E(\mathsf{N}_n)^k = \sum_{n \ge j_k, \ldots, j_1 \ge 0} E(\chi_I(\mathsf{S}_{j_k}) \cdots \chi_I(\mathsf{S}_{j_1})).$$

By combining all permutations of the same indices $j_1, \ldots, j_k$ we get

$$\mu_k(n) = k! \sum_{n \ge j_k \ge \cdots \ge j_1 \ge 0} E(\chi_I(\mathsf{S}_{j_k}) \cdots \chi_I(\mathsf{S}_{j_1}))$$

and

$$\mu_k(n) - \mu_k(n - 1) = k! \sum_{n \ge j_{k-1} \ge \cdots \ge j_1 \ge 0} E(\chi_I(\mathsf{S}_n)\chi_I(\mathsf{S}_{j_{k-1}}) \cdots \chi_I(\mathsf{S}_{j_1})).$$

Define transition probabilities on $\mathcal{B}_1$ by

$$p^{(j)}(B \mid x) = P(\mathsf{S}_{k+j} \in B \mid \mathsf{S}_k = x) = P(\mathsf{S}_j \in B - x).$$

So

$$\mu_k(n) - \mu_k(n-1) = k! \sum_{\substack{i_1, \ldots, i_k \geq 0, \\ i_1 + \cdots + i_k = n}} \int_I \cdots \int_I \prod_{m=1}^{k} p^{(i_m)}(dx_m \mid x_{m-1}),$$

where $x_0 \equiv 0$. If

$$p_r(B \mid x) = \sum_0^\infty r^n p^{(n)}(B \mid x)$$

where $p^{(0)}(B \mid x)$ is the point mass concentrated on $\{x\}$, and $0 \leq r < 1$, then, defining $\mu_k(-1) = 0$,

$$\text{(10.26)} \qquad \sum_{n=0}^{\infty} r^n(\mu_k(n) - \mu_k(n-1)) = k! \int_I \cdots \int_I \prod_{m=1}^{k} p_r(dx_m \mid x_{m-1}).$$

Proposition 10.27

$$\sqrt{1-r}\, p_r(I \mid x) \xrightarrow{uc} \frac{1}{\sigma\sqrt{2}}\, l_d(I)$$

for $x \in L_d$ *as* $r \uparrow 1$.

Proof. By Problem 1 of this chapter

$$\sigma\sqrt{2\pi n}\, p^{(n)}(I \mid x) \xrightarrow{uc} l_d(I)$$

for $x \in L_d$. Hence for all $x \in J \cap L_d$, J a finite interval, and for any $\epsilon > 0$, there exists n_0 such that for $n \geq n_0$,

$$\frac{(1-\epsilon)l_d(I)}{\sigma\sqrt{2\pi n}} \leq p^{(n)}(I \mid x) \leq \frac{(1+\epsilon)l_d(I)}{\sigma\sqrt{2\pi n}}.$$

By 10.24 we get uc convergence of $\sqrt{1-r}\, p_r(I \mid x)$ to $2\Gamma(\frac{3}{2})l_d(I)/\sigma\sqrt{2\pi}$.

Proposition 10.28

$$\lim_{r \uparrow 1} (1-r)^{k/2} \sum_{n=0}^{\infty} r^n(\mu_k(n) - \mu_k(n-1)) = k!\left(\frac{l_d(I)}{\sigma\sqrt{2}}\right)^k.$$

Proof. On the right-hand side of (10.26), look at the first factor of the integrand multiplied by $\sqrt{1-r}$, namely

$$\sqrt{1-r}\, p_r(I \mid x_{k-1}).$$

This converges uniformly for $x_{k-1} \in I \cap L_d$ to a constant. Hence we can pull it out of the integral. Now continue this way to get the result above.

With 10.28 in hand, apply Karamata's Tauberian theorem to conclude that

$$\text{(10.29)} \qquad \mu_k(n) \sim \frac{n^{k/2}}{\Gamma((k/2)+1)}\, k!\left(\frac{l_d(I)}{\sigma\sqrt{2}}\right)^k.$$

This almost completes the proof of

Theorem 10.30

$$\frac{\mathsf{N}_n}{\sqrt{n}} \xrightarrow{\mathfrak{D}} \frac{l_d(I)}{\sigma\sqrt{2}}\,\mathsf{X},$$

where

$$P(\mathsf{X} < x) = \frac{1}{\sqrt{\pi}}\int_0^x e^{-t^2/4}\,dt.$$

Proof. By (10.29),

$$E\left(\frac{\mathsf{N}_n}{\sqrt{n}}\right)^k \to \frac{k!}{\Gamma((k/2)+1)}\left(\frac{l_d(I)}{\sigma\sqrt{2}}\right)^k.$$

Let

$$m_k = E\mathsf{X}^k.$$

Use integration by parts to show that

$$m_{k+2} = 2(k+1)m_k,$$

and deduce from this that

$$\frac{k!}{\Gamma((k/2)+1)} = E\mathsf{X}^k.$$

Proposition 8.49 implies that the distribution is here uniquely determined by the moments. Therefore Theorem 8.48 applies. Q.E.D.

NOTES

The renewal theorem in the lattice case was stated and proved by Erdös, Feller, and Pollard [49] in 1949. Chung later pointed out that in this case the theorem follows from the convergence theorem for countable Markov chains due to Kolmogorov (see Chapter 7, Section 7). The theorem was gradually extended and in its present form was proved by Blackwell [6, 7]. New proofs continue to appear. There is an interesting recent proof due to Feller, see [59, Vol. II], which opens a new aspect of the theorem. The method of proof we use is adapted from Feller and Orey [60, 1961]. Charles Stone [132] by similar methods gets very accurate estimates for the rate of convergence of $H(B + x)$ as $x \to \infty$. A good exposition of the state of renewal theory as of 1958 is given by W. L. Smith [127]. One more result that is usually considered to be part of the renewal theorem concerns the case $E\mathsf{X} = \infty$. What is known here is that if the sums $\mathsf{S}_0, \mathsf{S}_1, \ldots$ are transient, then $H(I + y) \to 0$ as $y \to \pm\infty$ for all finite intervals I. Hence, in particular, if one of $E\mathsf{X}^+$, $E\mathsf{X}^-$ is finite, this result holds (see Feller [59, Vol. II, pp. 368 ff]).

Local limit theorems for lattice random variables have a long history. The original proof of the central limit theorem for coin-tossing was gotten by first estimating $P(\mathsf{S}_n = j)$ and thus used a local limit theorem to prove the tendency toward $\mathcal{N}(0, 1)$. More local limit theorems for the lattice case are in Gnedenko and Kolmogorov [62]. Essentially the theorem given in the text is due to Shepp [122]; the method of proof follows [133]. In its form in the text, the central limit theorem is not a consequence of the local theorem. But by very similar methods somewhat sharper results can be proved. For example, in the centered nonlattice case, for any interval I, let

$$\varphi(x) = \frac{1}{\sigma\sqrt{2\pi}}\, e^{-x^2/2\sigma^2}.$$

Stone proves that

$$\sup_x |\sqrt{n}P(\mathsf{S}_n \in I + x) - l(I)\varphi(x/\sqrt{n})| \to 0,$$

and the central limit theorem for centered nonlattice variables follows from this.

The occupation time theorem 10.30 was proven for normally distributed random variables by Chung and Kac [19], and in general, by Kallianpur and Robbins [84]. Darling and Kac [23] generalized their results significantly and simplified the proof by adding the Tauberian argument.

The occupation time problem for an infinite interval, say $(0, \infty)$, is considerably different. Then N_n becomes the number of positive sums among $\mathsf{S}_1, \ldots, \mathsf{S}_n$. The appropriate normalizing factor is n, and the famous arc sine theorem states

$$P\left(\frac{\mathsf{N}_n}{n} < x\right) \to \frac{2}{\pi} \text{ arc sine } \sqrt{x}.$$

See Spitzer's book [130], Chapter 4, for a complete discussion in the lattice case, or Feller [59, Vol. II, Chapter 12], for the general case.

CHAPTER 11

MULTIDIMENSIONAL CENTRAL LIMIT THEOREM AND GAUSSIAN PROCESSES

1. INTRODUCTION

Suppose that the objects under study are a sequence $\mathbf{X}_n$ of vector-valued variables $(X_1^{(n)}, \ldots, X_k^{(n)})$ where each $X_j^{(n)}$, $j = 1, \ldots, k$, is a random variable. What is a reasonable meaning to attach to

$$\mathbf{X}_n \xrightarrow{\mathfrak{D}} \mathbf{X},$$

where $\mathbf{X} = (X_1, \ldots, X_k)$? Intuitively, the meaning of convergence in distribution is that the probability that $\mathbf{X}_n$ is in some set $B \in \mathcal{B}_k$ converges to the probability that $\mathbf{X}$ is in B, that is,

$$P(\mathbf{X}_n \in B) \to P(\mathbf{X} \in B).$$

But when we attempt to make this hold for all $B \in \mathcal{B}_k$, difficulty is encountered in that $\mathbf{X}$ may be degenerate in part. In the one-dimensional case, the one-point sets to which the limit $\mathbf{X}$ assigned positive probability gave trouble and had to be excluded. In general, what can be done is to require that

$$P(\mathbf{X}_n \in B) \to P(\mathbf{X} \in B)$$

for all sets $B \in \mathcal{B}_k$ such that $P(\mathbf{X} \in bd(B)) = 0$.

The definition we use is directed at the problem from a different but equivalent angle. Let $\mathcal{E}_0$ be the class of all continuous functions on $R^{(k)}$ vanishing off of compact sets.

Definition 11.1. *The k-vectors* $\mathbf{X}_n$ *converge in distribution* (*or in law*) *to* $\mathbf{X}$ *if for every* $f(\mathbf{x}) \in \mathcal{E}_0$,

$$Ef(\mathbf{X}_n) \to Ef(\mathbf{X}).$$

This is written as $\mathbf{X}_n \xrightarrow{\mathfrak{D}} \mathbf{X}$ *or* $\mathcal{L}(\mathbf{X}_n) \to \mathcal{L}(\mathbf{X})$. *In terms of distribution functions,* $F_n \xrightarrow{\mathfrak{D}} F$ *if and only if*

$$\int f(\mathbf{x})F_n(d\mathbf{x}) \to \int f(\mathbf{x})F(d\mathbf{x})$$

for all $f \in \mathcal{E}_0$.

By considering continuous functions equal to one on some compact set and vanishing on the complement of some slightly larger compact set conclude that if $F_n \xrightarrow{\mathfrak{D}} F$, and $F_n \xrightarrow{\mathfrak{D}} G$, then $F = G$. Define as in the one-dimensional case:

Definition 11.2. *Let $\mathcal{N}_k$ be the set of all distribution functions on $R^{(k)}$. A set $\mathfrak{L} \subset \mathcal{N}_k$ is mass-preserving if for any $\epsilon > 0$, there is a compact set A such that $F(A^c) \le \epsilon$, for all $F \in \mathfrak{L}$.*

If $F_n \xrightarrow{\mathfrak{D}} F$, then $\{F_n\}$ is mass-preserving. From this, conclude that $F_n \xrightarrow{\mathfrak{D}} F$ if and only if for every bounded continuous function $f(\mathbf{x})$ on $R^{(k)}$,

$$\int f(\mathbf{x})F_n(d\mathbf{x}) \to \int f(\mathbf{x})F(d\mathbf{x}).$$

For any rectangle S such that $F(bd(S)) = 0$, approximate χ_S above and below by continuous functions to see that $F_n(S) \to F(S)$. Conversely, approximate $\int f\, dF_n$ by Riemann sums over rectangles such that $F(bd(S)) = 0$ to conclude that $F_n \xrightarrow{\mathfrak{D}} F$ is equivalent to

$$F_n(S) \to F(S), \qquad \text{if} \qquad F(bd(S)) = 0.$$

There are plenty of rectangles whose boundaries have probability zero, because if $P(\mathbf{X} \in S) = P(X_1 \in I_1, \ldots, X_k \in I_k)$, then

$$P(\mathbf{X} \in bd(S)) \le \sum_j P(X_j \in bd(I_j)).$$

By the same approximation as in 8.12, conclude that for $f(\mathbf{x})$ bounded, $\mathcal{B}_k$-measurable and with its discontinuity set having F-measure zero, that

$$\int f\, dF_n \to \int f\, dF.$$

From this, it follows that if B is in $\mathcal{B}_k$ and $F(bd(B)) = 0$, then $F_n(B) \to F(B)$.

Problem 1. Is $F_n \xrightarrow{\mathfrak{D}} F$ equivalent to requiring

$$F_n(x_1, \ldots, x_k) \to F(x_1, \ldots, x_k)$$

at every continuity point $(x_1, \ldots, x_k)$ of F? Prove or disprove.

2. PROPERTIES OF $\mathcal{N}_k$

The properties of k-dimensional probability measures are very similar to those of one-dimensional probabilities. The results are straightforward generalizations, and we deal with them sketchily. The major result is the generalization of the Helly-Bray theorem.

Theorem 11.3. *Let $\{F_n\} \subset \mathcal{N}_k$ be mass-preserving. Then there exists a subsequence F_{n_m} converging in distribution to some $F \in \mathcal{N}_k$.*

Proof. Here is a slightly different proof that opens the way for generalizations. Take $\{f_j\} \subset \mathcal{E}_0$ to be dense in $\mathcal{E}_0$ in the sense of uniform convergence. To verify that a countable set can be gotten with this property, look at the set of all polynomials with rational coefficients, then consider the set gotten by multiplying each of these by a function h_N which is one inside the k-sphere of radius N and zero outside the sphere of radius $N + 1$. Use diagonalization again; for every j, let I_j be an ordered subset of the positive integers such that $\int f_j \, dF_n$ converges as n runs through I_j, and $I_{j+1} \subset I_j$. Let n_m be the mth member of I_m; then the limit of $\int f_j \, dF_{n_m}$ exists for all j. Take $f \in \mathcal{E}_0$, $|f - f_j| \leq \epsilon$ so that

$$\left| \int f \, dF_{n_m} - \int f_j \, dF_{n_m} \right| \leq \epsilon,$$

hence $\lim \int f \, dF_{n_m}$ exists for all $f \in \mathcal{E}_0$. Because $\{F_n\}$ is mass-preserving, $\lim \int f \, dF_{n_m} = J(f)$ exists for all bounded continuous f. Denote the open rectangle $\{(x_1, \ldots, x_k);\ x_1 < y_1, \ldots, x_k < y_k\}$ by $S(\mathbf{y})$. Take g_n bounded and continuous such that $g_n \uparrow \chi_{S(\mathbf{y})}$. Then $J(g_n)$ is nondecreasing; call the limit $F(y_1, \ldots, y_k)$. The rest of the proof is the simple verification that $F(y_1, \ldots, y_k)$ is a distribution function, and that if $F\big(bd(S(\mathbf{y}))\big) = 0$, then $F_n\big(S(\mathbf{y})\big) \to F\big(S(\mathbf{y})\big)$.

Define sets $\mathcal{E}$ of $\mathcal{N}_k$-separating functions as in 8.14, prove as in 8.15 and 8.16 that $\{F_n\}$ mass-preserving, $\int f \, dF_n$ convergent, all $f \in \mathcal{E}$ imply the existence of an $F \in \mathcal{N}_k$ such that $F_n \xrightarrow{\mathfrak{D}} F$. Obviously, one sufficient condition for a class of functions to be $\mathcal{N}_k$-separating is that they be dense in $\mathcal{E}_0$ under *uc* convergence of uniformly-bounded sequences of functions.

Theorem 11.4. *The set of complex exponentials of the form*

$$\{\exp [i(u_1 x_1 + \cdots + u_k x_k)]\}, \quad \mathbf{u} \in R^{(k)}$$

is $\mathcal{N}_k$-separating.

Proof. The point here is that for any $f \in \mathcal{E}_0$, we can take n so large that $f(\mathbf{x}) = 0$ on the complement of $S_n = \{\mathbf{x};\ x_j \in [-n, +n], j = 1, \ldots, k\}$. Now approximate $f(\mathbf{x})$ uniformly on S_n by sums of terms of the form $\exp [\pi i(m_1 x_1 + \cdots + m_k x_k)/n]$, $m_1, \ldots, m_k$, integers. The rest goes through as in Theorem 8.24.

Definition 11.5. *For* $\mathbf{u}$, $\mathbf{x} \in R^{(k)}$, *let* $(\mathbf{u}, \mathbf{x}) = \sum_1^k u_i x_i$, *and define the characteristic function of the k-vector* $\mathbf{X} = (\mathrm{X}_1, \ldots, \mathrm{X}_k)$ *as* $f_{\mathbf{X}}(\mathbf{u}) = Ee^{i(\mathbf{u}, \mathbf{X})}$ *or of the distribution function* $F(\mathbf{x})$ *as*

$$f(\mathbf{u}) = \int e^{i(\mathbf{u}, \mathbf{x})} F(d\mathbf{x}).$$

The continuity theorem holds.

Theorem 11.6. *For $F_n \in \mathcal{N}_k$ having characteristic functions $f_n(\mathbf{u})$, if*

a) $\lim_n f_n(\mathbf{u})$ *exists for every* $\mathbf{u} \in R^{(k)}$,

b) $\lim_n f_n(\mathbf{u}) = h(\mathbf{u})$ *is continuous at the origin,*

then there is a distribution function $F \in \mathcal{N}_k$ such that $F_n \xrightarrow{\mathcal{D}} F$ and $h(\mathbf{u})$ is the characteristic function of F.

Proof. The only question at all is the analog of inequality 8.29. But this is simple. Observe that

$$\int \frac{1}{u^k}\left[\int_0^u \cdots \int_0^u \prod_1^k (1 - \cos v_j x_j)\, d\mathbf{v}\right] F(d\mathbf{x})$$
$$= \int \prod_1^k \left(1 - \frac{\sin u x_j}{u x_j}\right) F(d\mathbf{x}) \geq \frac{1}{\alpha^k} F(S_u^c),$$

where $S_u = \{|x_1| \leq 1/u, \ldots, |x_k| \leq 1/u\}$. For any function $g(v_1, \ldots, v_k)$ on $R^{(k)}$ define $T_j g$ to be the function

$$g(v_1, \ldots, v_{j-1}, 0, v_{j+1}, \ldots, v_k) - \tfrac{1}{2} g(v_1, \ldots, v_j, \ldots, v_k)$$
$$- \tfrac{1}{2} g(v_1, \ldots, -v_j, \ldots, v_k).$$

Then

$$\int \prod_1^k (1 - \cos v_j x_j) F(d\mathbf{x}) = T_k \cdots T_1 f(v_1, \ldots, v_k),$$

where f is the characteristic function of F. The function $T_k \cdots T_1 f(\mathbf{v})$ is continuous and is zero at the origin. Write the inequality as

$$F(S_u^c) \leq \alpha^k \cdot \frac{1}{u^n} \int_0^u \cdots \int_0^u T_k \cdots T_1 f(\mathbf{v})\, d\mathbf{v}.$$

Now, $f_n(\mathbf{u}) \to h(\mathbf{u})$ implies

$$\overline{\lim}\, F_n(S_u^c) \leq \alpha^k \cdot \frac{1}{u^n} \int_0^u \cdots \int_0^u T_k \cdots T_1 h(\mathbf{v})\, d\mathbf{v},$$

and everything goes through.

Problems

2. If $\mathsf{X}_1^{(n)}, \ldots, \mathsf{X}_k^{(n)}$ are independent for every n, and if for j fixed,

$$\mathsf{X}_j^{(n)} \xrightarrow{\mathcal{D}} \mathsf{X}_j,$$

prove that

$$(\mathsf{X}_1^{(n)}, \ldots, \mathsf{X}_k^{(n)}) \xrightarrow{\mathcal{D}} (\mathsf{X}_1, \ldots, \mathsf{X}_k)$$

and $\mathsf{X}_1, \ldots, \mathsf{X}_k$ are independent.

3. Let $\mathbf{X}^{(n)} \xrightarrow{\mathfrak{D}} \mathbf{X}$ and $\varphi_1, \ldots, \varphi_m$ be continuous functions on $R^{(k)}$. Show that

$$\left(\varphi_1(\mathbf{X}^{(n)}), \ldots, \varphi_m(\mathbf{X}^{(n)})\right) \xrightarrow{\mathfrak{D}} \left(\varphi_1(\mathbf{X}), \ldots, \varphi_m(\mathbf{X})\right).$$

4. Show that the conclusion of Problem 3 remains true if φ_k, $k = 1, \ldots, m$, continuous is replaced by $\varphi_k(\mathbf{x})$, $k = 1, \ldots, m$ a.s. continuous with respect to the distribution on $R^{(k)}$ given by $\mathbf{X} = (\mathsf{X}_1, \ldots, \mathsf{X}_k)$.

3. THE MULTIDIMENSIONAL CENTRAL LIMIT THEOREM

To use the continuity theorem, estimates on $f(\mathbf{u})$ are needed. Write

$$\|x\| = \sqrt{\sum_1^k x_j^2}.$$

Proposition 11.7. *Let $E\,\|\mathbf{X}\|^n < \infty$, then*

$$f(\mathbf{u}) = \sum_{m=0}^{n} \frac{i^m}{m!} E[(\mathbf{u}, \mathbf{X})^m] + \delta(\mathbf{u})\,\|\mathbf{u}\|^n,$$

where $\delta(\mathbf{u}) \to 0$ as $\mathbf{u} \to 0$.

Proof. Write

$$e^{i(\mathbf{u},\mathbf{x})} = \sum_j^n \frac{i^n}{m!}(\mathbf{u}, \mathbf{x})^m + \frac{(\mathbf{u}, \mathbf{x})^n}{n!}\varphi(\mathbf{u}, \mathbf{x}),$$

where

$$\varphi(\mathbf{u}, \mathbf{x}) = \cos[\theta_1(\mathbf{u}, \mathbf{x})] + i \sin[\theta_2(\mathbf{u}, \mathbf{x})] - 1,$$

θ_1, θ_2 real and $|\theta_1| \le 1$, $|\theta_2| \le 1$. By the Schwarz inequality we have $|(\mathbf{u}, \mathbf{x})| \le \|\mathbf{u}\|\,\|\mathbf{x}\|$. Thus

$$|E[(\mathbf{u}, \mathbf{X})^n \varphi(\mathbf{u}, \mathbf{X})]| \le \|\mathbf{u}\|^n\, E(\|\mathbf{X}\|^n\, |\varphi(\mathbf{u}, \mathbf{X})|).$$

The integrand is dominated by the integrable function $3\,\|\mathbf{X}\|^n$, and $\varphi(\mathbf{u}, \mathbf{X}) \to 0$ as $\mathbf{u} \to 0$ for all ω. Apply the bounded convergence theorem to get the result.

Definition 11.8. *Given a k-vector $(\mathsf{X}_1, \ldots, \mathsf{X}_k)$, $E\mathsf{X}_j = 0$, $j = 1, \ldots, k$. Define the $k \times k$ covariance matrix Γ by*

$$\Gamma_{ij} = E\mathsf{X}_i\mathsf{X}_j, \quad i,j = 1, \ldots, k.$$

Definition 11.9. *The vector $\mathbf{X}$, $E\mathsf{X}_j = 0$, $j = 1, \ldots, k$, is said to have a joint normal distribution $\mathcal{N}(0, \Gamma)$ if*

$$f_{\mathbf{X}}(\mathbf{u}) = \exp\left[-\tfrac{1}{2}\sum_{i,j}\Gamma_{ij}u_iu_j\right].$$

Theorem 11.10. *Let* $\mathbf{X}_1, \mathbf{X}_2, \ldots$ *be independent k-vectors having the same distribution with zero means and finite covariance matrix* Γ. *Then*

$$\frac{\mathbf{X}_1 + \cdots + \mathbf{X}_n}{\sqrt{n}} \xrightarrow{\mathcal{D}} \mathcal{N}(0, \Gamma).$$

Proof

$$E(\exp i[(\mathbf{u}, \mathbf{X}_1 + \cdots + \mathbf{X}_n)/\sqrt{n}]) = [E \exp [i(\mathbf{u}, \mathbf{X})/\sqrt{n}]^n,$$

where $\mathbf{X}$ has the same distribution as $\mathbf{X}_1, \mathbf{X}_2, \ldots$ By assumption, $EX_j^2 < \infty$, $j = 1, \ldots, k$, where $\mathbf{X} = (X_1, \ldots, X_k)$. Thus $E\,\|\mathbf{X}\|^2 < \infty$. By Proposition 11.7,

$$f(\mathbf{u}) = 1 - \tfrac{1}{2}E[(\mathbf{u}, \mathbf{X})^2] + \delta(\mathbf{u})\,\|\mathbf{u}\|^2,$$

so

$$Ee^{i(\mathbf{u},\mathbf{X})/\sqrt{n}} = f(\mathbf{u}/\sqrt{n}) = 1 - (1/2n)E[(\mathbf{u}, \mathbf{X})^2] + (1/n)\,\delta(\mathbf{u}/\sqrt{n})\,\|\mathbf{u}\|^2,$$

$$\begin{aligned}(f(\mathbf{u}/\sqrt{n}))^n &= \left[1 - \frac{1}{n}\left(\tfrac{1}{2}E[(\mathbf{u}, \mathbf{X})^2] + \delta(\mathbf{u}/\sqrt{n})\,\|\mathbf{u}\|^2\right)\right]^n \\ &\to e^{-1/2E[(\mathbf{u},\mathbf{X})^2]} = e^{-1/2\Sigma\Gamma_{ij}u_iu_j}.\end{aligned}$$

Once more, as long as second moments exist, the limit is normal. There are analogs here of Theorem 9.2 for the nonidentically distributed case which involve bounds on $E\,\|\mathbf{X}_n\|^3$. But we leave this; the tools are available for as many of these theorems as we want to prove.

4. THE JOINT NORMAL DISTRIBUTION

The neatest way of defining a joint normal distribution is

Definition 11.11. *Say that* $\mathbf{Y} = (Y_1, \ldots, Y_n)$, $EY_i = 0$, *has a joint normal (or joint Gaussian) distribution with zero means if there are k independent random variables* $\mathbf{X} = (X_1, \ldots, X_k)$, *each with* $\mathcal{N}(0, 1)$ *distribution, and a* $k \times n$ *matrix A such that*

$$\mathbf{Y} = \mathbf{X}A.$$

Obviously, the matrix A and vector $\mathbf{X}$ are not unique. Say that a set $Z_1, \ldots, Z_j$ of random variables is linearly independent if there are no real numbers $\alpha_1, \ldots, \alpha_j$ not all zero, such that

$$\alpha_1 Z_1 + \cdots + \alpha_j Z_j = 0 \quad \text{a.s.}$$

Then note that the minimal k, such that there is an A, $\mathbf{X}$ as in 11.11, with $\mathbf{Y} = \mathbf{X}A$, is the maximum number of linearly independent random variables in the set $Y_1, \ldots, Y_n$. If

$$\sum_{j=1}^{n} \alpha_j Y_j = 0 \quad \text{a.s.},$$

then

$$\sum_{j=1}^{n} \alpha_j \Gamma_{ij} = 0, \quad i = 1, \ldots, n,$$

so the minimum k is also given by the rank of the covariance matrix Γ. Throughout this section, take the joint normal distribution with zero means to be defined by 11.11. We will show that it is equivalent to 11.9.

If $\mathbf{Y} = \mathbf{X}A$, then the covariance matrix of $\mathbf{Y}$ is given by

$$EY_iY_j = E\left(\sum_k X_k a_{kj}\right)\left(\sum_m X_m a_{mj}\right) = \sum_k a_{ki}a_{kj}.$$

So

$$\Gamma = A^tA.$$

This is characteristic of covariance matrices.

Definition 11.12. *Call a square matrix M symmetric if $m_{ij} = m_{ji}$, nonnegative definite if $\boldsymbol{\alpha}M\boldsymbol{\alpha}^t \geq 0$ for all vectors $\boldsymbol{\alpha}$, where $\boldsymbol{\alpha}^t$ denotes the transpose of $\boldsymbol{\alpha}$.*

Proposition 11.13. *An $n \times n$ matrix Γ is the covariance matrix of some set of random variables $Y_1, \ldots, Y_n$ if and only if*

1) *Γ is symmetric nonnegative definite,*

equivalently,

2) *there is a matrix A such that*

$$\Gamma = A^tA.$$

Proof. One way is easy. Suppose $\Gamma_{ij} = EY_iY_j$; then obviously Γ is symmetric, and $\Sigma\alpha_i\Gamma_{ij}\alpha_j = E(\Sigma\alpha_iY_i)^2 \geq 0$. For the converse, we start with the well-known result (see [63], for example) that for Γ symmetric and nonnegative definite, there is an orthogonal matrix O such that

$$O\Gamma O^t = D,$$

where D is diagonal with nonnegative elements. Then taking B to be diagonal with diagonal elements the square roots of the corresponding elements D gives $B = D^tD$ so that

$$\Gamma = (O^tD^t)(DO).$$

Take $A = DO$, then $\Gamma = A^tA$. Now take $\mathbf{X}$ with components independent $\mathcal{N}(0, 1)$ variables, $\mathbf{Y} = \mathbf{X}A$ to get the result that the covariance matrix of $\mathbf{Y}$ is Γ.

If $\mathbf{Y}$ has joint normal distribution, $\mathbf{Y} = \mathbf{X}A$, then

$$(\mathbf{u}, \mathbf{Y}) = \sum_j u_jY_j = \sum_{j,k} u_jX_ka_{kj}.$$

Since the X_k are independent $\mathcal{N}(0, 1)$,

$$Ee^{i(\mathbf{u},\mathbf{Y})} = E \exp\left[i\sum_k X_k\left(\sum_j u_ja_{kj}\right)\right]$$

$$= \exp\left[-\tfrac{1}{2}\sum_k\left(\sum_j u_ja_{kj}\right)^2\right] = \exp\left[-\tfrac{1}{2}\mathbf{u}A^tA\mathbf{u}^t\right].$$

Hence the $\mathbf{Y}$ vector has a joint normal distribution in the sense of 11.9. Furthermore, if $\mathbf{Y}$ has the characteristic function $e^{-(1/2)\mathbf{u}\Gamma\mathbf{u}^t}$, by finding A such that $\Gamma = A^t A$, taking $\tilde{\mathbf{X}}$ to have independent $\mathcal{N}(0, 1)$ components and $\tilde{\mathbf{Y}} = \tilde{\mathbf{X}}A$, then we get $\tilde{\mathbf{Y}} \overset{\mathfrak{D}}{=} \mathbf{Y}$.

We can do better. Suppose $\mathbf{Y}$ has characteristic function $e^{-(1/2)\mathbf{u}\Gamma\mathbf{u}^t}$. Take O to be an orthogonal matrix such that $O\Gamma O^t = D$ is diagonal. Consider the vector $\mathbf{Z} = \mathbf{Y}O$.

$$E(Z_iZ_j) = E\left(\sum_{k,m} Y_k o_{ki} Y_m o_{mj}\right) = \sum_{k,m} o_{ki}\Gamma_{km} o_{mj}.$$

So

$$E(Z_iZ_j) = \begin{cases} 0, & i \neq j, \\ d_{jj}, & i = j. \end{cases}$$

Thus, the characteristic function of $\mathbf{Z}$ splits into a product, the $Z_1, Z_2, \ldots$ are independent by 8.33, and Z_j is $\mathcal{N}(0, d_{jj})$. Define $X_j \equiv 0$ if $d_{jj} = 0$, otherwise $X_j = Z_j/d_{jj}$. Then the nonzero X_j are independent $\mathcal{N}(0, 1)$ variables, and there is a matrix A such that $\mathbf{Y} = \mathbf{X}A$.

A fresh memory of linear algebra will show that all that has been done here is to get an orthonormal basis for the functions $Y_1, \ldots, Y_n$. This could be done for any set of random variables $Y_1, \ldots, Y_n$, getting random variables $X_1, \ldots, X_k$ such that $EX_j^2 = 1$, $EX_iX_j = 0$, $i \neq j$. But the variables $Y_1, \ldots, Y_n$, having a joint normal distribution and zero means, have their distribution completely determined by Γ with the pleasant and unusual property that $EY_iY_j = 0$ implies Y_i and Y_j are independent. Furthermore, if $I_1 = (i_1, \ldots, i_k)$, $I_2 = (j_1, \ldots, j_m)$ are disjoint subsets of $(1, \ldots, n)$ and $\Gamma_{ij} = 0, i \in I_1, j \in I_2$, then it follows that $(Y_{i_1}, \ldots, Y_{i_k})$ and $(Y_{j_1}, \ldots, Y_{j_m})$ are independent vectors.

We make an obvious extension to nonzero means by

Definition 11.14. *Say that* $\mathbf{Y} = (Y_1, \ldots, Y_n)$, *with* $\mathbf{m} = (EY_1, \ldots, EY_n)$, *has a joint normal distribution* $\mathcal{N}(\mathbf{m}, \Gamma)$ *if* $(Y_1 - EY_1, \ldots, Y_n - EY_n)$ *has the joint normal distribution* $\mathcal{N}(0, \Gamma)$.

Also,

Definition 11.15. *If* $\mathbf{Y} = (Y_1, \ldots, Y_n)$ *has the distribution* $\mathcal{N}(0, \Gamma)$, *the distribution is said to be nondegenerate if the* $Y_1, \ldots, Y_n$ *are linearly independent, or equivalently, if the rank of* Γ *is* n.

Problems

5. Show that if $Y_1, \ldots, Y_n$ are $\mathcal{N}(0, \Gamma)$, the distribution is nondegenerate if and only if there are n independent random variables $X_1, \ldots, X_n$, each $\mathcal{N}(0, 1)$, and an $n \times n$ matrix A such that $\det(A) \neq 0$, and $\mathbf{Y} = \mathbf{X}A$.

6. Let $Y_1, \ldots, Y_n$ have a joint normal nondegenerate distribution. Show that their distribution function has a density $f(\mathbf{y})$ given by

$$f(\mathbf{y}) = \frac{1}{(2\pi)^{n/2}\sqrt{\det(\Gamma)}} \exp\left[-\tfrac{1}{2}\mathbf{y}\Gamma^{-1}\mathbf{y}^t\right].$$

7. Show that if the random variables $Y_1, \ldots, Y_n$ have the characteristic function $\exp[-\frac{1}{2}\mathbf{u}H\mathbf{u}^t]$ for some $n \times n$ matrix H, then H is the covariance matrix of $\mathbf{Y}$ and $EY_j = 0, j = 1, \ldots, n$.

5. STATIONARY GAUSSIAN PROCESS

Definition 11.16. *A double ended process* $\ldots, X_{-1}, X_0, X_1, \ldots$ *is called Gaussian if every finite subset of variables has a joint normal distribution.*

Of course, this assumes that $E|X_n| < \infty$ for all n. When is a Gaussian zero-mean process $\mathbf{X}$ stationary? Take $\Gamma(m, n) = EX_nX_m$. Since the distribution of the process is determined by $\Gamma(m, n)$, the condition should be on $\Gamma(m, n)$.

Proposition 11.17. $\mathbf{X}$ *is stationary if and only if*

$$\Gamma(m, n) = \Gamma(m - n, 0), \quad \textit{all } m, n.$$

Proof. If X is stationary then $EX_mX_n = EX_{m-n}X_{n-n} = \Gamma(m - n, 0)$. Conversely, if true, then the characteristic function of $X_1, \ldots, X_n$ is

$$\exp\left(\frac{1}{2}\sum_{k,j=1}^{n} u_ju_k\Gamma(k - j, 0)\right).$$

But this is exactly the characteristic function of $X_{1+m}, \ldots, X_{n+m}$.

Use the notation (loose),

$$\Gamma(n) = \Gamma(n, 0)$$

and call $\Gamma(n)$ the covariance function. Call a function $M(n)$ on the integers nonnegative definite if for I any finite subset of the integers, and $\alpha_j, j \in I$, any real numbers,

$$\sum_{k,j \in I} \alpha_k\alpha_j M(k - j) \geq 0.$$

Clearly a covariance function is nonnegative definite. Just as in the finite case, given any symmetric nonnegative definite function $H(n)$ on the integers, we can construct a stationary zero-mean Gaussian process such that $EX_mX_n = H(m - n)$.

How can we describe the general stationary Gaussian process? To do this neatly, generalize a bit. Let $\ldots, X_{-1}, X_0, X_1, \ldots$ be a process of complex-valued functions $X_j = U_j + iV_j$, where U_j, V_j are random variables, and $EU_j = EV_j = 0$, all j. Call it a complex Gaussian process if any finite subset of the $\{U_n, V_n\}$ have a joint normal distribution, stationary if $EX_m\bar{X}_n = \Gamma(m - n)$. The covariance function of a complex Gaussian stationary process is Hermitian, $\Gamma(-n) = \bar{\Gamma}(n)$ and nonnegative definite in the sense that for a subset I of the integers, and α_j complex numbers,

$$\sum_{k,j\in I} \alpha_k\bar{\alpha}_j\Gamma(k - j) = E\left(\left|\sum_{k\in I}\alpha_k X_k\right|^2\right) \geq 0.$$

Consider a process that is a superposition of periodic functions with random amplitudes that are independent and normal. More precisely, let $\lambda_1, \ldots, \lambda_k$ be real numbers (called frequencies), and define

(11.18)
$$X_n = \sum_{j=1}^{k} e^{i\lambda_j n} Z_j,$$

where $Z_1, \ldots, Z_k$ are independent $\mathcal{N}(0, \sigma_j^2)$ variables. The $\{X_n\}$ process is a complex Gaussian process. Further,

$$\begin{aligned} EX_n\bar{X}_m &= \sum_j \sum_m e^{i\lambda_j n} e^{-i\lambda_l m} EZ_j Z_l \\ &= \sum_j \sigma_j^2 e^{i\lambda_j(n-m)}, \end{aligned}$$

so the process is stationary. The functions $e^{i\lambda_j n}$ are the periodic components with frequency λ_j, and we may as well take $\lambda_j \in [-\pi, +\pi)$. The formula (11.18) can be thought of as representing the $\{X_n\}$ process by a sum over frequency space λ, $\lambda \in [-\pi, +\pi)$. The main structural theorem for complex normal stationary processes is that every such process can be represented as an integral over frequency space, where the amplitudes of the various frequency components are, in a generalized sense, independent and normally distributed.

6. SPECTRAL REPRESENTATION OF STATIONARY GAUSSIAN PROCESSES

The main tool in the representation theorem is a representation result for covariance functions.

Herglotz Lemma 11.19 [70]. *$\Gamma(n)$ is a Hermitian nonnegative definite function on the integers if and only if there is a finite measure $F(B)$ on $\mathcal{B}_1([-\pi, +\pi))$ such that*

$$\Gamma(n) = \int e^{in\lambda} F(d\lambda).$$

Proof. One direction is quick. If

$$\Gamma(n) = \int e^{iu\lambda} F(d\lambda),$$

then $\Gamma(n)$ is Hermitian, and

$$\sum_{k,j\in I} \alpha_k\bar{\alpha}_k\Gamma(k-j) = \int |\Sigma\, \alpha_k e^{ik\lambda}|^2 F(d\lambda) \geq 0.$$

To go the other way, following Loève [108, p. 207], define

$$f_n(\lambda) = \frac{1}{2\pi}\sum_{k=-n+1}^{n-1}\left(1 - \frac{|k|}{n}\right)\Gamma(k)e^{-ik\lambda}$$

$$= \frac{1}{2\pi n}\sum_{j=1}^{n}\sum_{m=1}^{n}\Gamma(j-m)e^{-ij\lambda}e^{im\lambda} \geq 0.$$

Multiply both sides by e^{il} for $-n+1 \leq l \leq n-1$, and integrate over $[-\pi, +\pi]$ to get

$$\left(1 - \frac{|l|}{n}\right)\Gamma(l) = \int e^{il\lambda}f_n(\lambda)\,d\lambda,$$

$$\Gamma(0) = \int f_n(\lambda)\,d\lambda.$$

Define $F_n(d\lambda)$ as the measure on $[-\pi, +\pi]$ with density $f_n(\lambda)$, and take n' a subsequence such that $F_{n'} \xrightarrow{\mathfrak{D}} F$. Then

$$\Gamma(l) = \lim \int e^{il\lambda}F_{n'}(d\lambda)$$

$$= \int e^{il\lambda}F(d\lambda).$$

Now take the mass on the point $\{\pi\}$ and put it on $\{-\pi\}$ to complete the proof.

Note that the functions $\{e^{in\lambda}\}$ are separating on $[-\pi, \pi)$, hence $F(d\lambda)$ is uniquely determined by $\Gamma(n)$. For $\Gamma(n)$ the covariance function of a complex Gaussian stationary process $\mathbf{X}$, $F(d\lambda)$ is called the *spectral distribution function* of $\mathbf{X}$. To understand the representation of $\mathbf{X}$, an integral with respect to a random measure has to be defined.

Definition 11.20. *Let $\{\mathsf{Z}(\lambda)\}$ be a noncountable family of complex-valued random variables on $(\Omega, \mathcal{F}, P)$ indexed by $\lambda \in [-\pi, +\pi)$. For I an interval $[\lambda_1, \lambda_2)$, define $\mathsf{Z}(I) = \mathsf{Z}(\lambda_2) - \mathsf{Z}(\lambda_1)$. If the Riemann sums $\Sigma f(\lambda_k)\mathsf{Z}(I_k)$, $I_1, \ldots, I_n$ a disjoint partition of $[-\pi, +\pi)$ into intervals left-closed, right-open, $\lambda_k \in I_k$, converge in the second mean to the same random variable for any sequence of partitions such that* $\max_k |I_k| \to 0$, *denote this limit random variable by*

$$\int f(\lambda)\mathsf{Z}(d\lambda).$$

Now we can state

Theorem 11.21. *Let* $\mathbf{X}$ *be a complex Gaussian stationary process on* $(\Omega, \mathcal{F}, P)$ *with spectral distribution function* $F(d\lambda)$. *Then there exists a family* $\{\mathsf{Z}(\lambda)\}$ *of complex-valued random variables on* $(\Omega, \mathcal{F}, P)$ *indexed by* $\lambda \in [-\pi, +\pi)$ *such that*

i) *for any* $\lambda_1, \ldots, \lambda_m$, $\mathsf{Z}(\lambda_1), \ldots, \mathsf{Z}(\lambda_n)$ *have a joint normal distribution,*

ii) *for* I_1, I_2, *disjoint,* $E\mathsf{Z}(I_1)\overline{\mathsf{Z}(I_2)} = 0$,

iii) $E|\mathsf{Z}(I)|^2 = F(I)$, *for all intervals* I,

iv) $\mathsf{X}_n = \int e^{in\lambda}\mathsf{Z}(d\lambda)$, a.s. *all* n.

Proof. The most elegant way to prove this is to use some elementary Hilbert space arguments. Consider the space $\mathcal{L}(\mathbf{X})$ consisting of all finite linear combinations

$$\sum \alpha_k \mathsf{X}_k,$$

where the α_k are complex numbers. Consider the class of all random variables Y such that there exists a sequence $\mathsf{Y}_n \in \mathcal{L}$ with $\mathsf{Y}_n \xrightarrow{2} \mathsf{Y}$. On this class define an inner product $(\mathsf{Y}_1, \mathsf{Y}_2)$ by $E\mathsf{Y}_1\overline{\mathsf{Y}}_2$. Call random variables equivalent if they are a.s. equal. Then it is not difficult to check that the set of equivalence classes of random variables forms a complete Hilbert space $L_2(\mathbf{X})$ under the inner product $(\mathsf{Y}_1, \mathsf{Y}_2)$. Let $L_2(F)$ be the Hilbert space of all complex-valued $\mathcal{B}_1([-\pi, +\pi))$ measurable functions $f(\lambda)$ such that $\int |f(\lambda)|^2 F(d\lambda) < \infty$ under the inner product $(f, g) = \int f\bar{g}\, dF$ (take equivalence classes again).

To the element $\mathsf{X}_n \in L_2(\mathbf{X})$, correspond the function $e^{in\lambda} \in L_2(F)$. Extend this correspondence linearly,

$$\sum \alpha_k \mathsf{X}_k \leftrightarrow \sum \alpha_k e^{ik\lambda}.$$

Let $\mathcal{L}(F)$ be the class of all finite linear combinations $\sum \alpha_k e^{ik\lambda}$. Then

$$\mathsf{Y}_1, \mathsf{Y}_2 \in \mathcal{L}(\mathbf{X}), f, g \in \mathcal{L}(F),$$

and $\mathsf{Y}_1 \leftrightarrow f$, $\mathsf{Y}_2 \leftrightarrow g$ implies $\alpha\mathsf{Y}_1 + \beta\mathsf{Y}_2 \leftrightarrow \alpha f + \beta g$, and

$$(\mathsf{Y}_1, \mathsf{Y}_2) = E(\sum \alpha_j^{(1)}\mathsf{X}_j)\overline{(\sum \alpha_k^{(2)}\mathsf{X}_k)} = \sum \alpha_j^{(1)}\bar{\alpha}_k^{(2)}\Gamma(j-k)$$
$$= \int \sum \alpha_j^{(1)}\bar{\alpha}_k^{(2)} e^{i(j-k)\lambda}F(d\lambda) = \int f(\lambda)\bar{g}(\lambda)F(d\lambda) = (f, g).$$

If $\mathsf{Y}_n \in \mathcal{L}(\mathbf{X})$ and $\mathsf{Y}_n \xrightarrow{2} \mathsf{Y}$, then Y_n is Cauchy-convergent in the second mean; consequently so is the sequence $f_n \leftrightarrow \mathsf{Y}_n$. Hence there is an $f \in L_2(F)$ such that $f_n \xrightarrow{2} f$. Define $\mathsf{Y} \leftrightarrow f$; this can be checked to give a one-to-one correspondence between $L_2(F)$ and $L_2(\mathbf{X})$, which is linear and preserves inner products.

The function $\chi_{[-\pi,\xi)}(\lambda)$ is in $L_2(F)$; let $Z(\xi)$ be the corresponding element in $L_2(\mathbf{X})$. Now to check that the family $\{Z(\xi)\}$ has the properties asserted. Begin with (i). If real random variables $Y_k^{(n)} \xrightarrow{2} Y_k$, $k = 1, \ldots, m$, and $(Y_1^{(n)}, \ldots, Y_m^{(n)})$ has a joint normal distribution for each n, then $(Y_1, \ldots, Y_m)$ is joint normal. Because each element $\Gamma_{k,j}^{(n)} = EY_k^{(n)}Y_j^{(n)}$ of the covariance matrix converges to $\Gamma_{k,j} = EY_kY_j$; hence the characteristic function of $\mathbf{Y}_n$ converges to $e^{-(1/2)\mathbf{u}\Gamma\mathbf{u}^t}$, and this must be the characteristic function of $\mathbf{Y}$. Conclude that if $Y_1, \ldots, Y_m$ are in $L_2(\mathbf{X})$, then their real and imaginary components have joint normal distributions. Thus for $\xi_1, \ldots, \xi_m$ in $[-\pi, \pi)$, the real and imaginary components of $Z(\xi_1), \ldots, Z(\xi_m)$ have a joint normal distribution. For any interval $I = [\xi_1, \xi_2)$, $Z(I) \leftrightarrow \chi_I(\lambda)$. Hence for I_1, I_2 disjoint,

$$EZ(I_1)\overline{Z(I_2)} = \int \chi_{I_1}(\lambda)\chi_{I_2}(\lambda)F(d\lambda) = 0.$$

Also,

$$E\,|Z(I)|^2 = \int \chi_I^2(\lambda)F(d\lambda) = \int \chi_I(\lambda)F(d\lambda) = F(I).$$

Lastly, take $f(\lambda)$ to be a uniformly continuous function on $[-\pi, \pi)$. For a partition of $[-\pi, \pi)$ into disjoint intervals $I_1, \ldots, I_k$ left-closed, right-open, and $\lambda_k \in I_k$,

$$\Sigma f(\lambda_k)Z(I_k) \leftrightarrow \Sigma f(\lambda_k)\chi_{I_k}(\lambda).$$

The function on the right equals $f(\lambda_k)$ on the interval I_k, and converges uniformly to $f(\lambda)$ as $\max_k |I_k| \to 0$. So, in particular, it converges in the second mean to $f(\lambda)$. If $Y \leftrightarrow f(\lambda)$, then $\Sigma f(\lambda_k)Z(I_k) \xrightarrow{2} Y$. From the definition 11.20,

$$Y = \int f(\lambda)Z(d\lambda).$$

For $f(\lambda) = e^{in\lambda}$, the corresponding element is X_n, thus

$$X_n = \int e^{in\lambda}Z(d\lambda) \quad \text{a.s.} \qquad \text{Q.E.D.}$$

Proposition 11.22. *If $\{X_n\}$ is a real stationary Gaussian process, then the family $\{Z(\lambda)\}$, $\lambda \in [-\pi, +\pi)$ has the additional properties: If $\{Z_1(\lambda)\}$, $\{Z_2(\lambda)\}$ are the real and imaginary parts of the $\{Z(\lambda)\}$ process*

$$Z(\lambda) = Z_1(\lambda) + iZ_2(\lambda),$$

then

i) *For any two intervals I, J,*

$$EZ_1(I)Z_2(J) = 0.$$

ii) *For any two disjoint intervals I, J,*

$$EZ_1(I)Z_1(J) = 0, \qquad EZ_2(I)Z_2(J) = 0.$$

Proof. Write any intervals I, J as the union of the common part $I \cap J$ and nonoverlapping parts, and apply 11.21(ii) and (iii) to conclude that the imaginary part of $EZ(I)\overline{Z(J)}$ is zero. Therefore,

$$\textbf{(11.23)} \qquad EZ_1(I)Z_2(J) + EZ_2(I)Z_1(J) = 0.$$

Inspect the correspondence set up in the proof of 11.21 and notice that if $Y = \Sigma\, \alpha_k X_k$ and $Y \leftrightarrow f(\lambda)$, then $\overline{Y} = \Sigma\, \bar{\alpha}_k X_k$ corresponds to $\overline{f(-\lambda)}$. This extends to all $L_2(X)$ and $L_2(F)$. Hence, since $\chi_I(\lambda) \leftrightarrow Z(I)$, then $\chi_{-I}(\lambda) \leftrightarrow \overline{Z(I)}$, so

$$Z(-I) = \overline{Z(I)} \quad \text{a.s.},$$

where $-I = \{\lambda;\ -\lambda \in I\}$. From this,

$$Z_1(-I) = Z_1(I), \qquad Z_2(-I) = -Z_2(I).$$

Thus, if we change I to $-I$ in (11.23), the first term remains the same, and the second term changes sign; hence both terms are zero. For I, J disjoint, $EZ(I)\overline{Z(J)} = 0$ implies $EZ_1(I)Z_1(J) = EZ_2(I)Z_2(J)$. We can use the sign change again as above to prove that both sides are individually zero. Q.E.D.

For a real stationary Gaussian process with zero means, we can deduce from 11.22(i) that the processes $\{Z_1(\lambda)\}$, $\{Z_2(\lambda)\}$ are independent in the sense that all finite subsets

$$\{Z_1(I_j)\},\quad j = 1, \ldots, n, \qquad \{Z_2(I'_k)\},\quad k = 1, \ldots, m,$$

are independent. From 11.22 (ii), we deduce that for $I_1, \ldots, I_n$ disjoint intervals, the random variables $Z_1(I_1), \ldots, Z_1(I_n)$ are independent. Similarly for $Z_2(I_1), \ldots, Z_2(I_n)$.

7. OTHER PROBLEMS

The fact that $F(d\lambda)$ completely determines the distribution of a stationary Gaussian process with zero means leads to some compact results. For example, the process $\mathbf{X}$ is ergodic if and only if $F(d\lambda)$ assigns no mass to any one-point sets [111].

The correspondence between $L_2(\mathbf{X})$ and $L_2(F)$ was exploited by Kolmogorov [96, 97] and independently by Wiener [143] in a fascinating piece of analysis that leads to the solution of the prediction problem. The starting point is this: the best predictor in a mean-square sense of X_1 based on $X_0, X_{-1}, \ldots$ is $E(X_1 \mid X_0, X_{-1}, \ldots)$. But for a Gaussian process, there are constants $\alpha_k^{(n)}$ such that

$$E(X_1 \mid X_0, \ldots, X_{-n}) = \sum_{0}^{n} \alpha_k^{(n)} X_{-k}.$$

Because by taking the $\alpha_k^{(n)}$ such that

$$E\left[\left(X_1 - \sum_0^n \alpha_k^{(n)} X_{-k}\right) X_j\right] = 0, \quad j = 0, \ldots, -n,$$

or

$$\Gamma_{1,j} = \sum_0^n \alpha_k^{(n)} \Gamma_{-k,j}, \quad j = 0, \ldots, -n,$$

then $X_1 - \sum_1^n \alpha_k^{(n)} X_{-k}$ is independent of $(X_0, X_{-1}, \ldots, X_{-n})$, so that

$$E\left(X_1 - \sum_0^n \alpha_k^{(n)} X_{-k} \mid X_0, X_{-1}, \ldots, X_{-n}\right) = 0.$$

From the Martingale theorems, since $EX_1^2 < \infty$, it is easy to deduce that

$$E(X_1 \mid X_0, \ldots, X_{-n}) \xrightarrow{2} E(X_1 \mid X_0, \ldots).$$

Hence the best predictor is in the space $L_2(X_0, X_1, \ldots)$ generated by all linear combinations of $X_0, X_{-1}, \ldots$. By the isomorphism this translates into the problem: Let $L_2^-(F)$ be the space generated by all linear combinations of $e^{ik\lambda}$, $k = 0, -1, -2, \ldots$ Find the element $f(\lambda) \in L_2^-(F)$ which minimizes

$$\int |e^{i\lambda} - f(\lambda)|^2 \, F(d\lambda).$$

In a similar way, many problems concerning Gaussian processes translate over into interesting and sometimes well-known problems in functions of a real variable, in particular, usually in the area of approximation theory.

NOTES

Note that the representation theorem 11.21 for Gaussian processes depends only on the fact that $EX_n\bar{X}_m$ depends on the difference $n - m$. Define $\{X_n\}$, $n = 0, \pm 1, \ldots$ to be a complex process *stationary in the second order* if $\Gamma(n, m) = EX_n\bar{X}_m$ is a function of $n - m$. The only difference in the conclusion of 11.21 is that (i) is deleted. This representation theorem was proved by Cramér [21, 1942], and independently by Loève [107]. Since the work on the prediction problem in 1941–42 by Kolmogorov [96] and [97], and independently by Wiener [143], there has been a torrent of publications on second-order stationary processes and a sizeable amount on Gaussian processes. For a complete and rigorous treatment of these matters, refer to Doob's book [39]. For a treatment which is simpler and places more stress on applications, see Yaglom [144].

CHAPTER 12

STOCHASTIC PROCESSES AND BROWNIAN MOTION

1. INTRODUCTION

The natural generalization of a sequence of random variables $\{X_n\}$ is a collection of random variables $\{X_t\}$ indexed by a parameter t in some interval I. Such an object we will call a stochastic process.

Definition 12.1. *A stochastic process or continuous parameter process is a collection $\{X_t(\omega)\}$ of random variables on $(\Omega, \mathcal{F}, P)$ where t ranges over an interval $I \subset R^{(1)}$. Whenever convenient the notation $\{X(t, \omega)\}$ or simply $\{X(t)\}$ will be used.*

For fixed ω, what is produced by observing the values of $X(t, \omega)$ is a function $x(t)$ on I.

The most famous stochastic process and the most central in probability theory is Brownian motion. This comes up like so: let $X(t)$ denote one position coordinate of a microscopic particle undergoing molecular bombardments in a glass of water. Make the three assumptions given below.

Assumptions 12.2

1) *Independence*: $X(t + \Delta t) - X(t)$ *is independent of* $\{X(\tau)\}$, $\tau \le t$.

2) *Stationarity*: *The distribution of* $X(t + \Delta t) - X(t)$ *does not depend on* t.

3) *Continuity*: $$\lim_{\Delta t \downarrow 0} \frac{P(|X(t + \Delta t) - X(t)| \ge \delta)}{\Delta t} = 0, \quad \text{for all } \delta > 0.$$

This is the sense of the assumptions: (1) means that the change in position during time $[t, t + \Delta t]$ is independent of anything that has happened up to the time t. This is obviously only a rough approximation. Physically, what is much more correct is that the momentum imparted to the particle due to molecular bombardments during $[t, t + \Delta t]$ is independent of what has happened up to time t. This assumption makes sense only if the displacement of the particle due to its initial velocity at the beginning of the interval $[t, t + \Delta t]$ is small compared to the displacements it suffers as a result of molecular momentum exchange over $[t, t + \Delta t]$. From a model point of view this is the worst assumption of the three. Accept it for now; later we derive the so-called exact model for the motion in which (1) will be replaced. The second assumption is quite reasonable: It simply requires homogeneity in time; that the distribution of change over any time interval depend only on the length of the time interval, and not on the location of the origin of time. This corresponds to a model in which the medium is considered to be infinite in extent.

The third assumption is interesting. We want all the sample functions of our motion to be continuous. A model in which the particle took instantaneous jumps would be a bit shocking. Split the interval [0, 1] into n parts, $\Delta t = 1/n$. If the motion is continuous, then

$$h(\Delta t) = \sup_{1 \le k \le n} |\mathsf{X}(k\,\Delta t) - \mathsf{X}((k-1)\,\Delta t)|$$

must converge to zero as $\Delta t \to 0$. At a minimum, for any $\delta > 0$,

$$\lim_{\Delta t \downarrow 0} P\big(h(\Delta t) \ge \delta\big) = 0.$$

By (1) the variables $\mathsf{Y}_k = |\mathsf{X}(k\,\Delta t) - \mathsf{X}\big((k-1)\,\Delta t\big)|$ are independent; by (2), they all have the same distribution. Thus

$$\textbf{(12.3)} \quad P\big(h(\Delta t) \ge \delta\big) = 1 - P\Big(\sup_{1\le k \le n} \mathsf{Y}_k < \delta\Big) = 1 - [P(\mathsf{Y}_1 < \delta)]^n$$
$$= 1 - [1 - P(\mathsf{Y}_1 \ge \delta)]^n \simeq 1 - e^{-nP(\mathsf{Y}_1 \ge \delta)},$$

so that $P\big(h(\Delta t) \ge \delta\big) \to 0$ if and only if $nP(\mathsf{Y}_1 \ge \delta) \to 0$. This last is exactly

$$\lim_{n\to\infty} \frac{P(|\mathsf{X}(t + \Delta t) - \mathsf{X}(t)| \ge \delta)}{\Delta t} = 0.$$

Make the further assumption that $\mathsf{X}(0) \equiv 0$. This is not a restriction, but can be done by considering the process $\mathsf{X}(t) - \mathsf{X}(0)$, $t \ge 0$, which again satisfies (1), (2), and (3). Then

Proposition 12.4. *For any process* $\{\mathsf{X}(t)\}$, $t \ge 0$ *satisfying* 12.2. (1), (2), *and* (3) *with* $\mathsf{X}(0) \equiv 0$, $\mathsf{X}(t)$ *has a normal distribution with* $E\mathsf{X}(t) = \mu t$, $\sigma^2(\mathsf{X}(t)) = \sigma^2 t$.

Proof. For any t, let $\Delta t = t/n$, $\mathsf{Y}_k = \mathsf{X}(k\,\Delta t) - \mathsf{X}\big((k-1)\,\Delta t\big)$. Then $\mathsf{X}(t) = \mathsf{Y}_1 + \cdots + \mathsf{Y}_n$, where the $\mathsf{Y}_1, \ldots, \mathsf{Y}_n$ are independent and identically distributed. Therefore $\mathsf{X}(t)$ has an infinitely divisible law. Utilize the proof in (12.3) to show that $\mathsf{M}_n = \max_{1\le k\le n} |\mathsf{Y}_k|$ converges in probability to zero. By 9.6, $\mathsf{X}(t)$ has a normal distribution. Let $\varphi_1(t) = E\mathsf{X}(t)$. Then

$$\begin{aligned}\varphi_1(t+\tau) &= E\mathsf{X}(t+\tau)\\ &= E(\mathsf{X}(t+\tau) - \mathsf{X}(\tau)) + E\mathsf{X}(\tau)\\ &= \varphi_1(t) + \varphi_1(\tau).\end{aligned}$$

Let

$$\varphi_2(t) = \sigma^2(\mathsf{X}(t)),$$

so that

$$\begin{aligned}\varphi_2(t+\tau) &= E(\mathsf{X}(t+\tau) - \varphi_1(t+\tau))^2\\ &= E(\mathsf{X}(t+\tau) - \mathsf{X}(t) - \varphi_1(\tau) + \mathsf{X}(t) - \varphi_1(t))^2\\ &= \varphi_2(t) + \varphi_2(\tau).\end{aligned}$$

The fact is now that $\varphi_1(t)$ and $\varphi_2(t)$ are continuous. This follows from $\mathsf{X}(t+\tau) \xrightarrow{\mathfrak{D}} \mathsf{X}(t)$, as $\tau \to 0$, which for normal variables implies

$$E\mathsf{X}(t+\tau) \to E\mathsf{X}(t), \qquad \sigma^2(\mathsf{X}(t+\tau)) \to \sigma^2(\mathsf{X}(t)).$$

It is easy to show that any continuous solutions of the equation $\varphi(t+\tau) = \varphi(t) + \varphi(\tau)$ are linear.

Use the above heuristics to back into a definition of Brownian motion.

Definition 12.5. *Brownian motion is a stochastic process on* $[0, \infty)$ *such that* $\mathsf{X}(0) \equiv 0$ *and the joint distribution of*

$$\mathsf{X}(t_n), \ldots, \mathsf{X}(t_0), \qquad t_n > t_{n-1} > \cdots > t_0 \geq 0,$$

is specified by the requirement that $\mathsf{X}(t_k) - \mathsf{X}(t_{k-1}), k = 1, \ldots, n$ *be independent, normally distributed random variables with*

$$\begin{aligned}E(\mathsf{X}(t_k) - \mathsf{X}(t_{k-1})) &= (t_k - t_{k-1})\mu,\\ \sigma^2(\mathsf{X}(t_k) - \mathsf{X}(t_{k-1})) &= (t_k - t_{k-1})\sigma^2.\end{aligned}$$

This can be said another way. The random variables $\mathsf{X}(t_n), \ldots, \mathsf{X}(t_0)$ have a joint normal distribution with $E\mathsf{X}(t_k) = \mu t_k$ and

$$\begin{aligned}\Gamma(t_j, t_k) &= E(\mathsf{X}(t_j) - \mu t_j)(\mathsf{X}(t_k) - \mu t_k)\\ &= E[(\mathsf{X}(t_j) - \mathsf{X}(t_k) - \mu(t_j - t_k) + \mathsf{X}(t_k) - \mu t_k)(\mathsf{X}(t_k) - \mu t_k)]\\ &= \sigma^2 t_k, \quad \text{for } t_j > t_k,\end{aligned}$$

so that $\Gamma(t_j, t_k) = \sigma^2(\min(t_j, t_k))$.

Problem 1. Show that Brownian motion, as defined by 12.5, satisfies 12.2 (1), (2), and (3).

2. BROWNIAN MOTION AS THE LIMIT OF RANDOM WALKS

There are other ways of looking at Brownian motion. Consider a particle that moves to the right or left a distance Δx with probability $\frac{1}{2}$. It does this each Δt time unit. Let $Y_1, \ldots$ be independent, and equal $\pm\Delta x$ with probability $\frac{1}{2}$ each. The particle at time t has made $[t/\Delta t]$ jumps ($[z]$ indicates greatest integer $\leq z$). Thus the position of the particle is given by

$$D(t) = Y_1 + \cdots + Y_{[t/\Delta t]}.$$

The idea is that if Δx, $\Delta t \to 0$ in the right way, then $D(t)$ will approach Brownian motion in some way. To figure out how to let Δx, $\Delta t \to 0$, note that $ED^2(t) \simeq (\Delta x)^2 t/\Delta t$. To keep this finite and nonzero, Δx has to be of the order of magnitude of $\sqrt{\Delta t}$. For simplicity, take $\Delta x = \sqrt{\Delta t}$. Take $\Delta t = 1/n$, then the $Y_1, \ldots$ equal $\pm 1/\sqrt{n}$. Thus the $D(t)$ process has the same distribution as

$$X^{(n)}(t) = \frac{Z_1 + \cdots + Z_{[nt]}}{\sqrt{n}},$$

where $Z_1, \ldots$ are ± 1 with probability $\frac{1}{2}$.

Note that

$$X^{(n)}(t) = \sqrt{t}\,\frac{Z_1 + \cdots Z_{[nt]}}{\sqrt{nt}}$$

and apply the central limit theorem to conclude $X^{(n)}(t) \xrightarrow{\mathcal{D}} \mathcal{N}(0, t)$. In addition, it is no more difficult to show that all the joint distributions of $X^{(n)}(t)$ converge to those of Brownian motion. Therefore, Brownian motion appears as the limit of processes consisting of consecutive sums of independent, identically distributed random variables, and its study is an extension of the study of the properties of such sequences.

What has been done in 12.5 is to specify all the finite-dimensional distribution functions of the process. There is now the question again: Is there a process $\{X(t)\}$, $t \in [0, \infty)$ on $(\Omega, \mathcal{F}, P)$ with these finite-dimensional distributions? This diverts us into some foundational work.

3. DEFINITIONS AND EXISTENCE

Consider a stochastic process $\{X(t)\}$, $t \in I$ on $(\Omega, \mathcal{F}, P)$. For fixed ω, $X(t, \omega)$ is a real-valued function on I. Hence *denote by R^I the class of all real-valued functions $x(t)$ on I, and by $\mathbf{X}(\omega)$ the vector variable $\{X(t,\omega)\}$ taking values in R^I.*

Definition 12.6. *$\mathcal{F}(X(s), s \in J)$, $J \subset I$, is the smallest σ-field $\mathcal{F}$ such that all $X(s)$, $s \in J$, are $\mathcal{F}$-measurable.*

Definition 12.7. *A finite-dimensional rectangle in R^I is any set of the form*

$$S = \{x(\cdot) \in R^I;\ x(t_1) \in I_1, \ldots, x(t_n) \in I_n\},$$

where $I_1, \ldots, I_n$ are intervals. Let $\mathcal{B}_I$ be the smallest σ-field of subsets of R^I containing all finite dimensional rectangles.

For the understanding of what is going on here it is important to characterize $\mathcal{B}_I$. Say that a set $B \in \mathcal{B}_I$ has a countable base $T = \{t_j\}$ if it is of the form

$$B = \{x(\cdot);\ (x(t_1), x(t_2), \ldots) \in D\}, \quad D \in \mathcal{B}_\infty.$$

This means that B is a set depending only on the coordinates $x(t_1), x(t_2), \ldots$

Proposition 12.8. *The class $\mathcal{C}$ of all sets with a countable base forms a σ-field, hence $\mathcal{C} = \mathcal{B}_I$.*

Proof. Let $B_1, B_2, \ldots \in \mathcal{C}$, and take T_k as the base for B_k, then $T = \bigcup_k T_k$ is a base for all $B_1, B_2, \ldots$, and if $T = \{t_j\}$, each B_k may be written as

$$B_k = \{x(\cdot);\ (x(t_1), \ldots) \in D_k\}, \quad D_k \in \mathcal{B}_\infty.$$

Now it is pretty clear that any countable set combinations of the B_k produce a set with base T, hence a set in $\mathcal{C}$.

Corollary 12.9. *For $B \in \mathcal{B}_I$, $\{\omega;\ \mathbf{X}(\omega) \in B\} \in \mathcal{F}$.*

Proof. By the previous proposition there is a countable set $\{t_j\}$ such that

$$B = \{x(\cdot);\ (x(t_1), \ldots) \in D\}, \quad D \in \mathcal{B}_\infty.$$

Thus $\{\omega;\ \mathbf{X}(\omega) \in B\} = \{\omega;\ (X(t_1), \ldots) \in D\}$, and this in in $\mathcal{F}$ by 2.13.

Definition 12.10. *The finite-dimensional distribution functions of the process are given by*

$$F_{t_1 \cdots t_n}(x_1, \ldots, x_n) = P(X(t_1) < x_1, \ldots, X(t_n) < x_n), \; t_1 < t_2 < \cdots < t_n.$$

The notation $F_{\mathbf{t}}(\mathbf{x})$ may also be used.

Definition 12.11. *The distribution of the process is the probability $\hat{P}$ on $\mathcal{B}$ defined by*

$$\hat{P}(B) = P(\mathbf{X} \in B).$$

It is easy to prove that

Proposition 12.12. *Any two stochastic processes on I having the same finite dimensional distribution functions have the same distribution.*

Proof. If $X(t)$, $X'(t)$ are the two processes, it follows from 2.22 that $X(t_1)$, $X(t_2), \ldots$ and $X'(t_1), X'(t_2), \ldots$ have the same distribution. Thus if B is any set with base $\{t_1, t_2, \ldots\}$, there is a set $D \in \mathcal{B}_\infty$ such that

$$P(\mathbf{X} \in B) = P((X(t_1), \ldots) \in D) = P'((X'(t_1), \ldots) \in D) = P'(\mathbf{X}' \in B).$$

The converse is also true, namely, that starting with a consistent set of finite dimensional distribution functions we may construct a process having those distribution functions.

Definition 12.13. *Given a set of distribution functions*

$$\{F_{t_1 \cdots t_n}(x_1, \ldots, x_n)\},$$

defined for all finite subsets $\{t_1 < \cdots < t_n\}$ of I. They are said to be consistent if

$$\lim_{x_k \uparrow \infty} F_{t_1 \cdots t_n}(x_1, \ldots, x_n) = F_{t_1 \cdots \hat{t}_k \cdots t_n}(x_1, \ldots, \hat{x}_k, \ldots, x_n),$$

where the ^ denotes missing.

Theorem 12.14. *Given a set of consistent distribution functions as in* 12.13 *above, there is a stochastic process $\{\mathsf{X}(t)\}$, $t \in I$, such that*

$$F_{t_1 \cdots t_n}(x_1, \ldots, x_n) = P(\mathsf{X}(t_1) < x_1, \ldots, \mathsf{X}(t_n) < x_n).$$

Proof. Take $(\Omega, \mathcal{F})$ to be $(R^I, \mathcal{B}_I)$. Denote by T, T_1, T_2, etc., countable subsets of I, and by $\mathcal{B}_T$ all sets of the form

$$\{x(\cdot);\ (x(t_1), \ldots) \in D\}, \quad D \in \mathcal{B}_\infty, \quad \{t_j\} = T.$$

By the extension theorem 2.26, there is a probability P_T on $\mathcal{B}_T$ such that

$$P_T(x(\cdot);\ x(t_1) < x_1, \ldots, x(t_n) < x_n) = F_{t_1 \cdots t_n}(x_1, \ldots, x_n).$$

Take B any set in $\mathcal{B}_I$, then by 12.8 there is a T such that $B \in \mathcal{B}_T$. We would like to define P on $\mathcal{B}_I$ by $P(B) = P_T(B)$. To do this, the question is—is the definition well defined? That is, if we let $B \in \mathcal{B}_{T_1}$, $B \in \mathcal{B}_{T_2}$, is $P_{T_1}(B) = P_{T_2}(B)$? Now $B \in \mathcal{B}_{T_1 \cup T_2}$, hence it is sufficient to show that

$$T \subset T',\ B \in \mathcal{B}_T \Rightarrow P_T(B) = P_{T'}(B).$$

But $\mathcal{B}_T \subset \mathcal{B}_{T'}$, and $P_T = P_{T'}$ on all rectangles with base in T; hence $P_{T'}$ is an extension to $\mathcal{B}_{T'}$ of P_T on $\mathcal{B}_T$, so $P_T(B) = P_{T'}(B)$. Finally, to show P is σ-additive on $\mathcal{B}_I$, take $\{B_n\}$ disjoint in $\mathcal{B}_I$; then there is a T such that all $B_1, B_2, \ldots$ are in $\mathcal{B}_T$, hence so is $\cup B_k$. Obviously now, by the σ-additivity of P_T,

$$P(\cup B_k) = P_T(\cup B_k) = \sum_k P_T(B_k) = \sum_k P(B_k).$$

The probability space is now defined. Finish by taking

$$\mathsf{X}(t, x(\cdot)) = x(t).$$

4. BEYOND THE KOLMOGOROV EXTENSION

One point of the previous paragraph was that from the definition, the most complicated sets that could be guaranteed to be in $\mathcal{F}$ were of the form

$$\{\omega;\ (\mathsf{X}(t_1), \mathsf{X}(t_2), \ldots) \in D\}, \quad D \in \mathcal{B}_\infty.$$

Furthermore, starting from the distribution functions, the extension to $\mathcal{B}_I$ is unique, and the maximal σ-field to which the extension is unique is the completion, $\bar{\mathcal{B}}_I$, the class of all sets A such that A differs from a set in $\mathcal{B}_I$ by a subset of a set of probability zero (see Appendix A.10).

Now consider sets of the form

$$A_1 = \{\mathsf{X}(t, \omega) = 0 \quad \text{at least once for } t \in I\},$$

$$A_2 = \{|\mathsf{X}(t)| \le \alpha, \quad \text{all } t \in I\}.$$

These can be expressed as

$$A_1 = \bigcup_{t \in I} \{\mathsf{X}(t, \omega) = 0\},$$

$$A_2 = \bigcap_{t \in I} \{|\mathsf{X}(t, \omega)| \le \alpha\}.$$

If each $\mathsf{X}(t)$ has a continuous distribution, then A_1 is a noncountable union of sets of probability zero. A_2 is a noncountable intersection. Neither of A_1, A_2 depends on a countable number of coordinates because

$$A_1^c = \{\mathsf{X}(t, \omega) \ne 0 \quad \text{for all } t \in I\}.$$

Clearly, A_1^c does not contain any set of the form $\{(\mathsf{X}(t_1), \ldots) \in B\}$, $B \in \mathcal{B}_\infty$. Thus A_1^c is not of the form $\{\mathbf{X} \in B\}$, $B \in \mathcal{B}_I$, so neither is A_1. Similarly, A_2 contains no sets of the form $\{\mathbf{X} \in B\}$, $B \in \mathcal{B}_I$. This forces the unpleasant conclusion that if all we are given are the joint distribution functions of the process, there is no unique way of calculating the interesting and important probabilities that a process has a zero crossing during the time interval I or remains bounded below α in absolute value during I. (Unless, of course, these sets accidentally fall in $\bar{\mathcal{B}}_I$. See Problem 3 for an important set which is not in $\bar{\mathcal{B}}_I$.)

But a practical approach that seems reasonable is: Let

$$A_\epsilon = \left\{\inf_{t \in I} |\mathsf{X}(t)| < \epsilon\right\}$$

and hope that a.s. $A_1 = \lim_{\epsilon \downarrow 0} A_\epsilon$. To compute $P(A_\epsilon)$, for $I = [0, 1]$ say, compute

$$P_n = P\left(\inf_{k \le n} |\mathsf{X}(k/n)| < \epsilon\right),$$

and define $P(A_\epsilon) = \lim_n P_n$. Note that $\{\inf_{k \le n} |\mathsf{X}(k/n)| < \epsilon\} \in \mathcal{F}$, so its probability is well-defined and computable from the distribution functions. Finally, define $P(A_1) = \lim_{\epsilon \downarrow 0} P(A_\epsilon)$. This method of approximation is appealing. How to get it to make sense? We take this up in the next section.

Problems

2. Prove for Brownian motion that the fields

$$\mathcal{F}(\mathsf{X}(t),\, t \in [a, b]) \qquad \text{and} \qquad \mathcal{F}(\mathsf{X}(t) - \mathsf{X}(b),\, t \in [b, c]),$$

$0 \le a \le b < c \le \infty$, are independent.

3. Let

$$A = \{x(\cdot) \in R^I;\, x(\cdot) \text{ is } \mathcal{B}_1(I) \text{ measurable}\}.$$

Show by considering A and A^c, that A is never in $\bar{\mathcal{B}}_I$ for any probability P on $\mathcal{B}_I$.

5. EXTENSION BY CONTINUITY

We are going to insist that all the processes we deal with in this chapter have a very weak continuity property.

Definition 12.15. *Given a stochastic process $\{\mathsf{X}(t)\}$, $t \in I$, say that it is continuous in probability if for every $t \in I$, whenever $t_n \to t$, then $\mathsf{X}(t_n) \xrightarrow{P} \mathsf{X}(t)$.*

When is $\mathsf{X}(t, \omega)$ a continuous function of t? The difficulty here again is that the set

$$\{\omega;\ \mathsf{X}(t, \omega) \text{ continuous on } I\}$$

is not necessarily in $\mathcal{F}$. It certainly does not depend on only a countable number of coordinates. However, one way of getting around the problem is to take $T = \{t_j\}$ dense in I. The set

$$F = \{\omega;\ \mathsf{X}(\cdot, \omega) \text{ uniformly continuous on } T\}$$

is in $\mathcal{F}$. To see this more clearly, for $h > 0$ define

$$\mathsf{U}(h, \omega) = \sup_{|t_i - t_j| \le h} |\mathsf{X}(t_i, \omega) - \mathsf{X}(t_j, \omega)|, \qquad t_i, t_j \in T.$$

The function $\mathsf{U}(h, \omega)$ is the supremum over a countable set of random variables, hence is certainly a random variable. Furthermore, it is decreasing in h. If as $h \downarrow 0$, $\mathsf{U}(h, \omega) \xrightarrow{\text{a.s.}} 0$, then for almost every ω, $\mathsf{X}(t, \omega)$ is a uniformly continuous function on T. Let $C \in \mathcal{F}$ be the set on which $\mathsf{U}(h, \omega) \to 0$. Assume $P(C) = 1$. For $\omega \in C$, define $\tilde{\mathsf{X}}(t, \omega)$ to be the unique continuous function on I that coincides with $X(t, \omega)$ on T. For $t \in T$, obviously $\tilde{X}(t, \omega) = X(t, \omega)$. For $t \notin T$, $\omega \in C$, $\tilde{X}(t, \omega) = \lim_{t_j \to t} X(t_j, \omega)$. Define $\tilde{X}(t, \omega)$ to be anything continuous for $\omega \in C^c$, for example $X(t, \omega) \equiv 0$, all $t \in I$, $\omega \in C^c$. But note that $X(t_j) \xrightarrow{\text{a.s.}} \tilde{X}(t)$ for $t_j \in T$, $t_j \to t$, which implies that $X(t_j) \xrightarrow{P} \tilde{X}(t)$ and implies further that $\tilde{X}(t) = X(t)$ almost surely. When I is an infinite interval, then this same construction works if for any finite interval $J \subset I$, there is probability one that $X(\cdot)$ is uniformly continuous on $T \cap J$. Thus we have proved

Theorem 12.16. *If the process $\{X(t)\}$, $t \in I$, is continuous in probability, and if there is a countable set T dense in I such that*

$$P(\omega;\ X(t, \omega) \text{ is uniformly continuous on } T \cap J) = 1$$

for every finite subinterval $J \subset I$, then there is a process $\tilde{X}(t, \omega)$ such that $\tilde{X}(t, \omega)$ is a continuous function of $t \in I$ for every fixed ω, and for each t,

$$X(t, \omega) = \tilde{X}(t, \omega) \quad \text{a.s.}$$

The revised process $\{\tilde{X}(t, \omega)\}$ and the original process $\{X(t, \omega)\}$ *have the same distribution*, because for any countable $\{t_j\}$,

$$P(\omega;\ (X(t_1), \ldots) \neq (\tilde{X}(t_1), \ldots)) = 0.$$

Not only have the two processes the same distribution, so that they are indistinguishable probabilistically, but the $\{\tilde{X}(t)\}$ process is defined on the same probability space as the original process. The $\{\tilde{X}(t)\}$ process lends itself to all the computations and devices we wanted to use before. For example: for $I = [0, 1]$,

$$\tilde{A}_1 = \{\omega;\ \exists t \in I \text{ such that } \tilde{X}(t, \omega) = 0\},$$
$$\tilde{A}_\varepsilon = \{\omega;\ \exists t \in I \text{ such that } |\tilde{X}(t, \omega)| < \varepsilon\}.$$

It is certainly now true that

$$\tilde{A}_1 = \lim_{\epsilon \downarrow 0} \tilde{A}_\epsilon.$$

But take $\tilde{A}_{n,\epsilon}$ to be the set $\tilde{A}_{n,\epsilon} = \{\omega;\ \exists k \leq 2^n, \text{ such that } |\tilde{X}(k/2^n)| < \epsilon\}$. Then

$$\tilde{A}_{n,\epsilon} \in \mathcal{F},\ \tilde{A}_{n,\epsilon} \uparrow \tilde{A}_\epsilon,$$

which implies $\tilde{A}_\epsilon \in \mathcal{F}$, so, in turn, $\tilde{A}_1 \in \mathcal{F}$. Furthermore,

$$P(\tilde{A}_\epsilon) = \lim_n P(\tilde{A}_{n,\epsilon}).$$

Therefore, by slightly revising the original process, we arrive at a process having the same distribution, for which the reasonable approximation procedures we wish to use are valid and the various interesting sets are measurable.

Obviously not all interesting stochastic processes can be altered slightly so as to have all sample functions continuous. But the basic idea always is to pick and work with the smoothest possible version of the process.

Definition 12.17. *Given two processes* $\{X(t)\}$ *and* $\{\tilde{X}(t)\}$, $t \in I$, *on the same probability space* $(\Omega, \mathcal{F}, P)$. *They will be called versions of each other if*

$$P(X(t) \neq \tilde{X}(t)) = 0, \quad \text{all } t \in I.$$

Problems

4. Show that if a process $\{X(t)\}$, $t \in I$, is continuous in probability, then for any set $\{t_n\}$ dense in I, each $X(t)$ is measurable with respect to the completion of $\mathcal{F}(X(t_1), X(t_2), \ldots)$, or that each set of $\mathcal{F}(X(t), t \in I)$ differs from a set of $\mathcal{F}(X(t_1), \ldots)$ by a set of probability zero.

5. Conclude from the above that if $T_n \subset T_{n+1}$, $T_n \uparrow T$, T_n finite subsets of I, T dense in I, then for $J \subset I$, $A \in \mathcal{F}$,

$$P(A \mid X(t), t \in J) = \lim_{n \to \infty} P(A \mid X(t), t \in J \cap T_n), \quad \text{a.s.}$$

6. If $X(t)$, $\tilde{X}(t)$ are versions of each other for $t \in I$, and if both processes have all sample paths continuous on I, show that

$$P(X(t) = \tilde{X}(t) \quad \text{for all } t \in I) = 1.$$

7. If $\{X(t)\}$, $t \in I$, is a process all of whose paths are continuous on I, then show that the function $X(t, \omega)$ defined on $I \times \Omega$ is measurable with respect to the product σ-field $\mathcal{B}_1(I) \times \mathcal{F}$. [For I finite, let $I_1, \ldots, I_n$ be any partition of I, $t_k \in I_k$, and consider approximating $X(t)$ by the functions $\sum_k X(t_k)\chi_{I_k}(t)$.]

6. CONTINUITY OF BROWNIAN MOTION

It is easy to check that the finite-dimensional distributions given by 12.5 are consistent. Hence there is a process $\{X(t)\}$, $t \geq 0$ fitting them.

Definition 12.18. *Let* $X(t)$ *be a Brownian motion. If* $\mu \neq 0$ *it is said to be a Brownian motion with drift* μ. *If* $\mu = 0$, $\sigma^2 = 1$, *it is called normalized Brownian motion, or simply Brownian motion.*

Note that $(X(t) - \mu t)/\sigma$ is normalized. The most important single sample path property is contained in

Theorem 12.19. *For any Brownian motion* $X(t)$ *there is a dense set* T *in* $[0, \infty)$ *such that* $X(t)$ *is uniformly continuous on* $T \cap [0, a]$, $a < \infty$, *for almost every* ω.

In preparation for the proof of this, we need

Proposition 12.20. *Let T_0 be any finite collection of points, $0 = t_0 < t_1 < \cdots < t_n = \tau$; then for $\mathsf{X}(t)$ normalized Brownian motion*

$$P\left(\max_{t\in T_0} \mathsf{X}(t) > x\right) \le 2P(\mathsf{X}(\tau) > x),$$

$$P\left(\max_{t\in T_0} |\mathsf{X}(t)| > x\right) \le 2P(|\mathsf{X}(\tau)| > x).$$

Proof. Denote $\mathsf{Y}_k = \mathsf{X}(t_k) - \mathsf{X}(t_{k-1})$, and $\mathsf{j}^* = \{\text{first } j \text{ such that } \mathsf{X}(t_j) > x\}$. Then because $\mathsf{X}(t_n) - \mathsf{X}(t_j)$ has a distribution symmetric about the origin,

$$\tfrac{1}{2}P\left(\max_{t\in T_0} \mathsf{X}(t) > x\right) = \sum_{j=1}^{n} P(\mathsf{j}^* = j)P(\mathsf{X}(t_n) - \mathsf{X}(t_j) > 0).$$

Since $\{\mathsf{j}^* = j\} \in \mathcal{F}(\mathsf{X}(t_0), \ldots, \mathsf{X}(t_j))$, then

$$\tfrac{1}{2}P\left(\max_{t\in T_0} \mathsf{X}(t) > x\right) = \sum_{j=1}^{n} P(\mathsf{j}^* = j, \mathsf{X}(t_n) - \mathsf{X}(t_j) > 0)$$

$$\le \sum_{j=1}^{n} P(\mathsf{j}^* = j, \mathsf{X}(t_n) > x) \le P(\mathsf{X}(\tau) > x).$$

For the second inequality, use

$$\left\{\max_{t\in T_0} |\mathsf{X}(t)| > x\right\} = \left\{\max_{t\in T_0} \mathsf{X}(t) > x\right\} \cup \left\{\min_{t\in T_0} \mathsf{X}(t) < -x\right\}$$

and the fact that $-\mathsf{X}(t)$ is normalized Brownian motion.

Proof of 12.19. We show this for $a = 1$. Take $T_n = \{k/2^n;\ k = 0, \ldots, 2^n\}$, and $T = \bigcup_1^\infty T_n$. Define

$$\mathsf{U}_n = \sup_{\substack{t_j, t_k \in T \\ |t_j - t_k| \le 1/2^n}} |\mathsf{X}(t_j) - \mathsf{X}(t_k)|.$$

To show that $\mathsf{U}_n \to 0$ a.s., since $\mathsf{U}_n \in \downarrow$ it is sufficient to show that

$$\lim_n P(\mathsf{U}_n > \delta) = 0.$$

Let $I_k = [k/2^n, (k + 1)/2^n]$, and

$$\mathsf{Y}_k = \sup_{t\in I_k\cap T} |\mathsf{X}(t) - \mathsf{X}(k/2^n)|, \quad k = 0, 1, \ldots, 2^n - 1.$$

By the triangle inequality,

$$\mathsf{U}_n \le 3 \max_k \mathsf{Y}_k.$$

We show that $P\left(\max_k \mathsf{Y}_k > \delta\right) \to 0$. Use

$$P\left(\max_k \mathsf{Y}_k > \delta\right) = P\left(\bigcup_k \{\mathsf{Y}_k > \delta\}\right) \le \sum_0^{2^n-1} P(\mathsf{Y}_k > \delta).$$

The $\mathsf{Y}_1, \ldots$ are identically distributed, so

$$P\left(\max_k \mathsf{Y}_k > \delta\right) \le 2^n P(\mathsf{Y}_0 > \delta).$$

Note that

$$\mathsf{Y}_0 = \lim_N \max_{t \in I_0 \cap T_N} |\mathsf{X}(t)|.$$

By 12.20, for $N \ge n$,

$$P\left(\max_{t \in I_0 \cap T_N} |\mathsf{X}(t)| > \delta\right) \le 2P(|\mathsf{X}(2^{-n})| > \delta),$$

hence

$$P(\mathsf{Y}_0 > \delta) \le 2P(|\mathsf{X}(2^{-n})| > \delta).$$

Since Brownian motion satisfies $P\big(|\mathsf{X}(\Delta t)| > \delta\big)/\Delta t \to 0$ as $\Delta t \to 0$, then

$$2^n P\big(|\mathsf{X}(2^{-n})| > \delta\big) \to 0,$$

which proves the theorem.

Corollary 12.21. *There is a version of Brownian motion on* $[0, \infty)$ *such that all sample paths are continuous.*

Henceforth, we assume that the Brownian motion we deal with has all sample paths continuous.

Problems

8. Prove that for $\{\mathsf{X}(t)\}$ normalized Brownian motion on $[0, \infty)$, $P\big(\mathsf{X}(n\delta) \in J \text{ i.o.}\big) = 1$ for all intervals J such that $\|J\| > 0$, and fixed $\delta > 0$.

9. Define $\mathcal{T} = \bigcap_{t \ge 0} \mathcal{F}\big(\mathsf{X}(\tau), \tau \ge t\big)$, or $A \in \mathcal{T}$ if $A \in \mathcal{F}\big(\mathsf{X}(\tau), \tau \ge t\big)$ for all $t \ge 0$. Prove that $A \in \mathcal{T}$ implies $P(A) = 0$ or 1. [Apply a generalized version of 3.50.]

7. AN ALTERNATIVE DEFINITION

Normalized Brownian motion is completely specified by stating that it is a Gaussian process, i.e., all finite subsets $\{\mathsf{X}(t_1), \ldots, \mathsf{X}(t_n)\}$ have a joint normal distribution, $E\mathsf{X}(t) = 0$ and covariance $\Gamma(s, t) = \min(s, t)$. Since all sample functions are continuous, to specify Brownian motion it would only be necessary to work with a countable subset of random variables

$\{X(t)\}$, $t \in T$, and get the others as limits. This leads to the idea that with a proper choice of coordinates, $X(t)$ could be expanded in a countable coordinate system. Let $Y_1, Y_2, \ldots$ be independent and $\mathcal{N}(0, 1)$. Let $\varphi_k(t)$ be defined on a closed interval I such that $\sum_1^\infty |\varphi_k(t)|^2 < \infty$, all $t \in I$. Consider

$$Z(t) = \sum_1^\infty \varphi_k(t) Y_k.$$

Since

$$\sum_1^\infty E\,|\varphi_k(t) Y_k|^2 = \sum_1^\infty |\varphi_k(t)|^2 < \infty,$$

the sums converge a.s. for every $t \in I$, hence $Z(t)$ is well-defined for each t except on a set of probability zero. Furthermore, $Z(t_1), \ldots, Z(t_n)$ is the limit in distribution of joint normal variables, so that $\{Z(t)\}$ is a Gaussian process.

Note $EZ(t) = 0$, and for the $Z(t)$ covariance

(12.22) $$\Gamma(s, t) = \sum_1^\infty \varphi_k(t)\varphi_k(s), \quad (s, t) \in I \times I.$$

Hence if the $\varphi_k(t)$ satisfy

$$\min(s, t) = \sum_1^\infty \varphi_k(t)\varphi_k(s), \quad (s, t) \in I \times I,$$

then $Z(t)$ is normalized Brownian motion on I. I assert that on $I = [0, \pi]$,

(12.23) $$\min(s, t) = \frac{ts}{\pi} + \frac{2}{\pi}\sum_{k \ge 1} \frac{\sin kt \sin ks}{k^2}.$$

One way to verify this is to define a function of t on $[-\pi, \pi]$ for any $s \ge 0$, by

$$h_s(t) = \begin{cases} t, & |t| \le s; \\ -s, & t \le -s; \\ s, & t \ge s. \end{cases}$$

Denote the right-hand side of (12.23) by $g_s(t)$. The sum converges uniformly, hence $g_s(t)$ is continuous for all s, t. Simply check that for all integers k,

$$\int_{[-\pi,+\pi]} e^{ikt} h_s(t)\, dt = \int_{[-\pi,+\pi]} e^{ikt} g_s(t)\, dt,$$

and use the well-known fact that two continuous functions with the same Fourier coefficients are equal on $[-\pi, +\pi]$. Since $h_s(t) = \min(s, t)$, for $t \ge 0$, (12.23) results.

Proposition 12.24. *Let* $\mathsf{Y}_0, \mathsf{Y}_1, \ldots$ *be independent* $\mathcal{N}(0, 1)$, *then*

$$\mathsf{X}(t) = \frac{t}{\sqrt{\pi}}\mathsf{Y}_0 + \sqrt{\frac{2}{\pi}}\sum_{m\geq 1}\frac{\sin mt}{m}\mathsf{Y}_m$$

is normalized Brownian motion on $[0, \pi]$.

One way to prove the continuity of sample paths would be to define $\mathsf{X}^{(n)}(t)$ as the nth partial sum in 12.24, and to show that for almost all ω, the functions $x_n(t) = \mathsf{X}^{(n)}(t, \omega)$ converged uniformly on $[0, \pi]$. This can by shown true, at least for a subsequence $\mathsf{X}^{(n')}(t)$. See Ito and McKean [76, p. 22], for a proof along these lines.

8. VARIATION AND DIFFERENTIABILITY

The Brownian motion paths are extremely badly behaved for continuous functions. Their more obvious indices of bad behavior are given in this section: they are nowhere differentiable, and consequently of unbounded variation in every interval.

Theorem 12.25. *Almost every Brownian path is nowhere differentiable.*

Proof. We follow Dvoretski, Erdös, and Kakutani [42]. Fix $\beta > 0$, suppose that a function $x(t)$ has derivative $x'(s)$, $|x'(s)| < \beta$, at some point $s \in [0, 1]$; then there is an n_0 such that for $n > n_0$

(12.26) $$|x(t) - x(s)| \leq 2\beta\,|t - s|, \quad \text{if } |t - s| \leq 2/n.$$

Let $x(\cdot)$ denote functions on $[0, 1]$.

$$A_n = \{x(\cdot);\ \exists s \ \text{ such that } |x(t) - x(s)| \leq 2\beta\,|t - s|, \quad \text{if } |t - s| \leq 2/n\}.$$

The A_n increase with n, and the limit set A includes the set of all sample paths on $[0, 1]$ having a derivative at any point which is less than β in absolute value. If (12.26) holds, then, and we let k be the largest integer such that $k/n \leq s$, the following is implied:

$$y_k = $$
$$= \max\left(\left|x\left(\frac{k+2}{n}\right) - x\left(\frac{k+1}{n}\right)\right|, \left|x\left(\frac{k+1}{n}\right) - x\left(\frac{k}{n}\right)\right|, \left|x\left(\frac{k}{n}\right) - x\left(\frac{k-1}{n}\right)\right|\right)$$
$$\leq \frac{6\beta}{n}.$$

Therefore, if

$$B_n = \left\{x(\cdot);\ \text{at least one } y_k \leq \frac{6\beta}{n}\right\},$$

then $A_n \subset B_n$. Thus to show $P(A) = 0$, which implies the theorem, it is sufficient to get $\lim_n P(B_n) = 0$. But

$$B_n = \bigcup_{k=1}^{n-2} \left\{x(\cdot);\quad y_k \le \frac{6\beta}{n}\right\},$$

$$P(B_n) \le \sum_1^{n-2} P\left(\max\left(\left|X\left(\frac{k+2}{n}\right) - X\left(\frac{k+1}{n}\right)\right|, \left|X\left(\frac{k+1}{n}\right) - X\left(\frac{k}{n}\right)\right|,\right.\right.$$
$$\left.\left.\left|X\left(\frac{k}{n}\right) - X\left(\frac{k-1}{n}\right)\right|\right) \le \frac{6\beta}{n}\right)$$
$$\le nP\left(\max\left(\left|X\left(\frac{3}{n}\right) - X\left(\frac{2}{n}\right)\right|, \left|X\left(\frac{2}{n}\right) - X\left(\frac{1}{n}\right)\right|,\right.\right.$$
$$\left.\left.\left|X\left(\frac{1}{n}\right)\right|\right) \le \frac{6\beta}{n}\right)$$
$$= nP^3\left(\left|X\left(\frac{1}{n}\right)\right| \le \frac{6\beta}{n}\right) = n\left[\sqrt{\frac{n}{2\pi}}\int_{-6\beta/n}^{+6\beta/n} e^{-nx^2/2}\,dx\right]^3.$$

Substitute $nx = y$. Then

$$P(B_n) \le n\left[\frac{1}{\sqrt{2\pi n}}\int_{-6\beta}^{+6\beta} e^{-x^2/2n}\,dx\right]^3 \to 0.$$

Corollary 12.27. *Almost every sample path of $X(t)$ has infinite variation on every finite interval.*

Proof. If a sample function $X(t, \omega)$ has bounded variation on I, then it has a derivative existing almost everywhere on I.

A further result gives more information on the size of the oscillations of $X(t)$. Since $E\,|X(t + \Delta t) - X(t)|^2 = \Delta t$, as a rough estimate we would guess that $|X(t + \Delta t) - X(t)| \simeq \sqrt{\Delta t}$. Then for any fine partition $t_0, \ldots, t_n$ of the interval $[t, t + \tau]$,

$$\sum_1^n |X(t_k) - X(t_{k-1})|^2 \simeq \tau.$$

The result of the following theorem not only verifies this, but makes it surprisingly precise.

Theorem 12.28. *Let the partitions $\mathcal{T}_n$ of $[t, t + \tau]$, $\mathcal{T}_n = (t_0^{(n)}, \ldots, t_m^{(n)})$, $\|\mathcal{T}_n\| = \sup_k |t_{k+1}^{(n)} - t_k^{(n)}|$ satisfy $\|\mathcal{T}_n\| \to 0$. Then*

$$S_n = \sum_k |X(t_k^{(n)}) - X(t_{k-1}^{(n)})|^2 \xrightarrow{2} \tau.$$

If $\sum_1^\infty \|\mathcal{T}_n\| < \infty$, $S_n \xrightarrow{\text{a.s.}} \tau$.

Proof. Assume $t = t_0^{(n)}$, $t + \tau = t_m^{(n)}$, otherwise do some slight modification. Then $\tau = \sum_1^m (t_k - t_{k-1})$ (dropping the superscript), and

$$\begin{aligned} S_n - \tau &= \sum_1^m [(X(t_k) - X(t_{k-1}))^2 - (t_k - t_{k-1})] \\ &= \sum_1^m [(X(t_k) - X(t_{k-1}))^2 - E(X(t_k) - X(t_{k-1}))^2]. \end{aligned}$$

The summands are independent, with zero means. Hence

$$E(S_n - \tau)^2 = \sum_1^m E[(X(t_k) - X(t_{k-1}))^2 - (t_k - t_{k-1})]^2.$$

$(X(t_k) - X(t_{k-1}))^2/(t_k - t_{k-1})$ has the distribution of X^2, where X is $\mathcal{N}(0, 1)$. So

$$\begin{aligned} E(S_n - \tau)^2 &= E(X^2 - 1)^2 \cdot \sum_1^m (t_k - t_{k-1})^2 \\ &\leq E(X^2 - 1)^2 \cdot \|\mathcal{T}_n\| \, \tau, \end{aligned}$$

proving convergence in mean square. If $\Sigma \|\mathcal{T}_n\| < \infty$, then use the Borel-Cantelli lemma plus the Chebyshev inequality.

Theorem 12.28 holds more generally, with $S_n \xrightarrow{\text{a.s.}} \tau$ for any sequence $\mathcal{T}_n$ of partitions such that $\|\mathcal{T}_n\| \to 0$ and the $\mathcal{T}_n$ are successive refinements. (See Doob [39, pp. 395 ff.].)

9. LAW OF THE ITERATED LOGARITHM

Now for one of the most precise and well-known theorems regarding oscillations of a Brownian motion. It has gone through many refinements and generalizations since its proof by Khintchine in 1924.

Theorem 12.29. *For normalized Brownian motion*

$$\overline{\lim_{t \downarrow 0}} \frac{X(t)}{\sqrt{2t \log(\log 1/t)}} = 1 \quad \text{a.s.}$$

Proof. I follow essentially Lévy's proof. Let $\varphi(t) = \sqrt{2t \log(\log 1/t)}$.

1) For any $\delta > 0$,

$$P\left(\overline{\lim_{t \downarrow 0}}\, (X(t) - (1 + \delta)\varphi(t)) > 0\right) = 0.$$

Proof of (1). Take q any number in (0, 1), put $t_n = q^n$. The plan is to show that if C_n is the event

$$\{X(t) > (1 + \delta)\varphi(t) \quad \text{for at least one } t \in [t_{n+1}, t_n]\},$$

then $P(C_n \text{ i.o.}) = 0$. Define $\mathsf{M}(\tau) = \sup_{t \le \tau} \mathsf{X}(t)$ and use

$$C_n \subset \{\mathsf{M}(t_n) > (1+\delta)\varphi(t_{n+1})\},$$

valid since $\varphi(t)$ is increasing in t. Use the estimates gotten from taking limits in 12.20.

$$P\big(\mathsf{M}(t_n) > x\sqrt{t_n}\big) \le 2P\big(\mathsf{X}(t_n) > x\sqrt{t_n}\big) = \sqrt{\frac{2}{\pi}} \int_x^\infty e^{-z^2/2}\, dz$$

$$\le \sqrt{\frac{2}{\pi}} \frac{1}{x} \int_x^\infty z e^{-z^2/2}\, dz = \sqrt{\frac{2}{\pi}} \frac{e^{-x^2/2}}{x}.$$

Hence, letting $x_n = (1+\delta)\varphi(t_{n+1})/\sqrt{t_n}$

$$P\big(\mathsf{M}(t_n) > x_n\sqrt{t_n}\big) \le \sqrt{\frac{2}{\pi}} \frac{\exp\,[-x_n^2/2]}{x_n}.$$

and since

$$x_n = (1+\delta)\sqrt{2q \log\,[(n+1)\log 1/q]}$$
$$= \sqrt{2 \log c(n+1)^\lambda},$$

where $\lambda = q(1+\delta)^2$,

(12.30)
$$P\big(\mathsf{M}(t_n) > (1+\delta)\varphi(t_{n+1})\big)$$
$$\le c_1 \cdot \frac{1}{(n+1)^\lambda \sqrt{\log\,(n+1)}}.$$

For any δ, select q so that $q(1+\delta)^2 > 1$. Then the right-hand side of (12.30) is a term of a convergent sum and the first assertion is proved.

2) For any $\delta > 0$,

$$P\big(\overline{\lim_{t \downarrow 0}}\,(\mathsf{X}(t) - (1-\delta)\varphi(t)) > 0\big) = 1.$$

Proof of (2). Take q again in $(0, 1)$, $t_n = q^n$, let $\mathsf{Z}_n = \mathsf{X}(t_n) - \mathsf{X}(t_{n+1})$. The Z_n are independent. Suppose we could show that for $\epsilon > 0$,

$$P\big(\mathsf{Z}_n > (1-\epsilon)\varphi(t_n) \text{ i.o.}\big) = 1.$$

This would be easy in principle, because the independence of the Z_n allows the converse half of the Borel-Cantelli lemma to be used. On the other hand, from part one of this proof, because the processes $\{\mathsf{X}(t)\}$ and $\{-\mathsf{X}(t)\}$ have the same distribution,

$$P\big(\mathsf{X}(t_{n+1}) < -(1+\epsilon)\varphi(t_{n+1}) \text{ i.o.}\big) = 0$$

or

$$\mathsf{X}(t_{n+1}) \ge -(1+\epsilon)\varphi(t_{n+1})$$

holds for all n sufficiently large. From $X(t_n) = Z_n + X(t_{n+1})$ it follows that infinitely often

$$X(t_n) \geq (1 - \epsilon)\varphi(t_n) - (1 + \epsilon)\varphi(t_{n+1})$$
$$= \varphi(t_n)\left[1 - \epsilon - (1 + \epsilon)\frac{\varphi(t_{n+1})}{\varphi(t_n)}\right], \quad \text{a.s.}$$

Note that $\varphi(t_{n+1})/\varphi(t_n) \to \sqrt{q}$. Therefore, if we take ϵ, q so small that

$$1 - \epsilon - 2(1 + \epsilon)\sqrt{q} \geq 1 - \delta,$$

the second part would be established. So now, we start estimating:

$$P(Z_n > x\sqrt{(t_n - t_{n+1})}) = \frac{1}{\sqrt{2\pi}}\int_x^\infty e^{-z^2/2}\,dz \sim \frac{1}{x\sqrt{2\pi}}e^{-x^2/2}$$

as $x \to \infty$. Let

$$x_n = (1 - \epsilon)\frac{\varphi(t_n)}{\sqrt{t_n - t_{n+1}}} = \frac{(1 - \epsilon)}{\sqrt{1 - q}}\sqrt{2\log(n\log 1/q)}$$
$$= \sqrt{\alpha\log cn}, \qquad \alpha = 2(1 - \epsilon)^2/(1 - q).$$

Then

$$P(Z_n > (1 - \epsilon)\varphi(t_n)) \sim \frac{c}{\sqrt{2\pi}\,n^{\alpha/2}\sqrt{\log n}}.$$

By taking q even smaller, if necessary, we can get $\alpha < 2$. The right-hand side is then a term of a divergent series and the proof is complete. Q.E.D.

10. BEHAVIOR AT $t = \infty$

Let $Y_k = X(k) - X(k - 1)$. The Y_k are independent $\mathcal{N}(0, 1)$ variables, $X(n) = Y_1 + \cdots + Y_n$ is the sum of independent, identically distributed random variables. Thus $X(t)$ for t large has the magnitude properties of successive sums S_n. In particular,

Proposition 12.31

$$\frac{X(t)}{t} \xrightarrow{\text{a.s.}} 0 \quad \text{as} \quad t \to \infty.$$

Proof. Since $EY_k = 0$, we can use the strong law of large numbers to get $X(n)/n \xrightarrow{\text{a.s.}} 0$. Let

$$Z_k = \max_{0 \leq t \leq 1} |X(k + t) - X(k)|.$$

For $t \in [k, k + 1]$,

$$\left|\frac{X(t)}{t} - \frac{X(k)}{k}\right| \leq \frac{1}{k(k + 1)}|X(k)| + \frac{1}{k + 1}Z_k.$$

The first term $\xrightarrow{\text{a.s.}} 0$. Z_k has the same distribution as $\max_{0\le t\le 1}|\mathsf{X}(t)|$. By 12.20, $E\mathsf{Z}_k < \infty$. Now use Problem 10, Chapter 3, to conclude that $\mathsf{Z}_k/k \xrightarrow{\text{a.s.}} 0$.

This is the straightforward approach. There is another way which is surprising, because it essentially reduces behavior for $t \to \infty$ to behavior for $t \downarrow 0$. Define

$$\mathsf{X}^{(1)}(t) = t\mathsf{X}(1/t), \quad t > 0,$$
$$\equiv 0, \qquad t = 0.$$

Proposition 12.32. *$\mathsf{X}^{(1)}(t)$ is normalized Brownian motion on* $[0, \infty)$.

Proof. Certainly $\mathsf{X}^{(1)}(t)$ is Gaussian with zero mean. Also,

$$E\mathsf{X}^{(1)}(t)\mathsf{X}^{(1)}(s) = ts \min\left(\frac{1}{t}, \frac{1}{s}\right) = \min(s, t).$$

Now to prove 12.31 another way. The statement $\mathsf{X}(t)/t \to 0$ as $t \to \infty$ translates into $t\mathsf{X}(1/t) \to 0$ a.s. as $t \to 0$. So 12.31 is equivalent to proving that $\mathsf{X}^{(1)}(t) \to 0$ a.s. as $t \to 0$. If $\mathsf{X}^{(1)}(t)$ is a version of Brownian motion with all paths continuous on $[0, \infty)$, then trivially, $\mathsf{X}^{(1)}(t) \to 0$ a.s. at the origin. However, the continuity of $\mathsf{X}(t)$ on $[0, \infty)$ gives us only that the paths of $\mathsf{X}^{(1)}(t)$ are continuous on $(0, \infty)$. Take a version $\tilde{\mathsf{X}}^{(1)}(t)$ of $\mathsf{X}^{(1)}(t)$ such that all paths of $\tilde{\mathsf{X}}^{(1)}(t)$ are continuous on $[0, \infty)$. By Problem 5, almost all paths of $\tilde{\mathsf{X}}^{(1)}(t)$ and $\mathsf{X}^{(1)}(t)$ coincide on $(0, \infty)$. Since $\tilde{\mathsf{X}}^{(1)}(t) \to 0$ as $t \to 0$, this is sufficient.

By using this inversion on the law of the iterated logarithm we get

Corollary 12.33

$$\overline{\lim_{t\to\infty}} \frac{\mathsf{X}(t)}{\sqrt{2t \log(\log t)}} = 1.$$

Since $-\mathsf{X}(t)$ is also Brownian motion,

$$\underline{\lim_{t\to\infty}} \frac{\mathsf{X}(t)}{\sqrt{2t \log(\log t)}} = -1.$$

Therefore,

$$\overline{\lim_{t\to\infty}} \frac{|\mathsf{X}(t)|}{\sqrt{2t \log(\log t)}} = 1.$$

The similar versions of 12.29 *hold as $t \to 0$; for instance,*

(12.34)
$$\underline{\lim_{t\to 0}} \frac{\mathsf{X}(t)}{\sqrt{2t \log|\log t|}} = -1.$$

11. THE ZEROS OF $\mathsf{X}(t)$

Look at the set $T(\omega)$ of zeros of $\mathsf{X}(t, \omega)$ in the interval $[0, 1]$. For any continuous function, the zero set is closed. By (12.29) and (12.34), $T(\omega)$ is an infinite set a.s. Furthermore the Lebesgue measure of $T(\omega)$ is a.s. zero, because $l(T(\omega)) = l(t; \mathsf{X}(t) = 0) = \int_0^1 \chi_{\{0\}}(\mathsf{X}(t))\, dt$, so

$$El(T(\omega)) = E\int_0^1 \chi_{\{0\}}(\mathsf{X}(t))\, dt = \int_0^1 P(\mathsf{X}(t) = 0)\, dt = 0,$$

where the interchange of E and $\int_0^1 dt$ is justified by the joint measurability of $\mathsf{X}(t, \omega)$, hence of $\chi_{\{0\}}(\mathsf{X}(t, \omega))$. (See Problem 7.)

Theorem 12.35. *For almost all ω, $T(\omega)$ is a closed, perfect set of Lebesgue measure zero (therefore, noncountable).*

Proof. The remaining part is to prove that $T(\omega)$ has no isolated points. The idea here is that every time $\mathsf{X}(t)$ hits zero, it is like starting all over again and the law of the iterated logarithm guarantees a clustering of zeros starting from that point. For almost all paths, the point $t = 0$ is a limit point of zeros of $\mathsf{X}(t)$ from the right. For any point $a \geq 0$, let $\mathtt{t}^*$ be the position of the first zero of $\mathsf{X}(t)$ following $t = a$, that is,

$$\mathtt{t}^* = \inf\{t; \mathsf{X}(t) = 0, t \geq a\}.$$

Look at the process $\mathsf{X}^{(1)}(t) = \mathsf{X}(t + \mathtt{t}^*) - \mathsf{X}(\mathtt{t}^*)$. This is just looking at the Brownian process as though it started afresh at the time $\mathtt{t}^*$. Heuristically, what happens up to time $\mathtt{t}^*$ depends only on the process up to that time; starting over again at $\mathtt{t}^*$ should give a process that looks exactly like Brownian motion. If this argument can be made rigorous, then the set C_a of sample paths such that $\mathtt{t}^*$ is a limit point of zeros from the right has probability one. The intersection of C_a over all rational $a \geq 0$ has probability one, also. Therefore almost every sample path has the property that the first zero following any rational is a limit point of zeros from the right. This precludes the existence of any isolated zero. Therefore, the theorem is proved except for the assertion,

(12.36) $\mathsf{X}(t + \mathtt{t}^*) - \mathsf{X}(\mathtt{t}^*)$ is a Brownian motion.

The truth of (12.36) and its generalization are established in the next section. Suppose that it holds for more general random starting times. Then we could use this to prove

Corollary 12.37. *For any value a, the set $T^{(a)} = \{t;\ \mathsf{X}(t) = a, 0 \leq t \leq 1\}$ is, for almost all ω, either empty or a perfect closed set of Lebesgue measure zero.*

Proof. Let t^* be the first t such that $X(t) = a$, that is, $t^* = \inf\{t; X(t) = a\}$. If $t^* > 1$, then $T^{(a)}$ is empty. The set $\{t^* = 1\} \subset \{X(1) = a\}$ has probability zero. If $t^* < 1$, consider the process $X(t + t^*) - X(t^*)$ as starting out at the random time t^*. If this is Brownian motion, the zero set is perfect. But the zero set for this process is

$$T = \{t; X(t + t^*) - X(t^*) = 0, \quad 0 \le t \le 1\}$$
$$= \{t; X(t + t^*) = a, \quad 0 \le t \le 1\}.$$

Hence $T^{(a)} = T \cap [0, 1]$ is perfect a.s.

12. THE STRONG MARKOV PROPERTY

The last item needed to complete and round out this study of sample path properties is a formulation and proof of the statement: At a certain time t^*, where t^* depends only on the Brownian path up to time t^*, consider the motion as starting at t^*; that is, look at $X^{(1)}(t) = X(t + t^*) - X(t^*)$. Then $X^{(1)}(t)$ is Brownian motion and is independent of the path of the particle up to time t^*. Start with the observation that for $\tau \ge 0$, fixed,

(12.38) $$X^{(1)}(t) = X(t + \tau) - X(\tau), \quad t \ge 0$$

is Brownian motion and is independent of $\mathcal{F}(X(s), s \le \tau)$. Now, to have any of this make sense, we need:

Proposition 12.39. *If* $t^* \ge 0$ *is a random variable so is* $X(t^*)$.

Proof. For any $n > 0$, let

$$A_k^{(n)} = \left\{\frac{k-1}{n} \le t^* < \frac{k}{n}\right\},$$

$$X^{(n)}(t^*) = \sum_{k=1}^{\infty} X\left(\frac{k}{n}\right) \chi_{A_k^{(n)}}.$$

$X^{(n)}(t^*)$ is a random variable. On the set $\{t^* \le N\}$,

$$|X^{(n)}(t^*) - X(t^*)| \le \sup_{\substack{0 \le t \le N \\ 0 \le h \le 1/n}} |X(t + h) - X(t)|.$$

The right-hand side $\to 0$, so $X^{(n)}(t^*) \to X(t^*)$ everywhere.

Next, it is necessary to formulate the statement that the value of t^* depends only on the past of the process up to time t^*.

Definition 12.40. *For any process* $\{X(t)\}$ *a random variable* $t^* \ge 0$ *will be called a stopping time if for every* $t \ge 0$,

$$\{t^* \le t\} \in \mathcal{F}(X(\tau), \tau \le t).$$

The last step is to give meaning to "the part of the process up to time t^*."

Look at an example:

$$\mathsf{t}^* = \inf_t \{t; \mathsf{X}(t) \geq 1\},$$

so t^* is the first time that $\mathsf{X}(t)$ hits the point $x = 1$. It can be shown quickly that t^* is a stopping time. Look at sets depending on $\mathsf{X}(t)$, $0 \leq t \leq \mathsf{t}^*$; for example,

$$B = \left\{\omega; \inf_{t \leq \mathsf{t}^*} \mathsf{X}(t) \leq -1\right\}.$$

Note that for any $t \geq 0$, $B \cap \{\mathsf{t}^* \leq t\}$ depends only on the behavior of the sample path on $[0, t]$, that is,

$$B \cap \{\mathsf{t}^* \leq t\} \in \mathcal{F}(\mathsf{X}(\tau), \tau \leq t).$$

Generalizing from this we get

Definition 12.41. *The σ-field of events $B \in \mathcal{F}$ such that for every $t \geq 0$, $B \cap \{\mathsf{t}^* \leq t\} \in \mathcal{F}(\mathsf{X}(\tau), \tau \leq t)$ is called the σ-field generated by the process up to time t^* and denoted by $\mathcal{F}(\mathsf{X}(t), t \leq \mathsf{t}^*)$.*

This is all we need:

Theorem 12.42. *Let t^* be a stopping time, then*

$$\mathsf{X}^{(1)}(t) = \mathsf{X}(t + \mathsf{t}^*) - \mathsf{X}(\mathsf{t}^*), \quad t \geq 0$$

is normalized Brownian motion and

$$\mathcal{F}(\mathsf{X}^{(1)}(t), t \geq 0) \quad \text{is independent of} \quad \mathcal{F}(\mathsf{X}(t), t \leq \mathsf{t}^*).$$

Proof. If t^* takes on only a countable number of values $\{\tau_k\}$, then 12.42 is quick. For example, if

$$A_1, \ldots, A_j \in \mathcal{B}_1, \quad t_1, \ldots, t_j \geq 0, \qquad \text{and} \qquad B \in \mathcal{F}(\mathsf{X}(t), t \leq \mathsf{t}^*),$$

then

$$P(\mathsf{X}^{(1)}(t_1) \in A_1, \ldots, \mathsf{X}^{(1)}(t_j) \in A_j, B)$$
$$= \sum_k P(\mathsf{X}^{(1)}(t_1) \in A_1, \ldots, \mathsf{X}^{(1)}(t_j) \in A_j, \mathsf{t}^* = \tau_k, B).$$

Now,

$$\{\mathsf{X}^{(1)}(t_1) \in A_1, \ldots, \mathsf{t}^* = \tau_k, B\} = \{\mathsf{X}(t_1 + \tau_k) - \mathsf{X}(\tau_k) \in A_1, \ldots, \mathsf{t}^* = \tau_k, B\}.$$

Furthermore,

$$\{\mathsf{t}^* = \tau_k\} \cap B = \{\mathsf{t}^* = \tau_k\} \cap \{\mathsf{t}^* \leq \tau_k\} \cap B,$$

so is in $\mathcal{F}(\mathsf{X}(t), t \leq \tau_k)$. By (12.38), then,

$$P(\mathsf{X}^{(1)}(t_1) \in A_1, \ldots, \mathsf{t}^* = \tau_k, B)$$
$$= P(\mathsf{X}(t_1 + \tau_k) - \mathsf{X}(\tau_k) \in A_1, \ldots, \mathsf{X}(t_j + \tau_k) - \mathsf{X}(\tau_k) \in A_j) \cdot P(\mathsf{t}^* = \tau_k, B)$$
$$= P(\mathsf{X}(t_1) \in A_1, \ldots, \mathsf{X}(t_j) \in A_j) P(\mathsf{t}^* = \tau_k, B).$$

Summing over k, we find that

(12.43) $$P(\mathsf{X}^{(1)}(t_1) \in A_1, \ldots, \mathsf{X}^{(1)}(t_j) \in A_j, B) = P(\mathsf{X}(t_1) \in A_1, \ldots, \mathsf{X}(t_j) \in A_j) P(B).$$

This extends immediately to the statement of the theorem. In the general case, we approximate t^* by a discrete stopping variable. Define

$$\mathsf{t}_n^* = \begin{cases} \dfrac{k}{n}, & \dfrac{k-1}{n} < \mathsf{t}^* \leq \dfrac{k}{n}, \quad k > 1, \\[2ex] \dfrac{1}{n}, & 0 \leq \mathsf{t}^* \leq \dfrac{1}{n}, \quad k = 1. \end{cases}$$

Then t_n^* is a stopping time because for $k/n \leq t < (k + 1)/n$,

$$\{\mathsf{t}_n^* \leq t\} = \{\mathsf{t}^* \leq k/n\} \in \mathcal{F}(\mathsf{X}(s), \quad s \leq k/n).$$

But $\mathcal{F}(\mathsf{X}(s), s \leq k/n) \subset \mathcal{F}(\mathsf{X}(s), s \leq t)$. Also, I assert that

$$B \in \mathcal{F}(\mathsf{X}(t), t \leq \mathsf{t}^*) \Rightarrow B \in \mathcal{F}(\mathsf{X}(t), t \leq \mathsf{t}_n^*),$$

the latter because for $k/n \leq t < (k + 1)/n$,

$$B \cap \{\mathsf{t}_n^* \leq t\} = B \cap \{\mathsf{t}^* \leq k/n\} \in \mathcal{F}(\mathsf{X}(s), s \leq t).$$

Let $\mathsf{X}_n^{(1)}(t) = \mathsf{X}(t + \mathsf{t}_n^*) - \mathsf{X}(\mathsf{t}_n^*)$. By (12.43), for $B \in \mathcal{F}(\mathsf{X}(t), t \leq \mathsf{t}_n^*)$,

$$P(\mathsf{X}_n^{(1)}(t_1) < x_1, \ldots, \mathsf{X}_n^{(1)}(t_j) < x_j, B) = P(\mathsf{X}(t_1) < x_1, \ldots, \mathsf{X}(t_j) < x_j) P(B).$$

But, by the path continuity of $\mathsf{X}(t)$, $\mathsf{X}_n^{(1)}(t) \to \mathsf{X}^{(1)}(t)$ for every ω, t. This implies that at every point $(x_1, \ldots, x_j)$ which is a continuity point of the distribution function of $\mathsf{X}^{(1)}(t_1), \ldots, \mathsf{X}^{(1)}(t_j)$,

$$P(\mathsf{X}^{(1)}(t_1) < x_1, \ldots, \mathsf{X}^{(1)}(t_n) < x_n, B) = P(\mathsf{X}(t_1) < x_1, \ldots, \mathsf{X}(t_n) < x_n) P(B).$$

This is enough to ensure that equality holds for all $(x_1, \ldots, x_j)$. Extension now proves the theorem.

Problem 10. Prove that the variables

$$\mathsf{t}_1^* = \inf\{t; \mathsf{X}(t) = 0, t \geq a\},$$

$$\mathsf{t}_2^* = \inf\{t; \mathsf{X}(t) = b\},$$

are stopping times for Brownian motion.

NOTES

The motion of small particles suspended in water was noticed and described by Brown in 1828. The mathematical formulation and study was initiated by Bachelier [1, 1900], and Einstein [45, 1905], and carried on extensively from that time by both physicists and mathematicians. But rigorous discussion of sample path properties was not started until 1923 when Wiener [141] proved path continuity. Wiener also deduced the orthogonal expansion (12.24) in 1924 [142].

A good source for many of the deeper properties of Brownian motion is Lévy's books [103, 105], and in the recent book [76] by Ito and McKean. A very interesting collection of articles that includes many references to earlier works and gives a number of different ways of looking at Brownian motion has been compiled by Wax [139]. The article by Dvoretski, Erdös, and Katutani [42] gives the further puzzling property that no sample paths have any "points of increase."

The fact of nondifferentiability of the sample paths was discovered by Paley, Wiener, and Zygmund [115, 1933]. The law of the iterated logarithm for Brownian motion was proved by Khintchine [88, 1933]. The properties of the zero sets of its paths was stated by Lévy, who seemed to assume the truth of 12.42. This latter property was stated and proved by Hunt [74, 1956]. David Freedman suggested the proof given that no zeros of $\mathsf{X}(t)$ are isolated.

CHAPTER 13

INVARIANCE THEOREMS

1. INTRODUCTION

Let $S_1, S_2, \ldots$ be a player's total winnings in a fair coin-tossing game. A question leading to the famous arc sine theorem is: Let N_n be the number of times that the player is ahead in the first n games,

$$N_n = \{\text{number of } k;\ k \le n,\ S_k > 0\}.$$

The proportion of the time that the player is ahead in n games is $N_n/n = W_n$. Does a limiting distribution exist for W_n, and if so, what is it?

Reason this way: Define $Z^{(n)}(t) = S_{[nt]}$ as t ranges over the values $0 \le t \le 1$. Denote Lebesgue measure by l; then

$$W_n = l\{t;\ Z^{(n)}(t) > 0,\quad 0 \le t \le 1\}.$$

Now $Z^{(n)}(t) = S_{[nt]}$ does not converge to anything in any sense, but recall from Section 2 of the last chapter that the processes $X^{(n)}(t) = S_{[nt]}/\sqrt{n}$ have all finite dimensional distribution functions converging to those of normalized Brownian motion $X(t)$ as $n \to \infty$. We denote this by $X^{(n)}(\cdot) \xrightarrow{\mathcal{D}} X(\cdot)$. But W_n can also be written as

$$W_n = l\{t;\ X^{(n)}(t) > 0,\quad 0 \le t \le 1\}.$$

Of course, the big transition that we would like to make here would be to define

$$W = l\{t;\ X(t) > 0,\quad 0 \le t \le 1\},$$

so that W is just the proportion of time that a Brownian particle stays in the positive axis during $[0, 1]$, and then conclude that

$$W_n \xrightarrow{\mathcal{D}} W.$$

The general truth of an assertion like this would be a profound generalization of the central limit theorem. The transition from the obvious application of the central limit theorem to conclude that $X^{(n)}(\cdot) \xrightarrow{\mathcal{D}} X(\cdot)$ to get to

$\mathsf{W}_n \xrightarrow{\mathfrak{D}} \mathsf{W}$ is neither obvious or easy. But some general theorems of this kind would give an enormous number of limit theorems. For example, if we let

$$\mathsf{M}_n = \max (|\mathsf{S}_1|, \ldots, |\mathsf{S}_n|),$$

is it possible to find constants A_n such that $\mathsf{M}_n/A_n \xrightarrow{\mathfrak{D}} \mathsf{M}$, M nondegenerate? Again, write

$$\mathsf{M}_n = \max_{t \in [0,1]} |\mathsf{S}_{[nt]}| = \sqrt{n} \max_{t \in [0,1]} |\mathsf{X}^{(n)}(t)|.$$

Let

$$\mathsf{M} = \max_{t \in [0,1]} |\mathsf{X}(t)|,$$

then apply our nonexistent theorem to conclude

$$\mathsf{M}_n/\sqrt{n} \xrightarrow{\mathfrak{D}} \mathsf{M}.$$

This chapter will fill in the missing theorem. The core of the theorem is that successive sums of independent, identically distributed random variables with zero means and finite variances have the same distribution as Brownian motion with a random time index. The idea is not difficult—suppose $\mathsf{S}_1, \mathsf{S}_2, \ldots$ form the symmetric random walk. Define T_1 as the first time such that $|\mathsf{X}(t)| = 1$. By symmetry $P(\mathsf{X}(\mathsf{T}_1) = 1) = P(\mathsf{X}(\mathsf{T}_1) = -1) = \frac{1}{2}$. Define T_2 as the first time such that $|\mathsf{X}(t + \mathsf{T}_1) - \mathsf{X}(\mathsf{T}_1)| = 1$, and so on. But T_1 is determined only by the behavior of the $\mathsf{X}(t)$ motion up to time T_1, so that intuitively one might hope that the $\mathsf{X}(t + \mathsf{T}_1) - \mathsf{X}(\mathsf{T}_1)$ process would have the same distribution as Brownian motion, but be independent of $\mathsf{X}(t)$, $t \le \mathsf{T}_1$. To make this sort of construction hold, the strong Markov property is, of course, essential. But also, we need to know some more about the first time that a Brownian motion exits from some interval around the origin.

2. THE FIRST-EXIT DISTRIBUTION

For some set $B \in \mathcal{B}_1$, let

$$\mathsf{t}^*(B) = \inf \{t; \mathsf{X}(t) \in B^c\}$$

be the first exit time of the Brownian motion from the set B. In particular, the first time that the particle hits the point $\{a\}$ is identical with the first exit time of the particle from $(-\infty, a)$ if $a > 0$, or from (∞, a) if $a < 0$, and we denote it by t^*_a. Let $\mathsf{t}^*(a, b)$ be the first exit time from the interval (a, b). The information we want is the probability that the first exit of the particle from the interval (a, b), $a < 0 < b$, occurs at the point b, and the value of $E\mathsf{t}^*(a, b)$, the expected time until exit.

For normalized Brownian motion $\{X(t)\}$, use the substitution $X(t) = X(s) + (X(t) - X(s))$ to verify that for $s \le t$,

$$E(X(t) \mid X(\tau), \tau \le s) = X(s) \quad \text{a.s.}$$

$$E(X^2(t) - t \mid X(\tau), \tau \le s) = X^2(s) - s \quad \text{a.s.}$$

Hence the processes $\{X(t)\}$, $\{X^2(t) - t\}$ are martingales in the sense of

Definition 13.1. *A process $\{Y(t)\}$, $t \in I$, is a martingale if for all t in I, $E\,|Y(t)| < \infty$, and for any s, t in I, $s \le t$,*

$$E(Y(t) \mid Y(\tau), \tau \le s) = Y(s) \quad \text{a.s.}$$

Suppose that for stopping times t^* satisfying the appropriate integrability conditions, the generalization of 5.31 holds as the statement $EY(t^*) = EY(0)$. This would give, for a stopping time on Brownian motion,

(13.2) $$EX(t^*) = 0, \quad E(X^2(t^*) - t^*) = 0.$$

These two equations would give the parameters we want. If we take $t^* = t^*(a, b)$, (13.2) becomes

$$aP(X(t^*) = a) + bP(X(t^*) = b) = 0.$$

Solving, we get

(13.3) $$P(X(t^*) = b) = \frac{|a|}{|b| + |a|}, \qquad P(X(t^*) = a) = \frac{|b|}{|b| + |a|}.$$

The second equation of (13.2) provides us with

$$a^2P(X(t^*) = a) + b^2P(X(t^*) = b) = Et^*(a, b).$$

Using (13.3) we find that

(13.4) $$Et^*(a, b) = |a|\,|b|.$$

Rather than detour to prove the general martingale result, we defer proof until the next chapter and prove here only what we need.

Proposition 13.5. *For $t^* = t^*(a, b)$, $a < 0 < b$,*

$$EX(t^*) = 0, \; EX^2(t^*) = Et^*.$$

Proof. Let t^* be a stopping time taking values in a countable set $\{t_j\} \subset [0, \tau]$, $\tau < \infty$. Then

$$E(X(\tau) - X(t^*)) = \sum_j \int_{\{t^* = t_j\}} (X(\tau) - X(t_j))\, dP.$$

By independence, then, $E(\mathsf{X}(\tau) - \mathsf{X}(\mathsf{t}^*)) = 0$; hence $E\mathsf{X}(\mathsf{t}^*) = 0$. For the second equation, write

$$\begin{aligned} E\mathsf{X}(\tau)^2 &= E(\mathsf{X}(\tau) - \mathsf{X}(\mathsf{t}^*) + \mathsf{X}(\mathsf{t}^*))^2 \\ &= E(\mathsf{X}(\tau) - \mathsf{X}(\mathsf{t}^*))^2 + E\mathsf{X}^2(\mathsf{t}^*) + 2E(\mathsf{X}(\tau) - \mathsf{X}(\mathsf{t}^*))(\mathsf{X}(\mathsf{t}^*)). \end{aligned}$$

The last term is easily seen to be zero. The first term is

$$\sum_j \int_{\{\mathsf{t}^* = t_j\}} (\mathsf{X}(\tau) - \mathsf{X}(t_j))^2 \, dP$$

which equals

$$\sum_j (\tau - t_j) P(\mathsf{t}^* = t_j) = \tau - E\mathsf{t}^*.$$

Since $E\mathsf{X}(\tau)^2 = \tau$,

$$E\mathsf{X}^2(\mathsf{t}^*) = E\mathsf{t}^*.$$

Take $\mathsf{t}^{**} = \min(\mathsf{t}^*(a, b), \tau)$, and t_n^* a sequence of stopping times taking values in a countable subset of $[0, \tau]$ such that $\mathsf{t}_n^* \to \mathsf{t}^{**}$ everywhere. By the bounded convergence theorem, $E\mathsf{t}_n^* \to E\mathsf{t}^{**}$. Furthermore, by path continuity $\mathsf{X}(\mathsf{t}_n^*) \to \mathsf{X}(\mathsf{t}^{**})$, and

$$|\mathsf{X}(\mathsf{t}_n^*)|^2 \le \sup_{t \le \tau} |\mathsf{X}(t)|^2,$$

which is integrable by 12.20. Use the bounded convergence theorem again to get $E\mathsf{X}(\mathsf{t}_n^*) \to E\mathsf{X}(\mathsf{t}^{**})$, $E\mathsf{X}^2(\mathsf{t}_n^*) \to E\mathsf{X}^2(\mathsf{t}^{**})$. Write t^* for $\mathsf{t}^*(a, b)$, then

$$\int \mathsf{X}(\mathsf{t}^{**}) \, dP = \int_{\mathsf{t}^* \le \tau} \mathsf{X}(\mathsf{t}^*) \, dP + \int_{\mathsf{t}^* > \tau} \mathsf{X}(\tau) \, dP,$$

$$\int \mathsf{X}^2(\mathsf{t}^{**}) \, dP = \int_{\mathsf{t}^* \le \tau} \mathsf{X}^2(\mathsf{t}^*) \, dP + \int_{\mathsf{t}^* > \tau} \mathsf{X}^2(\tau) \, dP.$$

Note that $|\mathsf{X}(\mathsf{t}^*)| \le \max(|a|, |b|)$, and on $\mathsf{t}^* > \tau$, $|\mathsf{X}(\tau)| \le \max(|a|, |b|)$. Hence as $\tau \to \infty$,

$$E\mathsf{X}(\mathsf{t}^{**}) \to E\mathsf{X}(\mathsf{t}^*), \qquad E\mathsf{X}^2(\mathsf{t}^{**}) \to E\mathsf{X}^2(\mathsf{t}^*).$$

Since $\mathsf{t}^{**} \uparrow \mathsf{t}^*$, apply monotone convergence to get $\lim E\mathsf{t}^{**} = E\mathsf{t}^*$, completing the proof.

Problem 1. Use Wald's identity 5.34 on the sums

$$\mathsf{S}_n = \sum_1^n \big(\mathsf{X}(k\,\Delta t) - \mathsf{X}((k-1)\,\Delta t)\big),$$

and, letting $\Delta t \to 0$, prove that for $\mathsf{t}^* = \mathsf{t}^*(a, b)$, and any λ,

$$E \exp[\lambda \mathsf{X}(\mathsf{t}^*) - \tfrac{1}{2}\lambda^2 \mathsf{t}^*] = 1.$$

By showing that differentiation under the integral is permissible, prove 13.5.

3. REPRESENTATION OF SUMS

An important representation is given by

Theorem 13.6. *Given independent, identically distributed random variables* $Y_1, Y_2, \ldots, EY_1 = 0, EY_1^2 = \sigma^2 < \infty, S_n = Y_1 + \cdots + Y_n$, *there exists a probability space with a Brownian motion* $X(t)$ *and a sequence* $T_1, T_2, \ldots$ *of nonnegative, independent, identically distributed random variables defined on it such that the sequence* $X(T_1), X(T_1 + T_2), \ldots,$ *has the same distribution as* $S_1, S_2, \ldots,$ *and* $ET_1 = \sigma^2$.

Proof. Let (U_n, V_n), $n = 1, 2, \ldots$ be a sequence of identically distributed, independent random vectors defined on the same probability space as a Brownian motion $\{X(t)\}$ such that $\mathcal{F}(X(t), t \geq 0)$ and $\mathcal{F}(U_n, V_n, n = 1, 2, \ldots)$ are independent. This can be done by constructing $(\Omega_1, \mathcal{F}_1, P_1)$ for the Brownian motion, $(\Omega_2, \mathcal{F}_2, P_2)$ for the (U_n, V_n) sequence, and taking

$$(\Omega, \mathcal{F}, P) = (\Omega_1 \times \Omega_2, \mathcal{F}_1 \times \mathcal{F}_2, P_1 \times P_2).$$

Suppose $U_n \leq 0 \leq V_n$. Define

$$T_1 = \inf\{t; X(t) \in (U_1, V_1)^c\}.$$

Therefore the U_1, V_1 function as random boundaries. Note that T_1 is a random variable because

$$\{T_1 \leq t\} = \left\{\inf_{0 \leq \tau \leq t} X(\tau) \leq U_1\right\} \cup \left\{\sup_{0 \leq \tau \leq t} X(\tau) \geq V_1\right\}.$$

Further, $\mathcal{F}(X(t + \tau) - X(\tau), t \geq 0)$ is independent of $\mathcal{F}(X(s), s \leq \tau, U_1, V_1)$. By the same argument which was used to establish the strong Markov property,

$$X^{(1)}(t) = X(t + T_1) - X(T_1), \quad t \geq 0$$

is a Brownian motion and is independent of $X(T_1)$. Now define T_2 as the first exit time of $X^{(1)}(t)$ from (U_2, V_2). Then $X(T_1 + T_2) - X(T_1)$ has the same distribution as $X(T_1)$. Repeating this procedure we manufacture variables

$$X(T_1 + \cdots + T_k) - X(T_1 + \cdots + T_{k-1}),$$

which are independent and identically distributed. The trick is to select U_1, V_1 so that $X(T_1)$ has the same distribution as Y_1. For any random boundaries U_1, V_1, if $E\,|X(T_1)|^2 < \infty$, 13.5 gives

$$E(X(T_1) \mid U_1, V_1) = 0 \quad \text{a.s.},$$

$$E(X^2(T_1) \mid U_1, V_1) = E(T_1 \mid U_1, V_1) \quad \text{a.s.}$$

Hence

$$EX(T_1) = 0, \qquad EX^2(T_1) = ET_1.$$

Therefore if $X(T_1)$ has the same distribution as Y_1, then Y_1 must satisfy

$$E_1 Y = 0,$$

and automatically,

$$\sigma^2 = EY_1^2 = ET_1.$$

So $EY_1 = 0$ is certainly a necessary condition for the existence of random boundaries such that $X(T_1)$ has the same distribution as Y_1.

To show that it is also sufficient, start with the observation that if Y_1 takes on only two values, say $u < 0$, and $v > 0$ with probabilities p and q, then from $EY_1 = 0$, we have $pu + qv = 0$. For this distribution, we can take fixed boundaries $U_1 = u$, $V_1 = v$. Because, by 13.5, $EX(T_1) = 0$, $uP(X(T_1) = u) + vP(X(T_1) = v) = 0$, which implies that $P(X(T_1) = u) = p$. This idea can be extended to prove

Proposition 13.7. *For any random variable* Y, *such that* $EY = 0$, $EY^2 < \infty$ *there are random boundaries* $U \leq 0 \leq V$ *such that* $X(T) \stackrel{\mathfrak{D}}{=} Y$.

Proof. Assume first that the distribution of Y is concentrated on points $u_i \leq 0$, $v_i \geq 0$ with probability p_i, q_i such that (u_i, v_i) are pairs satisfying $u_i p_i + v_i q_i = 0$. Then take

$$P((U, V) = (u_i, v_i)) = p_i + q_i.$$

By the observation above,

$$P(X(T) = u_i \mid (U, V) = (u_i, v_i)) = \frac{p_i}{p_i + q_i}.$$

Therefore $X(T)$ has the same distribution as Y. Suppose now that the distribution of Y is concentrated on a finite set $\{y_j\}$ of points. Then it is easy to see that the pairs $u_i \leq 0$, $v_i \geq 0$ can be gotten such that Y assumes only values in $\{u_i\}$, $\{v_i\}$ and the pairs (u_i, v_i) satisfy the conditions above. (Note that u_i may equal u_j, $i \neq j$, and similarly the v_i are not necessarily distinct.) For Y having any distribution such that $EY = 0$, $EY^2 < \infty$, take $Y_n \xrightarrow{\mathfrak{D}} Y$ where Y_n takes values in a finite set of points, $EY_n = 0$. Define random boundaries U_n, V_n having stopping time T_n such that $X(T_n) \stackrel{\mathfrak{D}}{=} Y_n$. Suppose that the random vectors (U_n, V_n) have mass-preserving distributions. Then take $(U_{n'}, V_{n'}) \xrightarrow{\mathfrak{D}} (U, V)$. For these random boundaries and associated stopping time T, for $I \subset (0, \infty)$, use (13.3) to get

$$P(X(T) \in I \mid U, V) = \chi_I(V) \frac{|U|}{|U| + |V|} = g(U, V).$$

Similarly,

$$P(X(T_n) \in I \mid U_n, V_n) = g(U_n, V_n).$$

Hence if $P(\mathsf{Y} \in bd(I)) = P(\mathsf{V} \in bd(I)) = 0$, then

$$\begin{aligned} P(\mathsf{Y} \in I) &= \lim_n P(\mathsf{Y}_n \in I) \\ &= \lim_n Eg(\mathsf{U}_n, \mathsf{V}_n) \\ &= Eg(\mathsf{U}, \mathsf{V}) \\ &= P(\mathsf{X}(\mathsf{T}) \in I). \end{aligned}$$

The analogous proof holds for $I \subset (0, -\infty)$. To complete the proof we need therefore to show that the Y_n can be selected so that the $(\mathsf{U}_n, \mathsf{V}_n)$ have a mass-preserving set of distributions. Take $F(dy)$ to denote the distribution of Y. We can always select a nonempty finite interval (a, b) such that including part or all of the mass of $F(dy)$ at the endpoints of (a, b) in the integral we get

$$\int_a^b yF(dy) = 0.$$

Thus we can always take the distributions F_n of the Y_n such that

$$\int_{[a,b]} yF_n(dy) = 0.$$

In this case, the (u_i, v_i) pairs have the property that either both are in $[a, b]$ or both are outside of $[a, b]$. Since $E\mathsf{Y}^2 < \infty$, we can also certainly take the Y_n such that $E\mathsf{Y}_n^2 \le M < \infty$ for all n. Write

$$\begin{aligned} E\mathsf{Y}_n^2 &= E(E(\mathsf{X}^2(\mathsf{T}_n) \mid \mathsf{U}_n, \mathsf{V}_n)) = E(E(\mathsf{T}_n \mid \mathsf{U}_n, \mathsf{V}_n)) \\ &= E\,|\mathsf{U}_n\mathsf{V}_n|. \end{aligned}$$

But the function $|uv|$ goes to infinity as either $|u| \to \infty$ or $|v| \to \infty$ everywhere in the region $\{u < a, v > b\}$. This does it.

Problem 2. Let the distribution function of Y be $F(dy)$. Prove that the random boundaries (U, V) with distribution

$$\begin{aligned} P(\mathsf{U} \in du) &= F(du) \\ P(\mathsf{V} \in dv \mid \mathsf{U} = u) &= \alpha\,|v|\,g(u, v)F(dv), \end{aligned}$$

where $\alpha^{-1} = E\mathsf{Y}^+$ and $g(u, v)$ is zero or one as u and v have the same or opposite signs, give rise to an exit time T such that $\mathsf{X}(\mathsf{T}) \overset{\mathfrak{D}}{=} \mathsf{Y}$. (Here U and V can be both positive and negative.)

4. CONVERGENCE OF SAMPLE PATHS OF SUMS TO BROWNIAN MOTION PATHS

Now it is possible to show that in a very strong sense Brownian motion is the limit of random walks with smaller and smaller steps, or of normed sums of independent, identically distributed random variables. The random walk example is particularly illuminating. Let the walk $X^{(n)}(t)$ take steps of size $\pm 1/\sqrt{n}$ every $1/n$ time units. Using the method of the previous section, let T_1 be the first time that $X(t)$ changes by an amount $1/\sqrt{n}$, then T_2 the time until a second change of amount $1/\sqrt{n}$ occurs, etc.

The process

$$\tilde{X}^{(n)}(t) = X(T_1 + \cdots + T_{[nt]})$$

has the same distribution as the process $X^{(n)}(t)$. By definition,

$$T_1 + \cdots + T_{[nt]} = \text{time until } [nt] \text{ changes of magnitude } 1/\sqrt{n} \text{ have occurred along } X(t).$$

Therefore, up to time $T_1 + \cdots + T_{[nt]}$, the sum of the squares of the changes in $X(t)$ is approximately t. But by 12.28, this takes a length of time t. So, we would expect that $T_1 + \cdots + T_{[nt]} \to t$, hence that each sample path of the interpolated motion $\tilde{X}^{(n)}(t)$ would converge as a function of t to the corresponding path of Brownian motion. The convergence that does take place is uniform convergence. This holds, in general, along subsequences.

Theorem 13.8. *Let $Y_1, Y_2, \ldots$ be independent, identically distributed random variables, $EY_1 = 0$, $EY_1^2 = \sigma^2 < \infty$, $S_n = Y_1 + \cdots + Y_n$. Define the processes $X^{(n)}(t)$ by $S_{[nt]}/\sigma\sqrt{n}$. Then there are processes $\{\tilde{X}^{(n)}(t)\}$, for each n having the same distribution as $\{X^{(n)}(t)\}$, defined on a common probability space and a Brownian motion process $\{X(t)\}$ on the same space, such that for any subsequence $\{n_k\}$ increasing rapidly enough,*

$$\sup_{0\le t\le 1} |\tilde{X}^{(n_k)}(t) - X(t)| \to 0 \quad \text{a.s.}$$

Proof. Assume that $EY_1^2 = 1$. Let $(\Omega, \mathcal{F}, P)$ be constructed as in the representation theorem. For each n, consider the Brownian motion $X_n(t) = \sqrt{n}X(t/n)$. Construct $T_1^{(n)}, T_2^{(n)}, \ldots$ using the motion $X_n(t)$. Then the $\{S_k\}$ sequence has the same distribution as the $\{X_n(T_1^{(n)} + \cdots + T_k^{(n)})\}$ sequence. Thus the $X^{(n)}(t)$ process has the same distribution as

$$\tilde{X}^{(n)}(t) = X\left(\frac{T_1^{(n)} + \cdots + T_{[nt]}^{(n)}}{n}\right). \tag{13.9}$$

The sequence $T_1^{(n)}, T_2^{(n)}, \ldots$ for each n consists of independent, identically distributed random variables such that $ET_1^{(n)} = 1$, and $T_1^{(n)}$ has the same distribution for all n. The weak law of large numbers gives

$$\frac{T_1^{(n)} + \cdots + T_{[nt]}^{(n)}}{[nt]} \xrightarrow{P} 1,$$

so for n large, the random time appearing in (13.9) should be nearly t.

Argue that if it can be shown that

$$\mathsf{W}_n = \sup_{0 \le t \le 1} \left| \frac{\mathsf{T}_1^{(n)} + \cdots + \mathsf{T}_{[nt]}^{(n)}}{n} - t \right| \to 0 \quad \text{a.s.}$$

for n running through some subsequence, then the continuity of $\mathsf{X}(t)$ guarantees that along the same subsequence

$$\sup_{0 \le t \le 1} |\mathsf{X}^{(n)}(t) - \mathsf{X}(t)| \to 0 \quad \text{a.s.}$$

What can be easily proved is that $\mathsf{W}_n \xrightarrow{P} 0$. But this is enough because then for any subsequence increasing rapidly enough, $\mathsf{W}_{n_k} \xrightarrow{\text{a.s.}} 0$. Use

$$\bar{\mathsf{T}}_k^{(n)} = \mathsf{T}_k^{(n)} - 1,$$

so

$$E\bar{\mathsf{T}}_k^{(n)} = 0,$$

$$\mathsf{W}_n = \sup_{0 \le t \le 1} \left| \frac{\bar{\mathsf{T}}_1^{(n)} + \cdots + \bar{\mathsf{T}}_{[nt]}^{(n)}}{n} \right| + \frac{\boldsymbol{\theta}}{n}, \quad |\boldsymbol{\theta}| \le 1.$$

Ignore the second term, and write

$$\mathsf{W}_n \le \sup_{0 \le t \le 1} t \left| \frac{\bar{\mathsf{T}}_1^{(n)} + \cdots + \bar{\mathsf{T}}_{[nt]}^{(n)}}{[nt]} \right|.$$

For any ϵ, $0 < \epsilon < 1$, take

$$\mathsf{W}_n' = \epsilon \sup_{0 \le t \le \epsilon} \left| \frac{\bar{\mathsf{T}}_1^{(n)} + \cdots + \bar{\mathsf{T}}_{[nt]}^{(n)}}{[nt]} \right|$$

$$\le \epsilon \sup_{k \ge 1} \left| \frac{\bar{\mathsf{T}}_1^{(n)} + \cdots + \bar{\mathsf{T}}_k^{(n)}}{k} \right| = \epsilon \mathsf{M}^{(n)}.$$

The distribution of $\mathsf{M}^{(n)}$ is the same as $\mathsf{M}^{(1)}$. Now write

$$\mathsf{W}_n'' = \sup_{\epsilon \le t \le 1} \left| \frac{\bar{\mathsf{T}}_1^{(n)} + \cdots + \bar{\mathsf{T}}_{[nt]}^{(n)}}{[nt]} \right|$$

$$\le \sup_{k \ge [\epsilon n]} \left| \frac{\bar{\mathsf{T}}_1^{(n)} + \cdots + \bar{\mathsf{T}}_k^{(n)}}{k} \right|.$$

This bounding term has the same distribution as

$$\sup_{k \ge [\epsilon n]} \left| \frac{\bar{\mathsf{T}}_1^{(1)} + \cdots + \bar{\mathsf{T}}_k^{(1)}}{k} \right|.$$

The law of large numbers implies

$$\frac{\bar{\mathsf{T}}_1^{(1)} + \cdots + \bar{\mathsf{T}}_k^{(1)}}{k} \xrightarrow{\text{a.s.}} 0,$$

so that $\mathsf{W}_n'' \xrightarrow{P} 0$. Since $\mathsf{W}_n \leq \mathsf{W}_n' + \mathsf{W}_n''$, then for any $\epsilon > 0$, $x > 0$,

$$\overline{\lim}\, P(\mathsf{W}_n > x) \leq P(\epsilon \mathsf{M}^{(1)} > x).$$

Taking $\epsilon \downarrow 0$ now gives the result.

5. AN INVARIANCE PRINCIPLE

The question raised in the introduction to this chapter is generally this:

The sequence of processes $\mathsf{X}^{(n)}(\cdot)$ converges in distribution to Brownian motion $\mathsf{X}(\cdot)$, denoted $\mathsf{X}^{(n)}(\cdot) \xrightarrow{\mathfrak{D}} \mathsf{X}(\cdot)$, in the sense that for any $0 \leq t_1 < \cdots < t_k \leq 1$,

$$(\mathsf{X}^{(n)}(t_1), \ldots, \mathsf{X}^{(n)}(t_k)) \xrightarrow{\mathfrak{D}} (\mathsf{X}(t_1), \ldots, \mathsf{X}(t_k)).$$

Let $\mathsf{H}(x(\cdot))$ be defined on $R^{[0,1]}$. When is it true that

$$\mathsf{H}(\mathsf{X}^{(n)}(\cdot, \omega)) \xrightarrow{\mathfrak{D}} \mathsf{H}(\mathsf{X}(\cdot, \omega))?$$

There is no obvious handle. What is clear, however, is that we can proceed as follows: Let us suppose $\mathsf{X}^{(n)}(\cdot)$ and $\mathsf{X}(\cdot)$ are defined on the same space and take values in some subset $D \subset R^{[0,1]}$. On D define the sup-norm metric

$$\rho(x(\cdot), y(\cdot)) = \sup_{0 \leq t \leq 1} |x(t) - y(t)|,$$

and assume that $\mathsf{H}(x(\cdot))$ defined on D is continuous with respect to ρ. Then if the sample paths of the $\mathsf{X}^{(n)}(\cdot)$ processes converge uniformly to the corresponding paths of $\mathsf{X}(\cdot)$, that is, if

$$\sup_{0 \leq t \leq 1} |\mathsf{X}^{(n)}(t) - \mathsf{X}(t)| \to 0 \quad \text{a.s.},$$

then

$$\mathsf{H}(\mathsf{X}^{(n)}(\cdot)) \xrightarrow{\text{a.s.}} \mathsf{H}(\mathsf{X}(\cdot)).$$

But this is enough to give us what we want. Starting with the $\mathsf{X}^{(n)}(t) = \mathsf{S}_{[nt]}/\sigma\sqrt{n}$, we can construct $\tilde{\mathsf{X}}^{(n)}(\cdot)$ having the same distribution as $\mathsf{X}^{(n)}(\cdot)$ so that the $\tilde{\mathsf{X}}^{(n)}(t) \to \mathsf{X}(t)$ uniformly for $t \in [0,1]$ for n running through subsequences. Thus, $\mathsf{H}(\tilde{\mathsf{X}}^{(n)}(\cdot)) \xrightarrow{\text{a.s.}} \mathsf{H}(\mathsf{X}(\cdot))$, $n \in \{n_k\}$. This implies

$$\mathsf{H}(\mathsf{X}^{(n)}(\cdot)) \xrightarrow{\mathfrak{D}} \mathsf{H}(\mathsf{X}(\cdot)), \quad n \in \{n_k\}.$$

But this holding true for every subsequence $\{n_k\}$ increasing rapidly enough implies that the full sequence converges in distribution to $\mathsf{H}(\mathsf{X}(\cdot))$. Now to fasten down this idea.

Definition 13.10. *D is the class of all functions $x(t)$, $0 \leq t \leq 1$, such that $x(t-)$, $x(t+)$ exist for all $t \in (0, 1)$, and $x(t+) = x(t)$. Also, $x(0+) = x(0)$, $x(1-) = x(1)$. Define $\rho(x(\cdot), y(\cdot))$ on D by*

$$\sup_{0 \leq t \leq 1} |x(t) - y(t)|.$$

Definition 13.11. *For $\mathsf{H}(x(\cdot))$ defined on D, let G be the set of all functions $x(\cdot) \in D$ such that H is discontinuous at $x(\cdot)$ in the metric ρ. If there is a set $G_1 \in \mathcal{B}^{[0,1]}$ such that $G \subset G_1$ and for a normalized Brownian motion $\{\mathsf{X}(t)\}$, $P(\mathsf{X}(\cdot) \in G_1) = 0$, call H a.s. B-continuous.*

The weakening of the continuity condition on H to a.s. B-continuity is important. For example, the H that leads to the arc sine law is discontinuous at the set of all $x(\cdot) \in D$ such that

$$l\{t; x(t) = 0, \quad 0 \leq t \leq 1\} > 0.$$

(We leave this to the reader to prove as Problem 4.) But this set has probability zero in Brownian motion.

With these definitions, we can state the following special case of the "invariance principle."

Theorem 13.12. *Let H defined on D be a.s. B-continuous. Consider any process of the type*

$$\mathsf{X}^{(n)}(t) = \frac{\mathsf{S}_{[nt]}}{\sigma\sqrt{n}},$$

where the S_n are sums of independent, identically distributed random variables $\mathsf{Y}_1, \mathsf{Y}_2, \ldots$ with $E\mathsf{Y}_1 = 0$, $E\mathsf{Y}_1^2 = \sigma^2$. Assume that the $\mathsf{H}(\mathsf{X}^{(n)}(\cdot))$ are random variables. Then

$$\mathsf{H}(\mathsf{X}^{(n)}(\cdot)) \xrightarrow{\mathfrak{D}} \mathsf{H}(\mathsf{X}(\cdot)), \tag{13.13}$$

where $\{\mathsf{X}(t)\}$ is normalized Brownian motion.

Proof. Use 8.8; it is enough to show that any subsequence $\{n_k\}$ contains a subsequence $\{n_k'\}$ such that (13.13) holds along n_k'. Construct $\tilde{\mathsf{X}}_n(t)$ as in the proof of 13.8. Take n_k' any subsequence of n_k increasing rapidly enough. Then $\tilde{\mathsf{X}}^{(n_k')}(t)$ converges uniformly to $\mathsf{X}(t)$ for almost every ω, implying that (13.13) holds along the n_k' sequence.

There is a loose end in that $\mathsf{H}(\mathsf{X}(\cdot))$ was not assumed to be a random variable. However, since

$$\mathsf{H}(\tilde{\mathsf{X}}^{(n_k')}(\cdot)) \xrightarrow{\text{a.s.}} \mathsf{H}(\mathsf{X}(\cdot)),$$

the latter is a.s. equal to a random variable. Hence it is a random variable with respect to the completed probability space $(\Omega, \bar{\mathcal{F}}, \bar{P})$, and its distribution is well defined.

The reason that theorems of this type are referred to as *invariance principles* is that they establish convergence to a limiting distribution which does not depend on the distribution function of the independent summands $\mathsf{Y}_1, \mathsf{Y}_2, \ldots$ except for the one parameter σ^2. This gives the freedom to choose the most convenient way to evaluate the limit distribution. Usually, this is done either directly for Brownian motion or by combinatorial arguments for coin-tossing variables $\mathsf{Y}_1, \mathsf{Y}_2, \ldots$ In particular, see Feller's book [59, Vol. I], for a combinatorial proof that in fair coin-tossing, the proportion of times W_n that the player is ahead in the first n tosses has the limit distribution

$$\lim P(\mathsf{W}_n < x) = 2/\pi \text{ arc sine } \sqrt{x}.$$

Problems

3. Show that the function on D defined by

$$\mathsf{H}(x(\cdot)) = \sup_{0 \le t \le 1} |x(t)|$$

is continuous everywhere.

4. Show that the function on D defined by

$$\mathsf{H}(x(\cdot)) = l\{t;\, x(t) > 0, \quad 0 \le t \le 1\}$$

is continuous at $x(\cdot)$ if and only if $l\{t;\, x(t) = 0\} = 0$.

6. THE KOLMOGOROV-SMIRNOV STATISTICS

An important application of invariance is to an estimation problem. Let $\mathsf{Y}_1, \mathsf{Y}_2, \ldots$ be independent, identically distributed random variables with a continuous but unknown distribution function $F(x)$. The most obvious way to estimate $F(x)$ given n observations $\mathsf{Y}_1, \ldots, \mathsf{Y}_n$ is to put

$$\begin{aligned}\hat{\mathsf{F}}_n(x) &= \frac{1}{n}(\text{number of } \mathsf{Y}_k < x, \quad k = 1, \ldots, n) \\ &= \frac{1}{n}\sum_1^n \chi_{(-\infty,x)}(\mathsf{Y}_k).\end{aligned}$$

The law of large numbers guarantees

$$\hat{\mathsf{F}}_n(x) \xrightarrow{\text{a.s.}} F(x)$$

for fixed x. From the central limit theorem,

$$\sqrt{n}\,(\hat{\mathsf{F}}_n(x) - F(x)) \xrightarrow{\mathfrak{D}} \mathcal{N}(0, F(x)(1 - F(x))).$$

However, we will be more interested in uniform estimates:

$$\mathsf{D}_n^+ = \sqrt{n}\sup_x\big(\hat{\mathsf{F}}_n(x) - F(x)\big),$$

$$\mathsf{D}_n^- = \sqrt{n}\inf_x\big(\hat{\mathsf{F}}_n(x) - F(x)\big),$$

$$\mathsf{D}_n = \sqrt{n}\sup_x|\hat{\mathsf{F}}_n(x) - F(x)|,$$

and the problem is to show that D_n^+, D_n^-, D_n converge in distribution, and to find the limiting distribution.

Proposition 13.14. *Each of* D_n^+, D_n^-, D_n *has the same distribution for all continuous* $F(x)$.

Proof. Call $I = [a, b]$ an interval of constancy for $F(x)$ if $P(\mathsf{Y}_1 \in I) = 0$ and there is no larger interval containing I having this property. Let B be the union of all the intervals of constancy. Clearly, we can write

$$\mathsf{D}_n = \sqrt{n}\sup_{x\in B^c}|\hat{\mathsf{F}}_n(x) - F(x)|,$$

and similar equations for D_n^+ and D_n^-. For $x \in B^c$, the sets

$$\{\mathsf{Y}_k < x\}, \qquad \{F(\mathsf{Y}_k) < F(x)\}$$

are identical. Put $\mathsf{U}_k = F(\mathsf{Y}_k)$, and set

$$\hat{\mathsf{G}}_n(y) = \{\text{number of } \mathsf{U}_k < y, k = 1, \ldots, n\}.$$

Then

$$\mathsf{D}_n = \sqrt{n}\sup_{x\in B^c}|\hat{\mathsf{G}}_n(F(x)) - F(x)|$$
$$= \sqrt{n}\sup_x|\hat{\mathsf{G}}_n(F(x)) - F(x)|.$$

Since $F(x)$ maps $R^{(1)}$ onto $(0, 1)$ plus the points $\{0\}$ or $\{1\}$ possibly,

$$\mathsf{D}_n = \sqrt{n}\sup_{y\in(0,1)}|\hat{\mathsf{G}}_n(y) - y|$$
$$= \sqrt{n}\sup_{y\in[0,1]}|\hat{\mathsf{G}}_n(y) - y| \quad \text{a.s.},$$

the latter holding because $P(\mathsf{U}_1 = 0) = P(\mathsf{U}_1 = 1) = 0$. The distribution of U_1 is given by

$$P(\mathsf{U}_1 < y) = P\big(F(\mathsf{Y}_1) < y\big).$$

Put $x = \inf\{\xi; F(\xi) = y\}$. Then

$$P(\mathsf{U}_1 < y) = P\big(F(\mathsf{Y}_1) < F(x)\big) = P(\mathsf{Y}_1 < x) = F(x) = y.$$

Thus U_1 is uniformly distributed on $[0, 1]$, and D_n for arbitrary continuous

F has the same distribution as D_n for the uniform distribution. Similarly for D_n^-, and D_n^+.

Let $\mathsf{U}_1, \ldots, \mathsf{U}_n$ be independent random variables uniformly distributed on [0, 1]. The order statistics are defined as follows: $\mathsf{U}_1^{(n)}$ is the smallest, and so forth; $\mathsf{U}_n^{(n)}$ is the largest. The maximum of $|\hat{\mathsf{G}}_n(y) - y|$ or of $\hat{\mathsf{G}}_n(y) - y$ or $y - \hat{\mathsf{G}}_n(y)$ must occur at one of the jumps of $\hat{\mathsf{G}}_n(y)$. The jumps are at the points $\mathsf{U}_k^{(n)}$, and

$$\hat{\mathsf{G}}_n(\mathsf{U}_k^{(n)}) = \frac{k-1}{n}.$$

Since the size of the jumps is $1/n$, then to within $1/\sqrt{n}$,

$$\mathsf{D}_n = \sqrt{n} \max_{k \le n} \left| \mathsf{U}_k^{(n)} - \frac{k}{n} \right|,$$

$$\mathsf{D}_n^+ = \sqrt{n} \max_{k \le n} \left(\frac{k}{n} - \mathsf{U}_k^{(n)} \right),$$

$$\mathsf{D}_n^- = \sqrt{n} \min_{k \le n} \left(\frac{k}{n} - \mathsf{U}_k^{(n)} \right).$$

The fact that makes our invariance theorem applicable is that the $\mathsf{U}_1^{(n)}, \ldots, \mathsf{U}_n^{(n)}$ behave something like sums of independent random variables. Let $\mathsf{W}_1, \mathsf{W}_2, \ldots$ be independent random variables with the negative exponential distribution. That is,

$$P(\mathsf{W}_k > x) = e^{-x}, \quad x \ge 0.$$

Denote $\mathsf{Z}_n = \mathsf{W}_1 + \cdots + \mathsf{W}_n$; then

Proposition 13.15. *$\mathsf{U}_k^{(n)}$, $k = 1, \ldots, n$ have the same joint distribution as $\mathsf{Z}_k/\mathsf{Z}_{n+1}$, $k = 1, \ldots, n$.*

Proof. To show this, write (using a little symbolic freedom),

$$\begin{aligned} P(\mathsf{Z}_1 \in dx_1, \mathsf{Z}_2 \in dx_2, \ldots, \mathsf{Z}_{n+1} \in dx_{n+1}) \\ &= P(\mathsf{W}_1 \in dx_1, \mathsf{W}_2 \in dx_2 - x_1, \ldots, \mathsf{W}_{n+1} \in dx_{n+1} - x_n) \\ &= e^{-x_1} e^{-(x_2 - x_1)} \cdots e^{-(x_{n+1} - x_n)}\, dx_1 \cdots dx_{n+1} \\ &= e^{-x_{n+1}}\, dx_1 \cdots dx_{n+1}, \quad 0 \le x_1 \le x_2 \le \cdots \le x_{n+1}. \end{aligned}$$

Thus

$$P(\mathsf{Z}_1 \in dx_1, \ldots, \mathsf{Z}_n \in dx_n \mid \mathsf{Z}_{n+1} = x_{n+1}) = \begin{cases} n!\, x_{n+1}^{-n}\, dx_1 \cdots dx_n, & 0 \le x_1 \le x_2 \le \cdots \le x_{n+1}, \\ 0, & \text{otherwise}. \end{cases}$$

From this,

$$P\left(\frac{Z_1}{Z_{n+1}} \in dy_1, \ldots, \frac{Z_n}{Z_{n+1}} \in dy_n \mid Z_{n+1} = x_{n+1}\right)$$
$$= \begin{cases} n!\, dy_1 \cdots dy_n, & 0 \le y_1 \le \cdots \le y_n \le 1, \\ 0, & \text{otherwise.} \end{cases}$$

Therefore,

$$P\left(\frac{Z_1}{Z_{n+1}} \in dy_1, \ldots, \frac{Z_n}{Z_{n+1}} \in dy_n\right)$$
$$= \begin{cases} n!\, dy_1 \cdots dy_n, & 0 \le y_1 \le \cdots \le y_n \le 1, \\ 0, & \text{otherwise.} \end{cases}$$

On the other hand, for $0 \le y_1 \le \cdots \le y_n \le 1$,

$$P(U_1^{(n)} \in dy_1, \ldots, U_n^{(n)} \in dy_n) = \Sigma\, P(U_{l_1} \in dy_1, \ldots, U_{l_n} \in dy_n),$$

where the sum is over all permutations $(l_1, \ldots, l_n)$ of $(1, \ldots, n)$. Using independence this yields $n!\, dy_1 \cdots dy_n$.

Use this proposition to transform the previous expression for D_n into

$$D_n \overset{\mathfrak{D}}{=} \sqrt{n} \max_{k \le n} \left| \frac{Z_k}{Z_{n+1}} - \frac{k}{n} \right|,$$

with analogous expressions for D_n^+, D_n^-. Then

$$D_n \overset{\mathfrak{D}}{=} \frac{n}{Z_{n+1}} \max_{k \le n} \left| \frac{Z_k - k}{\sqrt{n}} - \frac{k}{n} \cdot \frac{Z_{n+1} - n}{\sqrt{n}} \right|.$$

Because $EW_1 = 1$, $\sigma^2(W_1) = 1$, it follows that $n/Z_{n+1} \xrightarrow{\text{a.s.}} 1$, and that $Z_k - k$ is a sum of independent, identically distributed random variables with first moment zero and second moment one. Put $S_k = Z_k - k$,

$$X^{(n)}(t) = \frac{S_{[nt]}}{\sqrt{n}},$$

and ignore the n/Z_{n+1} term and terms of order $1/\sqrt{n}$. Then D_n is given by

$$\sup_{0 \le t \le 1} |X^{(n)}(t) - tX^{(n)}(1)|.$$

Obviously,

$$\sup_{0 \le t \le 1} |x(t) - tx(1)|$$

is a continuous function in the sup-norm metric, so now applying the invariance principle, we have proved

Theorem 13.16

$$\mathsf{D}_n \xrightarrow{\mathcal{D}} \sup_{0\le t\le 1} |\mathsf{X}(t) - t\mathsf{X}(1)|,$$

$$\mathsf{D}_n^+ \xrightarrow{\mathcal{D}} \sup_{0\le t\le 1} (\mathsf{X}(t) - t\mathsf{X}(1)),$$

$$\mathsf{D}_n^- \xrightarrow{\mathcal{D}} \inf_{0\le t\le 1} (\mathsf{X}(t) - t\mathsf{X}(1)).$$

7. MORE ON FIRST-EXIT DISTRIBUTIONS

There is a wealth of material in the literature on evaluating the distributions of functions on Brownian motion. One method uses some transformations that carry Brownian motion into Brownian motion. A partial list of such transformations is

***Proposition* 13.17.** *If* $\mathsf{X}(t)$ *is normalized Brownian motion, then so is*

1) $-\mathsf{X}(t),\quad t \ge 0$ (*symmetry*),
2) $\mathsf{X}(t+\tau) - \mathsf{X}(t),\quad t \ge 0, \tau \ge 0$ fixed (*origin change*),
3) $t\mathsf{X}(1/t),\quad t \ge 0$ (*inversion*),
4) $(1/\sqrt{\alpha})\mathsf{X}(\alpha t),\quad t \ge 0, \alpha > 0$ (*scale change*),
5) $\mathsf{X}(T) - \mathsf{X}(T-t),\quad 0 \le t \le T$ fixed (*reversal*).

To get (4) and (5) just check that the processes are Gaussian with zero means and the right covariance.

We apply these transformations and the strong Markov property to get the distributions of some first exit times and probabilities. These are related to a number of important functions on Brownian motion. For example, for $x > 0$, if $\mathbf{t}_x^*$ is the first hitting time of the point $\{x\}$, then

$$P\left(\sup_{0\le\tau\le t} \mathsf{X}(\tau) < x\right) = P(\mathbf{t}_x^* > t). \tag{13.18}$$

To get the distribution of $\mathbf{t}_x^*$, let

$$\varphi(\lambda, x) = Ee^{-\lambda \mathbf{t}_x^*}, \quad \lambda \ge 0.$$

Take $x, y > 0$, note that $t_{x+y}^* = \mathbf{t}_x^* + \boldsymbol{\tau}_y^*$, where $\boldsymbol{\tau}_y^*$ is the first passage time of the process $\mathsf{X}(t) = \mathsf{X}(\mathbf{t}_x^* + t) - \mathsf{X}(\mathbf{t}_x^*)$ to the point y. By the strong Markov property, $\boldsymbol{\tau}_y^*$ has the same distribution as $\mathbf{t}_y^*$ and is independent of $\mathbf{t}_x^*$. Thus,

$$\varphi(\lambda, x+y) = \varphi(\lambda, x)\varphi(\lambda, y). \tag{13.19}$$

Since $\varphi(\lambda, x)$ is decreasing in x, and therefore well-behaved, 13.19 implies

(13.20) $$\varphi(\lambda, x) = e^{-g(\lambda)x}.$$

Now we can get more information by a scale change. Transformation 13.17(4) implies that a scale change in space by an amount $\sqrt{\alpha}$ changes time by a factor α. To be exact,

$$\begin{aligned} \mathbf{t}_x^* &= \inf\{t; \mathsf{X}(t) = x\} \\ &= \alpha \inf\{t/\alpha; (1/\sqrt{\alpha})\mathsf{X}(\alpha(t/\alpha)) = x/\sqrt{\alpha}\} \\ &= \alpha \inf\{t'; \mathsf{X}(t') = x/\sqrt{\alpha}\}. \end{aligned}$$

Therefore $\mathbf{t}_x^*$ has the same distribution as $\alpha \mathbf{t}_{x/\sqrt{\alpha}}^*$, yielding

(13.21) $$\begin{aligned} &\varphi(\lambda, x) = \varphi(\alpha\lambda, x/\sqrt{\alpha}), \\ &\Rightarrow g(\lambda) = c\sqrt{\lambda}, \\ &\Rightarrow \varphi(x, \lambda) = e^{-c\sqrt{\lambda}x}, \quad x > 0, \\ &\Rightarrow \varphi(x, \lambda) = e^{-c\sqrt{\lambda}|x|}, \quad \text{all } x, \quad \text{by symmetry.} \end{aligned}$$

Now $\varphi(x, \lambda)$ uniquely determines the distribution of t_x^*, so if we can get c, then we are finished. Unfortunately, there seems to be no very simple way to get c. Problem 5 outlines one method of showing that $c = \sqrt{2}$. Accept this for now, because arguments of this sort can get us the distribution of

$$\mathbf{t}^*(a, b) = \min(\mathbf{t}_a^*, \mathbf{t}_b^*), \qquad \{0\} \in (a, b).$$

Denote

$$\mathbf{t}^* = \min(\mathbf{t}_a^*, \mathbf{t}_b^*), \qquad A_a = \{\mathbf{t}_a^* < \mathbf{t}_b^*\}, \qquad A_b = \{\mathbf{t}_b^* < \mathbf{t}_a^*\},$$

so that on A_a, $\mathbf{t}_a^* = \mathbf{t}^*$, and on A_b, $\mathbf{t}_a^* = \mathbf{t}^* + \boldsymbol{\tau}^*$, where $\boldsymbol{\tau}^*$ is the additional time needed to get to $x = a$ once the process has hit $x = b$. So define

$$\boldsymbol{\tau}^* = \min\{t; \mathsf{X}(\mathbf{t}^* + t) - \mathsf{X}(\mathbf{t}^*) = a - b\}.$$

Put these together:

$$Ee^{-\lambda \mathbf{t}_a^*} = \int_{A_a} e^{-\lambda \mathbf{t}^*}\, dP + \int_{A_b} e^{-\lambda \mathbf{t}^*} e^{-\lambda \boldsymbol{\tau}^*}\, dP.$$

Now check that $A_b \in \mathcal{F}(\mathsf{X}(t), t \le \mathbf{t}^*)$. Since the variable $\boldsymbol{\tau}^*$ is independent of $\mathcal{F}(\mathsf{X}(t), t \le \mathbf{t}^*)$,

$$e^{-\sqrt{2\lambda}|a|} = \int_{A_a} e^{-\lambda \mathbf{t}^*}\, dP + e^{-\sqrt{2\lambda}(|a|+|b|)} \int_{A_b} e^{-\lambda \mathbf{t}^*}\, dP.$$

The same argument for $\mathbf{t}_b^*$ gives

(13.22) $$e^{-\sqrt{2\lambda}|b|} = e^{-\sqrt{2\lambda}(|b|+|a|)} \int_{A_a} e^{-\lambda \mathbf{t}^*}\, dP + \int_{A_b} e^{-\lambda \mathbf{t}^*}\, dP.$$

Now solve, to get

$$\textbf{(13.23)}\quad 1)\int_{A_a} e^{-\lambda t^*}\,dP = \frac{\sinh(\sqrt{2\lambda}\,|b|)}{\sinh(\sqrt{2\lambda}(|b| + |a|))},$$

$$2)\int_{A_b} e^{-\lambda t^*}\,dP = \frac{\sinh(\sqrt{2\lambda}\,|a|)}{\sinh(\sqrt{2\lambda}(|b| + |a|))}.$$

The sum of (13.23(1) and (2)) is $Ee^{-\lambda t^*}$, the Laplace transform of the distribution of t^*. By inverting this, we can get

$$P(a < X(\tau) < b, \text{ all } \tau \in [0, t]).$$

Very similar methods can be used to compute the probability that the Brownian motion ever hits the line $x = at + b$, $a \geq 0, b \geq 0$, or equivalently, exits from the open region with the variable boundary $x = at + b$. Let $p(a, b)$ be the probability that $X(t)$ ever touches the line $at + b$, $a \geq 0$, $b \geq 0$. Then

$$p(a, b_1 + b_2) = p(a, b_1)p(a, b_2),$$

the argument being that to get to $at + b_1 + b_2$, first the particle must get to $at + b_1$, but once it does, it then has to get to a line whose equation relative to its present position is $at + b_2$. To define this more rigorously, let τ^* be the time of first touching $at + b_1$; then $t^* = \min(\tau^*, s)$ is a stopping time. The probability that the process ever touches the line $at + b_1 + b_2$ and $\tau^* < s$ equals the probability that the process $X(t + t^*) - X(t^*)$ ever touches the line $at + b_2$ and $t^* < s$. By the strong Markov property, the latter probability is the product $p(a, b_2)P(\tau^* < s)$. Let $s \to \infty$ to get the result. Therefore, $p(a, b) = e^{-\psi(a)b}$. Take t_b^* to be the hitting time of the point b. Then

$$p(a, b) = P\left(\sup_{0 \leq t < \infty} (X(t + t_b^*) - X(t_b^*) - (at + at_b^*)) \geq 0\right).$$

Conditioning on t_b^* yields

$$p(a, b) = Ee^{-\psi(a)at_b^*},$$

(see 4.38). Use (13.21) to conclude that

$$p(a, b) = e^{-\sqrt{2a\psi(a)}\, b},$$

which leads to $2a\psi(a) = \psi^2(a)$, or $\psi(a) = 2a$. Thus

$$\textbf{(13.24)}\qquad p(a, b) = e^{-2ab}.$$

The probability,

$$P\left(\sup_t \frac{|X(t)|}{at + b} \geq 1\right),$$

of exiting from the two-sided region $|x| < at + b$ is more difficult to compute. One way is to first compute $Ee^{-\lambda t^*}$, where t^* is the first time of hitting $at + b$, and then imitate the development leading to the two-sided boundary distribution in (13.23). Another method is given in Doob [36].

The expression for $p(a, b)$ can be used to get the distribution of

$$\sup_{0 \leq t \leq 1} (X(t) - tX(1))$$

and therefore of $\lim_n P(D_n^+ < x)$. Let $Y(t) = X(t) - tX(1)$. Then $Y(t)$ is a Gaussian process with covariance

$$EY(t)Y(s) = s(1 - t), \qquad 0 \leq s \leq t \leq 1.$$

Consider the process

$$X^{(1)}(t) = (1 + t)Y\left(\frac{t}{1 + t}\right), \qquad t \geq 0.$$

Its covariance is $\min(s, t)$, so $X^{(1)}(t)$ is normalized Brownian motion. Therefore

$$\begin{aligned} P\left(\sup_{0 \leq t \leq 1} Y(t) \geq y\right) &= \left(\sup_{t \geq 0} \frac{X^{(1)}(t)}{1 + t} \geq y\right) \\ &= P\left(\sup_{t \geq 0} (X^{(1)}(t) - yt - y) \geq 0\right) \\ &= p(y, y) = e^{-2y^2}. \end{aligned}$$

The limiting distribution for D_n is similarly related to the probability of exiting from the two-sided region $\{|x| < y(1 + t)\}$.

Problems

5. Assuming

$$E \exp[-\lambda t_x^*] = \exp[-c\sqrt{\lambda}\,|x|],$$

find $Ee^{-\lambda t^*}$, where $t^* = t^*(-1, +1)$. Differentiating this with respect to λ, at $\lambda = 0$, find an expression for Et^* and compare this with the known value of Et^* to show that $c = \sqrt{2}$.

6. Use Wald's identity (see Problem 1) to get (13.23(1) and (2)) by using the equations for λ and $-\lambda$.

7. Using $E \exp[-\lambda t_x^*] = \exp[-\sqrt{2\lambda}\,|x|]$, prove that for $x \geq 0$,

$$P\left(\sup_{0 \leq \tau \leq t} X(\tau) \geq x\right) = 2P(X(t) \geq x).$$

8. For $S_1, S_2, \ldots$ sums of independent, identically distributed random variables with zero means and finite second moments, find normalizing constants so that the following random variables converge in distribution to a nondegenerate limit, and evaluate the distribution of the limit, or the Laplace transform of the limit distribution

a) $\max(S_1, \ldots, S_n)$, b) $\max(|S_1|, \ldots |S_n|)$, c) $S_1 + \cdots + S_n$.

8. THE LAW OF THE ITERATED LOGARITHM

Let S_n, $n = 1, 2, \ldots$, be successive sums of independent, identically distributed random variables $Y_1, Y_2, \ldots$, with

$$EY_1 = 0, \qquad EY_1^2 = \sigma^2 < \infty.$$

One version of the law of the iterated logarithm is

Theorem 13.25

$$\overline{\lim} \frac{S_n}{\sqrt{2\sigma^2 n \log(\log n)}} = 1 \quad \text{a.s.}$$

Strassen [134] noted recently that even though this is a strong limit theorem, it follows from an invariance principle, and therefore is a distant consequence of the central limit theorem. The result follows fairly easily from the representation theorem, 13.8. What we need is

Theorem 13.26. *There is a probability space with a Brownian motion* $X(t)$ *defined on it and a sequence* $\tilde{S}_n$, $n = 1, \ldots$, *having the same distribution as* S_n/σ, $n = 1, \ldots$, *such that*

$$\frac{\tilde{S}_{[t]} - X(t)}{\sqrt{2t \log(\log t)}} \to 0 \quad \text{a.s.} \tag{13.27}$$

as $t \to \infty$.

Proof. By 13.8, there is a sequence of independent, identically distributed, nonnegative random variables $T_1, T_2, \ldots, ET_1 = 1$ such that $X(T_1 + \cdots + T_n)$, $n = 1, 2, \ldots$, has the same distribution as S_n/σ, $n = 1, \ldots$ Therefore (13.27) reduces to proving that

$$\frac{X(T_1 + \cdots + T_{[t]}) - X(t)}{\varphi(t)} \to 0 \quad \text{a.s.,}$$

where $\varphi(t) = \sqrt{2t \log(\log t)}$. By the law of large numbers,

$$\frac{T_1 + \cdots + T_{[t]}}{t} \to 1 \quad \text{a.s.}$$

For any $\epsilon > 0$, there is an almost surely finite function $\mathsf{t}_0(\omega)$ such that for $t \geq \mathsf{t}_0(\omega)$,

$$\mathsf{T}_1 + \cdots + \mathsf{T}_{[t]} \in \left[\frac{t}{1+\epsilon}, t(1+\epsilon)\right].$$

Let

$$\mathsf{M}(t) = \sup_{\frac{t}{1+\epsilon} \leq \tau \leq (1+\epsilon)t} |\mathsf{X}(\tau) - \mathsf{X}(t)|.$$

Thus, for $t_k = (1+\epsilon)^k$, and $t_k \leq t \leq t_{k+1}$,

$$\begin{aligned} \mathsf{M}(t) &\leq \sup_{t_{k-1} \leq \tau \leq t_{k+2}} |\mathsf{X}(\tau) - \mathsf{X}(t)| \\ &\leq 2 \sup_{t_{k-1} \leq \tau \leq t_{k+2}} |\mathsf{X}(\tau) - \mathsf{X}(t_{k-1})|. \end{aligned}$$

In consequence, if we define.

$$\mathsf{M}_k = \sup_{t_{k-1} \leq \tau \leq t_{k+2}} |\mathsf{X}(\tau) - \mathsf{X}(t_{k-1})|,$$

$$\overline{\lim} \frac{\mathsf{M}(t)}{\varphi(t)} \leq 2\, \overline{\lim} \frac{\mathsf{M}_k}{\varphi(t_k)}.$$

By 12.20,

$$P\mathsf{M}(_k > x) \leq 2P(|\mathsf{X}(t_{k+2}) - \mathsf{X}(t_{k-1})| > x).$$

Write $t_{k+2} - t_{k-1} = \delta t_k$, where $\delta = (1+\epsilon)^2 - (1+\epsilon)^{-1}$. Then

$$\begin{aligned} P(\mathsf{M}_k \geq \sqrt{2\delta}\, \varphi(t_k)) &\leq 2P\left(\frac{|\mathsf{X}(t_{k+2}) - \mathsf{X}(t_{k-1})|}{\sqrt{t_{k+2} - t_{k-1}}} \geq 2\sqrt{\log(\log t_k)}\right) \\ &\leq \frac{2}{\sqrt{2\pi}} \exp[-2\log(\log t_k)] = \sqrt{\frac{2}{\pi}} \cdot \frac{1}{(\log t_k)^2} \\ &= \sqrt{\frac{2}{\pi}} \cdot \frac{1}{k^2(\log(1+\epsilon))^2}. \end{aligned}$$

Use Borel-Cantelli again, getting

$$P(\mathsf{M}_k \geq \sqrt{2\delta}\, \varphi(t_k) \text{ i.o.}) = 0$$

or

$$\overline{\lim} \frac{\mathsf{M}_k}{\varphi(t_k)} \leq \sqrt{2\delta} \quad \text{a.s.}$$

Going back,

$$\overline{\lim} \frac{|\mathsf{S}_{[t]} - \mathsf{X}(t)|}{\varphi(t)} \leq \sqrt{8\delta} \quad \text{a.s.}$$

Taking $\epsilon \downarrow 0$ gives $\delta \to 0$ which completes the proof.

9. A MORE GENERAL INVARIANCE THEOREM

The direction in which generalization is needed is clear. Let the $\{\mathsf{Y}^{(n)}(t)\}$, $t \in [0, 1], n = 0, 1, \ldots$ be a sequence of processes such that $\mathsf{Y}^{(n)}(\cdot) \xrightarrow{\mathfrak{D}} \mathsf{Y}^{(0)}(\cdot)$ in the sense that all finite dimensional distributions converge to the appropriate limit. Suppose that all sample functions of $\{\mathsf{Y}^{(n)}(t)\}$ are in D. Suppose also that some metric $\rho((\cdot), y(\cdot))$ is defined on D, and that in this metric H is a function on D a.s. continuous with respect to the distribution of $\mathsf{Y}^{(0)}(\cdot)$. Find conditions to ensure that

$$\mathsf{H}(\mathsf{Y}^{(n)}(\cdot)) \xrightarrow{\mathfrak{D}} \mathsf{H}(\mathsf{Y}^{(0)}(\cdot)).$$

This has been done for some useful metrics and we follow Skorokhod's strategy. The basic idea is similar to that in our previous work: Find processes $\tilde{\mathsf{Y}}^{(n)}(\cdot)$, $n = 0, 1, \ldots$ defined on a common probability space such that for each n, $\tilde{\mathsf{Y}}^{(n)}(\cdot)$ has the same distribution as $\mathsf{Y}^{(n)}(\cdot)$, and has all its sample functions in D. Suppose $\tilde{\mathsf{Y}}^{(n)}(\cdot)$ have the additional property that

$$\rho(\tilde{\mathsf{Y}}^{(n)}(\cdot), \tilde{\mathsf{Y}}^{(0)}(\cdot)) \xrightarrow{P} 0.$$

Then conclude, as in Section 5, that if $\mathsf{H}(\mathsf{Y}^{(n)}(\cdot))$ and $\mathsf{H}(\tilde{\mathsf{Y}}^{(n)}(\cdot))$ are random variables,

$$\mathsf{H}(\mathsf{Y}^{(n)}(\cdot)) \xrightarrow{\mathfrak{D}} \mathsf{H}(\mathsf{Y}^{(0)}(\cdot)).$$

The basic tool is a construction that yields the very general

Theorem 13.28. *Let $\{\mathsf{Y}^{(n)}(t)\}$, $t \in [0, 1]$, $n = 0, 1, \ldots$ be any sequence of processes such that $\mathsf{Y}^{(n)}(\cdot) \xrightarrow{\mathfrak{D}} \mathsf{Y}^{(0)}(\cdot)$. Then for any countable set $T \subset [0, 1]$, there are processes $\{\tilde{\mathsf{Y}}^{(n)}(t)\}$, $t \in T$, defined on a common space such that*

a) *For each n, $\{\tilde{\mathsf{Y}}^{(n)}(t)\}$, $t \in T$, and $\{\mathsf{Y}^{(n)}(t)\}$, $t \in T$, have the same distribution.*
b) *For every $t \in T$,*

$$\tilde{\mathsf{Y}}^{(n)}(t) \xrightarrow{\text{a.s.}} \tilde{\mathsf{Y}}^{(0)}(t)$$

as $n \to \infty$.

Proof. The proof of this is based on some simple ideas but is filled with technical details. We give a very brief sketch and refer to Skorokhod [124] for a complete proof.

First, show that a single sequence of random variables $\mathsf{X}_n \xrightarrow{\mathfrak{D}} \mathsf{X}_0$ can be replaced by $\tilde{\mathsf{X}}_n \xrightarrow{\text{a.s.}} \tilde{\mathsf{X}}_0$ on a common space with $\tilde{\mathsf{X}}_n \stackrel{\mathfrak{D}}{=} \mathsf{X}_n$. It is a bit surprising to go from $\xrightarrow{\mathfrak{D}}$ to $\xrightarrow{\text{a.s.}}$. But if, for example, X_n are fair coin-tossing random variables such that $\mathsf{X}_n \xrightarrow{\mathfrak{D}} \mathsf{X}_0$, then replace all X_n by the random variables $\tilde{X}_n$ on $((0, 1), \mathcal{B}((0, 1)), dx)$ defined by

$$\tilde{X}_n(x) = \begin{cases} -1, & x \in (0, \tfrac{1}{2}], \\ +1, & x \in (\tfrac{1}{2}, 1). \end{cases}$$

Not only does $\tilde{X}_n(x) \xrightarrow{\text{a.s.}} \tilde{X}_0(x)$, but $\tilde{X}_n(x) \equiv \tilde{X}_0(x)$. In general, take $(\tilde{\Omega}, \tilde{\mathcal{F}}, \tilde{P}) = ((0, 1), \mathcal{B}((0, 1)), dx)$. The device is simple: If $F_n(z)$, the distribution function of X_n, is continuous with a unique inverse, then take

$$\tilde{X}_n(x) = F_n^{-1}(x).$$

Consequently,

$$\begin{aligned} \tilde{P}(\tilde{X}_n(x) < z) &= l\{x; F_n^{-1}(x) < z\} \\ &= l\{x; x < F_n(z)\} \\ &= F_n(z). \end{aligned}$$

Since $X_n \xrightarrow{\mathcal{D}} X_0$, $F_n(z) \to F_0(z)$ for every z; thus $F_n^{-1}(x) \to F_0^{-1}(x)$, all x, or $\tilde{X}_n(x) \xrightarrow{\text{a.s.}} X_0(x)$. Because F_n may not have a unique inverse, define

$$\tilde{X}_n(x) = \inf_y \{y; F_n(y) > x\}.$$

Now verify that these variables do the job.

Generalize now to a sequence of process $\mathbf{X}_n = (X_1^{(n)}, \ldots)$ such that $\mathbf{X}_n \xrightarrow{\mathcal{D}} \mathbf{X}_0$. Suppose we have a nice 1-1 mapping θ: $R^{(\infty)} \leftrightarrow B$, $B \in \mathcal{B}_1$, such that θ, θ^{-1} are measurable $\mathcal{B}_\infty$, $\mathcal{B}_1(B)$ respectively, and such that the following holds:

$$\theta(\mathbf{X}_n) \xrightarrow{\mathcal{D}} \theta(\mathbf{X}_0).$$

Take Y_n, $n \geq 0$, random variables on a common space such that $Y_n \overset{\mathcal{D}}{=} \theta(\mathbf{X}_n)$ and $Y_n \xrightarrow{\text{a.s.}} Y_0$. Define $\tilde{\mathbf{X}}_n = \theta^{-1}(Y_n)$. It is easy to see that $\mathbf{X}_n$ and $\tilde{\mathbf{X}}_n$ have the same distribution. If θ^{-1} is smooth enough so that $Y_n \xrightarrow{\text{a.s.}} Y_0$ implies that every coordinate of $\theta^{-1}(Y_n)$ converges a.s. to the corresponding coordinate of $\theta^{-1}(Y_0)$, then this does it. To get such a θ, let $C_{n,k}$ be the set $\{x; P(X_k^{(n)} = x) > 0\}$. Let $C = \bigcup_{n,k} C_{n,k}$; C is countable. Take $\varphi(x)$: $R^{(1)} \leftrightarrow (0, 1)$ to be 1-1 and continuous such that $\varphi(C)$ contains no binary rationals. There is a 1-1 measurable mapping f: $(0, 1)^{(\infty)} \leftrightarrow (0, 1)$ constructed in Appendix A.47. The mapping θ: $R^{(\infty)} \leftrightarrow (0, 1)$ defined by

$$\theta(x_1, x_2, \ldots) = f(\varphi(x_1), \varphi(x_1), \varphi(x_2), \ldots)$$

has all the necessary properties.

Take one more step. The process $\{\tilde{X}^{(n)}(t)\}$, $t \in T$, having the same distribution as $\{X^{(n)}(t)\}$, $t \in T$, has the property that a.s. every sample function is the restriction to T of a function in D. Take T dense in $[0, 1]$, and for any $t \in [0, 1]$, $t \notin T$, define

$$\tilde{X}^{(n)}(t) = \lim_k \tilde{X}^{(n)}(t_k), \quad t_k \in T, t_k \downarrow t.$$

[Assume $\{1\} \in T$.] The processes $\{\tilde{X}^{(n)}(t)\}$, $t \in [0, 1]$, defined this way have all their sample functions in D, except perhaps for a set of probability zero. Furthermore, they have the same distribution as $X^{(n)}(\cdot)$ for each n.

Throwing out a set of probability zero, we get the statement: For each ω fixed, the sample paths $x_n(t) = \tilde{X}^{(n)}(t, \omega)$ are in D with the property that $x_n(t) \to x_0(t)$ for all $t \in T$. The extra condition needed is something to guarantee that this convergence on T implies that $\rho(x_n(\cdot), x_0(\cdot)) \to 0$ in the metric we are using. To illustrate this, use the metric

$$\rho(x(\cdot), y(\cdot)) = \sup_{0 \le t \le 1} |x(t) - y(t)|,$$

introduced in the previous sections. Other metrics will be found in Skorokhod's article referred to above. Define

$$\delta_n(h) = \sup_{|t-\tau| \le h} |x_n(t) - x_n(\tau)|, \qquad \delta(h) = \overline{\lim_n}\, \delta_n(h).$$

If $x_n(t) \to x_0(t)$ for $t \in T$, and $x_0(t) \in D$, then $\lim_{h \downarrow 0} \delta(h) = 0$ implies $\rho(x_n, x_0) \to 0$. Hence the following:

Theorem 13.29. *Under the above assumptions, if*

$$\text{(13.30)} \qquad \lim_{h \downarrow 0} \overline{\lim_n}\, P\left(\sup_{|t-\tau| \le h} |Y^{(n)}(t) - Y^{(n)}(\tau)| \ge \epsilon \right) = 0,$$

then

$$\sup_{0 \le t \le 1} |\tilde{Y}^{(n)}(t) - \tilde{Y}^{(0)}(t)| \xrightarrow{P} 0.$$

Proof. $\tilde{Y}^{(0)}(t)$ has continuous sample paths a.s. Because take $T_m = \{t_1, \ldots, t_m\} \subset T$. Then, letting

$$M_n = \sup_{|t_j - t_k| \le h} |\tilde{Y}^{(n)}(t_j) - \tilde{Y}^{(n)}(t_k)|,$$

we find that $M_n \xrightarrow{\text{a.s.}} M_0$ follows. For ϵ a continuity point in the distribution of M_0, $P(M_n \ge \epsilon) \to P(M_0 \ge \epsilon)$. Therefore,

$$P(M_0 \ge \epsilon) \le \overline{\lim_n}\, P\left(\sup_{|t-\tau| \le h} |\tilde{Y}^{(n)}(t) - \tilde{Y}^{(n)}(\tau)| \ge \epsilon \right).$$

Letting $T_m \uparrow T$ and using (13.30) implies the continuity.

Define random variables

$$\boldsymbol{\delta}_n(h) = \sup_{|t-\tau|\le h} |\tilde{\mathsf{Y}}^{(n)}(t) - \tilde{\mathsf{Y}}^{(n)}(\tau)|,$$

$$\mathsf{U}_n = \sup_{0\le t\le 1} |\tilde{\mathsf{Y}}^{(n)}(t) - \tilde{\mathsf{Y}}^{(0)}(t)|.$$

If $T_m = \{t_1, \ldots, t_m\} \subset T$ is such that

$$|t_{k+1} - t_k| \le h, \quad k = 1, \ldots, m-1,$$

then

$$\mathsf{U}_n \le \sup_{t\in T_m} |\tilde{\mathsf{Y}}^{(n)}(t) - \tilde{\mathsf{Y}}^{(0)}(t)| + \boldsymbol{\delta}_n(h) + \boldsymbol{\delta}_0(h).$$

The first term goes to zero a.s., leaving

$$\overline{\lim_n}\, P(\mathsf{U}_n > \epsilon) \le \overline{\lim_n}\, P(\boldsymbol{\delta}_n(h) > \epsilon/3) + P(\boldsymbol{\delta}_0(h) > \epsilon/3).$$

Take $h \downarrow 0$. Since U_n does not depend on h, the continuity of $\tilde{\mathsf{Y}}^{(0)}(\cdot)$, and (13.30) yields

$$\overline{\lim_n}\, P(\mathsf{U}_n > \epsilon) = 0 \Rightarrow \mathsf{U}_n \xrightarrow{P} 0.$$

Remark. Note that under (13.30), since $\tilde{\mathsf{Y}}^{(0)}(\cdot)$ has continuous sample paths a.s., then $\mathsf{Y}^{(0)}(t)$ has a version with all sample paths continuous.

The general theorems are similar; $\mathsf{Y}^{(n)}(\cdot) \xrightarrow{\mathcal{D}} \mathsf{Y}^{(0)}(\cdot)$ plus some equicontinuity condition on the $\mathsf{Y}^{(n)}(\cdot)$ gives

$$\rho(\tilde{\mathsf{Y}}^{(n)}(\cdot), \tilde{\mathsf{Y}}^{(0)}(\cdot)) \xrightarrow{P} 0.$$

Problem 9. For random variables having a uniform distribution on [0, 1], and $\hat{\mathsf{F}}_n(x)$, the sample distribution function defined in Section 6, use the multidimensional central limit theorem to show that

$$\{\hat{\mathsf{F}}_n(\xi) - \xi\} \xrightarrow{\mathcal{D}} \{\mathsf{Y}(\xi)\}, \quad \xi \in [0, 1],$$

where $\mathsf{Y}(\xi)$ is a Gaussian process with covariance

$$E\mathsf{Y}(\eta)\mathsf{Y}(\xi) = \eta(1 - \xi), \quad 0 \le \eta \le \xi \le 1.$$

Prove that (13.30) is satisfied by using 13.15 and the Skorokhod lemma 3.21.

NOTES

The invariance principle as applied to sums of independent, identically distributed random variables first appeared in the work of Erdös and Kac [47, 1946] and [48, 1947]. The more general result of 13.12 is due to Donsker [30, 1951]. The method of imitating the sums by using a Brownian motion evaluated at random times was developed by Skorokhod [126, 1961]. The possibility of using these methods on the Kolmogorov-Smirnov statistics was suggested by Doob [37] in a paper where he also evaluates the distribution of the limiting functionals on the Brownian motion. Donsker later [31, 1952] proved that Doob's suggested approach could be made rigorous. For some interesting material on the distribution of various functionals on Brownian motion, see Cameron and Martin [13], Kac [82], and Dinges [24].

Strassen's recent work [134, 1964] on the law of the iterated logarithm contains some fascinating generalizations of this law concerning the limiting fluctuations of Brownian motion. A relatively simple proof of the law of the iterated logarithm for coin-tossing is given by Feller [59, Vol. I]. A generalized version proved by Erdös [46, 1942] for coin-tossing, and extended by Feller [53, 1943] is: Let $\varphi(n)$ be a positive, monotonically increasing function, $\mathsf{S}_n = \mathsf{Y}_1 + \cdots + \mathsf{Y}_n$, $\mathsf{Y}_1, \ldots,$ independent and identically distributed random variables with mean zero and finite second moment. Then

$$P\left(\frac{\mathsf{S}_n}{\sigma\sqrt{n}} > \varphi(n) \quad \text{i.o.}\right)$$

equals zero or one, depending on whether

$$\sum_1^\infty \frac{\varphi(n)}{n} e^{-(1/2)\varphi^2(n)}$$

converges or diverges.

The general question concerning convergence of a sequence of processes $\mathsf{X}^{(n)}(\cdot) \xrightarrow{\mathfrak{D}} \mathsf{X}(\cdot)$ and related invariance results was dealt with in 1956 by Prokhorov [118] and by Skorokhod [124]. We followed the latter in Section 9.

The arc sine law has had an honorable history. Its importance in probability has been not so much in the theorem itself, as in the variety and power of the methods developed to prove it. For Brownian motion, it was derived by Paul Lévy [104, 1939]. Then Erdös and Kac [48, 1947] used an invariance argument to get it for sums of independent random variables. Then Sparre Andersen in 1954 [128] discovered a combinatorial proof that revealed the surprising fact that the law held for random variables whose second moments were not necessarily finite. Spitzer extended the combinatorial methods into entirely new areas [129, 1956]. For the latter, see particularly Spitzer's book [130], also the development by Feller [59, Vol. II]. Another interesting proof was given by Kac [82] for Brownian motion as a special case of a method that reduces the finding of distribution of functionals to related differential equations. There are at least three more proofs we know of that come from other areas of probability.

CHAPTER 14

MARTINGALES AND PROCESSES WITH STATIONARY, INDEPENDENT INCREMENTS

1. INTRODUCTION

In Chapter 12, Brownian motion was defined as follows:

1) $X(t + \tau) - X(t)$ is independent of everything up to time t,
2) The distribution of $X(t + \tau) - X(t)$ depends only on τ,
3) $$\lim_{\Delta t \downarrow 0} \frac{P(|X(\Delta t)| \geq \epsilon)}{\Delta t} = 0.$$

The third assumption involved continuity and had the eventual consequence that a version of Brownian motion was available with all sample paths continuous.

If the third assumption is dropped, then we get a class of processes satisfying (1) and (2) which have the same relation to Brownian motion as the infinitely divisible laws do to the normal law. In fact, examining these processes gives much more meaning to the representation for characteristic functions of infinitely divisible laws.

These processes cannot have versions with continuous sample paths, otherwise the argument given in Chapter 12 forces them to be Brownian motion. Therefore, the extension problem that plagued us there and that we solved by taking a continuous version, comes back again. We deal with this problem in the same way—we take the smoothest possible version available. Of the results available relating to smoothness of sample paths, one of the most general is for continuous parameter martingale processes. So first we develop the martingale theorems. With this theory in hand, we then prove that there are versions of any of the processes satisfying (1) and (2) above, such that all sample paths are continuous except for jumps. Then we investigate the size and number of jumps in terms of the distribution of the process, and give some applications.

2. THE EXTENSION TO SMOOTH VERSIONS

Virtually all the well-known stochastic processes $\{X(t)\}$, $t \in I$, can be shown to have versions such that all sample paths have only jump discontinuities. That is, the sample paths are functions $x(t)$ which have finite right- and left-hand limits $x(t-)$ and $x(t+)$ at all $t \in I$ for which these limits can be defined. This last phrase refers to endpoints. Make the convention that if t is in the interior of I, both $x(t-)$ and $x(t+)$ limits can be defined. At a closed right endpoint, only the $x(t-)$ limit can be defined. At a closed left endpoint only the $x(t+)$ limit can be defined. At open (including infinite) endpoints, neither limit can be defined. We specialize a bit more and define:

Definition 14.1. *$D(I)$ is the class of all functions $x(t)$, $t \in I$, which have only jump discontinuities and which are right-continuous; that is, $x(t+) = x(t)$ for all $t \in I$ such that $x(t+)$ is defined.*

Along with this goes

Definition 14.2. *A process $\{\mathsf{X}(t)\}$, $t \in I$ will be called continuous in probability from the right if whenever $\tau \downarrow t$,*

$$\mathsf{X}(\tau) \xrightarrow{P} \mathsf{X}(t).$$

We want to find conditions on the process $\{\mathsf{X}(t)\}$, $t \in I$ so that a version exists with all sample paths in $D(I)$. As with Brownian motion, start by considering the variables of the process on a set T countable and dense in I, with the convention that T includes any closed endpoints of I. In the case of continuous sample paths the essential property was that for I finite, any function defined and uniformly continuous on T had an extension to a continuous function on I. The analog we need here is

Definition 14.3. *A function $x(t)$ defined on T is said to have only jump discontinuities in I if the limits*

$$\lim_{s\uparrow t} x(s), \quad s \in T, \qquad \lim_{s\downarrow t} x(s), \quad s \in T$$

exist and are finite for all $t \in I$ where these limits can be defined.

Proposition 14.4. *If $x(t)$ defined on T has only jump discontinuities on I, then the function $\tilde{x}(t)$ defined on I by*

$$\tilde{x}(t) = \lim_{s\downarrow t} x(s), \quad s \in T$$

and $\tilde{x}(b) = x(b)$ for b a closed right endpoint of I is in $D(I)$.

Proof. Let $t_n \downarrow t$, t_n, $t \in I$, and take $s_n \in T$, $s_n > t_n > t$ such that $s_n \downarrow t$ and $x(s_n) - \tilde{x}(t_n) \to 0$. Since $x(s_n) \to \tilde{x}(t)$, this implies $\tilde{x}(t+) = \tilde{x}(t)$. Now take $t_n \uparrow t$, and $s_n \in T$ with $t_n < s_n < t$ and $\tilde{x}(t_n) - x(s_n) \to 0$. This shows that $\tilde{x}(t_n) \to \lim_{s\uparrow t} x(s)$, $s \in T$.

We use this to get conditions for the desired version.

Theorem 14.5. *Let the process $\{\mathsf{X}(t)\}$, $t \in I$, be continuous in probability from the right. Suppose that almost every sample function of the countable process $\{\mathsf{X}(t)\}$, $t \in T$, has only jump discontinuities on I. Then there is a version of $\{\mathsf{X}(t)\}$, $t \in I$, with all sample paths in $D(I)$.*

Proof. If for fixed ω, $\{X(t, \omega)\}$, $t \in T$, does not have only jump discontinuities on I, put $\tilde{X}(t, \omega) = 0$, all $t \in I$. Otherwise, define

$$\tilde{X}(t, \omega) = \lim_{s \downarrow t} X(s, \omega), \quad s \in T$$

and $\tilde{X}(b, \omega) = X(b, \omega)$ for b a closed right endpoint of I. By 14.4, the process $\{\tilde{X}(t)\}$, $t \in I$, so defined has all its sample paths in $D(I)$. For any $t \in I$ such that $s_n \downarrow t$, $s_n \in T$, $\tilde{X}(t) = \lim_n X(s_n)$ a.s. By the continuity in probability from the right, $X(s_n) \xrightarrow{P} X(t)$. Hence

$$\tilde{X}(t) = X(t) \quad \text{a.s.},$$

completing the proof.

Problem 1. For $x(t) \in D(I)$, J any finite closed subinterval of I, show that

1) $\sup_{t \in J} |x(t)| < \infty$,

2) for any $\delta > 0$, the set

$$\{t \in J; |x(t) - x(t-)| \geq \delta\}$$

is finite.

3) The set of discontinuity points of $x(t)$ is at most countable.

3. CONTINUOUS PARAMETER MARTINGALES

Definition 14.6. *A process* $\{X(t)\}$, $t \in I$, *is called a martingale* (MG) *if* $E|X(t)| < \infty$, *for all* $t \in I$, *and if for all* $s, t \in I$, $s \leq t$,

$$E(X(t) \mid X(\tau), \tau \leq s) = X(s) \quad \text{a.s.}$$

Call the process a submartingale (SMG) *if under the same conditions*

$$E(X(t) \mid X(\tau), \tau \leq s) \geq X(s) \quad \text{a.s.}$$

This definition is clearly the immediate generalization of the discrete parameter case. The basic sample path property is:

Theorem 14.7. *Let* $\{X(t)\}$, $t \in I$ *be a* SMG. *Then for* T *dense and countable in* I, *almost every sample function of* $\{X(t)\}$, $t \in T$, *has only jump discontinuities on* I.

Proof. It is sufficient to prove this for I a finite, closed interval $[r, \tau]$. Define

(14.8)
$$X^-(t-) = \varliminf_{s \uparrow t} X(s), \qquad X^+(t-) = \varlimsup_{s \uparrow t} X(s), \quad s \in T,$$

$$X^-(t+) = \varliminf_{s \downarrow t} X(s), \qquad X^+(t+) = \varlimsup_{s \downarrow t} X(s), \quad s \in T.$$

Of course, the limits for $r-$ and $\tau+$ are not defined. First we show that for almost every sample function the limits in (14.8) are finite for all $t \in I$. In fact,

(14.9)
$$\sup_{t \in T} |X(t)| < \infty \quad \text{a.s.}$$

To show this, take T_N finite subsets of T, $T_N \uparrow T$ and $r, \tau \in T_N$. By adding together both (1) and (2) of 5.13, deduce that

(14.10)
$$P\left(\sup_{t \in T_N} |X(t)| > x\right) \leq \frac{1}{x}(2E|X(\tau)| - EX(r)).$$

Letting $N \to \infty$ proves that (14.10) holds with T in place of T_N. Now take $x \to \infty$ to prove (14.9).

Now assume all limits in (14.8) are finite. If a sample path of $\{X(t)\}$, $t \in T$ does not have only jump discontinuities on I, then there is a point $v \in I$ such that either $X^-(v-) < X^+(v-)$ or $X^-(v+) < X^+(v+)$. For any two numbers $a < b$, let $D(a, b)$ be the set of all ω such that there exists a $v \in I$ with either

$$X^-(v-) < a < b < X^+(v-) \qquad \text{or} \qquad X^-(v+) < a < b < X^+(v+).$$

The union $\cup D(a, b)$ over all rational a, b; $a < b$, is then the set of all sample paths not having only jump discontinuities.

Take T_N finite subsets of T as above. Let $\boldsymbol{\beta}_N$ be the up-crossings of the interval $[a, b]$ by the SMG sequence $\{X(t_j)\}$, $t_j \in T_N$ (see Section 4, Chapter 5). Then $\boldsymbol{\beta}_N \uparrow \boldsymbol{\beta}$, where $\boldsymbol{\beta}$ is a random variable, possibly extended. The significant fact is

$$D(a, b) \subset \{\boldsymbol{\beta} = \infty\}.$$

Apply Lemma 5.17 to get

$$E\boldsymbol{\beta}_N \leq \frac{E(X(\tau) - a)^+}{b - a}$$

to conclude that $E\boldsymbol{\beta} < \infty$, hence $P(D(a, b)) = 0$. Q.E.D.

The various theorems concerning transformation of martingales by optional sampling and stopping generalize, if appropriate restrictions are imposed. See Doob [39, Chap. 7] for proofs under weak restrictions. We assume here that all the processes we work with have a version with all sample paths in $D(I)$.

Proposition 14.11. *Let* t^* *be a stopping time for a process* $\{X(t)\}$, $t \in I$, *having all sample paths right-continuous. Then* $X(t^*)$ *is a random variable.*

Proof. Approximate t^* by

$$t_n^* = \begin{cases} k/n, & (k-1)/n < t^* \leq k/n, \quad k > 1 \\ 1/n, & 0 \leq t^* \leq 1/n, \quad k = 1. \end{cases}$$

Then, $X(t_n^*)$ is a random variable. For n running through 2^m, $t_n^* \downarrow t^*$, so by right-continuity, $X(t_n^*) \to X(t^*)$ for every ω.

We prove, as an example, the generalization of 5.31.

Theorem 14.12. *Let* t^* *be a stopping time for the* SMG(MG) $\{X(t)\}$, $t \in [0, \infty)$. *If all the paths of the process are in* $D([0, \infty))$ *and if*

a) $E\,|X(t^*)| < \infty$, b) $\underline{\lim}_{t} \int_{\{t^* > t\}} |X(t)|\, dP = 0$,

then

$$EX(t^*) \underset{(=)}{\geq} EX(0).$$

Proof. Suppose first that t^* is uniformly bounded, $t^* \leq \tau$. Take $t_n^* \downarrow t^*$, $t_n^* \leq \tau$, but t_n^* taking on only a finite number of values. By 5.31,

$$EX(t_n^*) \underset{(=)}{\geq} EX(0).$$

and right-continuity implies $X(t_n^*) \to X(t^*)$ everywhere. Some sort of boundedness condition is needed now to conclude $EX(t_n^*) \to EX(t^*)$. Uniform integrability is sufficient, that is,

$$\sup_n \int_{\{|X(t_n^*)| > x\}} |X(t_n^*)|\, dP$$

goes to zero as $x \to \infty$. If $\{X(t)\}$ is a MG process, then $\{|X(t)|\}$ is a SMG. Hence by the optional sampling theorem, 5.10,

(14.13)
$$\int_{\{|X(t_n^*)| > x\}} |X(t_n^*)|\, dP \leq \int_{\{|X(t_n^*)| > x\}} |X(\tau)|\, dP.$$

Let

$$M = \sup_{0 \leq t \leq \tau} |X(t)|.$$

By the right-continuity and (14.9), $M < \infty$ a.s. Then the uniform integrability follows from $E\,|X(\tau)| < \infty$ and

$$\int_{\{|X(t_n^*)| > x\}} |X(\tau)|\, dP \leq \int_{\{M > x\}} |X(\tau)|\, dP.$$

But if $\{X(t)\}$ is not a MG, then use this argument: If the SMG $\{X(t)\}$ $t \geq 0$, were bounded below, say, $X(t) \geq \alpha$, all $t \leq \tau$ and ω, then for $x > |\alpha|$,

$$\int_{\{|X(t_n^*)| > x\}} |X(t_n^*)|\, dP = \int_{\{|X(t_n^*)| > x\}} X(t_n^*)\, dP$$
$$\leq \int_{\{|X(t_n^*)| > x\}} X(\tau)\, dP \leq \int_{\{|X(t^*)| > x\}} |X(\tau)|\, dP.$$

This gets us to (14.13) again. Proceed as above to conclude $E\mathsf{X}(\mathbf{t}^*) \geq E\mathsf{X}(0)$. In general, for α negative, take $\mathsf{Y}(t) = \max(\alpha, \mathsf{X}(t))$. Then $\{\mathsf{Y}(t)\}$, $t \geq 0$ is a SMG bounded below, so $E\mathsf{Y}(\mathbf{t}^*) \geq E\mathsf{Y}(0)$. Take $\alpha \to -\infty$ and note that

$$E\mathsf{X}(\mathbf{t}^*) - E\mathsf{Y}(\mathbf{t}^*) = \int_{\{\mathsf{X}(\mathbf{t}^*) \leq \alpha\}} (\mathsf{X}(\mathbf{t}^*) - \alpha)\, dP$$

goes to zero. Similarly for $E\mathsf{X}(0) - E\mathsf{Y}(0)$, proving the theorem for bounded stopping times. If $\mathbf{t}^*$ is not bounded, define the stopping time $\mathbf{t}^{**}$ as $\min(\mathbf{t}^*, \tau)$. Then

$$E\,|\mathsf{X}(t^{**})| \leq E\,|\mathsf{X}(\mathbf{t}^*)| + E\,|\mathsf{X}(\tau)| < \infty,$$

so $E\mathsf{X}(t^{**}) \geq E\mathsf{X}(0)$. But

$$E\mathsf{X}(\mathbf{t}^*) - E\mathsf{X}(\mathbf{t}^{**}) = \int_{\{\mathbf{t}^* > \tau\}} (\mathsf{X}(\mathbf{t}^*) - \mathsf{X}(\tau))\, dP.$$

The first term in this integral goes to zero as $\tau \to \infty$ because $E\,|\mathsf{X}(\mathbf{t}^*)| < \infty$ by hypothesis. For the second term, simply take a sequence $\tau_n \to \infty$ such that

$$\int_{\{\mathbf{t}^* > \tau_n\}} |\mathsf{X}(\tau_n)|\, dP \to 0.$$

For this sequence $E\mathsf{X}(\mathbf{t}^{**}) \to E\mathsf{X}(\mathbf{t}^*)$, completing the proof.

Problem 2. For Brownian motion $\{\mathsf{X}(t)\}$, $t \geq 0$, and $\mathbf{t}^* = \mathbf{t}^*(a, b)$, prove using 14.12 that $E\mathsf{X}(\mathbf{t}^*) = 0$, $E\mathsf{X}^2(\mathbf{t}^*) = E\mathbf{t}^*$.

4. PROCESSES WITH STATIONARY, INDEPENDENT INCREMENTS

Definition 14.14. *A process $\{\mathsf{X}(t)\}$, $t \in [0, \infty)$, has independent increments if for any t and $\tau > 0$, $\mathcal{F}(\mathsf{X}(t + \tau) - \mathsf{X}(t))$ is independent of $\mathcal{F}(\mathsf{X}(s), s \leq t)$.*

The stationary condition is that the distribution of the increase does not depend on the time origin.

Definition 14.15. *A process $\{\mathsf{X}(t)\}$, $t \in [0, \infty)$ is said to have stationary increments if $\mathcal{L}(\mathsf{X}(t + \tau) - \mathsf{X}(t))$, $\tau \geq 0$, does not depend on t.*

In this section we will deal with processes having independent, stationary increments, and we further normalize by taking $\mathsf{X}(0) \equiv 0$. Note that

$$\mathsf{X}(t) = (\mathsf{X}(t) - \mathsf{X}(t - \tau)) + (\mathsf{X}(t - \tau) - \mathsf{X}(t - 2\tau)) + \cdots + (\mathsf{X}(\tau) - \mathsf{X}(0)),$$

where $\tau = t/n$. Or

$$\mathsf{X}(t) = \mathsf{X}_1^{(n)} + \cdots + \mathsf{X}_n^{(n)},$$

where the $\mathsf{X}_k^{(n)}$, $k = 1, \ldots, n$ are independent and identically distributed. Ergo, $\mathsf{X}(t)$ must have an infinitely divisible distribution. Putting this formally, we have the following proposition.

Proposition 14.16. *Let* $\{X(t)\}$, $t \in [0, \infty)$ *be a process with independent, stationary increments; then* $X(t)$ *has an infinitely divisible distribution for every* $t \geq 0$.

It follows from $X(t + s) = (X(t + s) - X(s)) + (X(s) - X(0))$ that $f_t(u)$, the characteristic function of $X(t)$, satisfies the identity

$$f_{t+s}(u) = f_t(u)f_s(u). \tag{14.17}$$

If $f_t(u)$ had any reasonable smoothness properties, such as $f_t(u)$ measurable $\mathcal{B}_1$ in t for each u, then (14.17) would imply that $f_t(u) = [f_1(u)]^t$, $t \geq 0$. Unfortunately, a pathology can occur: Let $\varphi(t)$ be a real solution of the equation $\varphi(t + s) = \varphi(t) + \varphi(s)$, $t, s \geq 0$. Nonlinear solutions of this do exist [33]. They are nonmeasurable and unbounded in every interval. Consider the degenerate process $X(t) \equiv \varphi(t)$. This process has stationary, independent increments.

Starting with any process $\{X(t)\}$ such that $f_t(u) = [f_1(u)]^t$, then the process $X'(t) = X(t) + \varphi(t)$ has also stationary, independent increments, but $f'_t(u) \neq [f'_1(u)]^t$. This is the extent of the pathology, because it follows from Doob [39, pp. 407 ff.] that if $\{X(t)\}$ is a process with stationary, independent increments, then there is a function $\varphi(t)$, $\varphi(t + s) = \varphi(t) + \varphi(s)$ such that the process $\{X^{(1)}(t)\}$, $X^{(1)}(t) = X(t) - \varphi(t)$ has stationary, independent increments, $X^{(1)}(0) \equiv 0$, and $f_t^{(1)}(u)$ is continuous in t for every u. Actually, this is not difficult to show directly in this case (see Problem 3). A sufficient condition that eliminates this unpleasant case is given by

Proposition 14.18. *Let* $\{X(t)\}$ *be a process with stationary, independent increments such that* $f_t(u)$ *is continuous at* $t = 0$ *for every* u; *then* $\{X(t)\}$ *is continuous in probability and* $f_t(u) = [f_1(u)]^t$.

Proof. Fix u, then taking $s \downarrow 0$ in the equations $f_{t+s}(u) = f_t(u)f_s(u)$ and $f_t(u) = f_{t-s}(u)f_s(u)$ proves $f_t(u)$ continuous for all $t \geq 0$. The only continuous solutions of this functional equation are the exponentials. Therefore $f_t(u) = e^{t\varphi(u)}$. Evaluate this at $t = 1$ to get $f_1(u) = e^{\varphi(u)}$. Use $\lim_{s\downarrow 0} f_s(u) = 1$ to conclude $X(s) \xrightarrow{\mathfrak{D}} 0$ as $s \to 0$, implying $X(s) \xrightarrow{P} 0$. Since

$$\mathcal{L}(X(t) - X(t - s)) = \mathcal{L}(X(t + s) - X(t)) = \mathcal{L}(X(s)), \quad s \geq 0,$$

the proposition follows.

The converse holds.

Proposition 14.19. *Given the characteristic function* $f(u)$ *of an infinitely divisible distribution, there is a unique process* $\{X(t)\}$ *with stationary, independent increments,* $X(0) \equiv 0$, *such that* $f_t(u) = [f(u)]^t$.

Remark. In the above statement, uniqueness means that every process satisfying the given conditions has the same distribution.

Proof. All that is necessary to prove existence is to produce finite-dimensional consistent distribution functions. To specify the distribution of $X(t_1), X(t_2), \ldots, X(t_n)$, $t_1 < t_2 < \cdots < t_n$, define variables $Y_1, Y_2, \ldots, Y_n$ as being independent, Y_k having characteristic function $[f(u)]^{t_k - t_{k-1}}$, $t_0 = 0$. Define

$$\mathcal{L}(X(t_1), \ldots, X(t_n)) = \mathcal{L}(Y_1, Y_1 + Y_2, \ldots, Y_1 + Y_2 + \cdots + Y_n).$$

The consistency is obvious. By the extension theorem for processes, 12.14, there is a process $\{X(t)\}$, $t \geq 0$, having the specified distributions. For these distributions, $X(t + \tau) - X(t)$ is independent of the vector $(X(t_1), \ldots, X(t_n))$, $t_1, \ldots, t_n \leq t$. Thus the process $\{X(t)\}$ has independent increments. Further, the characteristic function of $X(t + \tau) - X(t)$ is $[f(u)]^\tau$, all $t \geq 0$, so implying stationarity of the increments. Of course, the characteristic function of $X(t)$ is $[f(u)]^t$. By construction, $X(0) = 0$ a.s. since $f_0(u) = 1$, so we take a version with $X(0) \equiv 0$. If there is any other such process $\{X(t)\}$ having characteristic function $[f(u)]^t$, clearly its distribution is the same.

Since there is a one-to-one correspondence between processes with stationary, independent increments, continuous in probability and characteristic functions of infinitely divisible distributions, we add the terminology: If

$$Ee^{iuX(t)} = e^{t\psi(u)},$$

call $\psi(u)$ the *exponent function* of the process.

Problems

3. Since $|f_t(u)| \leq 1$, all u, show that (14.17) implies that $|f_t(u)| = |f_1(u)|^t$. By (9.20), $\log f_t(u) = i\beta(t)u + \int \varphi(x, u)\gamma_t(dx)$. Use $|f_t(u)| = |f_1(u)|^t$ to show that $\gamma_t(R^{(1)})$ is uniformly bounded in every finite t-interval. Deduce that $\gamma_{t+s} \equiv \gamma_t + \gamma_s$; hence show that $\gamma_t \equiv t\gamma_1$. Now $\beta(t + s) = \beta(t) + \beta(s)$ by the unique representation. Conclude that $X^{(1)}(t) = X(t) - \beta(t)$ has a continuous characteristic function $f_t^{(1)}(u)$. (It is known that every measurable solution of $\varphi(t + s) = \varphi(t) + \varphi(s)$ is linear. Therefore if $\beta(t)$ is measurable, then it is continuous and $f_t(u)$ is therefore continuous.)

4. For a process with independent increments show that

$$\mathcal{F}(X(t + \tau) - X(t))$$

independent of $\mathcal{F}(X(s), s \leq t)$ for each $t, \tau \Rightarrow \mathcal{F}(X(t + \tau) - X(t), \tau \geq 0)$ independent of $\mathcal{F}(X(s), s \leq t)$.

5. For a process with stationary, independent increments, show that for $B \in \mathcal{B}_1$

1) $P(X(t + \tau) \in B \mid X(s), s \leq t) = P(X(t + \tau) \in B \mid X(t))$ a.s.,

2) $P(X(t + \tau) \in B \mid X(t) = x) = P(X(\tau) \in B - x)$ a.s. $P(X(t) \in dx)$.

5. PATH PROPERTIES

We can apply the martingale results to get this theorem:

Theorem 14.20. *Let $\{X(t)\}$, $t \geq 0$ be a process with stationary, independent increments continuous in probability. Then there is a version of $\{X(t)\}$ with all sample paths in $D([0, \infty))$.*

Remark. There are a number of ways to prove this theorem. It can be done directly, using Skorokhod's lemma (3.21) in much the same way as was done for path continuity in Brownian motion. But since we have martingale machinery available, we use a device suggested by Doob [39].

Proof. Take, as usual, $X(0) \equiv 0$. If $E|X(t)| < \infty$, all $t \geq 0$, we are finished. Because, subtracting off the means $EX(t)$ if necessary, assume $EX(t) \equiv 0$. Then $\{X(t)\}$, $t \geq 0$ is a martingale since for $0 \leq s \leq t$,

$$E(X(t) \mid X(\tau), \tau \leq s) = E(X(t) - X(s) \mid X(\tau), \tau \leq s) + X(s) = X(s) \quad \text{a.s.}$$

Simply apply the martingale path theorem (14.7), the continuity in probability, and 14.5.

If $E|X(t)|$ is not finite for all t, one interesting proof is the following: Take $\varphi(x)$ any continuous bounded function. The process on $[0, \tau]$ defined by

$$Y(t) = E(\varphi(X(\tau)) \mid X(s), s \leq t)$$

is a martingale. But (see Problem 5),

$$Y(t) = E(\varphi(X(\tau)) \mid X(t)) = \theta(\tau - t, X(t)),$$

where

$$\theta(t, x) = E\varphi(X(t) + x).$$

The plan is to deduce the path properties of $X(t)$ from the martingale path theorem by choosing suitable $\varphi(x)$. Take $\varphi(x)$ strictly increasing, $\varphi(+\infty) = \alpha$, $\varphi(-\infty) = -\alpha$, $E|\varphi(X(\tau))| = 1$. By the continuity in probability of $X(t)$, $\theta(t, x)$ is continuous in t. It is continuous and strictly increasing in x, hence jointly continuous in x and t. Thus,

$$y = \theta(\tau - t, x)$$

has an inverse

$$x = \Theta(t, y)$$

jointly continuous in (t, y), for $0 \leq t \leq \tau$, $|y| < \alpha$, and

$$X(t) = \Theta(t, Y(t)).$$

For T dense and countable in $[0, \tau]$, the martingale path theorem implies that

$\{Y(t)\}$, $t \in T$, has only jump discontinuities on $[0, \tau]$. Hence, for all paths such that $\sup_{t\in T} |Y(t)| < \alpha$, $\{X(t)\}$, $t \in T$, has only jump discontinuities on $[0, \tau]$. Now, to complete the proof, we need to show that $\sup_{t\in T} |X(t)| < \infty$, a.s. because for every sample function,

$$\sup_{t\in T} |X(t)| = \infty \Leftrightarrow \sup_{t\in T} |Y(t)| = \alpha.$$

Since $|Y(t)|$ is a SMG, apply 5.13 to conclude

$$P\left(\sup_{t\in T} |Y(t)| > \frac{\alpha}{2}\right) \leq \frac{2}{\alpha} E\,|Y(\tau)|.$$

Since

$$E\,|Y(\tau)| = E\,|\varphi(X(\tau))| = 1,$$

we get

$$P\left(\sup_{t\in T} |X(t)| = \infty\right) \leq \frac{2}{\alpha}.$$

Since α can be made arbitrarily large, conclude that

$$P\left(\sup_{t\in T} |X(t)| = \infty\right) = 0. \qquad \text{Q.E.D.}$$

In the rest of this chapter *we take all processes with stationary, independent increments to be continuous in probability with sample paths in $D([0, \infty))$.*

Problems

6. Prove, using Skorokhod's lemma 3.21 directly, that

$$\sup_{t\in T} |X(t)| < \infty \quad \text{a.s.}$$

7. For $\{X(t)\}$, $t \geq 0$, a process with stationary, independent increments with sample paths in $D([0, \infty))$, show that the strong Markov property holds; that is, if t^* is any stopping time, then

$$\{X(t + t^*) - X(t^*)\}, \quad t \geq 0,$$

has the same distribution as $\{X(t)\}$, $t \geq 0$ and is independent of $\mathcal{F}(X(t), t \leq t^*)$.

8. For $\{X(t)\}$, $t \geq 0$, a process continuous in probability with sample paths in $D([0, \infty))$, show that

$$P(|X(t) - X(t-)| > 0) = 0$$

for all $t \geq 0$.

6. THE POISSON PROCESS

This process stands at the other end of the spectrum from Brownian motion and can be considered as the simplest and most basic of the processes with stationary, independent increments. We get at it this way: A sample point ω consists of any countable collection of points of $[0, \infty)$ such that if $\mathsf{N}(I, \omega)$ is the number of points of ω falling into the interval I, then $\mathsf{N}(I, \omega) < \infty$ for all finite intervals I. Define $\mathcal{F}$ on Ω such that all $\mathsf{N}(I, \omega)$ are random variables, and impose on the probability P these conditions.

Conditions 14.21

a) *The number of points in nonoverlapping intervals is independent. That is,* $I_1, \ldots, I_k$ *disjoint* $\Rightarrow \mathsf{N}(I_1), \ldots, \mathsf{N}(I_k)$ *independent.*
b) *The distribution of the number of points in any interval I depends only on the length* $\|I\|$ *of I.*

A creature with this type of sample space is called a point process. Conditions (a) and (b) arise naturally under wide circumstances: For example, consider a Geiger counter held in front of a fairly sizeable mass of radioactive material, and let the points of ω be the successive registration times. Or consider a telephone exchange where we plot the times of incoming telephone calls over a period short enough so that the disparity between morning, afternoon, and nighttime business can be ignored. Define

(14.22) $$\mathsf{X}(t) = \mathsf{N}((0, t]).$$

Then by 14.21(a) and (b), $\mathsf{X}(t)$ is a process with stationary, independent increments. Now the question is: *Which one*? Actually, the prior part of this question is: *Is there one*? The answer is:

Theorem 14.23. *A process* $\mathsf{X}(t)$ *with stationary, independent increments has a version with all sample paths constant except for upward jumps of length one if and only if there is a parameter* $\lambda \geq 0$ *such that*

$$Ee^{iu\mathsf{X}(t)} = e^{\lambda t(e^{iu}-1)}.$$

Remark. By expanding, we find that $\mathsf{X}(t)$ has the Poisson distribution

$$P(\mathsf{X}(t) = n) = \frac{(\lambda t)^n}{n!} e^{-\lambda t}$$

and

$$P(\mathsf{N}(I) = n) = \frac{(\lambda \|I\|)^n}{n!} e^{-\lambda \|I\|}.$$

Proof. Let $\mathsf{X}(t)$ be a process with the given characteristic function, with paths in $D([0, \infty))$. Then

$$P(\mathsf{X}(t) = k) = \frac{(\lambda t)^k}{k!} e^{-\lambda t},$$

that is, $X(t)$ is concentrated on the nonnegative integers I^+, so that $P(X(t) \in I^+, \text{all } t \in T) = 1$ for any countable set T. Taking T dense in $[0, \infty)$ implies that the paths of the process are integer-valued, with probability one. Also, $\{X(t)\}$, $t \in T$, has nondecreasing sample paths, because $\mathcal{L}(X(t + \tau) - X(t)) = \mathcal{L}(X(\tau))$, and $X(\tau) \geq 0$, a.s. Therefore there is a version of $X(t)$ such that all sample paths take values in I^+ and jump upward only. I want to show that for this version,

$$P\left(\sup_{0 \leq t \leq 1} (X(t) - X(t-)) \leq 1\right) = 1.$$

By Problem 8, the probability that $X(t) - X(t-) > 0$ for t rational is zero. Hence, a.s.,

$$\sup_{0 \leq t \leq 1} (X(t) - X(t-)) = \lim_n \max_{1 \leq k \leq n} (X(k/n) - X(k - 1/n)).$$

But

$$\begin{aligned} P\left(\max_{1 \leq k \leq n} (X(k/n) - X(k - 1/n)) \leq 1\right) &= P(X(1/n)) \leq 1)^n \\ &= [e^{-\lambda/n}(1 + \lambda/n)]^n \\ &\to 1 \quad \text{as} \quad n \to \infty. \end{aligned}$$

To go the other way, let $X(t)$ be any process with stationary, independent increments, integer-valued, such that $X(t) - X(t-) = 0$ or 1. Take t_1^* to be the time until the first jump. It is a stopping time. Take t_2^* to be the time until the first jump of $X(t + t_1^*) - X(t_1^*)$. We know that t_2^* is independent of t_1^* with the same distribution. The time until the nth jump is $t_1^* + \cdots + t_n^*$,

$$P(t_1^* > t + \tau) = P(t_1^* > \tau, X(\xi) - X(\tau) = 0, \tau \leq \xi \leq t + \tau).$$

Now $\{t_1^* > \tau\} = \{t_1^* \leq \tau\}^c$, hence $\{t_1^* > \tau\}$ is in $\mathcal{F}(X(t), t \leq \tau)$. Therefore

$$\begin{aligned} P(t_1^* > t + \tau) &= P(t_1^* > \tau)P(X(\xi) - X(\tau) = 0, \tau \leq \xi \leq t + \tau) \\ &= P(t_1^* > \tau)P(t_1^* > t). \end{aligned}$$

This is the exponential equation; the solution is

$$P(t_1^* > t) = e^{-\lambda t}$$

for some $\lambda \geq 0$. To finish, write

$$P(X(t) = n) = P(t_1^* + \cdots + t_n^* < t) - P(t_1^* + \cdots + t_{n+1}^* < t).$$

Denote $Q_n(t) = P(t_1^* + \cdots + t_n^* > t)$, and note that

$$P(t_1^* + \cdots + t_{n+1}^* > t \mid t_1^* + \cdots + t_n^* = \xi) = \begin{cases} e^{-\lambda(t-\xi)}, & t \geq \xi, \\ 1, & t < \xi. \end{cases}$$

So we have

$$Q_{n+1}(t) = Q_n(t) - e^{-\lambda t}\int_0^t e^{\lambda\xi} Q_n(d\xi), \quad n \geq 1.$$

This recurrence relation and $Q_1(t) = e^{-\lambda t}$ gives

$$Q_{n+1}(t) = e^{-\lambda t}\left(1 + \lambda t + \cdots + \frac{(\lambda t)^n}{n!}\right),$$

leading to

$$P\big(\mathsf{X}(t) = n\big) = \frac{(\lambda t)^n}{n!} e^{-\lambda t}. \qquad \text{Q.E.D.}$$

7. JUMP PROCESSES

We can use the Poisson processes as building blocks. Let the jump points of a Poisson process $\mathsf{Y}(t)$ with parameter $\lambda > 0$ be $\mathsf{t}_1, \mathsf{t}_2, \ldots$ Construct a new process $\mathsf{X}(t)$ by assigning jump X_1 at time t_1, X_2 at time $\mathsf{t}_2, \ldots$, where $\mathsf{X}_1, \mathsf{X}_2, \ldots$ are independent, identically distributed random variables with distribution function $F(x)$, and $\mathcal{F}(\mathbf{X})$ is independent of $\mathcal{F}\big(\mathsf{Y}(t), t \geq 0\big)$. Then

Proposition 14.24. *$\mathsf{X}(t)$ is a process with stationary, independent increments, and*

$$E \exp\,[iu\mathsf{X}(t)] = \exp\left\{\lambda t \int [\exp\,(iux) - 1]\, dF(x)\right\}.$$

Proof. That $\mathsf{X}(t)$ has stationary, independent increments follows from the construction. Let $h(u) = \int e^{iux}\, dF$. Note that

$$E(\exp[iu\mathsf{X}(t)] \mid \mathsf{Y}(t) = n) = E \exp\,[iu(\mathsf{X}_1 + \cdots + \mathsf{X}_n)] = [h(u)]^n,$$

implying

$$Ee^{iu\mathsf{X}(t)} = \sum_0^\infty [h(u)]^n \frac{(\lambda t)^n e^{-\lambda t}}{n!} = e^{\lambda t(h(u)-1)}.$$

In Chapter 9, Section 4, it was pointed out that for a generalized Poisson distribution with jumps of size $x_1, x_2, \ldots$, each jump size contributes an independent component to the distribution. This is much more graphically illustrated here. Say that a process has a jump of size $B \in \mathcal{B}_1$ at time t if $\mathsf{X}(t) - \mathsf{X}(t-) \in B$. A process with stationary, independent increments and exponent function

$$\int (e^{iux} - 1)\mu(dx)$$

with $\mu(R^{(1)}) < \infty$ is of the type treated above, with $\lambda = \mu(R^{(1)})$, $F(B) = \mu(B)/\mu(R^{(1)})$. Therefore it has sample paths constant except for jumps. Let

$R^{(1)} - \{0\} = \bigcup_{k=1}^n B_k$, for disjoint $B_k \in \mathcal{B}_1$, $k = 1, \ldots, n$. Define processes $\{\mathsf{X}(B_k, t)\}$, $t \geq 0$, by

$$\mathsf{X}(B_k, t) = \sum_{\tau \leq t} \chi_{B_k}(\mathsf{X}(\tau) - \mathsf{X}(\tau -)).$$

Thus $\mathsf{X}(B_k, t)$ is the sum of the jumps of $\mathsf{X}(t)$ of size B_k up to time t. We need to show measurability of $\mathsf{X}(B_k, t)$, but this follows easily if we construct the process $\mathsf{X}(t)$ by using a Poisson process $\mathsf{Y}(t)$ and independent jump variables $\mathsf{X}_1, \mathsf{X}_2, \ldots$, as above. Then, clearly,

$$\mathsf{X}(t) = \sum_{k=1}^n \mathsf{X}(B_k, t).$$

Now we prove:

Proposition 14.25. *The processes* $\{\mathsf{X}(B_k, t)\}$, $t \geq 0$, *are independent of each other, and* $\{\mathsf{X}(B_k, t)\}$, $t \geq 0$ *is a process with stationary, independent increments and exponent function*

$$\int_{B_k} (e^{iux} - 1)\mu(dx).$$

Proof. It is possible to prove this directly, but the details are messy. So we resort to an indirect method of proof. Construct a sample space and processes $\{\tilde{\mathsf{X}}^{(k)}(t)\}$, $t \geq 0$, $n = 1, \ldots, n$ which are independent of each other and which have stationary, independent increments with exponent functions

$$\int_{B_k} (e^{iux} - 1)\mu(dx).$$

Each of these has the same type of distribution as the processes of 14.24, with $\lambda = \mu(B_k)$, $F_k(B) = \mu(B \cap B_k)/\mu(B_k)$. Hence $F_k(dx)$ is concentrated on B_k and $\{\tilde{\mathsf{X}}^{(k)}(t)\}$, $t \geq 0$, has jumps only of size B_k. Let

$$\tilde{\mathsf{X}}(t) = \sum_{k=1}^n \tilde{\mathsf{X}}^{(k)}(t).$$

Then $\{\tilde{\mathsf{X}}(t)\}$, $t \geq 0$ has stationary, independent increments and the same characteristic function for every t as $\mathsf{X}(t)$. Therefore $\{\tilde{\mathsf{X}}(t)\}$ and $\{\mathsf{X}(t)\}$ have the same distribution. But by construction,

$$\tilde{\mathsf{X}}^{(k)}(t) = \sum_{\tau \leq t} \chi_{B_k}(\tilde{\mathsf{X}}(t) - \tilde{\mathsf{X}}(t -)),$$

therefore $\{\tilde{\mathsf{X}}^{(k)}(t)\}$, $t \geq 0$ is defined on the $\{\tilde{\mathsf{X}}(t)\}$ process by the same function as $\{\mathsf{X}(B_k, t)\}$ on the $\{\mathsf{X}(t)\}$ process. Hence the processes $\{\mathsf{X}(B_k, t)\}$, $\{\mathsf{X}(t)\}$ have the same joint distribution as $\{\tilde{\mathsf{X}}^{(k)}(t)\}$, $\{\tilde{\mathsf{X}}(t)\}$, proving the proposition.

8. LIMITS OF JUMP PROCESSES

The results of the last section give some insight into the description of the paths of the general process. First, let $\{X(t)\}$, $t \geq 0$, be a process with stationary, independent increments, whose exponent function ψ has only an integral component:

$$\psi(u) = \int_{\{0\}^c} \left(e^{iux} - 1 - \frac{iux}{1 + x^2}\right) \mu\,(dx).$$

If $\mu(R^{(1)}) < \infty$, then the process is of the form $\{X(t) - \beta t\}$, where $\{X(t)\}$, $t \geq 0$ is of the type studied in the previous section, with sample paths constant except for isolated jumps. In the general case, μ assigns infinite mass to arbitrarily small neighborhoods of the origin. This leads to the suspicion that the paths for these processes have an infinite number of jumps of very small size. To better understand this, let D be any neighborhood of the origin, $\{X(D^c, t)\}$, $t \geq 0$, be the process of jumps of size greater than D, where we again define: *the process* $\{X(B, t)\}$, $t \geq 0$, *of jumps of* $\{X(t)\}$ *of size* $B \in \mathcal{B}_1$, $\{0\} \notin B$, *is given by*

$$X(B, t) = \sum_{\tau \leq t} \chi_B(X(\tau) - X(\tau-)). \tag{14.26}$$

Assuming that the results analogous to the last section carry over, $\{X(D^c, t)\}$ has exponent function

$$\int_{D^c} (e^{iux} - 1)\mu(dx).$$

Letting

$$\beta = \int_{D^c} \frac{x}{1 + x^2}\,\mu(dx),$$

we have that the exponent function of $X(D^c, t) - \beta t$ is given by

$$\psi_D(u) = \int_{D^c} \left(e^{iux} - 1 - \frac{iux}{1 + x^2}\right)\mu(dx).$$

Therefore, as D shrinks down to $\{0\}$, $\psi_D(u) \to \psi(u)$, and in some sense we expect that $X(t)$ is the limit of $X(D^c, t) - \beta t$. In fact, we can get a very strong convergence.

Theorem 14.27. *Let* $\{X(t)\}$, $t \geq 0$, *be a process with stationary, independent increments having only an integral component in its characteristic function. Take* $\{D_n\}$ *neighborhoods of the origin such that* $D_n \downarrow \{0\}$, *then* $\{X(D_n^c, t)\}$, $t \geq 0$ *is a process with stationary, independent increments and exponent function*

$$\int_{D_n^c} (e^{iux} - 1)\mu(dx).$$

Put

$$\beta_n = \int_{D_n^c} \frac{x}{1 + x^2} \mu(x).$$

For any $t_0 < \infty$,

$$\sup_{0 \le t \le t_0} |\mathsf{X}(t) - \mathsf{X}(D_n^c, t) + \beta_n t| \xrightarrow{P} 0$$

as $n \to \infty$.

Proof. Take $D_1 = R^{(1)}$, $B_n = D_{n+1} - D_n$. Construct a probability space on which there are processes $\{\tilde{\mathsf{Z}}_n(t)\}$, $t \geq 0$, with stationary, independent increments such that the processes are independent of each other, with exponent functions

$$\psi_n(u) = \int_{B_n} (e^{iux} - 1)\mu(dx).$$

Then 14.25 implies that the paths of $\tilde{\mathsf{Z}}_n(t)$ are constant except for jumps of size B_n. Denote

$$\tilde{\mathsf{X}}_n(t) = \sum_1^n \tilde{\mathsf{Z}}^{(k)}(t) \qquad \text{and} \qquad b_n = \int_{B_n} \frac{x}{1 + x^2} \mu(dx).$$

Consequently, $\tilde{\mathsf{X}}_n(t) - \beta_n t$ is the sum of the independent components $\tilde{\mathsf{Z}}_k(t) - b_k t$. Since the characteristic functions converge, for every t,

$$\tilde{\mathsf{X}}_n(t) - \beta_n t \xrightarrow{\mathfrak{D}}.$$

Because we are dealing with sums of independent random variables, 8.36 implies that for every t there is a random variable $\tilde{\mathsf{X}}(t)$ such that

$$\tilde{\mathsf{X}}_n(t) - \beta_n t \xrightarrow{\text{a.s.}} \tilde{\mathsf{X}}(t).$$

This implies that $\{\tilde{\mathsf{X}}(t)\}$, $t \geq 0$, is a process with stationary, independent increments having the same distribution as $\{\mathsf{X}(t)\}$, $t \geq 0$. We may assume that $\{\tilde{\mathsf{X}}(t)\}$, $t \geq 0$ has all its sample paths in $D([0, \infty))$. Take, for example, $t_0 = 1$, T any set dense in $[0, 1]$. For $\tilde{\mathsf{Y}}_n(t) = \tilde{\mathsf{X}}(t) - \tilde{\mathsf{X}}_n(t) + \beta_n t$,

$$\sup_{0 \le t \le 1} |\tilde{\mathsf{Y}}_n(t)| = \sup_{t \in T} |\tilde{\mathsf{Y}}_n(t)|.$$

Let T_N be finite subsets of T, $T_N \uparrow T$. Then $\sup_{t \in T} |\tilde{\mathsf{Y}}_n(t)| = \lim_n \sup_{t \in T_N} |\tilde{\mathsf{Y}}_n(t)|$. Because $\{\tilde{\mathsf{Y}}_n(t)\}$ is a process with stationary, independent increments, we can apply Skorokhod's lemma:

$$P\left(\sup_{t \in T_N} |\tilde{\mathsf{Y}}_n(t)| > 2y\right) \le \frac{1}{1 - C_N} P(|\tilde{\mathsf{Y}}_n(1)| > y),$$

where

$$C_N \le \sup_{0 \le t \le 1} P(|\tilde{\mathsf{Y}}_n(t)| > y).$$

Let $\psi_n(u)$ be the exponent function of $\tilde{Y}_n(t)$. Apply inequality 8.29 to write:

$$P(|\tilde{Y}_n(t)| > y) \leq \alpha y \int_0^{y^{-1}} |1 - e^{t\psi_n(u)}| \, du.$$

For $|e^z| \leq 1$, there is a constant $\gamma < \infty$ such that $|e^z - 1| < \gamma |z|$. Hence,

$$P(|\tilde{Y}_n(t)| > y) \leq \alpha \gamma y \int_0^{y^{-1}} |\psi_n(u)| \, du.$$

Since $\psi_n(u) \xrightarrow{uc} 0$, then $C_N \to 0$, leading to

$$\sup_{0 \leq t \leq 1} |\tilde{Y}_n(t)| \xrightarrow{P} 0.$$

Thus we can find a subsequence $\{n'\}$ such that a.s.

$$\tilde{X}_{n'}(t) - \beta_{n'} t \xrightarrow{uc} \tilde{X}(t).$$

Some reflection on this convergence allows the conclusion that the process $\tilde{X}_n(t)$ can be identified with $\tilde{X}(D_n^c, t)$. Therefore $\{\tilde{X}(t)\}$ and $\{\tilde{X}_n(t)\}$ have the same joint distribution as $\{X(t)\}$, $\{X(D_n^c, t\}$. Using this fact proves the theorem.

If a function $x(t) \in D([0, \infty))$ is, at all times t, the sum of all its jumps up to time t, then it is called a pure jump function. However, this is not well defined if the sum of the lengths of the jumps is infinite. Then $x(t)$ is the sum of positive and negative jumps which to some extent cancel each other out to produce $x(t)$, and the order of summation of the jumps up to time t becomes important. We define:

Definition 14.28. *If there is a sequence of neighborhoods* $D_n \downarrow \{0\}$ *such that*

$$\sum_{\tau \leq t} \chi_{D_n^c}(x(\tau) - x(\tau-)) \xrightarrow{uc} x(t),$$

then $x(t) \in D([0, \infty))$ *is called a pure jump function.*

Many interesting processes of the type studied in this section have the property that there is a sequence of neighborhoods $D_n \downarrow \{0\}$ such that

$$\beta = \lim_n \int_{D_n^c} \frac{x}{1 + x^2} \mu(dx)$$

exists and is finite. Under this assumption we have

Corollary 14.29. *Almost every sample path of* $\{X(t) - \beta t\}$, $t \geq 0$, *is a pure jump function.*

Another interesting consequence of this construction is the following: Take neighborhoods $D_n \downarrow \{0\}$ and take $\{X(D_n^c, t)\}$, as above, the process of jumps of size greater than D_n. Restrict the processes to a finite time interval $[0, t_0]$ and consider any event A such that for every n,

$$A \in \mathcal{F}(\mathsf{X}(t) - \mathsf{X}(D_n^c, t), 0 \leq t \leq t_0).$$

Then

Proposition 14.30. *$P(A)$ is zero or one.*

Proof. Let $B_n = D_{n+1} - D_n$; then the processes $\{\mathsf{X}(B_n, t)\}$, $t \in [0, t_0]$ are independent and A is measurable:

$$\mathcal{F}(\{\mathsf{X}(B_n, t)\}, \{\mathsf{X}(B_{n+1}, t)\}, \ldots, t \in [0, t_0])$$

for every n. Apply a slight generalization of the Kolmogorov zero-one law to conclude $P(A) = 0$ or 1.

The results of this section make it easier to understand why infinitely divisible laws were developed to use in the context of processes with independent increments earlier than in the central limit problem. The processes of jumps of different sizes proceed independently of one another, and the jump process of jumps of size $[x, x + \Delta x)$ contributes a Poisson component with exponent function approximately equal to

$$(e^{iux} - 1)\mu([x, x + \Delta x)).$$

The fact that the measure μ governs the number and the size of jumps is further exposed in the following problems, all referring to a process $\{\mathsf{X}(t)\}$, $t \geq 0$ with stationary, independent increments and exponent function having only an integral component.

Problems

9. Show that for any set $B \in \mathcal{B}_1$ bounded away from the origin, the process of jumps of size B, $\{\mathsf{X}(B, t)\}$, $t \geq 0$ has stationary, independent increments with exponent function

$$\int_B (e^{iux} - 1)\mu(dx),$$

and the processes $\{\mathsf{X}(t) - \mathsf{X}(B, t)\}$, $\{\mathsf{X}(B, t)\}$ are independent.

10. For B as above, show that the expected number of jumps of size B in $0 \leq t \leq 1$ is $\mu(B)$.

11. For B as above let t^* be the time of the first jump of size B. For $C \subset B$, $C \in \mathcal{B}_1$, prove that

$$P(\mathsf{X}(\mathsf{t}^*) - \mathsf{X}(\mathsf{t}^*-) \in C) = \mu(C)/\mu(B).$$

12. Show that except for a set of probability zero, either all sample functions of $\{\mathsf{X}(t)\}$, $t \in [0, t_0]$ have infinite variation or all have finite variation.

13. Show that for $\{\mathsf{X}(t)\}$, $t \geq 0$ as above, all sample paths have finite variation on every finite time interval $[0, t_0]$ if and only if

$$\int_{|x| \leq 1} |x| \, \mu(dx) < \infty.$$

[Take $t_0 = 1$. The function

$$\mathsf{V}_n = \sum_{k=1}^{2^n} |\mathsf{X}(k/2^n) - \mathsf{X}(k - 1/2^n)|$$

is monotonically nondecreasing. Now compute $Ee^{-\lambda \mathsf{V}_n}$ for $\lambda > 0$.]

9. EXAMPLES

Consider the first passage time $\mathbf{t}_\xi^*$ of a Brownian motion $\mathsf{X}(t)$ to the point ξ. Denote $\mathsf{Z}(\xi) = \mathbf{t}_\xi^*$. For $\xi_1, \xi_2 > 0$,

$$\mathbf{t}_{\xi_2+\xi_1}^* - \mathbf{t}_{\xi_1}^* = \inf \{t; \mathsf{X}(t + \mathbf{t}_{\xi_1}^*) - \mathsf{X}(\mathbf{t}_{\xi_1}^*) = \xi_2\}.$$

By the strong Markov property, $\mathsf{Z}(\xi_1 + \xi_2) - \mathsf{Z}(\xi_1)$ is independent of $\mathcal{F}(\mathsf{Z}(\xi), \xi \leq \xi_1)$ and is distributed as $\mathsf{Z}(\xi_2)$. Thus, to completely characterize $\mathsf{Z}(\xi)$, all we need is its characteristic function. From Chapter 13,

$$e^{iu\mathsf{Z}(\xi)} = e^{\xi\psi(u)},$$

$$\psi(u) = \begin{cases} - |u|^{1/2} e^{i\pi/4}, & u > 0, \\ - |u|^{1/2} e^{-i\pi/4}, & u < 0. \end{cases}$$

This is the characteristic function of a stable distribution with exponent $\frac{1}{2}$. The jump measure $\mu(dx)$ is given by

$$\mu(dx) = \begin{cases} 0, & x < 0, \\ c \dfrac{dx}{x^{3/2}}, & x > 0. \end{cases}$$

Doing some definite integrals gives $c = \frac{1}{2}\sqrt{\pi}$.

If the characteristic function of a process with stationary, independent increments is stable with exponent α, call the process *stable with exponent* α. For $0 < \alpha < 1$,

$$\int_0^\infty (e^{iux} - 1) \frac{dx}{x^{1+\alpha}}$$

exists. The processes with exponent functions

$$\psi(u) = m \int_0^\infty (e^{iux} - 1) \frac{dx}{x^{1+\alpha}}$$

are the limit of processes $\{X_n(t)\}$ with exponent functions

$$\psi_n(u) = m\int_{1/n}^{\infty}(e^{iux} - 1)\frac{dx}{x^{1+\alpha}}. \tag{14.31}$$

These latter processes have only upward jumps. Hence all paths of $\{X(t)\}$ are nondecreasing pure jump functions.

Stable processes of exponent ≥ 1 having nondecreasing sample paths do not exist. If $\mathcal{L}(X(t)) = \mathcal{L}(-X(t))$, the process is symmetric. Bochner [11] noted that it was possible to construct the symmetric stable processes from Brownian motion by a random time change. Take a sample space with a normalized Brownian motion $X(t)$ and a stable process $Z(t)$ defined on it such that $Z(t)$ has nondecreasing pure jump sample paths and $\mathcal{F}(X(t), t \geq 0)$, $\mathcal{F}(Z(t), t \geq 0)$ are independent.

Theorem 14.32. *If $Z(t)$ has exponent α, $0 < \alpha < 1$, then the process $Y(t) = X(Z(t))$ is a stable symmetric process of exponent 2α.*

Proof. The idea can be seen if we write

$$Y(t + \tau) - Y(t) = X((Z(t + \tau) - Z(t)) + Z(t)) - X(Z(t)).$$

Then, given $Z(t)$, the process $Y(t + \tau) - Y(t)$ looks just as if it were the $Y(\tau)$ process, independent of $Y(s)$, $s \leq t$.

For a formal proof, take $Z_n(t)$ the process of jumps of $Z(t)$ larger than $[0, 1/n)$. Its exponent function is

$$\psi_n(u) = m\int_{1/n}^{\infty}(e^{iux} - 1)\frac{dx}{x^{1+\alpha}}.$$

The jumps of $Z_n(t)$ occur at the jump times of a Poisson process with intensity

$$\lambda_n = m\int_{1/n}^{\infty}\frac{dx}{x^{1+\alpha}},$$

and the jumps have magnitude $Y_1, Y_2, \ldots$ independent of one another and of the jump times, and are identically distributed. Thus $X(Z_n(t))$ has jumps only at the jump times of the Poisson process. The size of the kth jump is

$$U_k = \begin{cases} X(Y_k + \cdots + Y_1) - X(Y_{k-1} + \cdots + Y_1), & k > 1, \\ X(Y_1), & k = 1. \end{cases}$$

By an argument almost exactly the same as the proof of the strong Markov property for Brownian motion, U_k is independent of $U_{k-1}, \ldots, U_1$ and has the same distribution as U_1. Therefore $X(Z_n(t))$ is a process with stationary, independent increments. Take $\{n'\}$ so that $Z_{n'}(t) \xrightarrow{uc} Z(t)$ a.s. Use continuity of

Brownian motion to get $X(Z_{n'}(t)) \xrightarrow{\text{a.s.}} X(Z(t))$ for every t. Thus $X(Z(t))$ is a process with stationary, independent increments. To get its characteristic function, write

$$E(e^{iuX(Z(1))} \mid Z(1) = z) = Ee^{iuX(z)} = e^{-zu^2/2}.$$

Therefore

$$Ee^{iuX(Z(1))} = Ee^{-(u^2/2)Z(1)} = e^{-c|u|^{2\alpha}}.$$

10. A REMARK ON A GENERAL DECOMPOSITION

Suppose $\{X(t)\}$, $t \geq 0$ is a process with stationary, independent increments and exponent function

$$\psi(u) = iu\beta - \frac{\sigma^2 u^2}{2} + \int_{\{0\}^c} \left(e^{iux} - 1 - \frac{iux}{1 + x^2}\right)\mu(dx).$$

Since $iu\beta - \sigma^2 u^2/2$ is the exponent function for a Brownian motion, a natural expectation is that $X(t) = X^{(1)}(t) + X^{(2)}(t)$, where $\{X^{(1)}(t)\}$, $t \geq 0$, is a Brownian motion with drift β and variance σ^2, and $\{X^{(2)}(t)\}$, $t \geq 0$, is a process with stationary, independent increments with exponent function having only an integral component, and that the two processes are independent. This is true, in fact, and can be proved by the methods of Sections 7 and 8. But as processes with stationary, independent increments appear in practice either as a Brownian motion, or a process with no Brownian component, we neglect the proof of this decomposition.

NOTES

For more material on processes with stationary, independent increments, see Doob [39, Chap. 8] and Paul Lévy's two books [103 and 105]. These latter two are particularly good on giving an intuitive feeling for what these processes look like. Of course, for continuous parameter martingales, the best source is Doob's book. The sample path properties of a continuous parameter martingale were given by Doob in 1951 [38], and applied to processes with independent increments.

Processes with independent increments had been introduced by de Finnett in 1929. Their sample path properties were studied by Lévy in 1934 [102]. He then proved Theorem 14.20 as generalized to processes with independent increments, not necessarily stationary. Most of the subsequent decomposition and building up from Poisson processes follows Lévy also, in particular [103, p. 93]. The article by Ito [75] makes this superposition idea more precise by defining an integral over Poisson processes.

CHAPTER 15

MARKOV PROCESSES, INTRODUCTION AND PURE JUMP CASE

1. INTRODUCTION AND DEFINITIONS

Markov processes in continuous time are, as far as definitions go, a straightforward extension of the Markov dependence idea.

Definition 15.1. *A process* $\{X(t)\}$, $t \geq 0$ *is called Markov with state space* $F \in \mathcal{B}_1$, *if* $X(t) \in F$, $t \geq 0$, *and for any* $B \in \mathcal{B}_1(F)$, $t, \tau \geq 0$.

$$\textbf{(15.2)} \quad P(X(t + \tau) \in B \mid X(s), s \leq t) = P(X(t + \tau) \in B \mid X(t)) \quad \text{a.s.}$$

To verify that a process is Markov, all we need is to have for any $t_{n+1} \geq t_n \geq \cdots \geq t_1 \geq 0$,

$$P(X(t_{n+1}) \in B \mid X(t_n), \ldots, X(t_1)) = P(X(t_{n+1}) \in B \mid X(t_n)) \quad \text{a.s.}$$

Since finite dimensional sets determine $\mathcal{F}(X(s), s \leq t)$, this extends to (15.2).

The Markov property is a statement about the conditional probability at the one instant $t + \tau$ in the future. But it extends to a general statement about the future, given the present and past:

Proposition 15.3. *If* $\{X(t)\}$ *is Markov, then for* $A \in \mathcal{F}(X(\tau), \tau \geq t)$,

$$P(A \mid X(s), s \leq t) = P(A \mid X(t)) \quad \text{a.s.}$$

The proof of this is left to the reader.

Definition 15.4. *By Theorem* 4.30, *for every* $t_2 > t_1$ *a version* $p_{t_2,t_1}(B \mid x)$ *of* $P(X(t_2) \in B \mid X(t_1) = x)$ *can be selected such that* $p_{t_2,t_1}(B \mid x)$ *is a probability on* $\mathcal{B}_1(F)$ *for* x *fixed, and* $\mathcal{B}_1(F)$ *measurable in* x *for* B *fixed. Call these a set of transition probabilities for the process.*

Definition 15.5. $\pi(\cdot)$ *on* $\mathcal{B}_1(F)$ *given by* $\pi(B) = P(X(0) \in B)$ *is the initial distribution for the process.*

The importance of transition probabilities is that the distribution of the process is completely determined by them if the initial distribution is specified.

This follows from

Proposition 15.6. *For* $B_n, \ldots, B_0 \in \mathcal{B}_1(F)$, $t_n > t_{n-1} > \cdots > t_0 = 0$,

$$(15.7)\quad P(\mathsf{X}(t_n) \in B_n, \ldots, \mathsf{X}(t_1) \in B_1, \mathsf{X}(t_0) \in B_0)$$
$$= \int_{B_n} \cdots \int_{B_1} \int_{B_0} p_{t_n,t_{n-1}}(dx_n \mid x_{n-1}) \cdots p_{t_1,t_0}(dx_1 \mid x_0)\pi(dx_0).$$

The proof is the same as for 7.3.

A special case of (15.7) are the Chapman-Kolmogorov equations. Reason this way: To get from x at time τ to B at time $t > \tau$, fix any intermediate time s, $t > s > \tau$. Then the probability of all the paths that go from x to B through the small neighborhood dy of y at time s is given (approximately) by $p_{t,s}(B \mid y)p_{s,\tau}(dy \mid x)$. So summing over dy gets us to the Chapman-Kolmogorov equations,

$$(15.8)\qquad p_{t,\tau}(B \mid x) = \int p_{t,s}(B \mid y)p_{s,\tau}(dy \mid x).$$

Actually, what is true is

Proposition 15.9. *The equation* (15.8) *holds* a.s. *with respect to* $P(\mathsf{X}(\tau) \in dx)$.

Proof. From (15.7), for every $C \in \mathcal{B}_1(F)$,

$$P(\mathsf{X}(t) \in B, \mathsf{X}(\tau) \in C) = \iint_C p_{t,s}(B \mid y)p_{s,\tau}(dy \mid x)P(\mathsf{X}(\tau) \in dx).$$

Taking the Radon derivative with respect to $P(\mathsf{X}(\tau) \in dx)$ gives the result.

One rarely starts with a process on a sample space $(\Omega, \mathcal{F}, P)$. Instead, consistent distribution functions are specified, and then the process constructed. For a Markov process, what gets specified are the transition probabilities and the initial distribution. Here there is a divergence from the discrete time situation in which the one-step transition probabilities $P(\mathsf{X}_{n+1} \in B \mid \mathsf{X}_n = x)$ determine all the multiple-step probabilities. There are no corresponding one-step probabilities here; the probabilities $\{p_{t,\tau}(B \mid x)\}$, $t > \tau \geq 0$ must all be specified, and they must satisfy among themselves at least the functional relationship (15.8).

2. REGULAR TRANSITION PROBABILITIES

Definition 15.10. *A set of functions* $\{p_{t,\tau}(B \mid x)\}$ *defined for all* $t > \tau \geq 0$, $B \in \mathcal{B}_1(F)$, $x \in F$ *is called a regular set of transition probabilities if*

1) $p_{t,\tau}(B \mid x)$ *is a probability on* $\mathcal{B}_1(F)$ *for* x *fixed and* $\mathcal{B}_1(F)$ *measurable in* x *for* B *fixed,*

2) *for every* $B \in \mathcal{B}_1(F)$, $x \in F$, $t > s > \tau \geq 0$,

$$p_{t,\tau}(B \mid x) = \int p_{t,s}(B \mid y) p_{s,\tau}(dy \mid x).$$

To be regular, then, a set of transition probabilities must satisfy the Chapman-Kolmogorov equations identically.

Theorem 15.11. *Given a regular set of transition probabilities and a distribution* $\pi(dx_0)$, *define probabilities on cylinder sets by: for* $t_n > t_{n-1} > \cdots > t_0 = 0$, $B_n, \ldots, B_0 \in \mathcal{B}_1(F)$,

(15.12) $P(\mathsf{X}(t_n) \in B_n, \ldots, \mathsf{X}(t_0) \in B_0)$

$$= \int_{B_n} \cdots \int_{B_0} p_{t_n, t_{n-1}}(dx_n \mid x_{n-1}) \cdots p_{t,t_0}(dx_1 \mid x_0) \pi(dx_0).$$

These are consistent and the resultant process $\{\mathsf{X}(t)\}$ *is Markov with the given functions* $p_{t,\tau}$, π *as transition probabilities and initial distribution.*

Proof. The first verification is consistency. Let $B_k = F$. The expression in $P(\mathsf{X}(t_n) \in B_n, \ldots)$ that involves x_k, t_k is an integration with respect to the probability defined for $B \in \mathcal{B}_1(F)$ by

$$\int_F p_{t_{k+1}, t_k}(B \mid x_k) p_{t_k, t_{k-1}}(dx_k \mid x_{k-1}).$$

By the Chapman-Kolmogorov equations, this is exactly

$$p_{t_{k+1}, t_{k-1}}(B \mid x_{k-1}),$$

which eliminates t_k, x_k and gives consistency. Thus, a bit surprisingly, the functional relations (15.8) are the key to consistency. Now extend and get a process $\{\mathsf{X}(t)\}$ on $(F^{[0,\infty)}, \mathcal{B}^{[0,\infty)}(F))$. To verify the remainder, let $A \in \mathcal{B}_n(F)$. By extension from the definition,

$$P(\mathsf{X}(t_n) \in B, (\mathsf{X}(t_{n-1}), \ldots, \mathsf{X}(t_0)) \in A)$$
$$= \int_{(x_{n-1}, \ldots, x_0) \in A} p_{t_n, t_{n-1}}(B \mid x_{n-1}) P(\mathsf{X}(t_{n-1}) \in dx_{n-1}, \ldots, \mathsf{X}(t_0) \in dx_0).$$

Now, take the Radon derivative to get the result.

One result of this theorem is that there are no functional relationships other than those following from the Chapman-Kolmogorov equations that transition probabilities must satisfy in general.

Another convention we make starting from a regular set of transition probabilities is this: Let $P_{(x,\tau)}(\cdot)$ be the probability on the space of paths $F^{[0,\infty)}$, gotten by "starting the process out at the point x at time τ." More specifically, let $P_{(x,\tau)}(\cdot)$ be the probability on $\mathcal{B}^{[0,\infty)}(F)$ extended from (15.12), where $t_0 = \tau$ and $\pi(dx_0)$ concentrated on the point $\{x\}$ are used. So $P_{(x,\tau)}(\cdot)$ is well-defined for all x, τ in terms of the transition probabilities only.

Convention. *For* $C \in \mathcal{F}(X(s), s \leq \tau)$, $A \in \mathcal{B}^{[0,\infty)}(F)$ *always use the version of* $P(X(\tau + \cdot) \in A \mid X(\tau) = x, C)$ *given by* $P_{(x,\tau)}(A)$.

Accordingly, we use the transition probabilities not only to get the distribution of the process but also, to manufacture versions of all important conditional probabilities. The point of requiring the Chapman-Kolmogorov equations to hold identically rather than a.s. is that if there is an x, B, $t > s > \tau \geq 0$, such that

$$p_{t,\tau}(B \mid x) \neq \int p_{t,s}(B \mid y) p_{s,\tau}(dy \mid x),$$

then these transition probabilities can not be used to construct "the process starting from x, τ."

Now, because we wish to study the nature of a Markov process as governed by its transition probabilities, no matter what the initial distribution, we enlarge our nomenclature. Throughout this and the next chapter when we refer to a Markov process $\{X(t)\}$ this will no longer refer to a single process. Instead, it will denote the totality of processes having the same transition probabilities but with all possible different initial starting points x at time zero. However, we will use only coordinate representation processes, so the measurability of various functions and sets will not depend on the choice of a starting point or initial distribution for the process.

3. STATIONARY TRANSITION PROBABILITIES

Definition 15.13. *Let* $\{p_{t,\tau}\}$ *be a regular set of transition probabilities. They are called stationary if for all* $t > \tau \geq 0$,

$$p_{t,\tau}(B \mid x) \equiv p_{t-\tau,0}(B \mid x).$$

In this case, the $p_t(B \mid x) = p_{t,0}(B \mid x)$ *are referred to as the transition probabilities for the process.*

Some simplification results when the transition probabilities are stationary. The Chapman-Kolmogorov equations become

$$p_{t+\tau}(B \mid x) = \int p_t(B \mid y) p_\tau(dy \mid x). \tag{15.14}$$

For any $A \in \mathcal{B}^{[0,\infty)}(F)$, $P_{(x,\tau)}(A) = P_{(x,0)}(A)$; the probabilities on the path-space of the process are the same no matter when the process is started. Denote for any $A \in \mathcal{B}^{[0,\infty)}(F)$, $f(\cdot)$ on $F^{[0,\infty)}$ measurable $\mathcal{B}^{[0,\infty)}(F)$,

$$P_x(A) = P_{(x,0)}(A), \tag{15.15}$$

$$E_x f(\cdot) = \int f(\mathbf{y}) P_x(d\mathbf{y}).$$

Assume also from now on that for any initial distribution, $\mathsf{X}(t) \xrightarrow{P} \mathsf{X}(0)$ as $t \to 0$. This is equivalent to the statement,

Definition 15.16. *Transition probabilities $p_t(B \mid x)$ are called standard if*

$$p_t(\cdot \mid x) \xrightarrow{\mathcal{D}} \delta_{\{x\}}(\cdot),$$

for all $x \in F$, where $\delta_{\{x\}}(\cdot)$ denotes the distribution with unit mass on $\{x\}$.

There is another property that will be important in the sequel. Suppose we have a stopping time t^* for the process $\mathsf{X}(t)$. The analog of the restarting property of Brownian motion is that the process $\mathsf{X}(t + \mathsf{t}^*)$, given everything that happened up to time t^*, has the same distribution as the process $\mathsf{X}(t)$ starting from the point $\mathsf{X}(\mathsf{t}^*)$.

Definition 15.17. *A Markov process $\{\mathsf{X}(t)\}$ with stationary transition probabilities is called strong Markov if for every stopping time t^* [see (12.40)], every starting point $x \in F$, and set $A \in \mathcal{B}^{[0,\infty)}(F)$,*

$$P_x(\mathsf{X}(\cdot + \mathsf{t}^*) \in A \mid \mathsf{X}(s), s \leq \mathsf{t}^*) = P_{\mathsf{X}(\mathsf{t}^*)}(\mathsf{X}(\cdot) \in A) \quad \text{a.s. } P_x.$$

Henceforth, call a stopping time for a Markov process a *Markov time.*

It's fairly clear from the definitions that for fixed $\tau > 0$,

$$\textbf{(15.18)} \qquad P_x(\mathsf{X}(\cdot + \tau) \in A \mid \mathsf{X}(\tau)) = P_{\mathsf{X}(\tau)}(\mathsf{X}(\cdot) \in A) \quad \text{a.s. } P_x.$$

In fact,

Proposition 15.19. *If t^* assumes at most a countable number of values $\{\tau_k\}$, then*

$$P_x(\mathsf{X}(\cdot + \mathsf{t}^*) \in A \mid \mathsf{X}(s), s \leq \mathsf{t}^*) = P_{\mathsf{X}(\mathsf{t}^*)}(\mathsf{X}(\cdot) \in A) \quad \text{a.s. } P_x.$$

Proof. Take $C \in \mathcal{F}(\mathsf{X}(s), s \leq \mathsf{t}^*)$. Let $\varphi(x)$ be any bounded measurable function, $\tilde{\varphi}(x) = E_x\varphi(\mathsf{X}(t))$. We prove the proposition first for A one-dimensional. Take φ the set indicator of A, then

$$\textbf{(15.20)} \qquad \int_C \varphi(\mathsf{X}(t + \mathsf{t}^*))\, dP_x = \sum_k \int_{C \cap \{\mathsf{t}^* = \tau_k\}} \varphi(\mathsf{X}(t + \tau_k))\, dP_x.$$

By definition, $C \cap \{\mathsf{t}^* = \tau_k\} \in \mathcal{F}(\mathsf{X}(s), s \leq \tau_k)$, so

$$\begin{aligned} \int_{C \cap \{\mathsf{t}^* = \tau_k\}} \varphi(\mathsf{X}(t + \tau_k))\, dP_x &= \int_{C \cap \{\mathsf{t}^* = \tau_k\}} E[\varphi(\mathsf{X}(t + \tau_k)) \mid \mathsf{X}(s), s \leq \tau_k]\, dP_x \\ &= \int_{C \cap \{\mathsf{t}^* = \tau_k\}} \tilde{\varphi}(\mathsf{X}(\tau_k))\, dP_x, \end{aligned}$$

which does it. The same thing goes for φ, a measurable function of many variables, and then for the general case.

It is not unreasonable to hope that the strong Markov property would hold in general. It doesn't! But we defer an example until the next chapter.

One class of examples of strong Markov processes with standard stationary transition probabilities are the processes with stationary, independent increments, where the sample paths are taken right-continuous. Let $X(t)$ be such a process. If at time t the particle is at the point x, then the distribution at time $t + \tau$ is gotten by adding to x an increment independent of the path up to x and having the same distribution as $X(\tau)$. Thus, these processes are Markov.

Problems

1. For $X(t)$, a process with stationary, independent increments, show that it is Markov with one set of transition probabilities given by

$$p_t(B \mid x) = P\big(X(t) \in B - x\big),$$

and that this set is regular and standard.

2. Show that processes with stationary, independent increments and right-continuous sample paths are strong Markov.

3. Show that the functions $P_x(A)$, $E_x f(\cdot)$ of (15.15) are $\mathcal{B}_1(F)$-measurable.

4. Show that any Markov process with standard stationary transition probabilities is continuous from the right in probability.

4. INFINITESIMAL CONDITIONS

What do Markov processes look like? Actually, what do their sample paths and transition probabilities look like? This problem is essentially one of connecting up global behavior with local behavior. Note, for example, that if the transition probabilities p_t are known for all t in any neighborhood of the origin, then they are determined for all $t > 0$ by the Chapman-Kolmogorov equations. Hence, one suspects that p_t would be determined for all t by specifying the limiting behavior of p_t as $t \to 0$. But, then, the sample behavior will be very immediately connected with the behavior of p_t near $t = 0$.

To get a feeling for this, look at the processes with stationary, independent increments. If it is specified that

$$P_x\big(|X(t) - x| > \epsilon\big) = o(t),$$

then the process is Brownian motion, all the transition probabilities are determined, and all sample paths are continuous. Conversely, if all sample paths are given continuous, the above limiting condition at $t = 0$ must hold.

At the other end, suppose one asks for a process $X(t)$ with stationary, independent increments having all sample paths constant except for isolated jumps. Then (see Section 6, Chapter 14) the probability of no jump in the time interval $[0, t]$ is given by $e^{-\lambda t}$, so

$$P_0(\mathsf{X}(t) = 0) = 1 - \lambda t + o(t).$$

If there is a jump, with magnitude governed by $F(x)$, then for $B \in \mathcal{B}_1$, $\{0\} \notin B$,

$$P_0(\mathsf{X}(t) \in B) = \lambda t F(B) + o(t).$$

Conversely, if there is a process $\mathsf{X}(t)$ with stationary, independent increments, and a λ, F such that the above conditions hold as $t \to 0$, it is easy to check that the process must be of the jump type with exponent function given by

$$\psi(u) = \lambda \int_{\{0\}^c} (e^{iux} - 1) F(dx).$$

In general now, let $\{\mathsf{X}(t)\}$ be any Markov process with stationary transition probabilities. Take $f(x)$ a bounded $\mathcal{B}_1$-measurable function on F. Consider the class of these functions such that the limit as $t \downarrow 0$ of

(15.21)
$$\frac{E_x f(\mathsf{X}(t)) - f(x)}{t}$$

exists for all $x \in F$. Denote the resulting function by $(Sf)(x)$. S is called the *infinitesimal operator* and summarizes the behavior of the transition probabilities as $t \to 0$. The class of bounded measurable functions such that the limit in (15.21) exists for all $x \in F$, we will call the domain of S, denoted by $\mathcal{D}(S)$. For example, for Poisson-like processes,

$$\begin{aligned} E_x f(\mathsf{X}(t)) &= E_0 f(\mathsf{X}(t) + x) \\ &= \lambda t \int_{\{0\}^c} f(y + x) F(dy) + f(x)[1 - \lambda t] + o(t). \end{aligned}$$

Define a measure $\mu(B; x)$ by

$$\mu(B; x) = \lambda F(B - x), \quad x \notin B, \qquad \text{and} \qquad \mu(\{x\}; x) = 0.$$

Then S for this process can be written as

$$(Sf)(x) = \int (f(y) - f(x)) \mu(dy; x).$$

In this example, no further restrictions on f were needed to make the limit as $t \to 0$ exist. Thus $\mathcal{D}(S)$ consists of all bounded measurable functions.

For Brownian motion, take $f(x)$ continuous and with a continuous, bounded second derivative. Write

$$f(y) = f(x) + (y - x) f'(x) + \frac{(y - x)^2}{2} (f''(x) + \delta_x(y - x)),$$

where

$$\delta_x(h) \leq \sup_{|\xi| \leq |h|} |f''(x + \xi) - f''(x)|.$$

From this,

$$E_x f(\mathsf{X}(t)) = E_0 f(\mathsf{X}(t) + x)$$
$$= f(x) + \tfrac{1}{2}(E_0 \mathsf{X}^2(t)) f''(x) + R(t).$$

Use

$$E\mathsf{X}^2(t) = t, \qquad R(t) \leq E\mathsf{X}^2(t)\delta_x(\mathsf{X}(t)) = o(t),$$

to get

$$(Sf)(x) = \frac{1}{2}\frac{d^2}{dx^2} f(x).$$

In this case it is not clear what $\mathfrak{D}(S)$ is, but it is certainly not the set of all bounded measurable functions. These two examples will be typical in this sense: The jumps in the sample paths contribute an integral operator component to S; the continuous nonconstant parts of the paths contribute a differential operator component.

Once the behavior of p_t near $t = 0$ is specified by specifying S, the problem of computing the transition probabilities for all $t > 0$ is present. S hooks into the transition probabilities in two ways. In the first method, we let the initial position be perturbed. That is, given $\mathsf{X}(0) = x$, we let a small time τ elapse and then condition on $\mathsf{X}(\tau)$. This leads to the *backwards equations*. In the second method, we perturb on the final position. We compute the distribution up to time t and then let a small time τ elapse. Figures 15.1 and 15.2 illustrate computing $P_x(\mathsf{X}(t + \tau) \in B)$.

Backwards Equations

$$P(\mathsf{X}(t + \tau) \in B \mid \mathsf{X}(0) = x) = E[P(\mathsf{X}(t + \tau) \in B \mid \mathsf{X}(\tau)) \mid \mathsf{X}(0) = x].$$

Letting $\varphi_t(x) = p_t(B \mid x)$, we can write the above as

$$\varphi_{t+\tau}(x) = E_x \varphi_t(\mathsf{X}(\tau)) \quad \text{or} \quad \varphi_{t+\tau}(x) - \varphi_t(x) = E_x \varphi_t(\mathsf{X}(\tau)) - \varphi_t(x).$$

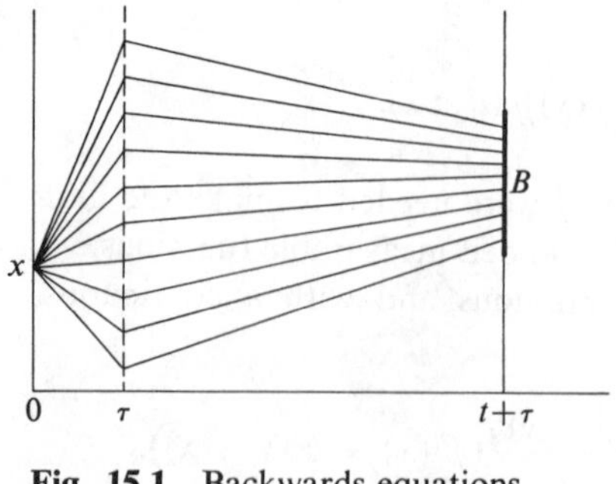

Fig. 15.1 Backwards equations.

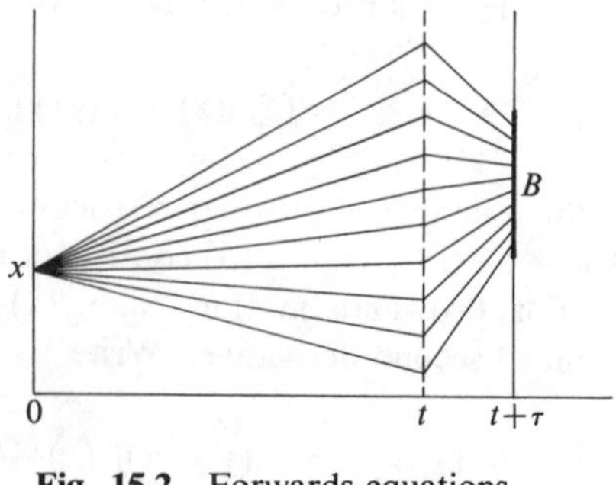

Fig. 15.2 Forwards equations.

Dividing by τ, letting $\tau \to 0$, if $p_t(B \mid x)$ is smooth enough in t, x, we find that

$$\text{(15.22)} \qquad \frac{\partial}{\partial t} p_t(B \mid x) = \big(Sp_t(B \mid \cdot)\big)(x),$$

that is, for Brownian motion, the backwards equations are

$$\frac{\partial}{\partial t} p_t(B \mid x) = \frac{1}{2} \frac{\partial^2}{\partial x^2} p_t(B \mid x).$$

For the Poisson-like processes,

$$\frac{\partial}{\partial t} p_t(B \mid x) = \int (p_t(B \mid y) - p_t(B \mid x))\mu(dy; x).$$

Forwards Equations. For any f for which Sf exists,

$$E_x f\big(\mathsf{X}(t + \tau)\big) = E_x\big[E_x\big(f(\mathsf{X}(t + \tau)) \mid \mathsf{X}(t)\big)\big]$$

or

$$\int f(y) p_{t+\tau}(dy \mid x) = \int \left[\int f(z) p_\tau(dz \mid y)\right] p_t(dy \mid x).$$

Subtract $E_x f\big(\mathsf{X}(t)\big)$ from both sides, divide by τ, and let $\tau \downarrow 0$. With enough smoothness,

$$\int f(y) \frac{\partial}{\partial t} p_t(dy \mid x) = \int (Sf)(y) p_t(dy \mid x).$$

Thus, if S has an adjoint S^*, the equations are

$$\text{(15.23)} \qquad \frac{\partial}{\partial t} p_t(B \mid x) = \big(S^* p_t(\cdot \mid x)\big)(B).$$

For Poisson-like processes,

$$\int (Sf)(y) p_t(dy \mid x) = \iint \big(f(z) - f(y)\big)\mu(dz; y)\, p_t(dy \mid x)$$

so the forwards equations are

$$\frac{\partial}{\partial t} p_t(B \mid x) = \int \bar{\mu}(B; y) p_t(dy \mid x),$$

where $\bar{\mu}(B; y) = \mu(B; y)$, $y \notin B$, and $\bar{\mu}(\{y\}; y) = -\mu(F; y)$. If $p_t(dy \mid x) \ll \mu(dy)$ for all t, x, then take S^* to be the adjoint with respect to $\mu(dy)$. For example, for Brownian motion $p_t(dy \mid x) \ll dy$. For all $f(y)$ with continuous second derivatives vanishing off finite intervals,

$$\int \frac{1}{2} f''(y) p_t(y \mid x)\, dy = \int f(y) \left(\frac{1}{2} \frac{\partial^2}{\partial y^2} p_t(y \mid x)\right) dy,$$

where $p_t(y \mid x)$ denotes (badly) the density of $p_t(dy \mid x)$ with respect to dy. Hence the forwards equation is

$$\frac{\partial}{\partial t} p_t(y \mid x) = \frac{1}{2} \frac{\partial^2}{\partial y^2} p_t(y \mid x).$$

The forwards or backwards equations, together with the boundary condition $p_t(\cdot \mid x) \xrightarrow{\mathcal{D}} \delta_{\{x\}}(\cdot)$ as $t \to 0$ can provide an effective method of computing the transition probabilities, given the infinitesimal conditions. But the questions regarding the existence and uniqueness of solutions are difficult to cope with. It is possible to look at these equations analytically, forgetting their probabilistic origin, and investigate their solutions. But the most illuminating approach is a direct construction of the required processes.

5. PURE JUMP PROCESSES

Definition 15.24. *A Markov process $\{X(t)\}$ will be called a pure jump process if, starting from any point $x \in F$, the process has all sample paths constant except for isolated jumps, and right-continuous.*

Proposition 15.25. *If $\{X(t)\}$ is a pure jump process, then it is strong Markov.*

Proof. By 15.19, for φ bounded and measurable $\mathcal{B}_k$, and

$$E_x\varphi(X(t_k), \ldots, X(t_0)) = \tilde{\varphi}(x),$$

we have for any $C \in \mathcal{F}(X(t), t \leq t_n^*)$, where t_n^* takes on only a countable number of values and $t_n^* \downarrow t^*$.

$$\int_C \varphi(X(t_k + t_n^*), \ldots, X(t_0 + t_n^*))\, dP_x = \int_C \tilde{\varphi}(X(t_n^*))\, dP_x.$$

Since $t^* \leq t_n^*$, this holds for all $C \in \mathcal{F}(X(t), t \leq t^*)$. Since all paths are constant except for jumps and right-continuous,

$$\tilde{\varphi}(X(t_n^*)) \to \tilde{\varphi}(X(t^*)),$$
$$\varphi(X(t_k + t_n^*), \ldots) \to \varphi(X(t_k + t^*), \ldots)$$

for every sample path. Taking limits in the above integral equality proves the proposition.

Assume until further notice that $\{X(t)\}$ is a pure jump process.

Definition 15.26. *Define the time T of the first jump as*

$$T = \inf\{t;\ X(t) \neq X(0)\}.$$

Proposition 15.27. *T is a Markov time.*

Proof. Let $\mathsf{T}_n = \inf\{k/2^n; \mathsf{X}(k/2^n) \neq \mathsf{X}(0)\}$. For every ω, $\mathsf{T}_n \downarrow \mathsf{T}$. Further,

$$\{\mathsf{T}_n \leq t\} = \bigcup_{k/2^n \leq t} \{\mathsf{X}(k/2^n) \neq \mathsf{X}(0)\} \in \mathcal{F}(\mathsf{X}(s), s \leq t.)$$

The sets $\{\mathsf{T}_n \leq t\}$ are monotonic increasing, and their limit is therefore in $\mathcal{F}(\mathsf{X}(s), s \leq t)$. If t is a binary rational, the limit is $\{\mathsf{T} \leq t\}$. If not, the limit is $\{\mathsf{T} < t\}$. But,

$$\{\mathsf{T} \leq \mathrm{t}\} = \{\mathsf{T} < t\} \cup \{\mathsf{X}(t) \neq \mathsf{X}(0)\}.$$

The basic structure of Markov jump processes is given by

Theorem 15.28. *Under P_x, T and $\mathsf{X}(\mathsf{T})$ are independent and there is a $\mathcal{B}_1$-measurable nonnegative function $\lambda(x)$ on F such that*

$$P_x(\mathsf{T} > t) = e^{-\lambda(x)t}.$$

Proof. Let $t + \mathsf{T}_1$ be the first exit time from state x past time t; that is,

$$t + \mathsf{T}_1 = \inf\{\tau; \mathsf{X}(\tau) \neq x, \tau > t\}.$$

Then, for $x \notin B$,

$$\begin{aligned} P_x(\mathsf{X}(\mathsf{T}) \in B, \mathsf{T} > t) &= P_x(\mathsf{X}(\mathsf{T}) \in B, \mathsf{T} > t, \mathsf{X}(t) = x) \\ &= P_x(\mathsf{X}(t + \mathsf{T}_1) \in B, \mathsf{T} > t, \mathsf{X}(t) = x) \\ &= P_x(\mathsf{X}(t + \mathsf{T}_1) \in B \mid \mathsf{T} > t, \mathsf{X}(t) = x) \\ &\qquad\qquad \cdot P_x(\mathsf{T} > t, \mathsf{X}(t) = x) \end{aligned}$$

if $P_x(\mathsf{T} > t) > 0$. Assume this for the moment. Then,

$$\{\mathsf{T} > t\} = \{\mathsf{T} \leq t\}^c \in \mathcal{F}(\mathsf{X}(s), s \leq t),$$

$$\begin{aligned} P_x(\mathsf{X}(t + \mathsf{T}_1) \in B \mid \mathsf{T} > t, \mathsf{X}(t) = x) &= P_x(\mathsf{X}(t + \mathsf{T}_1) \in B \mid \mathsf{X}(t) = x) \\ &= P_x(\mathsf{X}(\mathsf{T}) \in B). \end{aligned}$$

Going back, we have

$$P_x(\mathsf{X}(\mathsf{T}) \in B, \mathsf{T} > t) = P_x(\mathsf{X}(\mathsf{T}) \in B) P_x(\mathsf{T} > t).$$

There must be some $t_0 > 0$, such that $P_x(\mathsf{T} > t_0) > 0$. Take $0 < t \leq t_0$; then

$$\begin{aligned} P_x(\mathsf{T} > t + \tau) &= P_x(\mathsf{T} > t + \tau, \mathsf{T} > t, \mathsf{X}(t) = x) \\ &= P_x(\mathsf{T}_1 > \tau \mid \mathsf{X}(t) = x, \mathsf{T} > t) P_x(\mathsf{T} > t) \\ &= P_x(\mathsf{T} > \tau) P_x(\mathsf{T} > t). \end{aligned}$$

Therefore $\varphi(t) = P_x(\mathsf{T} > t)$ is a monotonic nonincreasing function satisfying $\varphi(t + \tau) = \varphi(t)\varphi(\tau)$ for $t \leq t_0$. This implies that there is a parameter $\lambda(x)$ such that

$$P_x(\mathsf{T} > t) = e^{-\lambda(x)t},$$

with $0 \le \lambda(x) < \infty$. If $\lambda(x) = 0$, then the state x is absorbing, that is, $P_x(\mathsf{X}(t) = x) = 1$, for all t. The measurability of $\lambda(x)$ follows from the measurability of $P_x(\mathsf{T} > t)$.

Corollary 15.29. *For a pure jump process, let*

$$P_x(\mathsf{T} > t) = e^{-\lambda(x)t}, \quad P_x(\mathsf{X}(\mathsf{T}) \in B) = p(B; x).$$

Then at every point x in F,

$$p_t(\{x\} \mid x) = 1 - \lambda(x)t + o(t),$$
$$p_t(B \mid x) = \big(\lambda(x)t + o(t)\big)\, p(B;\ x), \quad x \notin B.$$

Proof. Suppose that we prove that the probability of two jumps in time t is $o(t)$; then both statements are clear because

$$\begin{aligned} P_x(\mathsf{X}(t) = x) &= P_x(\mathsf{X}(t) = x, \text{ no more than one jump}) + o(t) \\ &= P_x(\mathsf{T} > t) + o(t) \\ &= 1 - \lambda(x)t + o(t). \end{aligned}$$

Similarly,

$$\begin{aligned} P_x(\mathsf{X}(t) \in B) &= P_x(\mathsf{X}(t) \in B, \text{ no more than one jump}) + o(t) \\ &= P_x(\mathsf{T} \le t, \mathsf{X}(\mathsf{T}) \in B) + o(t) \\ &= (\lambda(x)t + o(t))p(B; x). \end{aligned}$$

Remark. The reason for writing $o(t)$ in front of $p(B; x)$ is to emphasize that $o(t)/t \to 0$ uniformly in B.

We finish by

Proposition 15.30. *Let T_0 be the time of the first jump, $\mathsf{T}_0 + \mathsf{T}_1$ the time of the second jump. Then*

$$P_x(\mathsf{T}_0 + \mathsf{T}_1 \le t) = o(t).$$

Proof. $P_x(\mathsf{T}_0 + \mathsf{T}_1 \le t) \le P_x(\mathsf{T}_0 \le t, \mathsf{T}_1 \le t)$. Now,

$$P_x(\mathsf{T}_1 \le t \mid \mathsf{T}_0) = E_x[P_x(\mathsf{T}_1 \le t \mid \mathsf{T}_0, \mathsf{X}(\mathsf{T}_0)) \mid \mathsf{T}_0],$$

so

$$P_x(\mathsf{T}_1 \le t \mid \mathsf{T}_0) = \int (1 - e^{-\lambda(y)t}) p(dy; x).$$

This goes to zero as $t \to 0$, by the bounded convergence theorem. The following inequality

$$P_x(\mathsf{T}_0 + \mathsf{T}_1 \le t) \le (1 - e^{-\lambda(x)t}) \cdot \int (1 - e^{-\lambda(y)t}) p(dy\ ; x),$$

now proves the stated result.

Define a measure $\mu(dy; x)$ by

$$\mu(B; x) = \lambda(x)p(B; x), \qquad \mu(\{x\}, x) = 0.$$

The infinitesimal operator S is given, following 15.30, by

$$(Sf)(x) = \int (f(y) - f(x))\mu(dy; x), \tag{15.31}$$

and $\mathfrak{D}(S)$ consists of all bounded $\mathcal{B}_1(F)$ measurable functions. The following important result holds for jump processes.

Theorem 15.32. *$p_t(B \mid x)$ satisfies the backwards equations.*

Proof. First, we derive the equation

$$\begin{aligned} p_t(B \mid x) &= \chi_B(x)e^{-\lambda(x)t} \\ &\quad + \int_0^t \int_{\{x\}^c} p_{t-\tau}(B \mid y)\mu(dy; x)e^{-\lambda(x)\tau}\, d\tau. \end{aligned} \tag{15.33}$$

The intuitive idea behind (15.33) is simply to condition on the time and position of the first jump. To see this, write

$$P_x(\mathsf{X}(t) \in B) = P_x(\mathsf{X}(t) \in B, \mathsf{T} > t) + P_x(\mathsf{X}(t) \in B, \mathsf{T} \le t).$$

The first term is $\chi_B(x)e^{-\lambda(x)t}$. Reason that to evaluate the second term, if $\mathsf{T} \in d\tau$, and $\mathsf{X}(\mathsf{T}) \in dy$, then the particle has to get from y to B in time $t - \tau$. Hence the second term should be

$$\int_0^t \int p_{t-\tau}(B \mid y)P_x(\mathsf{X}(\mathsf{T}) \in dy)P_x(\mathsf{T} \in d\tau),$$

and this is exactly the second term in (15.33). A rigorous derivation could be given along the lines of the proof of the strong Markov property. But it is easier to use a method involving Laplace transforms which has wide applicability when random times are involved. We sketch this method: First note that since $\mathsf{X}(t)$ is jointly measurable in t, ω, $p_t(B \mid x) = E_x\chi_B(\mathsf{X}(t))$ is measurable in t. Define

$$\varphi_s(x) = \int_0^\infty e^{-st}p_t(B \mid x)\, dt = E_x\left[\int_0^\infty e^{-st}\chi_B(\mathsf{X}(t))\, dt\right], \quad s > 0.$$

Write

$$\int_0^\infty e^{-st}\chi_B(\mathsf{X}(t))\, dt = \int_0^{\mathsf{T}} e^{-st}\chi_B(\mathsf{X}(t))\, dt + \int_{\mathsf{T}}^\infty e^{-st}\chi_B(\mathsf{X}(t))\, dt.$$

The first term is

$$\chi_B(\mathsf{X}(0))\left(\frac{1 - e^{-s\mathsf{T}}}{s}\right).$$

The second is

$$e^{-s\mathsf{T}}\int_0^\infty e^{-st}\chi_B(\mathsf{X}(t+\mathsf{T}))\,dt.$$

By the strong Markov property,

$$E_x\left(e^{-s\mathsf{T}}\int_0^\infty e^{-st}\chi_B(\mathsf{X}(t+\mathsf{T}))\,dt \mid \mathsf{X}(\tau), \tau \le \mathsf{T}\right) = e^{-s\mathsf{T}}\varphi_s(\mathsf{X}(\mathsf{T})).$$

Hence

$$\varphi_s(x) = \chi_B(x)\cdot\frac{1}{s+\lambda(x)} + \frac{\lambda(x)}{s+\lambda(x)}\int \varphi_s(y)p(dy;x).$$

This is exactly the transform of (15.33), and invoking the uniqueness theorem for Laplace transforms (see [140]) gets (15.33) almost everywhere (dt). For $\{\mathsf{X}(t)\}$ a pure jump process, writing $p_t(B \mid x)$ as $E_x\chi_B(\mathsf{X}(t))$ makes it clear that $p_t(B \mid x)$ is right-continuous. The right side of (15.33) is obviously continuous in time; hence (15.33) holds identically.

Multiply (15.33) by $e^{\lambda(x)t}$ and substitute $t - \tau = \tau'$ in the second term:

$$e^{\lambda(x)t}p_t(B \mid x) = \chi_B(x) + \int_0^t\int_{\{x\}^c} e^{\lambda(x)\tau}p_\tau(B \mid y)\mu(dy;x)\,d\tau.$$

Hence $p_t(B \mid x)$ is differentiable, and

$$\frac{\partial}{\partial t}(e^{\lambda(x)t}p_t(B \mid x)) = e^{\lambda(x)t}\int_{\{x\}^c} p_t(B \mid y)\mu(dy;x).$$

An easy simplification gives the backwards equations.

The forwards equations are also satisfied. (See Chung [16, pp. 224 ff.], for instance.) In fact, most of the questions regarding the transition probabilities can be answered by using the representation

(15.34) $$\chi_B(\mathsf{X}(t)) = \sum_0^\infty \chi_{[\mathsf{R}_n,\mathsf{R}_{n+1})}(t)\chi_B(\mathsf{X}(\mathsf{R}_n)),$$

where R_n is the time of the nth jump, $\mathsf{R}_0 \equiv 0$, so

$$\mathsf{R}_n = \mathsf{T}_0 + \cdots + \mathsf{T}_{n-1}, \quad n > 0,$$

where the T_k are the first exit times after the kth jump. We make use of this in the next section to prove a uniqueness result for the backward equation.

Problem 5. Show that a pure jump process $\{\mathsf{X}(t)\}$, $t \ge 0$ is jointly measurable in (t, ω) with respect to $\mathcal{B}_1([0, \infty)) \times \mathcal{F}$.

6. CONSTRUCTION OF JUMP PROCESSES

In modeling Markov processes what is done, usually, is to prescribe infinitesimal conditions. For example: Let F be the integers, then a population model with constant birth and death rates would be constructed by specifying that in a small time Δt, if the present population size is j, the probability of increasing by one is $r_B j\,\Delta t$, where r_B is the birth rate. The probability of a decrease is $r_D j\,\Delta t$ where r_D is the death rate, and the probability of no change is $1 - r_D j\,\Delta t - r_B j\,\Delta t$. What this translates into is

$$\begin{aligned} p_t(j+1 \mid j) &= r_B jt + o(t),\ k \neq j, \\ p_t(j-1 \mid j) &= r_D jt + o(t), \\ p_t(j \mid j) &= 1 - (r_D + r_B)jt + o(t), \\ p_t(k \mid j) &= o(t), \quad \text{otherwise}. \end{aligned}$$

In general, countable state processes are modeled by specifying $q(k \mid j) \geq 0$, $q(j) = \sum_{k \neq j} q(k \mid j) < \infty$, such that

$$\begin{aligned} p_t(k \mid j) &= q(k \mid j)t + o(t), \quad k \neq j, \\ p_t(j \mid j) &= 1 - q(j)t + o(t). \end{aligned}$$

General jump processes are modeled by specifying finite measures $\mu(B; x)$, measurable in x for every $B \in \mathcal{B}_1(F)$, such that for every x,

$$\textbf{(15.35)} \qquad \begin{aligned} p_t(B \mid x) &= \mu(B; x)t + o(t), \quad x \notin B, \\ p_t(\{x\} \mid x) &= 1 - \mu(F; x)t + o(t). \end{aligned}$$

Now the problem is: Is there a unique Markov jump process fitting into (15.35)? Working backward from the results of the last section—we know that if there is a jump process satisfying (15.35), then it is appropriate to define

$$\lambda(x) = \mu(F; x), \qquad p(B; x) = \mu(B; x)/\mu(F; x), \qquad p(\{x\}; x) = 0,$$

and look for a process such that

$$P_x(\mathsf{T} > t) = e^{-\lambda(x)t}, \qquad P_x(\mathsf{X}(\mathsf{T}) \in B) = p(B; x).$$

Theorem 15.28 gives us the key to the construction of a pure jump process. Starting at x, we wait there a length of time T_0 exponentially distributed with parameter $\lambda(x)$; then independently of how long we wait, our first jump is to a position with distribution $p(dy; x)$. Now we wait at our new position y, time T_1 independent of T_0, with distribution parameter $\lambda(y)$, etc. Note that these processes are very similar to the Poisson-like processes with independent increments. Heuristically, they are a sort of patched-together assembly of such processes, in the sense that at every point x the process behaves at that point like a Poisson-like process with parameter $\lambda(x)$ and jump distribution given by $p(B; x)$.

At any rate, it is pretty clear how to proceed with the construction.

1) *The space structure* of this process is obtained by constructing a discrete Markov process $\mathsf{X}_0, \mathsf{X}_1, \mathsf{X}_2, \ldots$, moving under the transition probabilities $p(B; x)$, and starting from any point x.

2) *The time flow* of the process consists of slowing down or speeding up the rate at which the particle travels along the paths of the space structure.

For every n, $x \in F$, construct random variables $\mathsf{T}_n(x)$ such that

i) $P(\mathsf{T}_n(x) > t) = e^{-\lambda(x)t}$,

ii) $\mathsf{T}_n(x)$ are jointly measurable in ω, x,

iii) the processes $(\mathsf{X}_0, \ldots)$, $(\mathsf{T}_0(x), x \in F)$, $(\mathsf{T}_1(x), x \in F), \ldots$ are mutually independent.

The $\mathsf{T}_n(x)$ will serve as the waiting time in the state x after the nth jump. To see that joint measurability can be gotten, define on the probability space $([0, 1], \mathcal{B}([0, 1]), dz)$ the variables $\mathsf{T}(z, \lambda) = -(1/\lambda) \log z$. Thus $P(\mathsf{T}(z, \lambda) > t) = e^{-\lambda t}$. Now define $\mathsf{T}_0(x) = \mathsf{T}(z, \lambda(x))$ and take the cross-product space with the sample space for $\mathsf{X}_0, \mathsf{X}_1, \ldots$ Similarly, for $\mathsf{T}_1(x)$, $\mathsf{T}_2(x), \ldots$

For the process itself, proceed with

Definition 15.36. *Define variables as follows:*

$$\mathsf{R}_0 \equiv 0, \qquad \mathsf{R}_n = \mathsf{T}_0(\mathsf{X}_0) + \mathsf{T}_1(\mathsf{X}_1) + \cdots + \mathsf{T}_{n-1}(\mathsf{X}_{n-1}),$$

$$\mathsf{n}^*(t) = n, \quad \text{if} \quad \mathsf{R}_n \le t < \mathsf{R}_{n+1},$$

$$\mathsf{X}(t) = \mathsf{X}_{\mathsf{n}^*(t)}.$$

In this definition R_n functions as the time of the nth jump.

Theorem 15.37. *If* $\mathsf{n}^*(t)$ *is well-defined by* 15.36 *for all* t, *then* $\mathsf{X}(t)$ *is a pure jump Markov process with transition probabilities satisfying the given infinitesimal conditions.*

Proof. This is a straightforward verification. The basic point is that given $\mathsf{X}(t) = x$, and given, say, that we got to this space-time point in n steps, then the waiting time in x past t does not depend on how long has already been spent there; that is,

$$P(\mathsf{T}_n(x) > \tau + \xi \mid \mathsf{T}_n(x) > \xi) = e^{-\lambda(x)\tau}.$$

To show that the infinitesimal conditions are met, just show again that the probability of two jumps in time t is $o(t)$.

The condition that $\mathsf{n}^*(t)$ be well-defined is that

$$\mathsf{R}_\infty = \sum_0^\infty \mathsf{T}_k(\mathsf{X}_k) = \infty \quad \text{a.s. } P_x, \quad \text{all } x.$$

This is a statement that at most a finite number of jumps can occur in every finite time interval. If $P_x(\mathsf{R}_\infty < \infty) > 0$, there is no pure jump process that satisfies the infinitesimal conditions for all $x \in F$. However, even if $\mathsf{R}_\infty = \infty$ a.s., the question of uniqueness has been left open. Is there another Markov process, not necessarily a jump process, satisfying the infinitesimal conditions? The answer, in regard to distribution, is No. The general result states that if $P_x(\mathsf{R}_\infty < \infty) = 0$, all $x \in F$, then any Markov process $\{\mathsf{X}(t)\}$ satisfying the infinitesimal conditions (15.35) is a pure jump process and has the same distribution as the constructed process. For details of this, refer to Doob [39, pp. 266 ff.]. We content ourselves with the much easier assertion:

Proposition 15.38. *Any two pure jump processes having the same infinitesimal operator S have the same distribution.*

Proof. This is now almost obvious, because for both processes, T and $\mathsf{X}(\mathsf{T})$ have the same distribution. Therefore the sequence of variables $\mathsf{X}(\mathsf{T}_0)$, $\mathsf{X}(\mathsf{T}_0 + \mathsf{T}_1), \ldots$ has distribution governed by $p(B; x)$, and given this sequence, the jump times are sums of independent variables with the same distribution as the constructed variables $\{\mathsf{R}_n\}$.

Let $\{\mathsf{X}(t)\}$ be the constructed process. Whether or not $\mathsf{R}_\infty = \infty$ a.s., define $p_t^{(N)}(B \mid x)$ as the probability that $\mathsf{X}(t)$ reaches B in time t in N or fewer jumps. That is,

$$\begin{aligned} p_t^{(N)}(B \mid x) &= P_x\big(\mathsf{X}(t) \in B,\ \mathsf{R}_{N+1} > t\big) \\ &= \sum_{n=0}^{N} P_x(\mathsf{X}_n \in B,\ \mathsf{R}_n \leq t < \mathsf{R}_{n+1}). \end{aligned}$$

For $n \geq 1$, and $\tau < t$,

$$\begin{aligned} &P_x(\mathsf{X}_n \in B, \mathsf{T}_0 + \cdots + \mathsf{T}_{n-1} \leq t < \mathsf{T}_0 + \cdots + \mathsf{T}_n \mid \mathsf{X}_1 = y, \mathsf{T}_0 = \tau) \\ &\quad = P_x(\mathsf{X}_n \in B, \mathsf{T}_1 + \cdots + \mathsf{T}_{n-1} \leq t - \tau < \mathsf{T}_1 + \cdots + \mathsf{T}_n \mid \mathsf{X}_1 = y) \\ &\quad = P_y(\mathsf{X}_{n-1} \in B, \mathsf{T}_0 + \cdots + \mathsf{T}_{n-2} \leq t - \tau \\ &\qquad\qquad < \mathsf{T}_0 + \cdots + \mathsf{T}_{n-1}) \quad \text{a.s. } p(dy; x). \end{aligned}$$

The terms for $n \geq 1$ vanish for $\tau \geq t$, and the zero term is

$$P_x(\mathsf{X}_0 \in B,\ \mathsf{R}_1 > t) = \chi_B(x)e^{-\lambda(x)t};$$

hence integrating out X_1, T_0 gives

Proposition 15.39

$$p_t^{(N)}(B \mid x) = \chi_B(x)e^{-\lambda(x)t} + \int_0^t \int_{\{x\}^c} p_t^{(N-1)}(B \mid y)\mu(dy; x)e^{-\lambda(x)\tau}\, d\tau.$$

Letting $N \to \infty$ gives another proof that for a pure jump process, integral equation (15.33) and hence the backwards equations are satisfied.

Define

(15.40)
$$\bar{p}_t(B \mid x) = \lim_N p_t^{(N)}(B \mid x).$$

The significance of $\bar{p}_t(B \mid x) = \lim_N p_t^{(N)}(B \mid x)$ is that it is the probability of going from x to B in time t in a finite number of steps.

Proposition 15.41. *$\bar{p}_t(B \mid x)$ is the minimal solution of the backwards equations in the sense that if $q_t(x)$ is any other solution satisfying*

1) $$q_t(x) \geq 0,$$

2) $$q_t(x) \to \chi_B(x) \quad \text{as } t \to 0,$$

then

$$\bar{p}_t(B \mid x) \leq q_t(x).$$

Proof. The backwards equation is

$$\frac{\partial}{\partial \tau} q_\tau(x) = -\lambda(x)q_\tau(x) + \int_{\{x\}^c} q_\tau(y)\mu(dy; x).$$

Multiply by $e^{-\lambda(x)\tau}$, integrate from 0 to t, and we recover the integral equation

$$q_t(x) = \chi_B(x)e^{-\lambda(x)t} + \int_0^t \int_{\{x\}^c} q_\tau(y)\mu(dy; x)e^{-\lambda(x)\tau}\, d\tau.$$

Assume $q_t(x) \geq p_t^{(N)}(B \mid x)$. Then substituting this inequality in the integral on the right,

$$\begin{aligned} q_t(x) &\geq \chi_B(x)e^{-\lambda(x)t} + \int_0^t \int_{\{x\}^c} p_t^{(N)}(B \mid y)\mu(dy; x)e^{-\lambda(x)\tau}\, d\tau \\ &= p_t^{(N+1)}(B \mid x). \end{aligned}$$

By the nonnegativity of q_t,

$$q_t(x) \geq \chi_B(x)e^{-\lambda(x)t} = p_t^{(0)}(B \mid x).$$

Hence $q_t(x) \geq \lim_N p_t^{(N)}(B \mid x) = \bar{p}_t(B \mid x)$.

Corollary 15.42. *If $\{\mathsf{X}(t)\}$ is a pure jump process, equivalently, if $\mathsf{R}_\infty = \infty$ a.s. P_x, all $x \in F$, then $\{\bar{p}_t(B \mid x)\}$ are the unique set of transition probabilities satisfying the backwards equations.*

7. EXPLOSIONS

If there are only a finite number of jumps in every finite time interval, then everything we want goes through—the forwards and backwards equations are satisfied and the solutions are unique. Therefore it becomes important to be able to recognize from the infinitesimal conditions when the resulting process will be pure jump. The thing that may foul the process up is unbounded $\lambda(x)$. The expected duration of stay in state x is given by $E_x\mathsf{T} = 1/\lambda(x)$. Hence if $\lambda(x) \to \infty$ anywhere, there is the possibility that the particle will move from state to state, staying in each one a shorter period of time. In the case where F represents the integers, $\lambda(n)$ can go to ∞ only if $n \to \infty$. In this case, we can have infinitely many jumps only if the particle can move out to ∞ in finite time. This is dramatically referred to as the possibility of *explosions* in the process. Perhaps the origin of this is in a population explosion model with pure birth,

$$p_t(j+1 \mid j) = \lambda(j)t + o(t),$$
$$p_t(j \mid j) = 1 - \lambda(j)t + o(t).$$

Here the space structure is $p(n+1; n) = 1$; the particle must move one step to the right each unit. Hence, $\mathsf{X}_n = n$ if $\mathsf{X}_0 = 0$. Now

$$\mathsf{R}_n = \mathsf{T}_0(0) + \cdots + \mathsf{T}_{n-1}(n-1)$$

is the time necessary to move n steps. And

$$E_0\mathsf{R}_\infty = \sum_0^\infty \frac{1}{\lambda(n)}.$$

If this sum is finite, then $\mathsf{R}_\infty < \infty$ a.s. P_0. This is also sufficient for $P_j(\mathsf{R}_\infty < \infty) = 1$, for all $j \in F$. Under these circumstances the particle explodes out to infinity in finite time, and the theorems of the previous sections do not apply.

One criterion that is easy to derive is

Proposition 15.43. *A process satisfying the given infinitesimal conditions will be pure jump if and only if*

$$P_x\left(\sum_0^\infty \frac{1}{\lambda(\mathsf{X}_n)} < \infty\right) = 0, \quad \text{all } x \in F.$$

Proof. For $\sum_0^\infty \mathsf{T}_n$ a sum of independent, exponentially distributed random variables with parameters λ_n, $\sum_0^\infty \mathsf{T}_n < \infty$ a.s. iff $\sum_0^\infty 1/\lambda_n < \infty$. Because for $s \geq 0$,

$$Ee^{-s(\mathsf{T}_1+\cdots+\mathsf{T}_n)} = \frac{1}{\prod_1^n(1 + s/\lambda_k)}.$$

Verify that the infinite product on the right converges to a finite limit iff $\sum_0^\infty 1/\lambda_n < \infty$, and apply 8.36 and the continuity theorem given in Section 13, Chapter 8. Now note that given $(\mathsf{X}_0, \mathsf{X}_1, \ldots)$, R_∞ is a sum of such variables with parameters $\lambda(\mathsf{X}_n)$.

Corollary 15.44. *If* $\sup_{x \in F} \lambda(x) < \infty$, *then* $\{\mathsf{X}(t)\}$ *is pure jump.*

Note that for a pure birth process $\sum_0^\infty 1/\lambda(n) = \infty$ is both necessary and sufficient for the process to be pure jump. For F the integers, another obvious sufficient condition is that every state be recurrent under the space structure. Conversely, consider

$$E_j\left(\sum_0^\infty \frac{1}{\lambda(\mathsf{X}_n)}\right).$$

Let $\mathsf{N}(k)$ be the number of entries that the sequence $\mathsf{X}_0, \mathsf{X}_1, \ldots$ makes into state k. Then

$$E_j\left(\sum_0^\infty \frac{1}{\lambda(\mathsf{X}_n)}\right) = \sum_k \frac{E_j \mathsf{N}(k)}{\lambda(k)}.$$

If this is finite, then certainly there will be an explosion. Time-continuous birth and death processes are defined as processes on the integers such that each jump can be only to adjacent states, so that

$$p_t(j+1 \mid j) = b(j)t + o(t),$$
$$p_t(j-1 \mid j) = d(j)t + o(t),$$
$$p_t(j \mid j) = 1 - \big(b(j) + d(j)\big)t + o(t).$$

For a birth and death process with no absorbing states moving on the non-negative integers, $E_0(\mathsf{N}(k)) = 1 + M(k)$, where $M(k)$ is the expected number of returns to k given $\mathsf{X}_0 = k$. Then, as above,

$$E_0\left(\sum_0^\infty \frac{1}{\lambda(\mathsf{X}_n)}\right) = \sum_k \frac{1 + M(k)}{\lambda(k)}.$$

The condition that this latter sum be infinite is both necessary and sufficient for no explosions [12].

Another method for treating birth and death processes was informally suggested to us by Charles Stone. Let F be the nonnegative integers with no absorbing states. Let $\mathbf{t}_n^*$ be the first passage time from state 1 to state n, and $\boldsymbol{\tau}_n^*$ the first passage time from state n to state $n + 1$. The $\boldsymbol{\tau}_1^*, \boldsymbol{\tau}_2^*, \ldots$ are independent, $\mathbf{t}_n^* = \sum_1^{n-1} \boldsymbol{\tau}_k^*$.

Proposition 15.45. $\mathbf{t}_\infty^* = \lim_n \mathbf{t}_n^*$ *is finite* a.s. *if and only if* $\sum_1^\infty E\boldsymbol{\tau}_k^* < \infty$.

Proof. Let T_k be the duration of first stay in state k, then $\boldsymbol{\tau}_k^* \geq \mathsf{T}_k$. Further $\sum_1^\infty \mathsf{T}_k < \infty$ a.s. iff $\sum_1^\infty E\mathsf{T}_k < \infty$ or $\sum_1^\infty 1/\lambda(k) < \infty$. Hence if $\inf_k \lambda(k) = 0$, both $\sum_1^\infty \boldsymbol{\tau}_k^*$ and $\sum_1^\infty E\boldsymbol{\tau}_k^*$ are infinite. Now assume $\inf_k \lambda(k) = \delta > 0$. Given any succession of states $\mathsf{X}_0 = n, \mathsf{X}_1, \ldots, \mathsf{X}_m = n + 1$ leading from n to $n + 1$, $\boldsymbol{\tau}_k^*$ is a sum of independent, exponentially distributed random variables $\mathsf{T}_0 + \cdots + \mathsf{T}_m$, and $\sigma^2(\mathsf{T}_0 + \cdots + \mathsf{T}_m) = \sigma^2(\mathsf{T}_0) + \cdots + \sigma^2(\mathsf{T}_m) \leq \delta^{-1}E(\mathsf{T}_0 + \cdots + \mathsf{T}_m)$. Hence

$$\sigma^2(\boldsymbol{\tau}_k^*) \leq \delta^{-1}E\boldsymbol{\tau}_k^*.$$

If $\sum_1^\infty \boldsymbol{\tau}_k^*$ converges a.s., but $\sum_1^\infty E\boldsymbol{\tau}_k^* = \infty$, then for $0 < \epsilon < 1$,

$$P\left(\left|\sum_1^n (\boldsymbol{\tau}_k^* - E\boldsymbol{\tau}_k^*)\right| > \epsilon \sum_1^n E\boldsymbol{\tau}_k^*\right) \to 1.$$

Applying Chebyshev's inequality to this probability gives a contradiction which proves the proposition.

Problems

6. Show that $\alpha(k) = E\tau_k^*$ satisfies the difference equation

$$\alpha(k) = p(k)\cdot\frac{1}{\lambda(k)} + q(k)\left[\frac{1}{\lambda(k)} + \alpha(k-1) + \alpha(k)\right], \quad k \geq 1,$$

where

$$p(k) = b(k)/\lambda(k), \qquad q(k) = d(k)/\lambda(k),$$

or

$$p(k)\alpha(k) = \frac{1}{\lambda(k)} + q(k)\alpha(k-1), \quad k \geq 1.$$

Deduce conditions on $p(k)$, $q(k)$, $\lambda(k)$ such that there are no explosions. (See [87].)

7. Discuss completely the explosive properties of a birth and death process with $\{0\}$ a reflecting state and

$$b(n) = b\varphi(n), \qquad d(n) = d\varphi(n).$$

8. NONUNIQUENESS AND BOUNDARY CONDITIONS

If explosions are possible, then $\bar{p}_t(F \mid x) < 1$ for some x, t, and the process is not uniquely determined by the given infinitesimal conditions. The nature of the nonuniqueness is that the particle can *reach* points on some undefined "boundary" of F not included in F. Then to completely describe the process it is necessary to specify its evolution from these boundary points. This is seen most graphically when F is the integers. If $P_j(\mathsf{R}_\infty < \infty) > 0$ for some j, then we have to specify what the particle will do once it reaches ∞. One possible procedure is to add to F a state denoted $\{\infty\}$ and to specify transition probabilities from $\{\infty\}$ to $j \in F$. For example, we could make $\{\infty\}$ an absorbing state, that is, $p_t(\{\infty\} \mid \{\infty\}) = 1$. An even more interesting construction consists of specifying that once the particle reaches $\{\infty\}$ it instantaneously moves into state k with probability $Q(k)$. This is more interesting in that it is not necessary to adjoin an extra state $\{\infty\}$ to F.

To carry out this construction, following Chung [16], define

$$p_t^{(0)}(j \mid k) = \bar{p}_t(j \mid k) = P_k(\mathsf{X}(t) = j,\ \mathsf{R}_\infty \geq t).$$

Now look at the probability $p_t^{(1)}(j \mid k)$ that $k \to j$ in time t with exactly one passage to $\{\infty\}$. To compute this, suppose that $\mathsf{R}_\infty = \tau$; then the particle moves immediately to state l with probability $Q(l)$, and then must go from l to j in time $t - \tau$ with no further excursions to $\{\infty\}$. Hence, denoting $H_k(d\tau) = P_k(\mathsf{R}_\infty \in d\tau)$,

$$p_t^{(1)}(j \mid k) = \int_0^t \left[\sum_l p_{t-\tau}^{(0)}(j \mid l)Q(l)\right] H_k(d\tau).$$

Similarly, the probability $p^{(n)}(j \mid k)$, of $k \to j$ in time t with exactly n passages to $\{\infty\}$ is given by

$$p_t^{(n)}(j \mid k) = \int_0^t \left[\sum_l p_{t-\tau}^{(n-1)}(j \mid l) Q(l) \right] H_k(d\tau).$$

Now define

$$p_t(j \mid k) = \sum_{n=0}^{\infty} p_t^{(n)}(j \mid k).$$

Proposition 15.46. *$p_t(j \mid k)$ as defined above satisfies*

1) $\sum_j p_t(j \mid k) = 1$,
2) *the Chapman-Kolmogorov equations, and*
3) *the backwards equations.*

Proof. Left to reader.

Remark. $p_t(j \mid k)$ does not satisfy the forwards equations. See Chung [16, pp. 224 ff.].

The process constructed the above way has the property that

$$p_t(j \mid k) = p_t^{(0)}(j \mid k) + o(t).$$

This follows from noting that

$$p_t(j \mid k) = p_t^{(0)}(j \mid k) + \int_0^t \left(\sum_l p_{t-\tau}(j \mid l) Q(l) \right) H_k(d\tau).$$

The integral term is dominated by $P_k(\mathrm{R}_\infty < t)$. This is certainly less than the probability of two jumps in time t, hence is $o(t)$. Therefore, no matter what $Q(l)$ is, all these processes have the specified infinitesimal behavior. This leads to the observation (which will become more significant in the next chapter), that *if it is possible to reach a "boundary" point, then boundary conditions must be added to the infinitesimal conditions in order to specify the process.*

9. RESOLVENT AND UNIQUENESS

Although S with domain $\mathfrak{D}(S)$ does not determine the process uniquely, this can be fixed up with a more careful and restrictive definition of the domain of S. In this section the processes dealt with will be assumed to have standard stationary transition probabilities, but no restrictions are put on their sample paths.

Definition 15.47. *Say that functions $\varphi_t(x)$ converge boundedly pointwise to $\varphi(x)$ on some subset A of their domain as $t \to 0$ if*

i) $$\lim_{t \to 0} \varphi_t(x) = \varphi(x), \quad \text{all } x \in A,$$

ii) $$\sup_{x \in A} |\varphi_t(x)| \le M < \infty, \quad \text{for all } t \text{ sufficiently small.}$$

Denote this by $\varphi_t(x) \xrightarrow{bp} \varphi(x)$ *on* A.

Let $\mathfrak{L}$ be any class of bounded $\mathcal{B}_1(F)$-measurable functions. Then we use

Definition 15.48. $\mathfrak{D}(S, \mathfrak{L})$ *consists of all functions* $f(x)$ *in* $\mathfrak{L}$ *such that*

$$\frac{E_x(f(X(t))) - f(x)}{t}$$

converges boundedly pointwise on F *to a function in* $\mathfrak{L}$.

We plan to show that with an appropriate choice of $\mathfrak{L}$, that corresponding to a given S, $\mathfrak{D}(S, \mathfrak{L})$, there is at most one process. In the course of this, we will want to integrate functions of the type $E_x f(X(t))$, so we need

Proposition 15.49. *For* $f(x)$ *bounded and measurable on* F, $\varphi(x, t) = E_x f(X(t))$ *is jointly measurable in* (x, t), *with respect to* $\mathcal{B}_1(F) \times \mathcal{B}_1([0, \infty))$.

Proof. Take $f(x)$ bounded and continuous. Since $\{X(t)\}$ is continuous in probability from the right, the function $\varphi(x, t) = E_x f(X(t))$ is continuous in t from the right and $\mathcal{B}_1(F)$-measurable in x for t fixed. Consider the approximation

$$\varphi_n(x, t) = \sum_{k=0}^{\infty} \chi_{[k/2^n, k+1/2^n)}(t)\varphi(x, (k + 1)/2^n).$$

By the right-continuity, $\varphi_n(x, t) \to \varphi(x, t)$. But $\varphi_n(x, t)$ is jointly measurable, therefore so is $\varphi(x, t)$. Now consider the class of $\mathcal{B}_1(F)$-measurable functions $f(x)$ such that $|f(x)| \le 1$ and the corresponding $\varphi(x, t)$ is jointly measurable. This class is closed under pointwise convergence, and contains all continuous functions bounded by one. Hence it contains all bounded measurable functions bounded by one.

Definition 15.50. *The resolvent is defined as*

$$R_\lambda(B \mid x) = \int_0^\infty e^{-\lambda t} p_t(B \mid x)\, dt, \quad \lambda > 0$$

for any $B \in \mathcal{B}_1(F)$, $x \in F$.

It is easy to check that $R_\lambda(B \mid x)$ is a bounded measure on $\mathcal{B}_1(F)$ for fixed x. Furthermore, by 15.49 and the Fubini theorem, $R_\lambda(B \mid x)$ is $\mathcal{B}_1(F)$-measurable in x for $B \in \mathcal{B}_1(F)$ fixed. Denote, for f bounded and $\mathcal{B}_1(F)$-measurable,

$$(R_\lambda f)(x) = \int f(y) R_\lambda(dy \mid x) \quad \text{and} \quad (T_t f)(x) = \int f(y) p_t(dy \mid x).$$

Then, using the Fubini theorem to justify the interchange,

$$R_\lambda f = \int_0^\infty e^{-\lambda t}(T_t f)\,dt.$$

Take $\mathcal{G}$ to be the set of all bounded $\mathcal{B}_1(F)$-measurable functions $f(x)$ such that $(T_t f)(x) \to f(x)$ as $t \to 0$ for every $x \in F$. Note that $\mathfrak{D}(S, \mathcal{G}) \subset \mathcal{G}$.

Theorem 15.51. *If $f \in \mathcal{G}$, then $R_\lambda f$ is in $\mathfrak{D}(S, \mathcal{G})$ and*

1) $$(\lambda - S)(R_\lambda f) = f.$$

If f is in $\mathfrak{D}(S, \mathcal{G})$, then

2) $$R_\lambda((\lambda - S)f) = f.$$

Proof. If f is in $\mathcal{G}$, then since T_t and R_λ commute, the bounded convergence theorem can be applied to $T_t(R_\lambda f) = R_\lambda(T_t f)$ to establish $R_\lambda f \in \mathcal{G}$. Write

$$T_t(R_\lambda f) = T_t \int_0^\infty e^{-\lambda\tau}(T_\tau f)\,d\tau = \int_0^\infty e^{-\lambda\tau}(T_t(T_\tau f))\,d\tau.$$

The Chapman-Kolmogorov equations imply $T_t(T_\tau f) = T_{t+\tau} f$, so

$$T_t(R_\lambda f) = e^{\lambda t}\left[R_\lambda f - \int_0^t e^{-\lambda\tau}(T_\tau f)\,d\tau\right].$$

From this, denoting

$$S_t f = \frac{T_t f - f}{t},$$

we get

(15.52) $$S_t(R_\lambda f) = \left(\frac{e^{-\lambda t} - 1}{t}\right)R_\lambda f - \frac{e^{\lambda t}}{t}\int_0^t e^{-\lambda\tau}(T_\tau f)\,d\tau.$$

Using $\|f\| = \sup |f(x)|$, $x \in F$, we have

$$\|S_t(R_\lambda f)\| \le \frac{1 - e^{-\lambda t}}{\lambda t}\|f\| + \frac{e^{\lambda t}(1 - e^{-\lambda t})}{\lambda t}\|f\|;$$
$$\le c\,\|f\|, \quad 0 < t \le 1.$$

As t goes to zero, $(e^{-\lambda t} - 1)/t \to -\lambda$. As τ goes to zero, $(T_\tau f)(x) \to f(x)$. Using these in (15.52) completes the proof of the first assertion.

Now take f in $\mathfrak{D}(S, \mathcal{G})$; by the bounded convergence theorem,

$$R_\lambda((\lambda - S)f) = \lim_{t\to 0} R_\lambda((\lambda - S_t)f).$$

Note that R_λ and S_t commute, so

$$R_\lambda((\lambda - S)f) = \lim_{t\to 0}(\lambda - S_t)(R_\lambda f).$$

By part (1) of this theorem, $R_\lambda f$ is in $\mathfrak{D}(S, \mathcal{G})$, hence

$$\lim_{t\to 0} (\lambda - S_t)(R_\lambda f) = (\lambda - S)(R_\lambda f) = f.$$

The purpose of this preparation is to prove

Theorem 15.53. *There is at most one set of standard transition probabilities corresponding to given S, $\mathfrak{D}(S, \mathcal{G})$.*

Proof. Suppose there are two different sets, $p_t^{(1)}$ and $p_t^{(2)}$ leading to resolvents $R_\lambda^{(1)}$ and $R_\lambda^{(2)}$. For f in $\mathcal{G}$, let

$$g = (R_\lambda^{(1)} - R_\lambda^{(2)})f.$$

Then, by 15.51(1),

$$(\lambda - S)g = 0.$$

But $g \in \mathfrak{D}(S, \mathcal{G})$, so use 15.51(2) to get

$$R_\lambda^{(1)}((\lambda - S)g) = g.$$

Therefore g is zero. Thus for all $f \in \mathcal{G}$, $R_\lambda^{(1)} f = R_\lambda^{(2)} f$. Since $\mathcal{G}$ includes all bounded continuous functions, for any such function, and for all $\lambda > 0$,

$$\int_0^\infty e^{-\lambda t}(T_t^{(1)} f)\, dt = \int_0^\infty e^{-\lambda t}(T_t^{(2)} f)\, dt.$$

By the uniqueness theorem for Laplace transforms (see [140], for example) $(T_t^{(1)} f)(x) = (T_t^{(2)} f)(x)$ almost everywhere (dt). But both these functions are continuous from the right, hence are identically equal. Since bounded continuous functions separate, $p_t^{(1)}(B \mid x) \equiv p_t^{(2)}(B \mid x)$.

The difficulty with this result is in the determination of $\mathfrak{D}(S, \mathcal{G})$. This is usually such a complicated procedure that the uniqueness theorem 15.53 above has really only theoretical value. Some examples follow in these problems.

Problems

8. For the transition probabilities constructed and referred to in 15.46, show that a necessary condition for $f(j)$ to be in $\mathfrak{D}(S, \mathcal{G})$ is

$$\sum_j f(j) Q(j) = 0.$$

9. Show that for any $\lambda', \lambda'' > 0$,

$$R_{\lambda''} f = R_{\lambda'} f + (\lambda' - \lambda'') R_{\lambda'}(R_{\lambda''} f).$$

Use this to show that the set $\mathfrak{R}_\lambda$ consisting of all functions $\{R_\lambda f\}$, f in $\mathcal{G}$, does not depend on λ. Use 15.51 to show that $\mathfrak{R}_\lambda = \mathfrak{D}(S, \mathcal{G})$.

10. For Brownian motion, show that the resolvent has a density $r_\lambda(y \mid x)$ with respect to dy given by

$$r_\lambda(y \mid x) = (1/\sqrt{2\lambda})e^{-\sqrt{2\lambda}|y-x|}.$$

11. Let $\mathcal{C}$ be the class of all bounded continuous functions on $R^{(1)}$. Use the identity in Problem 9, and the method of that problem to show that $\mathfrak{D}(S, \mathcal{C})$ for Brownian motion consists of all functions f in $\mathcal{C}$ such that $f''(x)$ is in $\mathcal{C}$.

12. For a pure jump process, show that if $\sup \lambda(x) < \infty$, then $\mathfrak{D}(S, \mathcal{G})$ consists of all bounded $\mathcal{B}_1(F)$-measurable functions.

10. ASYMPTOTIC STATIONARITY

Questions concerning the asymptotic stationarity of a Markov process $\{X(t)\}$ can be formulated in the same way as for discrete time chains. In particular,

Definition 15.54. *$\bar{\pi}(dx)$ on $\mathcal{B}_1(F)$ will be called a stationary initial distribution for the process if for every $B \in \mathcal{B}_1(F)$ and $t > 0$,*

$$\bar{\pi}(B) = \int p_t(B \mid x)\bar{\pi}(dx).$$

Now ask, when do the probabilities $p_t(B \mid x)$ converge as $t \to \infty$ to some stationary distribution $\bar{\pi}(B)$ for all $x \in F$? Interestingly enough, the situation here is less complicated than in discrete time because there is no periodic behavior. We illustrate this for $X(t)$ a pure jump process moving on the integers. Define the times between successive returns to state k by

$$t_1^* = \inf \{t; X(t) = k, \exists\, \tau < t \text{ such that } X(\tau) \neq k\},$$
$$t_2^* = \inf \{t; X(t + t_1^*) = k, \exists\, t_1^* < \tau < t + t_1^* \text{ such that } X(\tau) \neq k\},$$

and so forth. By the strong Markov property,

Proposition 15.55. *If $P_k(t_1^* < \infty) = 1$, then the $t_1^*, t_2^*, \ldots$ are independent and identically distributed.*

If $P_k(t_1^* < \infty) < 1$, then the state k is called *transient*.

To analyze the asymptotic behavior of the transition probabilities, use

$$\{X(t) = k\} = \bigcup_{n=0}^{\infty} \{t_1^* + \cdots + t_n^* \le t,\ t_1^* + \cdots + t_n^* + T_n(k) > t\},$$

where $T_n(k)$ is the duration of stay in state k after the nth return. It is independent of $t_1^*, \ldots, t_n^*$, and $P_k(T_n(k) > t) = e^{-\lambda(k)t}$. So

$$p_t(k \mid k) = e^{-\lambda(k)t} + \sum_{n=1}^{\infty} P_k(t_1^* + \cdots + t_n^* \le t,\ t_1^* + \cdots + t_n^* + T_n(k) > t).$$

Put

$$R(t) = \sum_{1}^{\infty} P_k(\mathsf{t}_1^* + \cdots + \mathsf{t}_n^* \le t);$$

then

$$p_t(k \mid k) = e^{-\lambda(k)t} + \int_0^t P_k(\mathsf{T}_0(k) > \xi)R(t - d\xi).$$

Argue that t_1^* is nonlattice because

$$\mathsf{t}_1^* = \mathsf{T}_0(k) + \boldsymbol{\tau}_1^*,$$

where $\boldsymbol{\tau}_1^*$ is the time from the first exit to the first return. By the strong Markov property, $\mathsf{T}_0(k)$, $\boldsymbol{\tau}_1^*$ are independent. Finally, note that $\mathsf{T}_0(k)$ has a distribution absolutely continuous with respect to Lebesgue measure; hence so does t_1^*.

Now apply the renewal theorem 10.8. As $t \to \infty$,

$$R(t - d\xi) \xrightarrow{w} \frac{d\xi}{E\mathsf{t}_1^*}.$$

Conclude that

$$\lim_t p_t(k \mid k) = \frac{1}{E\mathsf{t}_1^*}\int_0^\infty e^{-\lambda(k)t}\,dt.$$

Hence

Proposition 15.56. *Let* T *be the first exit time from state* k, t_1^* *the time of first return. If* $E\mathsf{t}_1^* < \infty$, *then*

$$p_t(k \mid k) \to \frac{E\mathsf{T}}{E\mathsf{t}_1^*} = \pi(k). \tag{15.57}$$

The following problems concern the rest of the problem of asymptotic convergence.

Problems

13. Let all the states communicate under the space structure given by $p(j; k)$, and let the expected time of first return be finite for every state. Show that

1) For any k, j,

$$\lim_t p_t(j \mid k) = \pi(j),$$

where $\pi(j)$ is defined by (15.57).

2) If $\tilde{\pi}(k)$ is the stationary initial distribution under $p(j; k)$, that is,

$$\tilde{\pi}(j) = \sum_k p(j; k)\tilde{\pi}(k),$$

then

$$\pi(j) = \alpha\tilde{\pi}(j)/\lambda(j),$$

where α is normalizing constant.

14. Show that if k is a transient state, then $p_t(k \mid k)$ goes to zero exponentially fast.

15. Show that π is a stationary initial distribution for a process having standard stationary transition probabilities if and only if

$$\int g(y)\pi(dy) = 0$$

for all g such that there exists an f in $\mathfrak{D}(S, \mathfrak{G})$ with $g = Sf$.

NOTES

K. L. Chung's book [16] is an excellent reference for the general structure of time-continuous Markov chains with a countable state space. Even with this simple a state space, the diversity of sample path behavior of processes with standard stationary transition probabilities is dazzling. For more general state spaces and for jump processes in particular, see Doob's book [39, Chap. 6]. For a thoroughly modern point of view, including discussion of the strong Markov property and the properties of S, $\mathfrak{D}(S, \mathfrak{G})$ and the resolvent, see Dynkin [44, especially Vol. I].

The fundamental work in this field started with Kolmogorov [93, 1931]. The problems concerning jump processes were treated analytically by Pospišil [117, 1935–1936] and Feller in 1936, but see Feller [52] for a fuller treatment. Doeblin [26, 1939] had an approach closer in spirit to ours. Doob [33 in 1942] carried on a more extended study of the sample path properties. The usefulness of the resolvent and the systematic study of the domain of S were introduced by Feller [57, 1952]. His idea was that the operators $\{T_t\}$, ≥ 0, formed a semi-group, hence methods for analyzing semi-groups of operators could be applied to get useful results.

There is an enormous literature on applications of pure jump Markov processes, especially for those with a countable state space. For a look at some of those, check the books by Bharucha-Reid [3], T. E. Harris [67], N. T. J. Bailey [2], and T. L. Saaty [120]. An extensive reference to both theoretical and applied sources is the Bharucha-Reid book.

CHAPTER 16

DIFFUSIONS

1. THE ORNSTEIN-UHLENBECK PROCESS

In Chapter 12, the Brownian motion process was constructed as a model for a microscopic particle in liquid suspension. We found the outstanding nonreality of the model was the assumption that increments in displacement were independent—ignoring the effects of the velocity of the particle at the beginning of the incremental time period. We can do better in the following way:

Let $V(t)$ be the *velocity* of a particle of mass m suspended in liquid. Let $\Delta V = V(t + \Delta t) - V(t)$, so that $m\,\Delta V$ is the change in momentum of the particle during time Δt. The basic equation is

$$m\,\Delta V = -\beta V\,\Delta t + \Delta M. \tag{16.1}$$

Here $-\beta V$ is the viscous resistance force, so $-\beta V\,\Delta t$ is the loss in momentum due to viscous forces during Δt. ΔM is the momentum transfer due to molecular bombardment of the particle during time Δt.

Let $M(t)$ be the momentum transfer up to time t. Normalize arbitrarily to $M(0) = 0$. Assume that

i) $M(t + \Delta t) - M(t)$ is independent of $\mathcal{F}(M(\tau), \tau \le t)$,
ii) the distribution of ΔM depends only on Δt,
iii) $M(t)$ is continuous in t.

The third assumption may be questionable if one uses a hard billiard-ball model of molecules. But even in this case we reason that the jumps of $M(t)$ would have to be quite small unless we allowed the molecules to have enormous velocities. At any rate (iii) is not unreasonable as an approximation.

But (i), (ii), (iii) together characterize $M(t)$ as a Brownian motion. The presence of drift in $M(t)$ would put a $\mu\,\Delta t$ term on the right-hand side of (16.1). Such a term corresponds to a constant force field, and would be useful, for example, in accounting for a gravity field. However, we will assume no constant force field exists, and set $EM(t) = 0$. Put $EM^2(t) = \sigma^2 t$; hence $M(t) = \sigma X(t)$, where $X(t)$ is normalized Brownian motion. Equation (16.1) becomes

$$m\,\Delta V = -\beta V\,\Delta t + \sigma\,\Delta X. \tag{16.2}$$

Doing what comes naturally, we divide by Δt, let $\Delta t \to 0$ and produce the Langevin equation

$$\text{(16.3)} \qquad m\frac{d\mathsf{V}}{dt} = -\beta\mathsf{V} + \sigma\frac{d\mathsf{X}}{dt}.$$

The difficulty here is amusing: We know from Chapter 12 that $d\mathsf{X}/dt$ exists nowhere. So (16.3) makes no sense in any orthodox way. But look at this: Write it as

$$\frac{d}{dt}(e^{\alpha t}\mathsf{V}(t)) = \gamma e^{\alpha t}\frac{d\mathsf{X}(t)}{dt},$$

where $\alpha = \beta/m$, $\gamma = \sigma/m$. Assume $\mathsf{V}(0) = 0$ and integrate from 0 to t to get

$$e^{\alpha t}\mathsf{V}(t) = \gamma\int_0^t e^{\alpha\tau}\,d\mathsf{X}(\tau).$$

Do an integration by parts on the integral,

$$e^{\alpha t}\mathsf{V}(t) = \gamma e^{\alpha t}\mathsf{X}(t) - \gamma\alpha\int_0^t \mathsf{X}(\tau)e^{\alpha\tau}\,d\tau.$$

Now the integral appearing is for each ω just the integral of a continuous function and makes sense. Thus the expression for $\mathsf{V}(t)$ given by

$$\mathsf{V}(t) = \gamma\int_0^t e^{\alpha(t-\tau)}\,d\mathsf{X}(\tau)$$

can be well defined by this procedure, and results in a process with continuous sample paths.

To get a more appealing derivation, go back to (16.2). Write it as

$$\Delta(e^{\alpha t}\mathsf{V}) = \gamma e^{\alpha t}\Delta\mathsf{X} + \delta(\Delta t),$$

where $\delta(\Delta t) = o(\Delta t)$ because by (16.2), $\mathsf{V}(t)$ is continuous and bounded in every finite interval. By summing up, write this as

$$e^{\alpha t}\mathsf{V}(t) \simeq \gamma\sum_{k=0}^{n-1} e^{\alpha t_k^{(n)}}(\mathsf{X}(t_{k+1}^{(n)}) - \mathsf{X}(t_k^{(n)})),$$

where $0 = t_0^{(n)} < \cdots < t_n^{(n)} = t$ is a partition $\mathcal{T}_n$ of $[0, t]$. If the limit of the right-hand side exists in some decent way as $\|\mathcal{T}_n\| \to 0$, then it would be very reasonable to define $e^{\alpha t}\mathsf{V}(t)$ as this limit. Replace the integration by parts in the integral by a similar device for the sum,

$$\sum_{k=0}^{n-1} e^{\alpha t_k}(\mathsf{X}(t_{k+1}) - \mathsf{X}(t_k)) = e^{\alpha t}\mathsf{X}(t) - \sum_{k=0}^{n-1}(e^{\alpha t_{k+1}} - e^{\alpha t_k})\mathsf{X}(t_{k+1}).$$

The second sums are the Riemann-Stieltjes sums for the integral $\int_0^t \mathsf{X}(\tau)\,d(e^{\alpha\tau})$.

For every sample path, they converge to the integral. Therefore:

Definition 16.4. *The Ornstein-Uhlenbeck process* $\mathsf{V}(t)$ *normalized to be zero at* $t = 0$, *is defined as*

$$\mathsf{V}(t) = \gamma \int_0^t e^{-\alpha(t-\tau)}\, d\mathsf{X}(\tau),$$

where the integral is the limit of the approximating sums for every path.

Proposition 16.5. $\mathsf{V}(t)$ *is a Gaussian process with covariance*

$$E\mathsf{V}(s)\mathsf{V}(t) = \rho(e^{-\alpha|s-t|} - e^{-\alpha(s+t)}),$$

where $\rho = \gamma^2/2\alpha$.

Proof. That $\mathsf{V}(t)$ is Gaussian follows from its being the limit of sums $\sum_k \varphi(t_k)\, \Delta_k\mathsf{X}$, where the $\Delta_k\mathsf{X} = \mathsf{X}(t_{k+1}) - \mathsf{X}(t_k)$ are independent, normally-distributed random variables. To get $\Gamma(s, t)$, take $s > t$, put

$$0 = t_0 < \cdots < t_n = t < t_{n+1} < \cdots < t_m = s.$$

Write

$$\mathsf{V}(s) \simeq \gamma \sum_0^m e^{-\alpha(s-t_k)}\, \Delta_k\mathsf{X}, \qquad \mathsf{V}(t) \simeq \gamma \sum_0^n e^{-\alpha(t-t_k)}\, \Delta_k\mathsf{X}.$$

Use $E(\Delta_k\mathsf{X})(\Delta_j\mathsf{X}) = 0$ if $k \neq j$, $E(\Delta_k\mathsf{X})^2 = t_{k+1} - t_k$, to get

$$E\mathsf{V}(s)\mathsf{V}(t) \simeq \gamma^2 \sum_0^n e^{-\alpha(s+t)+2\alpha t_k}(t_{k+1} - t_k).$$

Going to the limit, we get

$$E\mathsf{V}(s)\mathsf{V}(t) = \gamma^2 e^{-\alpha(s+t)} \int_0^t e^{2\alpha\tau}\, d\tau.$$

As $t \to \infty$, $E\mathsf{V}(t)^2 \to \rho$, so $\mathsf{V}(t) \xrightarrow{\mathfrak{D}} \mathsf{V}(\infty)$, where $\mathsf{V}(\infty)$ is $\mathcal{N}(0, \rho)$. What if we start the process with this limiting distribution? This would mean that the integration of the Langevin equation would result in

$$e^{\alpha t}\mathsf{V}_1(t) - \mathsf{V}_1(0) = \gamma \int_0^t e^{-\alpha(t-\tau)}\, d\mathsf{X}(\tau).$$

Define the stationary Ornstein-Uhlenbeck process by

$$\mathsf{Y}(t) = \mathsf{V}(t) + e^{-\alpha t}\mathsf{V}_1(0), \tag{16.6}$$

where $\mathsf{V}_1(0)$ is $\mathcal{N}(0, \rho)$ and independent of $\mathcal{F}(\mathsf{V}(t), t \geq 0)$.

Proposition 16.7. $\mathsf{Y}(t)$ *is a stationary Gaussian process with covariance* $\Gamma(s, t) = \rho e^{-\alpha|s-t|}$.

Proof. Direct computation.

Remark. Stationarity has not been defined for continuous parameter processes, but the obvious definition is that all finite-dimensional distributions remain invariant under a time shift. For *Gaussian processes* with zero means, stationarity is equivalent to $\Gamma(s, t) = \varphi(|s - t|)$.

The additional important properties of the Ornstein-Uhlenbeck process are:

Proposition 16.8. $\mathsf{Y}(t)$ *is a Markov process with stationary transition probabilities having all sample paths continuous.*

Proof. Most of 16.8 follows from the fact that for $\tau \geq 0$, $\mathsf{Y}(t + \tau) - e^{-\alpha t}\mathsf{Y}(\tau)$ is independent of $\mathcal{F}(\mathsf{Y}(s), s \leq \tau)$. To prove this, it is necessary only to check the covariance

$$E[(\mathsf{Y}(t + \tau) - e^{-\alpha t}\mathsf{Y}(\tau))\mathsf{Y}(s)] = \Gamma(t + \tau, s) - e^{-\alpha t}\Gamma(\tau, s)$$
$$= 0, \qquad s \leq \tau.$$

Now,

$$P(\mathsf{Y}(t + \tau) \in A \mid \mathsf{Y}(\tau) = x, \mathsf{Y}(s), s \leq \tau)$$
$$= P(\mathsf{Y}(t + \tau) - e^{-\alpha t}\mathsf{Y}(\tau) \in A - e^{-\alpha t}x \mid \mathsf{Y}(\tau) = x, \mathsf{Y}(s), s \leq \tau)$$
$$= P(\mathsf{Y}(t + \tau) - e^{-\alpha t}\mathsf{Y}(\tau) \in A - e^{-\alpha t}x).$$

The random variable $\mathsf{Y}(t + \tau) - e^{-\alpha t}\mathsf{Y}(\tau)$ is normal with mean zero, and

$$E(\mathsf{Y}(t + \tau) - e^{-\alpha t}\mathsf{Y}(\tau))^2$$
$$= E[(\mathsf{Y}(t + \tau) - e^{-\alpha t}\mathsf{Y}(\tau))(\mathsf{Y}(t + \tau))] = \rho(1 - e^{-2\alpha t}).$$

Thus $p_t(\cdot \mid x)$ has the distribution of

$$\mathcal{N}(e^{-\alpha t}x, \rho(1 - e^{-2\alpha t})).$$

The continuity of paths follows from the definition of $\mathsf{V}(t)$ in terms of an integral of $\mathsf{X}(t)$.

Problems

1. Show that if a process is Gaussian, stationary, Markov, and continuous in probability, then it is of the form $\mathsf{Y}(t) + c$, where $\mathsf{Y}(t)$ is an Ornstein-Uhlenbeck process.

2. Let $\mathbf{Z}$ be a vector-valued random variable taking values in $R^{(m)}$, $m \geq 2$. Suppose that the components of $\mathbf{Z}$, $(\mathsf{Z}_1, \mathsf{Z}_2, \ldots, \mathsf{Z}_m)$, are independent and identically distributed with a symmetric distribution. Suppose also that the components have the same property under all other orthogonal coordinate systems gotten from the original one by rotation. Show that $\mathsf{Z}_1, \ldots, \mathsf{Z}_m$ are $\mathcal{N}(0, \sigma^2)$.

Remark. The notable result of this problem is that any model for Brownian motion in three dimensions leads to variables normally distributed providing the components of displacement of velocity along the different axes are independent and identically distributed (symmetry is not essential, see Kac [79]) irrespective of which orthogonal coordinate system is selected. However, it does not follow from this that the process must be Gaussian.

2. PROCESSES THAT ARE LOCALLY BROWNIAN

In the spirit of the Langevin approach of the last section, if $\mathsf{Y}(t)$ is Brownian motion with drift μ, variance σ^2, then write

$$\Delta\mathsf{Y} = \mu\,\Delta t + \sigma\,\Delta\mathsf{X}.$$

The same integration procedure as before would then result in a process $\mathsf{Y}(t)$ which would be, in fact, Brownian motion with parameters μ, σ. To try to get more general Markov processes with continuous paths, write

(16.9)
$$\Delta\mathsf{Y} = \mu(\mathsf{Y})\,\Delta t + \sigma(\mathsf{Y})\,\Delta\mathsf{X}.$$

As before $\mathsf{X}(t)$ is normalized Brownian motion. $\mathsf{Y}(t)$ should turn out to be Markov with continuous paths and stationary transition probabilities. Argue this way: $\mu(\mathsf{Y})\,\Delta t$ is a term approximately linear in t, but except for this term $\Delta\mathsf{Y}$ is of the order of $\Delta\mathsf{X}$, hence $\mathsf{Y}(t)$ should be continuous. Further, assume $\mathsf{Y}(t)$ is measurable $\mathcal{F}\big(\mathsf{X}(\tau), \tau \le t\big)$. Then the distribution of $\Delta\mathsf{Y}$ depends only on $\mathsf{Y}(t)$, $\big($through $\mu(\mathsf{Y})$ and $\sigma(\mathsf{Y})\big)$, and on $\Delta\mathsf{X}$, which is independent of $\mathcal{F}\big(\mathsf{Y}(\tau), \tau \le t\big)$ with distribution depending only on Δt.

Roughly, a process satisfying (16.9) is *locally Brownian.* Given that $\mathsf{Y}(t) = y$, it behaves for the next short time interval as though it were a Brownian motion with drift $\mu(y)$, variance $\sigma(y)$. Therefore, we can think of constructing this kind of process by patching together various Brownian motions. Note, assuming $\Delta\mathsf{X}$ is independent of $\mathsf{Y}(t)$,

$$E\big(\Delta\mathsf{Y} \mid \mathsf{Y}(t) = y\big) = \mu(y)\,\Delta t,$$
$$E\big((\Delta\mathsf{Y})^2 \mid \mathsf{Y}(t) = y\big) = \sigma^2(y)E(\Delta\mathsf{X})^2 + o(\Delta t) = \sigma^2(y)\,\Delta t + o(\Delta t).$$

Of course, the continuity condition is also satisfied,

$$P\big(|\Delta\mathsf{Y}| > \epsilon \mid \mathsf{Y}(t) = y\big) = o(\Delta t).$$

Define the truncated change in Y by

$$\Delta_\epsilon\mathsf{Y} = \begin{cases} \Delta\mathsf{Y} & \text{if } |\Delta\mathsf{Y}| \le \epsilon, \\ 0, & \text{otherwise.} \end{cases}$$

As a first approximation to the subject matter of this chapter, I will say that we are going to look at Markov processes $\mathsf{Y}(t)$ taking values in some interval

F, with stationary transition probabilities p_t satisfying for every $\epsilon > 0$, and y in the interior of F,

$$P(|\Delta \mathsf{Y}| > \epsilon \mid \mathsf{Y}(t) = y) = o(\Delta t),$$

(16.10)
$$E(\Delta_\epsilon \mathsf{Y} \mid \mathsf{Y}(t) = y) = \mu(y)\,\Delta t + o(\Delta t),$$

$$E((\Delta_\epsilon \mathsf{Y})^2 \mid \mathsf{Y}(t) = y) = \sigma^2(y)\,\Delta t + o(\Delta t),$$

and having continuous sample paths. Conditions (16.10) are the infinitesimal conditions for the process. A Taylor expansion gives

Proposition 16.11. *Let $f(x)$ be bounded with a continuous second derivative. If $p_t(dy \mid x)$ satisfies* (16.10), *then $(Sf)(x)$ exists for every point x in the interior of F and equals*

$$\tfrac{1}{2}\sigma^2(x)\frac{d^2f(x)}{dx^2} + \mu(x)\frac{df(x)}{dx}.$$

Thus,

Proposition 16.12. *If the transition probabilities satisfy* (16.10), *and have densities $p_t(y \mid x)$ with a continuous second derivative for $x \in \operatorname{int}(F)$, then*

(16.13)
$$\frac{\partial}{\partial t} p_t(y \mid x) = \tfrac{1}{2}\sigma^2(x)\frac{\partial^2}{\partial x^2} p_t(y \mid x) + \mu(x)\frac{\partial}{\partial x} p_t(y \mid x),\ x \in \operatorname{int}(F).$$

Proof. This is the backwards equation.

Problem 3. Show by direct computation that the transition probabilities for the Ornstein-Uhlenbeck process satisfy

$$\frac{\partial}{\partial t} p_t(y \mid x) = \tfrac{1}{2}\sigma^2 \frac{\partial^2}{\partial x^2} p_t(y \mid x) - \alpha x \frac{\partial}{\partial x} p_t(y \mid x).$$

3. BROWNIAN MOTION WITH BOUNDARIES

For $\mathsf{X}(t)$ a locally Brownian process as in the last section, the infinitesimal operator S is defined for all interior points of F by 16.11. Of course, this completely defines S if F has only interior points. But if F has a closed boundary point, the definition of S at this point is not clear. This problem is connected with the question of what boundary conditions are needed to uniquely solve the backwards equation (16.13). To illuminate this problem a bit, we consider two examples of processes where F has a finite closed boundary point.

Definition 16.14. *Use $\mathsf{X}_0(t)$ to denote normalized Brownian motion on $R^{(1)}$, $p_t^{(0)}(dy \mid x)$ are its transition probabilities.*

The examples will be concerned with Brownian motion restricted to the interval $F = [0, \infty)$.

Example 1. Brownian motion with an absorbing boundary. Take $F = [0, \infty)$. The Brownian motion $\mathsf{X}(t)$ starting from $x > 0$ with absorption at $\{0\}$ is defined by

$$\mathsf{X}(t) = \mathsf{X}_0(\min(t, \mathsf{t}_0^*)),$$

where $\mathsf{X}_0(t)$ is started from x.

It is not difficult to check that $\mathsf{X}(t)$ is Markov with stationary transition probabilities. To compute these rigorously is tricky. Let $A \subset (0, \infty)$, $A \in \mathcal{B}_1$, and consider $P_x(\mathsf{X}_0(t) \in A, \mathsf{t}_0^* < t)$. The set $\{\mathsf{X}_0(t) \in A, \mathsf{t}_0^* < t\}$ consists of all sample paths that pass out of $(0, \infty)$ at least once and then come back in and get to A by time t. Let $A^\perp$ be the reflection of A around the point $x = 0$. Argue that after hitting $\{0\}$ at time $\tau < t$ it is just as probable (by symmetry) to get to $A^\perp$ by time t as it is to get to A, implying

(16.15) $$P_x(\mathsf{X}_0(t) \in A, \mathsf{t}_0^* < t) = P_x(\mathsf{X}_0(t) \in A^\perp, \mathsf{t}_0^* < t).$$

This can be proven rigorously by approximating t_0^* by stopping times that take only a countable number of values. We assume its validity. Proceed by noting that $\{\mathsf{X}_0(t) \in A^\perp\} \subset \{\mathsf{t}_0^* < t\}$ so that

$$\begin{aligned} P_x(\mathsf{X}_0(t) \in A^\perp, \mathsf{t}_0^* < t) &= P_x(\mathsf{X}_0(t) \in A^\perp) \\ &= p_t^{(0)}(A^\perp \mid x). \end{aligned}$$

Now,

$$\begin{aligned} P_x(\mathsf{X}(t) \in A) &= P_x(\mathsf{X}_0(t) \in A, \mathsf{t}_0^* \geq t) \\ &= P_x(\mathsf{X}_0(t) \in A) - P_x(\mathsf{X}_0(t) \in A, \mathsf{t}_0^* < t) \\ &= p_t^{(0)}(A \mid x) - p_t^{(0)}(A^\perp \mid x). \end{aligned}$$

The density for $p_t^{(0)}(A^\perp \mid x)$ is $p_t^{(0)}(-y \mid x)$. Thus

(16.16) $$p_t(y \mid x) = p_t^{(0)}(y \mid x) - p_t^{(0)}(-y \mid x).$$

Example 2. Brownian motion with a reflecting boundary. Define the Brownian motion $\mathsf{X}(t)$ on $F = [0, \infty)$ with a reflecting boundary at $\{0\}$ to be

$$\mathsf{X}(t) = |\mathsf{X}_0(t)|,$$

where we start the motion from $x > 0$. What this definition does is to take all parts of the $\mathsf{X}_0(t)$ path below $x = 0$ and reflect them in the $x = 0$ axis getting the $\mathsf{X}(t)$ path.

Proposition 16.17. *$\mathsf{X}(t)$ is Markov with stationary transition probability density*

$$p_t(y \mid x) = p_t^{(0)}(y \mid x) + p_t^{(0)}(-y \mid x).$$

Proof. Take $A \in \mathcal{B}_1([0, \infty))$, $x \geq 0$. Consider the probabilities

$$P(\mathsf{X}(t + \tau) \in A \mid \mathsf{X}_0(t) = x, \mathsf{X}_0(s), s \leq t),$$
$$P(\mathsf{X}(t + \tau) \in A \mid \mathsf{X}_0(t) = -x, \mathsf{X}_0(s), s \leq t).$$

Because $\mathsf{X}_0(t)$ is Markov, these reduce to

$$P(|\mathsf{X}_0(t + \tau)| \in A \mid \mathsf{X}_0(t) = x),$$
$$P(|\mathsf{X}_0(t + \tau)| \in A \mid \mathsf{X}_0(t) = -x).$$

These expressions are equal. Hence

$$\begin{aligned} P(\mathsf{X}(t + \tau) \in A \mid \mathsf{X}(t) = x, \mathsf{X}(s), s \leq t) &\\ = E(P(\mathsf{X}(t + \tau) \in A \mid |\mathsf{X}_0(t)| = x, \mathsf{X}_0(s), s \leq t) \mid \mathsf{X}(s), s \leq t) &\\ = P(|\mathsf{X}_0(t + \tau)| \in A \mid \mathsf{X}_0(t) = x) &\\ = p_t^{(0)}(A \mid x) + p_t^{(0)}(A^{\perp} \mid x). & \end{aligned}$$

In both examples, $\mathsf{X}(t)$ equals the Brownian motion $\mathsf{X}_0(t)$ until the particle reaches zero. Therefore, in both cases, for $x > 0$ and f bounded and continuous on $[0, \infty)$,

$$|E_x f(\mathsf{X}(t)) - E_x f(\mathsf{X}_0(t))| = o(t).$$

As expected, in the interior of F, then

$$(Sf)(x) = \frac{1}{2} \frac{d^2 f(x)}{dx^2}$$

for functions with continuous second derivatives. Assume that the limits $f'(0+), f''(0+)$, as $x \downarrow 0$ of the first and second derivatives, exist. In the case of a reflecting boundary at zero, direct computation gives

$$E_0 f(\mathsf{X}(t)) - f(0) = \sqrt{\frac{t}{2\pi}}\, f'(0+) + \frac{t}{2} f''(0+) + o(t).$$

Thus, $(Sf)(0)$ does not exist unless $f'(0+) = 0$. If $f'(0+) = 0$, then $(Sf)(0) = \frac{1}{2} f''(0+)$, so not only is $(Sf)(x)$ defined at $x = 0$, but it is also continuous there.

If $\{0\}$ is absorbing, then for any f, $(Sf)(0) = 0$. If we want $(Sf)(x)$ to be continuous at zero, we must add the restriction $f''(0+) = 0$.

Does the backwards equation (16.13) have the transition probabilities of the process as its unique solution? Even if we add the restriction that we will consider only solutions which are densities of transition probabilities of Markov processes, the examples above show that the solution is not unique. However, note that in the case of absorption

$$\lim_{x \downarrow 0} p_t(y \mid x) = 0$$

for all $t, y > 0$. Intuitively this makes sense, because the probability starting from x of being absorbed at zero before hitting the point y goes to one as $x \to 0$. For reflection, use the symmetry to verify that

$$\lim_{x \downarrow 0} \frac{\partial}{\partial x} p_t(y \mid x) = 0.$$

If either of the above boundary conditions are imposed on the backwards equation, it is possible to show that there is a unique solution which is a set of transition probability densities.

Reflection or absorption is not the only type of behavior possible at boundary points. Odd things can occur, and it was the occurrence of some of these eccentricities which first prompted Feller's investigation [56] and eventually led to a complete classification of boundary behavior.

Problems

4. Show that the process $\mathsf{X}(t)$ defined on $[0, 1]$ by folding over Brownian motion,

$$\mathsf{X}(t) = |\mathsf{X}_0(t) - 2n|$$

if $|\mathsf{X}_0(t) - 2n| \leq 1$, is a Markov process with stationary transition probabilities such that

$$\frac{\partial}{\partial x} p_t(y \mid x)\bigg|_{0+} = \frac{\partial}{\partial x} p_t(y \mid x)\bigg|_{1-} = 0.$$

Evaluate $(Sf)(x)$ for $x \in (0, 1)$. For what functions does $(Sf)(x)$ exist at the endpoints? [This process is called Brownian motion with two reflecting boundaries.]

5. For Brownian motion on $[0, \infty)$, either absorbing or reflecting, evaluate the density $r_\lambda(y \mid x)$ of the resolvent $R_\lambda(dy \mid x)$. For $\mathcal{C}$ the class of all bounded continuous functions on $[0, \infty)$, show that

a) For absorbing Brownian motion,

$$\mathfrak{D}(S, \mathcal{C}) = \{f \in \mathcal{C};\ f'' \in \mathcal{C},\ f''(0+) = 0\}.$$

b) For reflecting Brownian motion,

$$\mathfrak{D}(S, \mathcal{C}) = \{f \in \mathcal{C};\ f'' \in \mathcal{C},\ f'(0+) = 0\}.$$

[See Problems 10 and 11, Chapter 15. Note that in (a), $R_\lambda(dy \mid 0)$ assigns all of its mass to $\{0\}$.]

4. FELLER PROCESSES

The previous definitions and examples raise a host of interesting and difficult questions. For example:

1) Given the form of S, that is, given $\sigma^2(x)$ and $\mu(x)$ defined on int(F), and certain boundary conditions. Does there exist a unique Markov process with continuous paths having S as its infinitesimal operator and exhibiting the desired boundary behavior?

2) If the answer to (1) is Yes, do the transition probabilities have a density $p_t(y \mid x)$? Are these densities smooth enough in x so that they are a solution of the backwards equations? Do the backwards equations have a unique solution?

The approach that will be followed is similar to that of the previous chapter. Investigate first the properties of the class of processes we want to construct—try to simplify their structure as far as possible. Then work backwards from the given infinitesimal conditions to a process of the desired type.

To begin, assume the following.

Assumption 16.18(a). *F is an interval closed, open, or half of each, finite or infinite. $\mathsf{X}(t)$ is a Markov process with stationary transition probabilities such that starting from any point $x \in F$, all sample paths are continuous.*

The next step would be to prove that 16.18(a) implies that the strong Markov property holds. *This is not true.* Consider the following counter example: Let $F = R^{(1)}$. Starting from any $x \neq 0$, $\mathsf{X}(t)$ is Brownian motion starting from that point. Starting from $x = 0$, $\mathsf{X}(t) \equiv 0$. Then, for $x \neq 0$,

$$P_x(\mathsf{X}(t + \mathsf{t}_0^*) \in B) = P_0(\mathsf{X}_0(t) \in B) .$$

But if $\mathsf{X}(t)$ were strong Markov, since t_0^* is a Markov time,

$$P_x(\mathsf{X}(t + \mathsf{t}_0^*) \in B) = P_{\mathsf{X}(\mathsf{t}_0^*)}(\mathsf{X}(t) \in B) = \chi_B(0).$$

The pathology in this example is that starting from the point $x = 0$ gives distributions drastically different from those obtained by starting from any point $x \neq 0$. When you start going through the proof of the strong Markov property, you find that it is exactly this large change in the distribution of the process when the initial conditions are changed only slightly that needs to be avoided. This recalls the concept of stability introduced in Chapter 8.

Definition 16.19. *Call the transition probabilities stable if for any sequence of initial distributions $\pi_n(\cdot) \xrightarrow{\mathfrak{D}} \pi(\cdot)$, the corresponding probabilities satisfy*

$$P_{\pi_n}(\mathsf{X}(t) \in \cdot) \xrightarrow{\mathfrak{D}} P_\pi(\mathsf{X}(t) \in \cdot)$$

for any $t > 0$. Equivalently, for all $\varphi(x)$ continuous and bounded on F, $E_x\varphi(\mathsf{X}(t))$ is continuous on F.

Assume, in addition to 16.18(a),

Assumption 16.18(b). *The transition probabilities of* $\mathsf{X}(t)$ *are stable.*

Definition 16.20. *A process satisfying* 16.18(a),(b) *will be called a Feller process.*

Now we can carry on.

Theorem 16.21. *A Feller process has the strong Markov property.*

Proof. Let $\varphi(x)$ be bounded and continuous on F, then $\tilde{\varphi}(x) = E_x\varphi(\mathsf{X}(t))$ is likewise. Let t^* be a Markov time, t_n^* a sequence of Markov times such that $\mathsf{t}_n^* \downarrow \mathsf{t}^*$ and t_n^* takes on only a countable number of values. From (15.20), for $C \in \mathcal{F}(\mathsf{X}(t), t \leq \mathsf{t}^*)$,

$$\int_C \varphi(\mathsf{X}(t + \mathsf{t}_n^*))\, dP_x = \int_C \tilde{\varphi}(\mathsf{X}(\mathsf{t}_n^*))\, dP_x.$$

The path continuity and the continuity of φ, $\tilde{\varphi}$ give

$$\varphi(\mathsf{X}(t + \mathsf{t}_n^*)) \xrightarrow{\text{a.s.}} \varphi(\mathsf{X}(t + \mathsf{t}^*)),$$
$$\tilde{\varphi}(\mathsf{X}(t_n^*)) \xrightarrow{\text{a.s.}} \tilde{\varphi}(\mathsf{X}(\mathsf{t}^*)).$$

Thus, for φ continuous,

$$E_x\big(\varphi(\mathsf{X}(t + \mathsf{t}^*)) \mid \mathsf{X}(\tau), \tau \leq \mathsf{t}^*\big) = E_{\mathsf{X}(\mathsf{t}^*)}\big(\varphi(\mathsf{X}(t))\big).$$

The continuous functions separate, thus, for any $B \in \mathcal{B}_1(F)$,

$$P_x\big(\mathsf{X}(t + \mathsf{t}^*) \in B \mid \mathsf{X}(\tau), \tau \leq \mathsf{t}^*\big) = P_{\mathsf{X}(\mathsf{t}^*)}\big(\mathsf{X}(t) \in B\big).$$

To extend this, let $\varphi(x_1, \ldots, x_k)$ on $F^{(k)}$ equal a product $\varphi_1(x_1) \cdots \varphi_k(x_k)$, where $\varphi_1, \ldots, \varphi_k$ are bounded and continuous on F. It is easy to check that

$$E_x\varphi\big(\mathsf{X}(t_k), \ldots, \mathsf{X}(t_1)\big) = \tilde{\varphi}(x)$$

is continuous in x. By the same methods as in (15.20), get

$$\int_C \varphi(\mathsf{X}(t_k + \mathsf{t}_n^*), \ldots, \mathsf{X}(t_1 + \mathsf{t}_n^*))\, dP_x = \int_C \tilde{\varphi}(\mathsf{X}(\mathsf{t}_n^*))\, dP_x,$$

conclude that

$$\int_C \varphi(\mathsf{X}(t_k + \mathsf{t}^*), \ldots, \mathsf{X}(t_1 + \mathsf{t}^*))\, dP_x = \int_C \tilde{\varphi}(\mathsf{X}(\mathsf{t}^*))\, dP_x,$$

and now use the fact that products of bounded continuous functions separate probabilities on $\mathcal{B}_k(F)$.

Of course, this raises the question of how much stronger a restriction stability is than the strong Markov property. The answer is—Not much!

To go the other way, it is necessary that the state space have something like an indecomposability property—that every point of F can be reached from every interior point.

Definition 16.22. *The process is called regular if for every $x \in \text{int}(F)$ and $y \in F$,*

$$P_x(t_y^* < \infty) > 0.$$

Theorem 16.23. *If $X(t)$ is regular and strong Markov, then its transition probabilities are stable.*

Proof. Let $x < y < z$, $x \in \text{int}(F)$. Define $\bar{t}_y = \min(t_y^*, s)$, $\bar{t}_z = \min(t_z^*, s)$, for $s > 0$ such that $P_x(t_z^* \leq s) > 0$. These are Markov times. Take $\varphi(x)$ bounded and continuous, $\tilde{\varphi}(x) = E_x\varphi(X(t))$. By the strong Markov property,

$$\begin{aligned} E_x\varphi(X(t + \bar{t}_y)) &= E_x[E_x(\varphi(X(t + \bar{t}_y)) \mid X(\tau), \tau \leq \bar{t}_y)] \\ &= E_x\tilde{\varphi}(X(\bar{t}_y)), \\ E_x\varphi(X(t + \bar{t}_z)) &= E_x\tilde{\varphi}(X(\bar{t}_z)). \end{aligned}$$

Suppose that on the set $\{t_z^* < \infty\}$, $t_y^* \uparrow t_z^*$ a.s. P_x as $y \uparrow z$, implying $\bar{t}_y \uparrow \bar{t}_z$ a.s. So $E_x\varphi(X(t + \bar{t}_y)) \to E_x\varphi(X(t + \bar{t}_z))$. The right-hand sides of the above equations are

$$\tilde{\varphi}(y)P_x(t_y^* \leq s) + \int_{\{t_y^* > s\}} \tilde{\varphi}(X(s))\, dP_x$$

and

$$\tilde{\varphi}(z)P_x(t_z^* \leq s) + \int_{\{t_z^* > s\}} \tilde{\varphi}(X(s))\, dP_x.$$

For s a continuity point of $P_x(t_z^* \leq s)$, $P_x(t_y^* \leq s) \to P_x(t_z^* \leq s)$, and the sets $\{t_y^* > s\} \uparrow \{t_z^* > s\}$. The conclusion is that $\tilde{\varphi}(y) \to \tilde{\varphi}(z)$.

The final part is to get $t_y^* \uparrow t_z^*$ as $y \uparrow z$. Let $t_z^*(\omega) < \infty$ and denote $\tau = \lim t_y^*$, as $y \uparrow z$. On the set $\{t_y^* < \infty\}$, $X(t_y^*) = y$. By the path continuity, $X(\tau) = z$, so $\tau = t_z^*$. By varying x to the left and right of points, 16.23 results.

From here on, *we work only with regular processes.*

Problem 6. Let $\mathcal{C}$ be the class of all bounded continuous functions on F. For a Feller process show that if $f \in \mathcal{C}$, then $R_\lambda f$ is in $\mathcal{C}$. Deduce that $\mathfrak{D}(S, \mathcal{C}) \subset \mathfrak{D}(S, \mathcal{G})$. Prove that there is at most one Feller process corresponding to a given S, $\mathfrak{D}(S, \mathcal{C})$. (See Theorem 15.53.)

5. THE NATURAL SCALE

For pure jump processes, the structure was decomposable into a space structure, governed by a discrete time Markov chain and a time rate which determined how fast the particle moved through the paths of the space structure. Regular Feller processes can be decomposed in a very similar way. The idea is clearer when it is stated a bit more generally. Look at a path-continuous Markov process with stable stationary transition probabilities taking values in n-dimensional space, $\mathbf{X}(t) \in R^{(n)}$. Consider a set $B \in \mathcal{B}_n$; let $\mathfrak{t}^*(B)$ be the first exit time from B, then using the nontrivial fact that $\mathfrak{t}^*(B)$ is measurable, define probabilities on the Borel subsets of the boundary of B by

$$Q_x(C) = P_x\big(\mathbf{X}(\mathfrak{t}^*(B)) \in C\big), \qquad C \subset bd(B).$$

The $Q_x(C)$ are called exit distributions and specify the location of $\mathbf{X}(t)$ upon first leaving B. Suppose two such processes have the same exit distributions for all open sets. Then we can prove that under very general conditions they differ from each other only by a random change of time scale [10]. Thus the exit distributions characterize the space structure of the process. To have the exit distributions make sense, it is convenient to know that the particle does exit a.s. from the set in question. Actually, we want and get much more than this.

Theorem 16.24. *Let J be a finite open interval such that $\bar{J} \subset F$. Then*

$$\sup_{x \in J} E_x \mathfrak{t}^*(J) < \infty.$$

Proof. First we need

Lemma 16.25. *Let $\sup_{x\in J} P_x\big(\mathfrak{t}^*(J) > t\big) = \alpha < 1$ for some $t > 0$. Then*

$$E_x \mathfrak{t}^*(J) \le \frac{t}{1-\alpha}.$$

Proof. Let $\alpha_n = \sup_{x\in J} P_x\big(\mathfrak{t}^*(J) > nt\big)$. Write

$$P_x\big(\mathfrak{t}^*(J) > (n+1)t\big) \le \alpha_n P_x\big(\mathfrak{t}^*(J) > (n+1)t \mid \mathfrak{t}^*(J) > nt\big).$$

Let $\mathfrak{t}_n^*(J)$ be the first exit time from J, starting from time nt. Then

$$P_x\big(\mathfrak{t}^*(J) > (n+1)t\big) \le \alpha_n P_x\big(\mathfrak{t}_n^*(J) > t \mid \mathfrak{t}^*(J) > nt\big).$$

Since $\{t^*(J) > nt\} \in \mathcal{F}(X(\tau), \tau \le nt)$, then

$$P_x\big(\mathfrak{t}^*(J) > (n+1)t\big) \le \alpha\alpha_n.$$

Hence $\alpha_n \le \alpha^n$, and

$$E_x \mathfrak{t}^*(J) = \int_0^\infty P_x\big(\mathfrak{t}^*(J) > \tau\big)\, d\tau$$

$$= \sum_{n=0}^\infty \int_{nt}^{(n+1)t} P_x\big(\mathfrak{t}^*(J) > \tau\big)\, d\tau$$

$$\le t \sum_{n=0}^\infty \alpha^n = \frac{t}{1-\alpha}.$$

To finish the proof of 16.24, let $\bar{J} = [a, b]$, pick $y \in J$. By regularity, there exists a t and α such that $P_y(t_a^* > t) \le \alpha < 1$, $P_y(t_b^* > t) \le \alpha < 1$. For $y \le x < b$,

$$P_x(t^*(J) > t) \le P_x(t_b^* > t) \le P_y(t_b^* > t),$$

and for $a < x \le y$,

$$P_x(t^*(J) > t) \le P_x(t_a^* > t) \le P_y(t_a^* > t).$$

Apply the lemma now to deduce the theorem.

Remark. Note that the lemma has wider applicability than the theorem. Actually, it holds for all intervals J, finite or infinite.

In one dimension the relevant exit probabilities are:

Definition 16.26. *For any open interval $J = (a, b)$ such that $P_x(t^*(J) < \infty) = 1$, $x \in J$, define*

$$p^+(x, J) = P_x(X(t^*(J)) = b),$$

$$p^-(x, J) = 1 - p^+(x, J).$$

Theorem 16.27. *There exists a continuous, strictly increasing function $u(x)$ on F, unique up to a linear transformation, such that for $\bar{J} \subset F$, $J = (a, b)$,*

(16.28)
$$p^+(x, J) = \frac{u(x) - u(a)}{u(b) - u(a)}.$$

Proof. For $J \subset I$, note that exiting right from I starting from $x \in J$ can occur in two ways:

i) Exit right from J starting from x, then exit right from I starting from b.
ii) Exit left from J starting from x, then exit right from I starting from a.

Use the strong Markov property to conclude that for $x \in J$,

(16.29)
$$p^+(x, I) = p^+(x, J)p^+(b, I) + p^-(x, J)p^+(a, I)$$

or

$$p^+(x, J) = \frac{p^+(x, I) - p^+(a, I)}{p^+(b, I) - p^+(a, I)}.$$

If F were bounded and closed, then we could take $u(x) = p^+(x, \text{int}(F))$ and satisfy the theorem. In general, we have to use extension. Take I_0 to be a bounded open interval, such that if F includes any one of its endpoints, then $\bar{I}_0$ includes that endpoint. Otherwise I_0 is arbitrary. Define $u(x)$ on $I_0 = (x_1, x_2)$ as $p^+(x, I_0)$. By the equation above, for $I_0 \subset I_1$, $x \in I_0$,

(16.30)
$$u(x) = \frac{p^+(x, I_1) - p^+(x_1, I_1)}{p^+(x_2, I_1) - p^+(x_1, I_1)}.$$

Define an extension $u^{(1)}(x)$ on I_1 by the right-hand side of (16.30). Suppose another interval I_2 is used to get an extension, say $I_1 \subset I_2$. Then for $x \in I_2$, we would have

$$u^{(2)}(x) = \frac{p^+(x, I_2) - p^+(x_1, I_2)}{p^+(x_2, I_2) - p^+(x_1, I_2)}.$$

For $x \in I_1$, (16.29) gives

$$p^+(x, I_2) = c_1 p^+(x, I_1) + c_2.$$

Substitute this into (16.30) to conclude that $u^{(1)}(x) = u^{(2)}(x)$ on I_1. Thus the extensions are unique. Continuing this way, we can define $u(x)$ on $\text{int}(F)$ so that (16.28) is satisfied.

It is increasing; otherwise there exists a finite open J, $J \subset F$, and $x \in J$, such that $p^+(x, J) = 0$. This contradicts regularity. Extend u, by taking limits, to endpoints of F included in F. Now let J_n be open, $J_n \uparrow J = (a, b)$. I assert that $\mathsf{t}^*(J_n) \uparrow \mathsf{t}^*(J)$. Because $\mathsf{t}^*(J_n) \leq \mathsf{t}^*(J)$, by monotonicity $\mathsf{t}^* = \lim_n \mathsf{t}^*(J_n)$ exists, and by continuity $\mathsf{X}(\mathsf{t}^*) = a$ or b. For any $\epsilon > 0$ and n sufficiently large,

$$p^+(x, J_n) = P_x\big(\mathsf{X}(\mathsf{t}^*(J_n)) \geq b - \epsilon\big).$$

Since $\mathsf{X}(\mathsf{t}^*(J_n)) \to \mathsf{X}(\mathsf{t}^*(J))$ a.s., taking limits of the above equation gives

$$p^+(x, J_n) \to p^+(x, J).$$

By taking either $a_n \downarrow a$ or $b_n \uparrow b$ we can establish the continuity. The fact that $u(x)$ is unique up to a linear transformation follows from (16.28).

We will say that a process is on its natural scale if it has the same exit distributions as Brownian motion. From (13.3),

Definition 16.31. *A process is said to be on its natural scale if for every* $J = (a, b)$, $\bar{J} \subset F$,

$$p^+(x, J) = \frac{x - a}{b - a},$$

that is, if $u(x) = x$ *satisfies* (16.28).

The distinguishing feature of the space structure of normalized Brownian motion is that it is *driftless*. There is as much tendency to move to the right as to the left. More formally, if J is any finite interval and x_0 its midpoint, then for normalized motion, by symmetry, $p^+(x_0, J) = \frac{1}{2}$. We generalize this to

Proposition 16.32. *A process is on its natural scale if and only if for every finite open* J, $\bar{J} \subset F$, x_0 *the midpoint of* J,

$$p^+(x_0, J) = \tfrac{1}{2}.$$

Proof. Consider n points equally spaced in J,

$$a = x_1 < \cdots < x_n = b.$$

Starting from x_k, the particle next hits x_{k-1} or x_{k+1} with equal probability, so $p^+(x_k, (x_{k-1}, x_{k+1})) = \frac{1}{2}$. Therefore, the particle behaves like a symmetric random walk on the points of the partition. From Chapter 7, Section 10,

$$p^+(x_k, J) = \frac{x_k - a}{b - a}.$$

The continuity established in Theorem 16.27 completes the proof of 16.32.

Let the state space undergo the transformation $\tilde{x} = u(x)$. Equivalently, consider the process

(16.33) $$\tilde{X}(t) = u(X(t)).$$

If $X(t)$ is a regular Feller process, then so is $\tilde{X}(t)$. The importance of this transformation is:

Proposition 16.34. *$\tilde{X}(t)$ is on its natural scale.*

Proof. Let $\tilde{J} = (\tilde{a}, \tilde{b})$, $\tilde{a} = u(a)$, $\tilde{b} = u(b)$, $J = (a, b)$. For the $\tilde{X}(t)$ process, with $\tilde{x} = u(x)$,

$$\begin{aligned} p^+(\tilde{x}, \tilde{J}) &= P_{\tilde{x}}\big(\tilde{X}(t^*(\tilde{J})) = \tilde{b}\big) \\ &= P_x\big(X(t^*(J)) = b\big) \\ &= \frac{u(x) - u(a)}{u(b) - u(a)} \\ &= \frac{\tilde{x} - \tilde{a}}{\tilde{b} - \tilde{a}}. \end{aligned}$$

For any regular Feller process then, a simple space transformation gives another regular Feller process having the same space structure as normalized Brownian motion. *Therefore, we restrict attention henceforth to this type and examine the time flow.*

Remark. The reduction to natural scale derived here by using the transformation $\tilde{x} = u(x)$ does not generalize to Feller processes in two or more dimensions. Unfortunately, then, a good deal of the theory that follows just does not generalize to higher dimensions.

6. SPEED MEASURE

The functions $m(x, J) = E_x t^*(J)$ for open intervals J, determine how fast the process moves through its paths. There is a measure $m(dx)$ closely associated with these functions. Define, for $J = (a, b)$ finite,

$$(16.35) \qquad G_J(x, y) = \begin{cases} \dfrac{2(x-a)(b-y)}{b-a}, & x, y \in J, \quad x \leq y, \\ \dfrac{2(y-a)(b-x)}{b-a}, & x, y \in J, \quad x \geq y, \\ 0, & \text{otherwise.} \end{cases}$$

Then,

Theorem 16.36. *Let $X(t)$ be on its natural scale. Then there is a unique measure $m(dx)$ defined on $\mathcal{B}_1(\text{int } F)$, $m(B) < \infty$ for B bounded, $\bar{B} \subset \text{int}(F)$, such that for finite open J, $\bar{J} \subset F$,*

$$m(x, J) = \int G_J(x, y) m(dy).$$

Proof. The proof will provide some justification for the following terminology.

Definition 16.37. *$m(dx)$ is called the speed measure for the process.*

Consider (a, b) partitioned by points $a = x_0 < x_1 < \cdots < x_n = b$, where $x_k = a + k\delta$. Define $J_k = (x_{k-1}, x_{k+1})$. Note that $m(x_k, J_k)$ gives some quantitative indication of how fast the particle is traveling in the vicinity of x_k. Consider the process only at the exit times from one of the J_k intervals. This is a symmetric random walk $Z_0, Z_1, \ldots$ moving on the points $x_0, \ldots, x_n$. Let $n(x_j; x_k)$ be the expected number of times the random walk visits the point x_k, starting from x_j before hitting x_0 or x_n. Then

$$(16.38) \qquad m(x_j, J) = \sum_{k=1}^{n-1} n(x_j; x_k) m(x_k, J_k).$$

This formula is not immediately obvious, because $t^*(I)$ and $X(t^*(I))$ are not, in general, independent. Use this argument: Let t_N^* be the time taken for the transition $Z_N \to Z_{N+1}$. For $x \in \{x_0, \ldots, x_n\}$,

$$E_x t_N^* = E_x(E_x(t_N^* \mid Z_N)) = \sum_{k=1}^{n-1} P_x(Z_N = x_k) m(x_k, J_k).$$

Sum over N, noting that

$$n(x_j; x_k) = E_{x_j}\left(\sum_{N=0}^{\infty} \chi_{\{x_k\}}(Z_N)\right) = \sum_{N=0}^{\infty} P_{x_j}(Z_N = x_k).$$

This function was evaluated in Problem 14, Chapter 7, with the result,

$$n(x_j; x_k) = \frac{1}{\delta} G_J(x_j, x_k).$$

Defining $\hat{m}(dx)$ as a measure that gives mass $m(x_k, J_k)/\delta$ to the point x_k, we get

$$m(x_j, J) = \int G_J(x_j, y)\hat{m}(dy).$$

Now to get $m(dx)$ defined on all int(F). Partition F by successive refinements $\mathfrak{I}^{(n)}$ having points a distance δ_n apart, with $\delta_n \to 0$. Define the measure m_n as assigning mass

$$m(x_k, (x_{k-1}, x_{k+1}))/\delta_n$$

to all points $x_k \in \mathfrak{I}^{(n)}$ which are not endpoints of the partition. For a, b, $x_j \in \mathfrak{I}^{(n)}$, $J = (a, b)$, $\bar{J} \subset F$,

$$m(x_j, J) = \int G_J(x_j, y)m_n(dy). \tag{16.39}$$

For any finite interval I such that $\bar{I} \subset \text{int}(F)$, (16.39) implies that $\overline{\lim}_n\, m_n(I) < \infty$. Use this fact to conclude that there exists a subsequence $m_{n'} \xrightarrow{w} m$, where m is a measure on $\mathcal{B}_1(\text{int } F)$ (see 10.5). Furthermore, for any $J = (a, b)$, and $x \in J$, where a, b, x are in $\bigcup_n \mathfrak{I}^{(n)}$,

$$m(x, J) = \int G_J(x, y)m(dy). \tag{16.40}$$

For any arbitrary finite open interval J, $\bar{J} \subset F$, take $J_1 \subset J$ where J_1 has endpoints in $\bigcup_n \mathfrak{I}^{(n)}$ and pass to the limit as $J_1 \uparrow J$, to get (16.40) holding for J and any $x \in \bigcup_n \mathfrak{I}^{(n)}$. To extend this to arbitrary x, we introduce an identity: If $I \subset J$, $I = (a, b)$, $x \in I$, then the strong Markov property gives

$$m(x, J) = m(x, I) + p^+(x, I)m(b, J) + p^-(x, I)m(a, J). \tag{16.41}$$

Take J finite and open, $\bar{J} \subset F$, $x \in J$, $y < z < x$ and $y, z \in \bigcup \mathfrak{I}^{(n)} \cap J$. Use (16.41) to write

$$m(z, J) = m(z, (y, x)) + p^+(z, (y, x))m(x, J) + p^-(z, (y, x))m(y, J). \tag{16.42}$$

Take $z \uparrow x$. By (16.40), $m(z, (y, x)) \to 0$, so

$$\lim_{z \uparrow x} m(z, J) = m(x, J).$$

Since the integral $\int G_J(x, y)m(dy)$ is continuous in x, then (16.40) holds for all $x \in J$. But now the validity of (16.40) and the fact that the set of functions $\{G_J(x, y)\}$, a, b, $x \in \text{int}(F)$, are separating on int(F) imply that $m(dy)$ is unique, and the theorem follows.

One question left open is the assignment of mass to closed endpoints of F. This we defer to Section 7.

Problems

7. If $\mathsf{X}(t)$ is not on its natural scale, show that $m(dx)$ can still be defined by

$$m(x, J) = \int G_J(x, y)m(dy)$$

by using the definition

$$G_J(x, y) = \begin{cases} \dfrac{2(u(x) - u(a))(u(b) - u(y))}{u(b) - u(a)}, & x, y \in J, \quad x \le y, \\ \dfrac{2(u(y) - u(a))(u(b) - u(x))}{u(b) - u(a)}, & x, y \in J, \quad x \ge y, \\ 0, & \text{otherwise.} \end{cases}$$

8. For $\mathsf{X}(t)$ Brownian motion with zero drift, show that for $J = (a, b)$,

$$E_x \mathsf{t}^*(J) = \frac{(x - a)(b - x)}{\sigma^2}.$$

Use this to prove

$$m(dy) = \frac{1}{\sigma^2}\, dy.$$

[Another way to see the form of $m(dy)$ for Brownian motion is this: $m_n(dy)$ assigns equal masses to points equally spaced. The only possible limit measure of measures of this type is easily seen to be a constant multiple $c\,dy$ of Lebesgue measure. If c_0 is the constant for normalized Brownian motion, a scale transformation of $\mathsf{X}(t)$ shows that $c = c_0/\sigma^2$.]

9. For $x \in \text{int}(F)$ and J_n open neighborhoods of x such that $J_n \downarrow \{x\}$, show that $\mathsf{t}^*(J_n) \xrightarrow{\text{a.s.}} 0$. Deduce that $\mathsf{t}^*(\{x\}) = 0$ a.s. P_x.

10. For $f(x)$ a bounded continuous function on F, J any finite open interval such that $\bar{J} \subset F$, prove that

$$E_x\left[\int_0^{\mathsf{t}^*(J)} f(\mathsf{X}(t))\, dt\right] = \int_J G_J(x, y) f(y) m(dy).$$

[Use the approximation by random walks on partitions.]

7. BOUNDARIES

This section sets up classifications that summarize the behavior at the boundary points of F. If F is infinite in any direction, say to the right, call $+\infty$ the boundary point on the right; similarly for $-\infty$.

For a process on its natural scale the speed measure $m(dx)$ defined on $\text{int}(F)$ will to a large extent determine the behavior of the process at the boundaries of F. For example, we would expect that knowing the speed measure in the vicinity of a boundary point b of F would tell us whether the process would ever hit b, hence whether b was an open or closed endpoint of F. Here, by open endpoint, we mean $b \notin F$, by closed, $b \in F$. In fact, we have

Proposition 16.43. *If b is a finite endpoint of F, then $b \notin F$ if and only if for $J \subset F$ any nonempty open neighborhood with b as endpoint,*

$$\int_J |b - y| \, m(dy) = \infty.$$

Remark. A closed endpoint of F is sometimes called *accessible*, for obvious reasons.

Proof. Assume b is a right endpoint. If $b \in F$, then there exists a t such that for $c \in \text{int}(F), P_c(\mathbf{t}_b^* > t) = \alpha < 1$. Let $J = (c, b)$. Then $P_x(\mathbf{t}^*(J) > t) \leq \alpha$, all $x \in J$. Use Lemma 16.25 to get $m(x, J) < \infty$, all $x \in J$. For $b \in F, \bar{J} \subset F$; hence 16.36 holds:

$$\begin{aligned} m(x, J) &= \int_J G_J(x, y) m(dy) \\ &\geq \frac{2(x - c)}{b - c} \int_{(x,b)} |b - y| \, m(dy). \end{aligned}$$

Therefore the integral over (x, b) is finite. Conversely, if the integral is finite for one $c \in \text{int}(F)$, then as $z \uparrow b$, if $b \notin F$, and $x > c$,

$$E_x \mathbf{t}_c^* = \lim m(x, (c, z)) = \int_J G_J(x, y) m(dy), \quad J = (c, b).$$

The left-hand side of this is nondecreasing as $x \uparrow b$. But the integral equals

$$\frac{2(b - x)}{b - c} \int_{(c,x)} (y - c) m(dy) + \frac{2(x - c)}{b - c} \int_{[x,b)} (b - y) m(dy),$$

which goes to zero as $x \uparrow b$.

For open boundary points b, there is two-way classification.

Definition 16.44. *Let $x \in \text{int}(F)$, $y \in \text{int}(F)$, $y \to b$ monotonically. Call b*

natural *if for all $t > 0$, $\lim_{y \to b} P_y(\mathbf{t}_x^* < t) = 0$;*

entrance *if there is a $t > 0$ such that $\lim_{y \to b} P_y(\mathbf{t}_x^* < t) > 0$.*

A natural boundary behaves like the points at ∞ for Brownian motion. It takes the particle a long time to get close to such a boundary point and then a long time to get away from it. An entrance boundary has the odd property that it takes a long time for the particle to get out to it but not to get away from it.

Proposition 16.45. *Let b be an open boundary point.*

a) *If b is finite, it must be natural.*

b) *If b is infinite, it is natural if and only if for $J \subset F$ any open interval with b as an endpoint,*

$$\int_J |y|\, m(dy) = \infty.$$

Proof. Take b finite and right-hand, say. If b is an entrance point, then for $J = (c, b)$, use Lemma 16.25 to get $m(x, J) \leq M < \infty$ for all $x \in J$. This implies, if we take $J_1 \uparrow J$ and use the monotone convergence theorem, that

$$\int_J (b - y)m(dy) < \infty.$$

This is impossible, by 16.43, since b is an open endpoint. For $b = \infty$, check that

$$\lim_{\uparrow \infty} m(x, (a, c)) = 2\int_a^x (y - a)\, dm + 2(x - a)\int_x^\infty dm.$$

If $b = \infty$ is entrance, then there is an a such that for $J = (a, \infty)$

$$P_x(\mathrm{t}^*(J) > t) = P_x(\mathrm{t}_a^* > t) \leq 1 - \alpha < 1$$

for all $x \in J$. Use Lemma 16.25 to get

$$\int_a^x (y - a)\, dm + (x - a)\int_x^\infty dm \leq M < \infty, \quad x \in J.$$

Taking $x \to \infty$ proves that $\int_J |y|\, dm < \infty$. Conversely, if the integral is finite for $J = (a, \infty)$, then there exists an $M < \infty$ such that for all $a < x < c < \infty$, $m(x, (a, c)) \leq M$. Assume $b = \infty$ is not an entrance boundary. For any $0 < \epsilon < \frac{1}{4}$, take c, x such that $a < x < c$ and

$$P_x(\mathrm{t}_a^* \leq 2M) \leq \epsilon, \qquad P_x(\mathrm{t}_c^* \leq 2M) \leq \epsilon.$$

Then

$$m(x, (a, c)) \geq 2MP_x(\mathrm{t}^*((a, c)) \geq 2M) \geq 2M(1 - 2\epsilon) > M,$$

and this completes the proof.

There is also a two-way classification of closed boundary points b. This is connected with the assignment of speed measure $m(\{b\})$ on these points. Say b is a closed left-hand endpoint. We define a function $G_J(x, y)$ for all intervals of the form $J = [b, c)$, and $x, y \in J$ as follows: Let $\overset{\perp}{J}$ be the reflection of J around b, that is, $\overset{\perp}{J} = (b - (c - b), b]$ and $\overset{\perp}{y}$ the reflection of y around b, $\overset{\perp}{y} = b - (y - b)$. Then define

$$G_J(x, y) = G_{J \cup \overset{\perp}{J}}(x, y) + G_{J \cup \overset{\perp}{J}}(x, \overset{\perp}{y}), \tag{16.46}$$

where $G_{J\cup \overset{\perp}{J}}$ is defined by (16.35). This definition leads to an extension of (16.40).

Theorem 16.47. *It is possible to define $m(\{b\})$ so that for all finite intervals $J = [b, c), J \subset F$,*

$$m(x, J) = \int_J G_J(x, y)m(dy).$$

Remark. Note that if $\int_{(b,c)} m(dy) = \infty$, then no matter how $m(\{b\})$ is assigned, $m(x, J) = \infty$. This leads to

Definition 16.48. *Let b be a closed boundary point, $J \subset F$ any finite open interval with b as one endpoint such that $\bar{J}$ is smaller than F. Call b*

$$\begin{array}{rl} \text{regular} & \textit{if} \quad m(J) < \infty, \\ \text{exit} & \textit{if} \quad m(J) = \infty. \end{array}$$

Proof of 16.47. The meaning of the definition of G_J will become more intuitive in the course of this proof. Partition any finite interval $[b, c)$ by equally spaced points $b = x_0 < x_1 < \cdots < x_n = c$, a distance δ apart. Define $J_k = (x_{k-1}, x_{k+1})$, $k = 1, \ldots, n-1$, $J_0 = [x_0, x_1)$. By exactly the same reasoning as in deriving 16.38,

$$m(x_j, J) = \sum_{k=0}^{n-1} n^{(r)}(x_j; x_k)m(x_k, J_k),$$

where $n^{(r)}(x_j; x_k)$ is the expected number of visits to x_k of a random walk $\mathsf{Z}_0, \mathsf{Z}_1, \ldots$ on $x_0, \ldots, x_n$ starting from x_j, with reflection at x_0 and absorption at x_n. Construct a new state space $x_{-n}, \ldots, x_{-1}, x_0, \ldots, x_n$, where $x_{-k} = \overset{\perp}{x}_k$, and consider a symmetric random walk $\overset{\perp}{\mathsf{Z}}_0, \overset{\perp}{\mathsf{Z}}_1, \ldots$ on this space with absorption at x_{-n}, x_n. Use the argument that $\mathsf{Z}_k = |\overset{\perp}{\mathsf{Z}}_k - x_0| + x_0$ is the reflected random walk we want, to see that for $x_k \neq x_0$,

$$n^{(r)}(x_j; x_k) = \overset{\perp}{n}(x_j; x_k) + \overset{\perp}{n}(x_j; x_{-k}),$$

where $\overset{\perp}{n}(x_j, x_k)$ refers to the $\overset{\perp}{\mathsf{Z}}$ random walk, and

$$n^{(r)}(x_j; x_0) = \overset{\perp}{n}(x_j; x_0).$$

Hence, for $G_J(x, y)$ as defined in (16.46),

$$n^{(r)}(x_j; x_k) = \begin{cases} \dfrac{1}{\delta} G_J(x_j, x_k), & k \neq 0, \\[2ex] \dfrac{1}{2\delta} G_J(x_j, x_0), & k = 0. \end{cases}$$

For $m(x_k, J_k)$, $k \geq 1$, use the expression $\int_{J_k} G_{J_k}(x_k, y)m(dy)$. Then

$$m(x_j, J) = \frac{1}{2\delta} G_J(x_j, x_0)m(x_0, J_0) + \int_{(b,c)} \left(\sum_{k=1}^{n-1} \frac{1}{\delta} G_J(x_j, x_k)G_{J_k}(x_k, y) \right) m(dy).$$

The integrand converges uniformly to $G_J(x_j, y)$ as $\delta \to 0$. Hence for b not an exit boundary, the second term of this converges to

$$\int_{J-\{b\}} G_J(x_j, y)m(dy)$$

as we pass through successive refinements such that $\delta \to 0$. Therefore, $m(x_0, J_0)/2\delta$ must converge to a limit, finite or infinite. Define $m(\{b\})$ to be this limit. Then for x_j any point in the successive partitions,

$$m(x_j, J) = \int_J G_J(x_j, y)m(dy).$$

For $m(\{b\}) < \infty$, extend this to all $x \in J$ by an argument similar to that in the proof of 16.36. For $m(\{b\}) = \infty$, show in the same way that $m(x, J) = \infty$ for all $x \in J$.

Note that the behavior of the process is specified at all boundary points except regular boundary points by $m(dy)$ on $\mathcal{B}_1(\text{int } F)$. Summarizing graphically, for b a right-hand endpoint,

Classification	Type
$\text{int}(F) \mathrel{\substack{\nrightarrow \\ \nleftarrow}} b$	natural boundary
$\text{int}(F) \mathrel{\substack{\nrightarrow \\ \leftarrow}} b$	entrance boundary
$\text{int}(F) \mathrel{\substack{\rightarrow \\ \nleftarrow}} b$	exit boundary
$\text{int}(F) \rightleftarrows b$	regular boundary

The last statement ($\rightleftarrows$ for regular boundary points) needs some explanation. Consider reflecting and absorbing Brownian motion on $[b, \infty)$ as described in Section 3. Both of these processes have the same speed measure $m(dy) = dy$ on (b, ∞), and b is a regular boundary point for both of them. They differ in the assignment of $m(\{b\})$. For the absorbing process, obviously $m(\{b\}) = \infty$; for the reflecting process $m(\{b\}) < \infty$. Hence, in terms of m on $\text{int}(F)$ it is possible to go $\text{int}(F) \to b$ and $b \to \text{int}(F)$. Of course, the latter is ruled out if $m(\{b\}) = \infty$, so $\rightleftarrows$ should be understood above to mean "only in terms of m on $\text{int}(F)$."

Definition 16.49. *A regular boundary point b is called*

absorbing	*if* $m(\{b\}) = \infty$,
slowly reflecting	*if* $0 < m(\{b\}) < \infty$,
instantaneously reflecting	*if* $m(\{b\}) = 0$.

See Problem 11 for some interesting properties of a slowly reflecting boundary.

Problems

11. Show, by using the random walk approximation, that

$$(c - b)m(\{b\}) = E_b(l\{t; \mathsf{X}(t) = b, t \leq \mathsf{t}_c^*\}).$$

Conclude that if $m(\{b\}) > 0$, then $\mathsf{X}(t)$ spends a positive length of time at point $b \Rightarrow p_t(\{b\} \mid b) > 0$ for some t. Show also that for almost all sample paths, $\{t; \mathsf{X}(t) = b\}$ contains no intervals with positive length and no isolated points.

12. For b an entrance boundary, $J = (b, c)$, $c \in \text{int}(F)$, show that

$$m(b, J) = \int_J (c - y)m(dy),$$

where $m(b, J) = \lim m(x, J)$, $x \to b$.

13. For b a regular boundary point, $J = [b, c)$, $J \subset F$, show that

$$m(b, J) = \int_J |y - c| \, m(dy).$$

14. For b an exit boundary, and any $J = (a, b]$, $x \in J$, show that $m(x, J) = \infty$ (see the proof of 16.47). Use 16.25 to conclude that $P_b(\mathsf{t}_x^* < \infty) = 0$. Hence, deduce that $p_t(\{b\} \mid b) = 1$, $t \geq 0$.

8. CONSTRUCTION OF FELLER PROCESSES

Assume that a Feller process is on its natural scale. Because $\mathsf{X}(t)$ has the same exit probabilities as Brownian motion $\mathsf{X}_0(t)$, we should be able to construct a process with the same distribution as $\mathsf{X}(t)$ by expanding or contracting the time scale of the Brownian motion, depending on the current position of the particle. Suppose $m(dx)$ is absolutely continuous with respect to Lebesgue measure,

$$m(dx) = V(x)\, dx,$$

where $V(x)$ is continuous on F. For J small,

$$m(x, J) \simeq V(x)m^{(0)}(x, J),$$

where $m^{(0)}(x, J)$ is the expected time for $\mathsf{X}_0(t)$ to exit from J. So if it takes Brownian motion time Δt to get out of J, it takes $\mathsf{X}(t)$ about $V(x)\,\Delta t$ to get out of J. Look at the process $\mathsf{X}_0(\mathsf{T}(t))$, where $\mathsf{T}(t) = \mathsf{T}(t, \omega)$ is for every ω an increasing function of time. If we want this process to look like $\mathsf{X}(t)$, then when t changes by $V(x)\,\Delta t$, we must see that T changes by the amount Δt. We get the differential equation

$$\frac{d\mathsf{T}}{dt} = \frac{1}{V(x)}.$$

We are at the point $x = \mathsf{X}_0(\mathsf{T}(t))$, so

$$\frac{d\mathsf{T}}{dt} = \frac{1}{V(\mathsf{X}_0(\mathsf{T}))} \quad \text{or} \quad V(\mathsf{X}_0(\mathsf{T}))\, d\mathsf{T} = dt.$$

Integrating, we get

$$\int_0^{\mathsf{T}} V(\mathsf{X}_0(\xi))\, d\xi = t.$$

Hence,

Definition 16.50. *Take $\mathsf{X}_0(t)$ to be Brownian motion on F, instantaneously reflecting at all finite endpoints. Denote*

$$\mathsf{I}(r) = \int_0^r V(\mathsf{X}_0(\xi))\, d\xi.$$

Define $\mathsf{T}(t)$ to be the solution of $\mathsf{I}(r) = t$, that is,

$$t = \int_0^{\mathsf{T}(t)} V(\mathsf{X}_0(\xi))\, d\xi.$$

Remark. Because $m(J) > 0$ for all open neighborhoods J, $\{x;\ V(x) = 0\}$ can contain no open intervals. But almost no sample function of $\mathsf{X}_0(t)$ can have any interval of constancy. Hence $\mathsf{I}(\tau)$ is a strictly increasing continuous function of τ, and so is $\mathsf{T}(t)$. Further, note that for every t, $\mathsf{T}(t)$ is a Markov time for $\{\mathsf{X}_0(t)\}$.

Theorem 16.51. $\mathsf{X}(t) = \mathsf{X}_0(\mathsf{T}(t))$ *is a Feller process on natural scale with speed measure*

$$m(dx) = V(x)\, dx.$$

Proof. For any Markov time t^*, $\mathsf{X}_0(\mathsf{t}^*)$ is measurable $\mathcal{F}(\mathsf{X}_0(s), s \le \mathsf{t}^*)$. Further, it is easy to show for any two Markov times $\mathsf{t}_1^* \le \mathsf{t}_2^*$, that

$$\mathcal{F}(\mathsf{X}_0(s), s \le \mathsf{t}_1^*) \subset \mathcal{F}(\mathsf{X}_0(s), s \le \mathsf{t}_2^*).$$

This implies that $\mathsf{X}(\tau) = \mathsf{X}_0(\mathsf{T}(\tau))$ is measurable $\mathcal{F}(\mathsf{X}_0(s), s \le \mathsf{T}(t))$, for any $\tau \le t$. So

$$\mathcal{F}(\mathsf{X}(s), s \le t) \subset \mathcal{F}(\mathsf{X}_0(s), s \le \mathsf{T}(t)).$$

Hence $P_x(\mathsf{X}(t+\tau) \in B \mid \mathsf{X}(s), s \le t)$ equals the expectation of

$$P_x(\mathsf{X}(t+\tau) \in B \mid \mathsf{X}_0(s), s \le \mathsf{T}(t)),$$

given $\mathcal{F}(\mathsf{X}(s), s \le t)$. To evaluate $\mathsf{T}(t+\tau)$, write

$$t + \tau = \int_0^{\mathsf{T}(t+\tau)} V(\mathsf{X}_0(\xi))\, d\xi = \int_0^{\Delta} V\big(\mathsf{X}_0(\xi + \mathsf{T}(t))\big)\, d\xi + t,$$

where $\Delta = \mathsf{T}(t+\tau) - \mathsf{T}(t)$. Thus Δ is the solution of $\mathsf{I}_1(r) = \tau$, where

$$\mathsf{I}_1(r) = \int_0^r V\big(\mathsf{X}_0(\xi + \mathsf{T}(t))\big)\, d\xi.$$

Now $X(t + \tau) = X_0(T(t + \tau)) = X_0(\Delta + T(t))$. Because $I_1(r)$ is the same function on the process $X_0(\cdot + T(t))$ as $I(r)$ is on $X_0(\cdot)$, the strong Markov property applies:

$$P_x\big(X_0(\Delta + T(t)) \in B \mid X_0(s), s \le T(t)\big)$$
$$= P_{X_0(T(t))}\big(X_0(T(\tau)) \in B\big) = P_{X(t)}\big(X(\tau) \in B\big).$$

The proof that $X(t)$ is strong Markov is only sketched. Actually, what we will prove is that the transition probabilities for the process are stable. Let $\varphi(x)$ be bounded and continuous on F. Denote by t_y^* the first passage time of $X_0(t)$ to y. As $y \to x$, $t_y^* \to 0$ a.s. P_x. Hence, by path continuity, as $y \to x$,

$$E_x\varphi\big(X(t + I(t_y^*))\big) \to E_x\varphi(X(t)).$$

$T(t + I(t_y^*))$ is defined as the solution of

$$t + I(t_y^*) = \int_0^T V(X_0(\xi))\, d\xi.$$

Thus $\Delta = T(t + I(t_y^*)) - t_y^*$ is the solution of

$$t = \int_0^\Delta V(X_0(\xi + t_y^*))\, d\xi.$$

Use the strong Markov property of $X_0(t)$ to compute

$$E_x\varphi\big(X(t + I(t_y^*))\big) = E_x\varphi(X_0(\Delta + t_y^*)) = E_y\varphi(X(t)).$$

Thus $E_x\varphi(X(t))$ is continuous.

Denote by $t_0^*(J)$ the first exit time of $X_0(t)$ from J, $t^*(J)$ the exit time for $X(t)$. The relationship is

(16.52) $\quad t_0^*(J) = T(t^*(J)) \quad$ or $\quad t^*(J) = \displaystyle\int_0^{t_0^*(J)} V(X_0(\xi))\, d\xi.$

Taking expectations, we get

$$m(x, J) = E_x \int_0^{t_0^*(J)} V(X_0(\xi))\, d\xi.$$

By Problem 10, the latter integral is

$$\int_J G_J(x, y)V(y)m_0(dy) = \int_J G_J(x, y)V(y)\, dy.$$

Thus, the process has the asserted speed measure.

To show $X(t)$ is on its natural scale, take $J = (a, b)$,

$$P_x\big(X(t^*(J)) = b\big) = P_x\Big(X_0\big(T(t^*(J))\big) = b\Big) = P_x\big(X_0(t_0^*(J)) = b\big).$$

If $m(dx)$ is not absolutely continuous, the situation is much more difficult. The key difficulty lies in the definition of $I(r)$. So let us attempt to transform the given definition of $\mathsf{l}(r)$ above into an expression not involving $V(x)$: Let $\mathsf{L}(t, J)$ be the time that $\mathsf{X}_0(\xi)$ spends in J up to time t,

$$\mathsf{L}(t, J) = l\{\xi;\ \mathsf{X}_0(\xi) \in J,\ \xi \leq t\}.$$

Then $\mathsf{l}(r)$ can be written approximately as

$$\mathsf{l}(r) = \sum_k V(x_k)\mathsf{L}(t, J_k) = \sum_k \frac{\mathsf{L}(t, J_k)}{\|J_k\|} m(J_k).$$

Suppose that

$$\lim_{J \downarrow \{y\}} \frac{L(t, J)}{\|J\|} = \mathfrak{l}^*(t, y)$$

exists for all y, t. Then assuming the limit exists in some nice way in y, we get the important alternative expression for $\mathsf{l}(r)$,

$$\textbf{(16.53)} \qquad \mathsf{l}(r) = \int \mathfrak{l}^*(r, y) m(dy).$$

That such a function $\mathfrak{l}^*(t, x)$ exists having some essential properties is the consequence of a remarkable theorem due to Trotter [136].

Theorem 16.54. *Let $\mathsf{X}_0(t)$ be unrestricted Brownian motion on $R^{(1)}$. Almost every sample path has the property that there is a function $\mathfrak{l}^*(t, y)$ continuous on $\{(t, y);\ t \geq 0,\ y \in R^{(1)}\}$ such that for all $B \in \mathcal{B}_1$,*

$$l\{\xi;\ \mathsf{X}_0(\xi) \in B,\ \xi \leq t\} = \int_B \mathfrak{l}^*(t, y)\, dy.$$

Remarks. $\mathfrak{l}^*(t, y)$ is called local time for the process. It has to be nondecreasing in t for y fixed. Because of its continuity, the limited procedure leading up to (16.53) is justified. The proof of this theorem is too long and technical to be given here. See Ito and McKean [76, pp. 63 ff.] or Trotter [136].

Assume the validity of 16.54. Then it is easy to see that local time $\mathfrak{l}_1^*(t, y)$ exists for the Brownian motion with reflecting boundaries. For example, if $x = 0$ is a reflecting boundary, then for $B \in \mathcal{B}_1([0, \infty))$,

$$l\{\xi;\ |\mathsf{X}_0(\xi)| \in B,\ \xi \leq t\} = \int_B \mathfrak{l}_1^*(t, y)\, dy,$$

where $\mathfrak{l}_1^*(t, y) = \mathfrak{l}^*(t, y) + \mathfrak{l}^*(t, -y)$.

Definition 16.55. *Take $\mathsf{X}_0(t)$ to be Brownian motion on F, instantaneously reflecting at all closed endpoints, with local time $\mathfrak{l}^*(t, y)$. Let $m(dx)$ be any measure on $\mathcal{B}_1(F)$ such that $0 < m(J) < \infty$ for all finite open intervals J with $\bar{J} \subset \operatorname{int}(F)$. Denote*

$$\mathsf{l}(r) = \int \mathfrak{l}^*(r, y) m(dy),$$

and let $\mathsf{T}(t)$ be the solution of $\mathsf{l}(r) = t$.

Since $m(F)$ is not necessarily finite, then along some path there may be an r such that $\mathsf{l}(r) = \infty$; hence $\mathsf{l}(s) = \infty$, $s \geq r$. But if $\mathsf{l}(r) < \infty$, then $\mathsf{l}(s)$ is continuous on $0 \leq s \leq r$. Furthermore, it is strictly increasing on this range. Otherwise there would be an s, t, $0 \leq t < s \leq r$ such that $\mathfrak{l}^*(t, y) = \mathfrak{l}^*(s, y)$ for all $y \in F$. Integrate this latter equality over F with respect to dy to get the contradiction $s = t$. Thus, $\mathsf{T}(t)$ will always be well defined except along a path such that there is an r with $\mathsf{l}(r) < \infty$, $\mathsf{l}(r+) = \infty$ (by the monotone convergence theorem $\mathsf{l}(r)$ is always left-continuous). If this occurs, define $\mathsf{T}(t) \equiv r$ for all $t \geq \mathsf{l}(r)$. With this added convention, $\mathsf{T}(t)$ is continuous and strictly increasing unless it becomes identically constant.

Theorem 16.56. *$\mathsf{X}_0(\mathsf{T}(t))$ is a Feller process on natural scale with speed measure $m(dy)$.*

Proof. That $\mathsf{X}(t) = \mathsf{X}_0(\mathsf{T}(t))$ is on natural scale is again obvious. To compute the speed measure of $\mathsf{X}(t)$, use (16.52), $\mathfrak{t}_0^*(J) = \mathsf{T}(\mathfrak{t}^*(J))$ or $\mathsf{l}(\mathfrak{t}_0^*(J)) = \mathfrak{t}^*(J)$. Hence

$$m(x, J) = \int E_x \mathfrak{l}^*(\mathfrak{t}_0^*(J), y) m(dy).$$

The integrand does not depend on $m(dy)$. By the definition and continuity properties of local time the integral

$$\int_I E_x \mathfrak{l}^*(\mathfrak{t}_0^*(J), y)\, dy, \qquad I \subset J$$

is the expected amount of time the Brownian particle starting from x spends in the interval I before it exits from J. We use Problem 10 to deduce that this expected time is also given by

$$\int_I G_J(x, y)\, dy.$$

The verification that $E_x \mathfrak{l}^*(\mathfrak{t}_0^*(J), y)$ is continuous in y leads to its identification with $G_J(x; y)$, and proves the assertion regarding the speed measure of the process. Actually, the identity of Problem 10 was asserted to hold only for the interior of F, but the same proof goes through for J including a closed endpoint of F when the extended definition of $G_J(x, y)$ is used.

The proof in 16.51 that $\mathsf{X}(t)$ is strong Markov is seen to be based on two facts:

a) $\mathsf{T}(t)$ is a Markov time for every $t \geq 0$;
b) $\mathsf{T}(t + \tau) = \Delta + \mathsf{T}(t)$, where Δ is the solution of $\mathsf{l}_1(r) = \tau$, and $\mathsf{l}_1(r)$ is the integral $\mathsf{l}(r)$ based on the process $\mathsf{X}_0(\cdot + \mathsf{T}(t))$.

It is easy to show that (a) is true in the present situation. To verify (b), observe that $\mathfrak{l}^*(t, y)$ is a function of y and the sample path for $0 \leq \xi \leq t$.

Because

$$l\{\xi;\, X_0(\xi) \in B,\, \xi \le r + s\}$$
$$= l\{\xi;\, X_0(\xi) \in B,\, \xi \le r\} + l\{\xi;\, X_0(\xi + r) \in B,\, 0 \le \xi \le s\},$$

it follows that

$$\mathfrak{l}^*(r + s, y) = \mathfrak{l}^*(r, y) + \mathfrak{l}^{**}(s, y),$$

where $\mathfrak{l}^{**}(s, y)$ is the function $\mathfrak{l}^*(s, y)$ evaluated along the sample path $X_0(r + \xi)$, $0 \le \xi \le s$. Therefore

$$\mathsf{l}(r + s) = \mathsf{l}(r) + \mathsf{l}_1(s),$$

where $\mathsf{l}_1(s)$ is $\mathsf{l}(s)$ evaluated along the path $X_0(r + \xi)$, $0 \le \xi \le s$. The rest goes as before.

Many details in the above proof are left to the reader to fill in. An important one is the examination of the case in which $\mathsf{l}(r) = \infty$ for finite r. This corresponds to the behavior of $X(t)$ at finite boundary points. In particular, if at a closed boundary point b the condition of accessibility (16.43) is violated, then it can be shown that the constructed process never reaches b. Evidently, for such measures $m(dy)$, the point b should be deleted from F. With this convention we leave to the reader the proof that the constructed process $X(t)$ is regular on F.

Problem 15. Since $\mathfrak{l}^*(t, 0)$ is nondecreasing, there is an associated measure $\mathfrak{l}^*(dt, 0)$. Show that $\mathfrak{l}^*(dt, 0)$ is concentrated on the zeros of $X_0(t)$. That is, prove

$$\int_{\{\xi;\, X_0(\xi)=0,\, \xi \le t\}} \mathfrak{l}^*(d\xi, 0) = \mathfrak{l}^*(t, 0).$$

(This problem illustrates the fact that $\mathfrak{l}^*(t, 0)$ is a measure of the time the particle spends at zero.)

9. THE CHARACTERISTIC OPERATOR

The results of the last section show that corresponding to every speed measure $m(dx)$ there is at least one Feller process on natural scale. In fact, there is only one. Roughly, the reasoning is that by breaking F down into smaller and smaller intervals, the distribution of time needed to get from one point to another becomes determined by the expected times to leave the small subintervals, hence by $m(dx)$. But to make this argument firm, an excursion is needed. This argument depends on what happens over small space intervals. The operator S which determines behavior over small time intervals can be computed by allowing a fixed time t to elapse, averaging over $X(t)$, and then taking $t \to 0$. Another approach is to fix the terminal space positions and average over the time it takes the particle to get to these terminal space positions. Take $x \in J$, J open, and define

$$(16.57) \qquad (U_J f)(x) = \frac{E_x f(X(t^*(J))) - f(x)}{E_x t^*(J)}.$$

Let $J \downarrow \{x\}$, if $\lim\, (U_J f)(x)$ exists, then see whether, under some reasonable conditions, the limit will equal $(Sf)(x)$.

Definition 16.58. *Let $x \in F$. For any neighborhood J of x open relative to F, suppose a function $\varphi(J)$ is defined. Say that*

$$\lim_{J \downarrow \{x\}} \varphi(J) = \alpha$$

if for every system J_n of such neighborhoods, $J_n \downarrow \{x\}$, $\lim_n \varphi(J_n) = \alpha$.

Theorem 16.59. *Let $f \in \mathfrak{D}(S, \mathcal{C})$, where $\mathcal{C}$ consists of all bounded continuous functions on F. Then*

$$\lim_{J \downarrow \{x\}} (U_J f)(x) = (Sf)(x)$$

for all $x \in F$.

Proof. The proof is based on an identity due to Dynkin.

Lemma 16.60. *For any Markov time t^* such that $E_x t^* < \infty$, for $f \in \mathfrak{D}(S, \mathcal{C})$ and $g = Sf$,*

$$E_x f(X(t^*)) - f(x) = E_x \int_0^{t^*} g(X(t))\, dt.$$

Proof. For any bounded measurable $h(x)$ consider $f(x) = (R_\lambda h)(x)$ and write

$$(16.61) \qquad \begin{aligned} E_x e^{-\lambda t^*} f(X(t^*)) &= E_x \int_0^\infty e^{-\lambda(t+t^*)} h(X(t + t^*))\, dt \\ &= E_x \int_{t^*}^\infty e^{-\lambda t} h(X(t))\, dt \\ &= f(x) - E_x \int_0^{t^*} e^{-\lambda t} h(X(t))\, dt. \end{aligned}$$

For f in $\mathfrak{D}(S, \mathcal{C})$, take $h = (\lambda - S)f$. Then by 15.51(2), $f = R_\lambda h$ and (16.61) becomes

$$Ee^{-\lambda t^*} f(X(t^*)) - f(x) = E_x \int_0^{t^*} e^{-\lambda t} g(X(t))dt - \lambda E_x \int_0^{t^*} e^{-\lambda t} f(X(t))dt,$$

where the last integral exists for all $\lambda \geq 0$ by the boundedness of $f(x)$ and $E_x t^* < \infty$. Taking $\lambda \to 0$ now proves the lemma.

To finish 16.59, if x is an absorbing or exit endpoint, then both expressions are zero at x. For any other point x, there are neighborhoods J

of x, open relative to F, such that $E_x \mathfrak{t}^*(J) < \infty$. Now g is continuous at x, so take J sufficiently small so that $|g(X(t)) - g(x)| \leq \epsilon$, $t \leq \mathfrak{t}^*(J)$. Then by the lemma,

$$E_x f(X(\mathfrak{t}^*(J))) - f(x) = (g(x) \pm \epsilon) E_x \mathfrak{t}^*(J).$$

Definition 16.62. *Define* $(Uf)(x)$ *for any measurable function* f *as*

$$\lim_{J \downarrow \{x\}} (U_J f)(x)$$

wherever this limit exists. Denote $f \in \mathfrak{D}(U, I)$ *if the limit exists for all* $x \in I$.

Corollary 16.63. *If* $f \in \mathfrak{D}(S, \mathcal{C})$, *then* $f \in \mathfrak{D}(U, F)$.

There are a number of good reasons why U is usually easier to work with than S. For our purposes an important difference is that U is expressible in terms of the scale and speed measure. For any Feller process we can show that $(Uf)(x)$ is very nearly a second-order differential operator. There is a simple expression for this when the process is on natural scale.

Theorem 16.64. *Let* $X(t)$ *be on its natural scale,* $f \in \mathfrak{D}(S, \mathcal{C})$. *Then*

$$(Uf)(x) = \frac{1}{2} \frac{d}{dm} \frac{d}{dx} f(x)$$

in the following sense for $x \in \text{int}(F)$,

i) $f'(x)$ *exists except perhaps on the countable set of points* $\{x;\ m(\{x\}) > 0\}$.
ii) *For* $J = (x_1, x_2)$ *finite, such that* $f'(x_1), f'(x_2)$ *exist,*

$$f'(x_2) - f'(x_1) = 2 \int_J g(x) m(dx),$$

where $g(x) = (Sf)(x)$.

Proof. Use 16.60 and Problem 10 to get

$$E_x f(X(\mathfrak{t}^*(J))) - f(x) = \int_J G_J(x, y) g(y) m(dy).$$

Take $J = (x - h_1, x + h_2)$ and use the appropriate values for $p^+(x, J)$ to get the following equations.

$$\frac{f(x + h_2) - f(x)}{h_2} - \frac{f(x) - f(x - h_1)}{h_1} = \frac{h_2 + h_1}{h_1 h_2} \int G_J(x, y) g(y) m(dy).$$

$$\frac{h_2 + h_1}{2 h_2 h_1} G_J(x, y) = \begin{cases} \dfrac{y - (x - h_1)}{h_1}, & x - h_1 \leq y \leq x, \\[2ex] \dfrac{(x + h_2) - y}{h_2}, & x < y \leq x + h_2. \end{cases}$$

Taking h_1, $h_2 \downarrow 0$ separately we can show that both right- and left-hand derivatives exist. When h_1, h_2 both go to zero, the integrand converges to zero everywhere except at $y = x$. Use bounded convergence to establish (i).

By substituting $x + h_2 = b$, $x - h_1 = a$, form the identity,

$$(x - a)f(b) + (b - x)f(a) - f(x)(b - a) = (b - a)\int G_J(x, y)g(y)m(dy).$$

Substituting in $x + h$ for x and subtracting we get

$$f(b) - f(a) - (b - a)\,\frac{f(x + h) - f(x)}{h}$$
$$= (b - a)\int \frac{G_J(x + h, y) - G_J(x, y)}{h}\,g(y)m(dy).$$

The integrand is bounded for all h, and

$$\lim_{h \downarrow 0}(b - a)\,\frac{G_J(x + h, y) - G_J(x, y)}{h} = \begin{cases} -2(y - a), & y < x, \\ 2(b - y), & y > x. \end{cases}$$

Therefore $g(x)m(\{x\}) = 0$ implies that

$$f(b) - f(a) - (b - a)f'(x) = -2\int_a^x (y - a)g\,dm + 2\int_x^b (b - y)g\,dm.$$

Take $a < x_1 < x_2 < b$, such that

$$g(x_1)m(\{x_1\}) = g(x_2)m(\{x_2\}) = 0.$$

Use both x_1 and x_2 in the equation above and subtract

$$(b - a)\big(f'(x_2) - f'(x_1)\big) = 2\int_{x_1}^{x_2} (y - a)g\,dm + 2\int_{x_2}^{x_1} (b - y)g\,dm$$
$$= 2(b - a)\int_{x_1}^{x_2} g\,dm. \qquad \text{Q.E.D.}$$

There is an interesting consequence of this theorem in an important special case:

Corollary 16.65. *Let $m(dx) = V(x)\,dx$ on $\mathcal{B}(\text{int } F)$ where $V(x)$ is continuous on* int(F). *Then $f \in \mathfrak{D}(S, \mathfrak{C})$ implies that $f''(x)$ exists and is continuous on* int(F), *and for $x \in$* int(F),

$$(Uf)(x) = \frac{1}{2V(x)}\,\frac{d^2f(x)}{2}.$$

Problems

16. If b is a closed left endpoint, $f \in \mathfrak{D}(S, \mathfrak{C})$, show that

$$b \text{ reflecting} \Rightarrow f'_+(b) = m(\{b\})(Sf)(b),$$

where $f'_+(b)$ is the one-sided derivative

$$\lim_{h \downarrow 0} \frac{f(b+h) - f(b)}{h}, \quad b + h \in F.$$

Show also that

$$b \text{ absorbing or exit} \Rightarrow (Sf)(b) = 0.$$

17. If $\mathsf{X}(t)$ is not on its natural scale, use the definition

$$g(x) = \left(\frac{df}{du}\right)(x) \Leftrightarrow f(x_2) - f(x_1) = \int_{(x_1,x_2)} g(x)\, du(x).$$

Show that for $f \in \mathfrak{D}(S, \mathfrak{C})$, $x \in \text{int}(F)$,

$$(Uf)(x) = \frac{1}{2}\left(\frac{d}{dm}\frac{df}{du}\right)(x).$$

10. UNIQUENESS

U is given by the scale and speed. But the scale and speed can also be recovered from U.

Proposition 16.66. *$p^+(x, J)$ satisfies*

$$(Up^+)(x) = 0, \quad x \in J;$$

and $m(x, J)$ satisfies

$$(Um)(x) = -1, \quad x \in J,$$

for J open and finite, $\bar{J} \subset F$.

Proof. Let $I \subset J$, $I = (x_1, x_2)$. Then for $x \in I$,

$$E_x p^+(\mathsf{X}(t^*(I)), J) = p^+(x_2, J)p^+(x, I) + p^+(x_1, J)p^-(x, I) = p^+(x, J),$$

$$E_x m(\mathsf{X}(t^*(I), J)) = m(x_2, J)p^+(x, I) + m(x_1, J)p^-(x, I) = m(x, J) - m(x, I).$$

For b a closed reflecting endpoint and $J = [b, c)$, even simpler identities hold and 16.66 is again true for J of this form.

To complete the recovery of m, p^+ from U, one needs to know that the solutions of the equations in 16.66 are unique, subject to the conditions m, p^+ continuous on $J = (a, b)$, and $p^+(b-, J) = 1$, $p^+(a+, J) = 0$, $m(a+, J) = 0$, $m(b-, J) = 0$.

Proposition 16.67. *Let $J = (a, b)$. If $f(x)$ is continuous on $\bar{J}$, then*

$$(Uf)(x) = 0, \quad x \in J, \qquad f(a) = f(b) = 0 \Rightarrow f(x) \equiv 0, \quad x \in J.$$

Proof. This is based on the following minimum principle for U:

Lemma 16.68. *If $h(x)$ is a continuous function in some neighborhood of $x_0 \in F$, if $h \in \mathfrak{D}(U, \{x_0\})$, and if $h(x)$ has a local minimum at x_0, then*

$$(Uh)(x_0) \geq 0.$$

This is obvious from the expression for U. Now suppose there is a function $\varphi(x)$ continuous on $\bar{J}$ such that $\varphi > 0$ on J and $U\varphi = \lambda\varphi$ on J, $\lambda > 0$. Suppose that $f(x)$ has a minimum in J, then for $\epsilon > 0$ sufficiently small, $f - \epsilon\varphi$ has a minimum in J, but

$$U(f - \epsilon\varphi) = -\lambda\epsilon\varphi < 0 \quad \text{on } J.$$

By the minimum principle, f cannot have a minimum. Similarly it cannot have a maximum, thus $f \equiv 0$. All that is left to do is get φ. This we will do shortly in Theorem 16.69 below.

This discussion does not establish the fact that U determines $m(\{b\})$ at a closed, reflecting endpoint b, because if $J = [b, c)$, then $m(b, J) \neq 0$. In this case suppose there are two solutions, f_1, f_2 of $Uf = -1$, $x \in J$, such that both are continuous on $\bar{J}$, and $f_1(c) = f_2(c) = 0$, $f_1(b) = \alpha$, $f_2(b) = \beta > \alpha$. Form

$$f(x) = f_2(x) - \frac{\beta}{\alpha} f_1(x) \qquad \text{so} \qquad Uf = \frac{\beta}{\alpha} - 1 > 0.$$

The minimum principle implies that $f(x) \leq 0$ on J. But $(Uf)(b) > 0$ implies there is a number d, $b < d < c$, such that $f(d) - f(b) > 0$. Hence $\beta = \alpha$, $f_1(x) = f_2(x)$ on J.

Theorem 16.69. $U\varphi = \lambda\varphi$, $x \in \operatorname{int}(F)$, $\lambda > 0$, *has two continuous solutions $\varphi_+(x)$, $\varphi_-(x)$ such that*

i) $\varphi_+(x) > 0$, $\varphi_-(x) > 0$, $x \in \operatorname{int}(F)$,
ii) *$\varphi_+(x)$ is strictly increasing, $\varphi_-(x)$ is strictly decreasing,*
iii) *at closed endpoints b, φ_+ and φ_- have finite limits as $x \to b$, at open right (left) endpoints $\varphi_+(\varphi_-) \to \infty$,*
iv) *any other continuous solution of $U\varphi = \lambda\varphi$, $x \in \operatorname{int}(F)$ is a linear combination of φ_+, φ_-.*

Proof. The idea is simple in the case $F = [a, b]$. Here we could take

$$\varphi_+(x) = E_x e^{-\lambda t_b^*}, \qquad \varphi_-(x) = E_x e^{-\lambda t_a^*}$$

and show that these two satisfy the theorem. In the general case, we have to use extensions of these two functions. Denote for $x < z$,

$$\theta_+(x, z) = E_x e^{-\lambda t^*_z},$$
$$\theta_-(z, x) = E_z e^{-\lambda t^*_x}.$$

For $x < y < z$, the time from x to z consists of the time from x to y plus the time from y to z. Use the strong Markov property to get

(16.70) $\theta_+(x, z) = \theta_+(x, y)\theta_+(y, z), \qquad \theta_-(z, x) = \theta_-(z, y)\theta_-(y, x).$

Now, in general, t^*_x and t^*_z are extended random variables and therefore not stopping times. But, by truncating them and then passing to the limit, identities such as (16.70) can be shown to hold. Pick z_0 in F to be the closed right endpoint if there is one, otherwise arbitrary. Define, for $x \le z_0$,

$$\varphi_+(x) = E_x e^{-\lambda t^*_{z_0}}.$$

Now take $z_1 > z_0$, and for $x \le z_1$ define

$$\varphi_+(x) = E_x e^{-\lambda t^*_{z_1}} / E_{z_0} e^{-\lambda t^*_{z_1}}.$$

Use (16.70) to check that the φ_+ as defined for $x \le z_1$ is an extension of φ_+ defined for $x \le z_0$.

Continuing this way we define $\varphi_+(x)$ on F. Note that $\varphi_+ > 0$ on int(F), and is strictly increasing. Now define φ_- in an analogous way. As $w \uparrow y$ or $w \downarrow y$, $t^*_w \to t^*_y$ a.s. P_x on the set $\{t^*_y < \infty\}$. Use this fact applied to (16.70) to conclude that φ_+ and φ_- are continuous. If the right-hand endpoint b is open, then lim E_x exp $[-\lambda t^*_z] = 0$ as $z \to b$, otherwise b is accessible. For $z > z_0$, the definition of φ_+ gives

$$\varphi_+(z) = 1/E_{z_0} e^{-\lambda t^*_z},$$

so $\varphi_+(z) \to +\infty$ as $z \to b$.

For y arbitrary in F, let $\varphi(x) = E_x e^{-\lambda t^*_y}$, then assert that

Proposition 16.71. *$\varphi \in \mathfrak{D}(U, F - \{y\})$, and for $x \ne y$,*

$$U\varphi = \lambda\varphi.$$

Proof. Take $h(x) \in \mathfrak{C}$ such that $h(x) = 0$ for all $x \le y$. Then $f = R_\lambda h$ is in $\mathfrak{D}(S, \mathfrak{C})$. By truncating, it is easy to show that the identity (16.61) holds for the extended stopping variable t^*_y. For $x \le y$, (16.61) now becomes

$$E_x e^{-\lambda t^*_y} f(X(t^*_y)) = f(x)$$

or

$$f(y)\varphi(x) = f(x), \quad x \le y.$$

By 15.51, $(\lambda - S)f = h$, so

$$(Uf)(x) = \lambda f(x), \quad x \le y.$$

But for $x < y$,

$$(Uf)(x) = f(y)(U\varphi)(x),$$

which leads to $U\varphi = \lambda\varphi$. An obviously similar argument does the trick for $x > y$.

Therefore φ_+, φ_- are solutions of $U\varphi = \lambda\varphi$ in int(F). Let φ be any solution of $U\varphi = \lambda\varphi$ in int(F). For $x_1 < x_2$ determine constants c_1, c_2 so that

$$\varphi(x_2) = c_1\varphi_+(x_2) + c_2\varphi_-(x_2),$$

$$\varphi(x_1) = c_1\varphi_+(x_1) + c_2\varphi_-(x_1).$$

This can always be done if $\varphi_+(x_2)\varphi_-(x_1) - \varphi_-(x_2)\varphi_+(x_1) \neq 0$. The function $D(x) = \varphi_+(x)\varphi_-(x_1) - \varphi_-(x)\varphi_+(x_1)$ is strictly increasing in x, so $D(x_1) = 0 \Rightarrow D(x_2) > 0$. The function $\tilde{\varphi}(x) = \varphi(x) - c_1\varphi_+(x) - c_2\varphi_-(x)$ satisfies $U\tilde{\varphi} = \lambda\tilde{\varphi}$, and $\tilde{\varphi}(x_1) = \tilde{\varphi}(x_2) = 0$. The minimum principle implies $\tilde{\varphi}(x) = 0$, $x_1 \le x \le x_2$. Thus

$$\varphi(x) = c_1\varphi_+(x) + c_2\varphi_-(x), \quad x_1 \le x \le x_2.$$

This must hold over int(F). For if c_1', c_2' are constants determined by a larger interval (x_1', x_2'), then

$$(c_1' - c_1)\varphi_+(x) + (c_2' - c_2)\varphi_-(x) = 0, \quad x_1 \le x \le x_2.$$

This is impossible unless φ_+, φ_- are constant over (x_1, x_2). Hence $c_1' = c_1$, $c_2' = c_2$.

We can now prove the uniqueness.

Theorem 16.72. *For $f(x)$ bounded and continuous on F, $g(x) = (R_\lambda f)(x)$, $\lambda > 0$, is in $\mathfrak{D}(U, F)$ and is the unique bounded continuous solution of*

$$(Ug)(x) = \lambda g(x) + f(x), \quad \text{all } x.$$

Proof. That $g(x) \in \mathfrak{D}(U, F)$ follows from 16.63, and 15.51. Further,

$$Sg = \lambda g + f \Rightarrow Ug = \lambda g + f.$$

Suppose that there were two bounded continuous solutions g_1, g_2 on F. Then $\varphi(x) = g_1(x) - g_2(x)$ satisfies

$$U\varphi = \lambda\varphi, \quad x \in F.$$

By 16.69,

$$\varphi(x) = c_1\varphi_+(x) + c_2\varphi_-(x), \quad x \in \text{int}(F).$$

If F has two open endpoints, φ cannot be bounded unless $c_1 = c_2 = 0$.

Therefore, take F to have at least one closed endpoint b, say to the left. Assume that $\varphi(b) \geq 0$; otherwise, use the solution $-\varphi(x)$. If $\varphi(b) > 0$, then

$$(U\varphi)(b) > 0 \Rightarrow \varphi(b + h) > \varphi(b)$$

for all h sufficiently small. Then $\varphi(x)$ can never decrease anywhere, because if it did, it would have a positive maximum in int(F), contradicting the minimum principle. If $F = [b, c)$, then $\varphi(x) = c_2\varphi_-(x)$. Since $\varphi_-(b) = 1$, the case $\varphi(b) = 0$ leads to $\varphi(x) \equiv 0$. But $\varphi(b) > 0$ is impossible because $\varphi_-(x)$ is decreasing. Finally, look at $F = [b, c]$. If $\varphi(c) > 0$ or if $\varphi(c) < 0$, then an argument similar to the above establishes that $\varphi(x)$ must be decreasing on F or increasing on F, respectively. In either case, $\varphi(b) \geq 0$ is impossible. The only case not ruled out is $\varphi(b) = \varphi(c) = 0$. Here the minimum principle gives $\varphi(x) \equiv 0$ and the theorem follows.

Corollary 16.73. *There is exactly one Feller process having a given scale function and speed measure.*

Proof. It is necessary only to show that the transition probabilities $\{p_t(B \mid x)\}$ are uniquely determined by $u(x)$, $m(dx)$. This will be true if $E_x f(\mathsf{X}(t))$ is uniquely determined for all bounded continuous f on F. The argument used to prove uniqueness in 15.51 applies here to show that $E_x f(\mathsf{X}(t))$ is completely determined by the values of $(R_\lambda f)(x)$, $\lambda > 0$. But $g(x, \lambda) = (R_\lambda f)(x)$ is the unique bounded continuous solution of $Ug = \lambda g + f$, $x \in F$.

11. $\varphi_+(x)$ AND $(\varphi_- x)$

These functions, (which depend also on λ) have a central role in further analytic developments of the theory of Feller processes.

For example, let J be any finite interval, open in F, with endpoints x_1, x_2. The first passage time distributions from J can be specified by the two functions

$$\theta^+(\lambda, x) = \int_{A_+} e^{-\lambda t^*(J)}\, dP_x, \qquad \theta^-(\lambda, x) = \int_{A_-} e^{-\lambda t^*(J)}\, dP_x,$$

where $A_+ = \{\mathsf{X}(t^*(J)) = x_2\}$, A_- is the other exit set.

Theorem 16.74 (*Darling and Siegert*)

$$\theta^+(\lambda, x) = \frac{\varphi_+(x_1)\varphi_-(x) - \varphi_+(x)\varphi_-(x_1)}{\varphi_+(x_1)\varphi_-(x_2) - \varphi_+(x_2)\varphi_-(x_1)},$$

$$\theta^-(\lambda, x) = \frac{\varphi_-(x_2)\varphi_+(x) - \varphi_-(x)\varphi_+(x_2)}{\varphi_+(x_1)\varphi_-(x_2) - \varphi_+(x_2)\varphi_-(x_1)}.$$

Remark. Since these expressions are invariant under linear transformations, we could just as well take instead of φ_+, φ_- any two linearly independent solutions g_+, g_- of $Ug = \lambda g$ such that $g_+ \uparrow, g_- \downarrow$.

Proof. Use the strong Markov property to write

$$E_x e^{-\lambda t^*_{x_2}} = \int_{A_+} e^{-\lambda t^*(J)}\, dP_x + (E_{x_1} e^{-\lambda t^*_{x_2}}) \int_{A_-} e^{-\lambda t^*(J)}\, dP_x,$$

$$E_x e^{-\lambda t^*_{s_1}} = (E_{x_2} e^{-\lambda t^*_{s_1}}) \int_{A} e^{-\lambda t^*(J)}\, dP_x + \int_{A_-} e^{-\lambda t^*(J)}\, dP_x.$$

Let $h^+(x) = E_x e^{-\lambda t^*_{x_2}}$, $h^-(x) = E_x e^{-\lambda t^*_{x_1}}$, and solve the above equations to get

$$\theta^+(\lambda, x) = \frac{h^+(x) - h^+(x_1)h^-(x)}{1 - h^-(x_2)h^+(x_1)}.$$

Since $h^-(x_1) = 1$, $h^+(x_2) = 1$, this has the form of the theorem. By construction $h^-(x)$, $h^+(x)$ are constant multiples of $\varphi_-(x)$, $\varphi_+(x)$.

More important, in determining the existence of densities of the transition probabilities $p_t(B \mid x)$ and their differentiability is the following sequence of statements:

Theorem 16.75

i) $\dfrac{d}{dx}\varphi_+(x)$, $\dfrac{d}{dx}\varphi_-(x)$ *both exist except possibly at a countable number of points.*

ii) $\mathfrak{Z} = \varphi_- \dfrac{d\varphi_-}{dx} - \varphi_+ \dfrac{d\varphi_-}{dx}$ *is constant except possibly where* $\dfrac{d}{dx}\varphi_+$ *or* $\dfrac{d}{dx}\varphi_-$ *do not exist.*

iii) $R_\lambda(dy \mid x) \ll m(dy)$ *on* $\mathcal{B}_1(F)$, *and putting* $r_\lambda(y, x) = \dfrac{dR_\lambda}{dm}$, *we have*

$$r_\lambda(y, x) = \begin{cases} \dfrac{1}{\mathfrak{Z}}\varphi_+(y)\varphi_-(x), & y \le x, \\ \dfrac{1}{\mathfrak{Z}}\varphi_-(y)\varphi_+(x), & y \ge x, \end{cases}$$

or equivalently,

$$(R_\lambda f)(x) = \int f(y) r_\lambda(y, x) m(dy)$$

for $r_\lambda(y, x)$ *as above.*

The proof, which I will not give (see Ito and McKean, pp. 149 ff.), goes like this: Show that $U\varphi_+$, $U\varphi_-$ have the differential form given in 16.64. Then statement (ii) comes from the fact that

$$\int_{(x_1,x_2)} (\varphi_- U\varphi_+ - \varphi_+ U\varphi_-)\, dm = 0.$$

Statement (iii) comes from showing directly that the function

$$g(x) = \int f(y) r_\lambda(y, x) m(dy)$$

satisfies $Ug = \lambda g + f$ for all continuous bounded f.

The statement 16.75(iii) together with some eigenvalue expansions finally shows that $p_t(dy \mid x) \ll m(dy)$ and that $p_t(y \mid x)$, *the density with respect to* $m(dy)$ exists and is a symmetric function of y, x. Also, the same development shows that $p_t(y \mid x) \in \mathfrak{D}(S, \mathfrak{C})$ and that

$$\frac{\partial}{\partial t} p_t(y \mid x) = (Sp_t(y \mid \cdot))(x).$$

Unfortunately, the proofs of these results require considerably more analytic work. It is disturbing that such basic things as existence of densities and the proofs that $p_t(y \mid x)$ are sufficiently differentiable to satisfy the backwards equations lie so deep.

Problem 18. If regular Feller process has a stationary initial distribution $\pi(dx)$ on F, then show that $\pi(dx)$ must be a constant multiple of the speed measure $m(dx)$. [Use 16.75(iii).]

12. DIFFUSIONS

There is some disagreement over what to call a diffusion. We more or less follow Dynkin [44].

Definition 16.76. *A diffusion is a Feller process on F such that there exist functions* $\sigma^2(x)$, $\mu(x)$ *defined and continuous on* int(F), $\sigma^2(x) > 0$ *on* int(F), *with*

i) $\displaystyle \frac{1}{t} P_x(|X(t) - x| \geq \epsilon) \xrightarrow{bp} 0,$

ii) $\displaystyle \frac{1}{t} \int_{|X(t)-x|<\epsilon} (X(t) - x)\, dP_x \xrightarrow{bp} \mu(x),$

iii) $\displaystyle \frac{1}{t} \int_{|X(t)-x|<\epsilon} (X(t) - x)^2\, dP_x \xrightarrow{bp} \sigma^2(x),$

where the convergence ($\xrightarrow{bp}$) *is bounded pointwise on all finite intervals* J, $\bar{J} \subset$ int(F).

So a diffusion is pretty much what we started in with—a process that is locally Brownian. Note

Proposition 16.77. *Let $f(x)$ have a continuous second derivative on* int(F). *For a diffusion,* $f \in \mathfrak{D}(U, \text{int}(F))$ *and*

$$(Uf)(x) = \tfrac{1}{2}\sigma^2(x)\frac{d^2f}{dx^2} + \mu(x)\frac{df}{dx}.$$

Proof. Define $f(x) = \tilde{f}(x)$ in some neighborhood J of x such that $\tilde{f}(x)$ is bounded and has continuous second derivative on int(F) and vanishes outside a compact interval I contained in the interior of F. On I^c prove that $(S_t\tilde{f})(x) \xrightarrow{bp} 0$. Apply a Taylor expansion now to conclude that $\tilde{f} \in \mathfrak{D}(S, \mathcal{C})$, and that $(S\tilde{f})(x)$ is given by the right-hand side above.

The scale is given in terms of μ, σ by

Proposition 16.78. *For a diffusion, the scale function $u(x)$ is the unique (up to a linear transformation) solution on* int(F) *of*

$$\textbf{(16.79)} \qquad \tfrac{1}{2}\sigma^2(x)\frac{d^2u(x)}{dx^2} + \mu(x)\frac{du(x)}{dx} = 0.$$

Proof. Equation (16.79) has the solution

$$\textbf{(16.80)} \qquad \frac{du_0(x)}{dx} = \exp\left[-\int_{x_0}^{x}\frac{2\mu(z)}{\sigma^2(z)}\,dz\right]$$

for an arbitrary x_0. This $u_0(x)$ has continuous second derivative on int(F). Thus by 16.77 $(Uu_0)(x) = 0$, $x \in \text{int}(F)$, and $u_0(x)$ is a scale function.

Now that the scale is determined, transform to natural scale, getting the process

$$\tilde{X}(t) = u(X(t)).$$

Proposition 16.81. $\tilde{X}(t)$ *is a diffusion with zero drift and*

$$\tilde{\sigma}^2(y) = \sigma^2(x)[u'(x)]^2, \quad y = u(x).$$

Proof. Proposition 16.81 follows from

Proposition 16.82. *Let $X(t)$ be a diffusion on F, $w(x)$ a function continuous on F such that $|w'(x)| \neq 0$ on* int(F), *$w''(x)$ continuous on* int(F). *Then $\tilde{X}(t) = w(X(t))$ is a diffusion. If $\tilde{x} = w(x)$,*

$$\tilde{\mu}(\tilde{x}) = \tfrac{1}{2}\sigma^2(x)w''(x) + \mu(x)w'(x),$$

$$\tilde{\sigma}^2(\tilde{x}) = \sigma^2(x)[w'(x)]^2.$$

Proof. The transformations of the drift and variance here come from

$$\frac{E_x w(\mathsf{X}(t)) - w(x)}{t} \to Sw(x),$$

$$\frac{E_x\big(w(\mathsf{X}(t)) - w(x)\big)^2}{t} = \frac{E_x\big(w^2(\mathsf{X}(t)) - w^2(x)\big)}{t} - 2w(x) \cdot \frac{E_x w(\mathsf{X}(t)) - w(x)}{t}$$

$$\to Sw^2 - 2w \cdot Sw.$$

What needs to be shown is that for the revised process, $\xrightarrow{bp} \tilde{\mu}(\tilde{x})$, $\xrightarrow{bp} \tilde{\sigma}^2(\tilde{x})$ takes place on finite intervals in the interior of $\tilde{F}$. This is a straightforward verification, and I omit it.

By this time it has become fairly apparent from various bits of evidence that the following should hold:

Theorem 16.83. *For a diffusion with zero drift the speed measure is given on* $\mathcal{B}_1(\text{int } F)$ *by*

$$m(dx) = \frac{1}{\sigma^2(x)}\, dx.$$

Proof. Take f to be zero off a compact interval $I \subset \text{int}(F)$, with a continuous second derivative. There are two expressions, given by 16.77 and 16.64, for $(Uf)(x)$. Equating these,

$$\sigma^2(x)\frac{d^2f(x)}{dx^2} = \left(\frac{d}{dm}\frac{df}{dx}\right)(x)$$

or,

$$f'(x_2) - f'(x_1) = \int_{(x_1,x_2)} \sigma^2(y) f''(y) m(dy)$$

$$= \int_{(x_1,x_2)} f''(y)\, dy$$

which implies the theorem.

These results show that $\mu(x)$, $\sigma^2(x)$ determine the scale function and the speed measure completely on int(F). Hence, specifying m on any and all regular boundary points completely specifies the process. The basic uniqueness result 16.73 guarantees at most one Feller process with the given scale and speed. Section 8 gives a construction for the associated Feller process. What remains is to show that the process constructed is a diffusion.

Theorem 16.84. *Let* $m(dx) = V(x)\,dx$ *on* int(F), $V(x)$ *continuous and positive on* int(F). *The process on natural scale with this speed measure is a diffusion with* $\mu(x) = 0$, $\sigma^2(x) = 1/V(x)$.

Proof. Use the representation $X(t) = X_0(T(t))$, where $T(t)$ is the random time change of Section 8. The first step in the proof is: Let the interval $J_\epsilon = (x - \epsilon,\ x + \epsilon)$, take $J \subset \text{int}(F)$ finite such that for any $x \in J$, $J_\epsilon \subset I \subset \text{int}(F)$. Then

i°) $$(1/t)P_x(t^*(J_\epsilon) \leq t) \xrightarrow{bp} 0, \quad x \in J.$$

Proof of (i°). $\{t^*(J_\epsilon) \leq t\} = \{t_0^*(J_\epsilon) \leq T(t)\}$. In the latter set, by taking inverses we get

$$\{t_0^*(J_\epsilon) \leq T(t)\} = \left\{\int_0^{t_0^*(J_\epsilon)} V(X_0(\xi))\, d\xi \leq t\right\}.$$

But letting $M = \inf_I V(y)$, we find that, denoting $J_\epsilon^0 = (-\epsilon, \epsilon)$,

$$P_x\left(\int_0^{t_0^*(J_\epsilon)} V(X_0(\xi))\, d\xi \leq t\right) \leq P_x(Mt_0^*(J_\epsilon) \leq t) = P_0(t_0^*(J_\epsilon^0) \leq t/M).$$

Since $P_x(|X(t) - x| \geq \epsilon) \leq P_x(t^*(J_\epsilon) \leq t)$, condition (i) of the diffusion definition 16.76 is satisfied.

To do (ii) and (iii): By (i°) it is sufficient to prove

$$\frac{1}{t}\int_{\{t<t^*(J_\epsilon)\}} (X(t) - x)\, dP_x \xrightarrow{bp} 0, \qquad \frac{1}{t}\int_{\{t<t^*(J_\epsilon)\}} (X(t) - x)^2\, dP_x \xrightarrow{bp} \frac{1}{V(x)},$$

for x in finite J, $\bar{J} \subset \text{int}(F)$.

Let $\tau_\epsilon^* = \min\,(t, t^*(J_\epsilon))$. Then we will show that

ii°) $$E_x X(\tau_\epsilon^*) = x, \qquad E_x X^2(\tau_\epsilon^*) = x^2 + E_x T(\tau_\epsilon^*).$$

Proof of (ii°). Use (16.52); that is (dropping the ϵ),

$$t_0^*(J) = T(t^*(J)),$$
$$X(\tau^*) = X_0(T(\tau^*)) = X_0(\min\,(T(t), t_0^*(J))).$$

But $\tau_0^* = \min\,(T(t), t_0^*(J))$ is a stopping time for Brownian motion. Since $X_0(t)$, $X_0^2(t) - t$ are martingales, (ii°) follows.

Use (ii°) as follows:

$$\int_{\{t<t^*(J_\epsilon)\}} X(t)\, dP_x = \int_{\{t<t^*(J_\epsilon)\}} X(\tau_\epsilon^*)\, dP_x = x - \int_{\{t\geq t^*(J_\epsilon)\}} X(\tau_\epsilon^*)\, dP_x$$

Hence, we get the identity

$$\int_{\{t<t^*(J_\epsilon)\}} (X(t) - x)\, dP_x = -\int_{\{t\geq t^*(J_\epsilon)\}} (X(t^*(J_\epsilon)) - x)\, dP_x.$$

This latter integral is in absolute value less than $\epsilon P_x(\mathsf{t}^*(J_\epsilon) \le t)$. Apply (i°) now.

For (iii), write

$$\int_{\{t<\mathsf{t}^*(J_\epsilon)\}} (\mathsf{X}(t) - x)^2 \, dP_x = \int_{\{t<\mathsf{t}^*(J_\epsilon)\}} (\mathsf{X}(\boldsymbol{\tau}_\epsilon^*) - x)^2 \, dP_x$$

$$= E_x(\mathsf{X}(\boldsymbol{\tau}_\epsilon^*) - x)^2 - \int_{\{t\ge\mathsf{t}^*(J_\epsilon)\}} (\mathsf{X}(\boldsymbol{\tau}_\epsilon^*) - x)^2 \, dP_x.$$

By (ii°),

$$E_x(\mathsf{X}(\boldsymbol{\tau}_\epsilon^*) - x)^2 = E_x\mathsf{T}(\boldsymbol{\tau}_\epsilon^*).$$

The remaining integral above is bounded by $\epsilon^2 P_x(\mathsf{t}^*(J_\epsilon) \le t)$, hence gives no trouble. To evaluate $E_x\mathsf{T}(\boldsymbol{\tau}_\epsilon^*)$, write

$$t \ge \int_0^{\mathsf{T}(\boldsymbol{\tau}_\epsilon^*)} V(\mathsf{X}_0(\xi)) \, d\xi \ge M\mathsf{T}(\boldsymbol{\tau}_\epsilon^*),$$

so that

$$\frac{1}{t} E_x\mathsf{T}(\boldsymbol{\tau}_\epsilon^*) \le \frac{1}{M},$$

for all t and $x \in J$. Further, since $\mathsf{T}(t) \to 0$ as $t \to 0$,

$$\frac{t}{\mathsf{T}(t)} = \frac{1}{\mathsf{T}(t)} \int_0^{\mathsf{T}(t)} V(\mathsf{X}_0(\xi)) \, d\xi \to V(\mathsf{X}_0(0)) \quad \text{a.s.}$$

as $t \to 0$. Since $\mathsf{T}(\boldsymbol{\tau}_\epsilon^*)/t$ is bounded by M, the bounded convergence theorem can be applied to get

$$\lim_{t\to 0} \frac{1}{t} E_x \mathsf{T}(\boldsymbol{\tau}_\epsilon^*) = \frac{1}{V(x)}$$

for every $x \in J$.

These results help us to characterize diffusions. For example, now we know that of all Feller processes on natural scale, the diffusion processes are those such that $dm \ll dx$ on int(F) and dm/dx has a continuous version, positive on int(F). A nonscaled diffusion has the same speed measure property, plus a scale function with nonvanishing first derivative and continuous second derivative.

Problem 19. Show that for a diffusion, the functions φ_+, φ_- of Section 10 are solutions in int(F) of

$$\tfrac{1}{2}\sigma^2 \frac{d^2\varphi}{dx^2} + \mu \frac{d\varphi}{dx} = \lambda\varphi.$$

NOTES

Most of the material in this chapter was gleaned from two recent books on this subject, Dynkin [44], of which an excellent English translation was published in 1965, and Ito, McKean [76, 1965].

The history of the subject matter is very recent. The starting point was a series of papers by Feller, beginning in 1952 [57], which used the semigroup approach. See also [58]. Dynkin introduced the idea of the characteristic operator in [43, 1955], and subsequently developed the theory using the associated concepts. The idea of random time substitutions was first exploited in this context by Volkonskiĭ [138, 1958]. The construction of the general process using local time was completed by Ito and McKean [76].

The material as it stands now leaves one a little unhappy from the pedagogic point of view. Some hoped-for developments would be: (1) a simple proof of the local time theorem (David Freedman has shown me a simple proof of all the results of the theorem excepting the continuity of $\mathfrak{l}^*(t, y)$); (2) a direct proof of the unique determination of the process by scale function and speed measure to replace the present detour by means of the characteristic operator; (3) a simplification of the proofs of the existence of densities and smoothness properties for the transition probabilities.

Charles Stone has a method of getting Feller processes as limits of birth and death processes which seems to be a considerable simplification both conceptually and mathematically. Most of this is unpublished. However, see [131] for a bit of it.

The situation in two or more dimensions is a wilderness. The essential property in one dimension that does not generalize is that if a path-continuous process goes from x to y, then it has to pass through all the points between x and y. So far, the most powerful method for dealing with diffusions in any number of dimensions is the use of stochastic integral equations (see Doob [39, Chap. VI], Dynkin [44, Chap. XI]) initiated by Ito. The idea here is to attempt a direct integration to solve the equation

$$\Delta \mathsf{Y} = \mu(\mathsf{Y})\,\Delta t + \sigma^2(\mathsf{Y})\,\Delta \mathsf{X}$$

and its multidimensional analogs.

APPENDIX

ON MEASURE AND FUNCTION THEORY

The purpose of this appendix is to give a brief review, with very few proofs, of some of the basic theorems concerning measure and function theory. We refer for the proofs to Halmos [64] by page number.

1. MEASURES AND THE EXTENSION THEOREM

For Ω a set of points ω, define

Definition A.1. *A class $\mathcal{F}$ of subsets of Ω is a field if A, $B \in \mathcal{F}$ implies A^c, $A \cup B$, $A \cap B$ are in $\mathcal{F}$. The class $\mathcal{F}$ is σ-field if it is a field, and if, in addition, $A_n \in \mathcal{F}$, $n = 1, 2, \ldots$ implies $\bigcup_1^\infty A_n \in \mathcal{F}$.*

Notation A.2. *We will use A^c for the complement of A, $A - B$ for $A \cap B^c$, $\emptyset$ for the empty set, $A \,\triangle\, B$ for the symmetric set difference*

$$(A - B) \cup (B - A).$$

Note

Proposition A.3. *For any class $\mathcal{C}$ of subsets of Ω, there is a smallest field of subsets, denoted $\mathcal{F}_0(\mathcal{C})$, and a smallest σ-field of subsets, denoted $\mathcal{F}(\mathcal{C})$, containing all the sets in $\mathcal{C}$.*

Proof. The class of all subsets of Ω is a σ-field containing $\mathcal{C}$. Let $\mathcal{F}(\mathcal{C})$ be the class of sets A such that A is in every σ-field that contains $\mathcal{C}$. Check that $\mathcal{F}(\mathcal{C})$ so defined is a σ-field and that if $\mathcal{F}$ is a σ-field, $\mathcal{C} \subset \mathcal{F}$, then $\mathcal{F}(\mathcal{C}) \subset \mathcal{F}$. For fields a finite construction will give $\mathcal{F}_0(\mathcal{C})$.

If A_n is a sequence of sets such that $A_n \subset A_{n+1}$, $n = 1, 2, \ldots$ and $A = \cup A_n$, write $A_n \uparrow A$. Similarly, if $A_{n+1} \subset A_n$, $A = \cap A_n$, write $A_n \downarrow A$. *Define a monotone class of subsets $\mathcal{C}$ by: If $A_n \in \mathcal{C}$, and $A_n \uparrow A$ or $A_n \downarrow A$ then $A \in \mathcal{C}$.*

Monotone Class Theorem A.4 *(Halmos, p. 27). The smallest monotone class of sets containing a field $\mathcal{F}_0$ is $\mathcal{F}(\mathcal{F}_0)$.*

Definition A.5. *A finitely additive measure μ on a field $\mathcal{F}$ is a real-valued (including $+\infty$), nonnegative function with domain $\mathcal{F}$ such that for $A, B \in \mathcal{F}$, $A \cap B = \emptyset$,*

$$\mu(A \cup B) = \mu(A) + \mu(B).$$

This extends to: If $A_1, \ldots, A_n \in \mathcal{F}$ are pairwise disjoint, $A_i \cap A_j = \emptyset$, $i \neq j$, then

$$\mu\left(\bigcup_1^n A_j\right) = \sum_1^n \mu(A_j).$$

Whether or not the sets $A_1, \ldots, A_n \in \mathcal{F}$ are disjoint,

$$\mu\left(\bigcup_1^n A_j\right) \leq \sum_1^n \mu(A_j).$$

Definition A.6. *A σ-additive measure (or just measure) on a σ-field $\mathcal{F}$ is a real-valued ($+\infty$ included), nonnegative function with domain $\mathcal{F}$ such that for $A_1, \ldots \in \mathcal{F}$, $A_i \cap A_j = \emptyset$, $i \neq j$,*

$$\mu\left(\bigcup_j A_j\right) = \sum_j \mu(A_j).$$

We want some finiteness:

Definition A.7. *A measure (finitely or σ-additive) on a field $\mathcal{F}_0$ is σ-finite if there are sets $A_k \in \mathcal{F}_0$ such that $\bigcup_k A_k = \Omega$ and for every k, $\mu(A_k) < \infty$.*

We restrict ourselves henceforth to σ-finiteness! The extension problem for measures is: Given a finitely additive measure μ_0 on a field $\mathcal{F}_0$, when does there exist a measure μ on $\mathcal{F}(\mathcal{F}_0)$ agreeing with μ_0 on $\mathcal{F}_0$? A measure has certain continuity properties:

Proposition A.8. *Let μ be a measure on the σ-field $\mathcal{F}$. If $A_n \downarrow A$, $A_n \in \mathcal{F}$, and if $\mu(A_n) < \infty$ for some n, then*

$$\lim_n \mu(A_n) = \mu(A).$$

Also, if $A_n \uparrow A$, $A_n \in \mathcal{F}$, then

$$\lim_n \mu(A_n) = \mu(A).$$

This is called continuity from above and below. Certainly, if μ_0 is to be extended, then the minimum requirement needed is that μ_0 be continuous on its domain. Call μ_0 *continuous from above at* $\emptyset$ if whenever $A_n \in \mathcal{F}_0$, $A_n \downarrow \emptyset$, and $\mu_0(A_n) < \infty$ for some n, then

$$\lim_n \mu_0(A_n) = 0.$$

Carathéodory Extension Theorem A.9. *If μ_0 on $\mathcal{F}_0$ is continuous from above at $\emptyset$, then there is a unique measure μ on $\mathcal{F}(\mathcal{F}_0)$ agreeing with μ_0 on $\mathcal{F}_0$ (see Halmos, p.* 54).

Definition A.10. *A measure space is a triple $(\Omega, \mathcal{F}, \mu)$ where $\mathcal{F}$, μ are a σ-field and measure. The completion of a measure space, denoted by $(\Omega, \bar{\mathcal{F}}, \bar{\mu})$, is gotten by defining $A \in \bar{\mathcal{F}}$ if there are sets A_1, A_2 in $\mathcal{F}$, $A_1 \subset A \subset A_2$ and $\mu(A_2 - A_1) = 0$. Then define $\bar{\mu}(A) = \mu(A_1)$.*

$\bar{\mathcal{F}}$ is the largest σ-field for which unique extension under the hypothesis of A.9 holds. That is, $\bar{\mu}$ is the only measure on $\bar{\mathcal{F}}$ agreeing with μ_0 on $\mathcal{F}_0$ and

Proposition A.11. *Let $B \notin \bar{\mathcal{F}}$, $\mathcal{F}_1$ the smallest σ-field containing both B and $\bar{\mathcal{F}}$. Then there is an infinity of measures on $\mathcal{F}_1$ agreeing with μ_0 on $\mathcal{F}_0$. (See Halmos, p.* 55 *and p.* 71, *Problem* 3).

Note that $\mathcal{F}(\mathcal{F}_0)$ depends only on $\mathcal{F}_0$ and not on μ_0, but that $\bar{\mathcal{F}}(\mathcal{F}_0)$ depends on μ_0. The measure μ on $\mathcal{F}(\mathcal{F}_0)$, being a unique extension, must be approximable in some sense by μ_0 on $\mathcal{F}_0$. One consequence of the extension construction is

Proposition A.12 (*Halmos, p.* 56). *For every $A \in \mathcal{F}(\mathcal{F}_0)$, and $\epsilon > 0$, there is a set $A_0 \in \mathcal{F}_0$ such that*

$$\mu(A \,\triangle\, A_0) \le \epsilon.$$

We will designate a space Ω and a σ-field $\mathcal{F}$ of subsets of Ω as a *measurable space* $(\Omega, \mathcal{F})$. If $F \subset \Omega$, denote by $\mathcal{F}(F)$ the σ-field of subsets of F of the form $A \cap F$, $A \in \mathcal{F}$, and take the complement relative to F.

Some important measurable spaces are

$R^{(1)}$	the real line
$\mathcal{B}_1$	the smallest σ-field containing all intervals
$R^{(k)}$	k-dimensional Euclidean space
$\mathcal{B}_k$	the smallest σ-field containing all k-dimensional rectangles
$R^{(\infty)}$	the space of all infinite sequences $(x_1, x_2, \ldots)$ of real numbers
$\mathcal{B}_\infty$	the smallest σ-field containing all sets of the form $\{(x_1, x_2, \ldots);\ x_1 \in I_1, \ldots, x_n \in I_n\}$ for any n where $I_1, \ldots, I_n$ are any intervals
R^I	the space of all real-valued functions $x(t)$ on the interval $I \subset R^{(1)}$
$\mathcal{B}_I$	the smallest σ-field containing all sets of the form $\{x(\cdot) \in R^I;\ x(t_1) \in I_1, \ldots, x(t_n) \in I_n\}$ for any $t_1, \ldots, t_n \in I$ and intervals $I_1, \ldots, I_n$

Definition A.13. *The class of all finite unions of disjoint intervals in* $R^{(1)}$ *is a field. Take* μ_0 *on this field to be length. Then* μ_0 *is continuous from above at* $\emptyset$ *(Halmos, pp.* 34 ff.). *The extension of length to* $\mathcal{B}_1$ *is Lebesgue measure, denoted by* l *or by* dx.

Henceforth, if we have a measure space $(\Omega, \mathcal{F}, \mu)$ and a statement holds for all $\omega \in \Omega$ with the possible exception of $\omega \in A$, where $\mu(A) = 0$, we say that the statement holds *almost everywhere* (a.e.)

2. MEASURABLE MAPPINGS AND FUNCTIONS

Definition A.14. *Given two spaces* Ω, R, *and a mapping* $\mathbf{X}(\omega)$: $\Omega \to R$, *the inverse image of a set* $B \subset R$ *is defined as*

$$\mathbf{X}^{-1}B = \{\omega \in \Omega; \mathbf{X}(\omega) \in B\}.$$

Denote this by $\{\mathbf{X} \in B\}$. *The taking of inverse images preserves all set operations; that is,*

$$\left\{\mathbf{X} \in \bigcup_\lambda B_\lambda\right\} = \bigcup_\lambda \{\mathbf{X} \in B_\lambda\},$$

$$\left\{\mathbf{X} \in \bigcap_\lambda B_\lambda\right\} = \bigcap_\lambda \{\mathbf{X} \in B_\lambda\},$$

$$\{\mathbf{X} \in B^c\} = \{\mathbf{X} \in B\}^c.$$

Definition A.15. *Given two measurable spaces* $(\Omega, \mathcal{F})$, $(R, \mathcal{B})$. *A mapping* $\mathbf{X}$: $\Omega \to R$ *is called measurable if the inverse of every set in* $\mathcal{B}$ *is in* $\mathcal{F}$.

Proposition A.16 (*See* 2.29). *Let* $\mathcal{C} \subset \mathcal{B}$ *such that* $\mathcal{F}(\mathcal{C}) = \mathcal{B}$. *Then* $\mathbf{X}$: $\Omega \to R$ *is measurable if the inverse of every set in* $\mathcal{C}$ *is in* $\mathcal{F}$.

Definition A.17. X: $\Omega \to R^{(1)}$ *will be called a measurable function if it is a measurable map from* $(\Omega, \mathcal{F})$ *to* $(R^{(1)}, \mathcal{B}_1)$.

From A.16 it is sufficient that $\{\mathsf{X} < x\} \in \mathcal{F}$ for all x in a set dense in $R^{(1)}$. Whether or not a function is measurable depends on both Ω and $\mathcal{F}$. Refer therefore to measurable functions on $(\Omega, \mathcal{F})$ as $\mathcal{F}$-measurable functions.

Proposition A.18. *The class of* $\mathcal{F}$*-measurable functions is closed under pointwise convergence. That is, if* $\mathsf{X}_n(\omega)$ *are each* $\mathcal{F}$*-measurable, and* $\lim_n \mathsf{X}_n(\omega)$ *exists for every* ω, *then* $\mathsf{X}(\omega) = \lim_n \mathsf{X}_n(\omega)$ *is* $\mathcal{F}$*-measurable.*

Proof. Suppose $\mathsf{X}_n(\omega) \downarrow \mathsf{X}(\omega)$ for each ω; then $\{\mathsf{X} < x\} = \bigcup_n \{\mathsf{X}_n < x\}$. This latter set is in $\mathcal{F}$. In general, if $\mathsf{X}_n(\omega) \to \mathsf{X}(\omega)$, take

$$\mathsf{Y}_n(\omega) = \sup_{m \geq n} \mathsf{X}_m(\omega).$$

Then $\{\mathsf{Y}_n > y\} = \bigcup_{m \geq n} \{\mathsf{X}_m > y\}$ which is in $\mathcal{F}$. Then Y_n is $\mathcal{F}$-measurable and $\mathsf{Y}_n \downarrow \mathsf{X}$.

Define an *extended measurable function* as a function $\mathsf{X}(\omega)$ which takes values in the extended real line $R^{(1)} \cup \{\infty\}$ such that $\{\mathsf{X} \in B\} \in \mathcal{F}$ for every $B \in \mathcal{B}_1$.

By the argument above, if X_n are $\mathcal{F}$-measurable, then $\overline{\lim}\, \mathsf{X}_n$, $\underline{\lim}\, \mathsf{X}_n$ are extended $\mathcal{F}$-measurable, hence the set

$$\{\lim \mathsf{X}_n \text{ exists}\} = \{\underline{\lim}\, \mathsf{X}_n = \overline{\lim}\, \mathsf{X}_n\} \cap \{|\underline{\lim}\, \mathsf{X}_n| < \infty\}$$

is in $\mathcal{F}$.

Proposition A.19 (*See* 2.31). *If* $\mathbf{X}$ *is a measurable mapping from* $(\Omega, \mathcal{F})$ *to* $(R, \mathcal{B})$ *and* φ *is a* $\mathcal{B}$*-measurable function, then* $\varphi(\mathbf{X})$ *is an* $\mathcal{F}$*-measurable function.*

The *set indicator* of a subset $A \subset \Omega$ is the function

$$\chi_A(\omega) = \begin{cases} 1, & \omega \in A, \\ 0, & \omega \in A^c. \end{cases}$$

A *simple function* is any finite linear combination of set indicators,

$$g(\omega) = \sum_{k=1}^{n} \alpha_k \chi_{A_k}(\omega)$$

of sets $A_k \in \mathcal{F}$.

Proposition A.20. *The class of* $\mathcal{F}$*-measurable functions is the smallest class of functions containing all simple functions and closed under pointwise convergence.*

Proof. For any $n \geq 0$ and $\mathsf{X}(\omega)$ a measurable function, define sets $A_k = \{\mathsf{X} \in [k/n, k + 1/n)\}$ and consider

$$\mathsf{X}_n = \sum_{k=-n^2}^{n^2} \frac{k}{n} \chi_{A_k}(\omega).$$

Obviously $\mathsf{X}_n \to \mathsf{X}$.

For any measurable mapping $\mathbf{X}$ from $(\Omega, \mathcal{F})$ to $(R, \mathcal{B})$, denote by $\mathcal{F}(\mathbf{X})$ *the* σ*-field of inverse images of sets in* $\mathcal{B}$. Now we prove 4.9.

Proposition A.21. *If* Z *is an* $\mathcal{F}(\mathbf{X})$*-measurable function, then there is a* $\mathcal{B}$*-measurable function* θ *such that* $\mathsf{Z} = \theta(\mathbf{X})$.

Proof. Consider the class of functions $\varphi(\mathbf{X})$, $\mathbf{X}$ fixed, as φ ranges over the $\mathcal{B}$-measurable functions. Any set indicator $\chi_A(\omega)$, $A \in \mathcal{F}(\mathbf{X})$, is in this class, because $A = \{\mathbf{X} \in B\}$ for some $B \in \mathcal{B}$. Hence

$$\chi_A(\omega) = \chi_B(\mathbf{X}).$$

Now the class is closed under addition, so by A.20 it is sufficient to show it closed under pointwise convergence. Let $\varphi_n(\mathbf{X}) \to \mathsf{Y}$, φ_n $\mathcal{B}$-measurable.

Let $B = \{\lim_n \varphi_n \text{ exists}\}$. Then $B \in \mathcal{B}$, and $\Omega = \{\mathbf{X} \in B\}$. Define

$$\varphi = \begin{cases} \lim_n \varphi_n, & \text{on } B \\ 0, & \text{on } B^c. \end{cases}$$

Obviously, then $\mathsf{Y} = \varphi(\mathbf{X})$.

We modify the proof of A.20 slightly to get 2.38;

Proposition A.22. *Consider a class $\mathfrak{L}$ of $\mathcal{F}$-measurable functions having the properties*

i) $\mathsf{X}, \mathsf{Y} \in \mathfrak{L}$, $\alpha, \beta \geq 0 \Rightarrow \alpha\mathsf{X} + \beta\mathsf{Y} \in \mathfrak{L}$,
ii) $\mathsf{X}_n \in \mathfrak{L}$, $\mathsf{X}_n \uparrow \mathsf{X} \Rightarrow \mathsf{X} \in \mathfrak{L}$,
iii) *for every $A \in \mathcal{F}$, $\chi_A(\omega) \in \mathfrak{L}$;*

then $\mathfrak{L}$ includes all nonnegative $\mathcal{F}$-measurable functions.

Proof. For $\mathsf{X} \geq 0$, $\mathcal{F}$-measurable, let $\mathsf{X}_n = \sum_{k=0}^{n^2} k/n\chi_{A_k}(\omega)$ where $A_k = \{\mathsf{X} \in [k/n, k + 1/n)\}$. Then certainly $\mathsf{X}_n \in \mathfrak{L}$, and $\mathsf{X}_{n'} \uparrow \mathsf{X}$ if we take n' the subsequence $\{2^m\}$.

3. THE INTEGRAL

Take $(\Omega, \mathcal{F}, \mu)$ to be a measure space. Let $\mathsf{X}(\omega) \geq 0$ be a nonnegative $\mathcal{F}$-measurable function. To define the integral of X let $\mathsf{X}_n \geq 0$ be simple functions such that $\mathsf{X}_n \uparrow \mathsf{X}$.

Definition A.23. *The integral $\int \mathsf{X}_n \, d\mu$ of the nonnegative simple function $\mathsf{X}_n = \Sigma\, \alpha_k \chi_{A_k}(\omega)$, $\alpha_k \geq 0$, is defined by $\Sigma\, \alpha_k \mu(A_k)$.*

For $\mathsf{X}_n \uparrow \mathsf{X}$, it is easy to show that $\int \mathsf{X}_{n+1} \, d\mu \geq \int \mathsf{X}_n \, d\mu \geq 0$.

Definition A.24. *Define $\int \mathsf{X} \, d\mu$ as $\lim_n \int \mathsf{X}_n \, d\mu$. Furthermore, the value of this limit is the same for all sequences of nonnegative simple functions converging up to X (Halmos, p. 101).*

Note that this limit may be infinite. For any $\mathcal{F}$-measurable function X, suppose that $\int |\mathsf{X}| \, d\mu < \infty$. In this case define $\mathcal{F}$-measurable functions

$$\mathsf{X}^+(\omega) = \begin{cases} \mathsf{X}(\omega), & \mathsf{X}(\omega) \geq 0, \\ 0, & \mathsf{X}(\omega) < 0, \end{cases}$$

$$\mathsf{X}^-(\omega) = \begin{cases} -\mathsf{X}(\omega), & \mathsf{X}(\omega) \leq 0, \\ 0, & \mathsf{X}(\omega) > 0. \end{cases}$$

Note $|\mathsf{X}| = \mathsf{X}^+ + \mathsf{X}^-$.

Definition A.25. *If* $\int |X|\, d\mu < \infty$, *define*

$$\int X\, d\mu = \int X^+\, d\mu - \int X^-\, d\mu;$$

we may sometimes use the notation

$$\int X(\omega)\mu(d\omega).$$

The elementary properties of the integral are: If the integrals of X and Y exist,

i) $X \geq Y \Rightarrow \int X\, d\mu \geq \int Y\, d\mu,$

ii) $\int (\alpha X + \beta Y)\, d\mu = \alpha \int X\, d\mu + \beta \int Y\, d\mu,$

iii) $A, B \in \mathcal{F}, A \cap B = \emptyset \Rightarrow \int_{A \cup B} X\, d\mu = \int_A X\, d\mu + \int_B X\, d\mu.$

Some nonelementary properties begin with

Monotone Convergence Theorem A.26 (*Halmos, p.* 112). *For* $X_n \geq 0$ *non-negative* $\mathcal{F}$*-measurable functions,* $X_n \uparrow X$, *then*

$$\lim \int X_n\, d\mu = \int X\, d\mu.$$

From this comes the

Fatou Lemma A.27. If $X_n \geq 0$, *then*

$$\int \underline{\lim}\, X_n\, d\mu \leq \underline{\lim} \int X_n\, d\mu.$$

Proof. To connect up with A.26 note that for $X_1, \ldots, X_n$ arbitrary, non-negative $\mathcal{F}$-measurable functions,

$$\int \inf (X_1, \ldots, X_n)\, d\mu \leq \int X_k\, d\mu, \quad k = 1, \ldots, n.$$

Hence, by taking limits

$$\int \inf_n X_n\, d\mu \leq \inf_n \int X_n\, d\mu.$$

Let $Y_n = \inf_{m \geq n} X_m$; then

$$\int Y_n\, d\mu \leq \inf_{m \geq n} \int X_n\, d\mu.$$

Since $Y_n \geq 0$, and $Y_n \uparrow \underline{\lim}\, X_n$, apply A.26 to complete the proof.

Another useful convergence result is:

Bounded Convergence Theorem A.28 (2.44). *Let* $\mathsf{X}_n \to \mathsf{X}$ *pointwise, where the* X_n *are* $\mathcal{F}$*-measurable functions such that there is an* $\mathcal{F}$*-measurable function* Z *with* $|\mathsf{X}_n| \le \mathsf{Z}$, *all* n, ω, *and* $\int \mathsf{Z}\, d\mu < \infty$. *Then* (*see Halmos, p.* 110)

$$\lim_n \int \mathsf{X}_n \, d\mu = \int \mathsf{X}\, d\mu.$$

From these convergence theorems can be deduced the *σ-additivity* of an integral: For $\{B_n\}$ disjoint, $B_n \in \mathcal{F}$, and $\int |\mathsf{X}|\, d\mu < \infty$,

$$\int_{\cup B_n} \mathsf{X}\, d\mu = \sum \int_{B_n} \mathsf{X}\, d\mu.$$

Also, if $\int |\mathsf{X}|\, d\mu < \infty$, then the integral is absolutely continuous. That is, for every $\epsilon > 0$, there exists a $\delta > 0$ such that if $A \in \mathcal{F}$ and $\mu(A) \le \delta$, then

$$\int_A \mathsf{X}\, d\mu \le \epsilon.$$

4. ABSOLUTE CONTINUITY AND THE RADON-NIKODYM THEOREM

Consider a measurable space $(\Omega, \mathcal{F})$ and two measures μ, ν on $\mathcal{F}$.

Definition A.29. *Say that* ν *is absolutely continuous with respect to* μ, *denoted* $\nu \ll \mu$, *if* $A \in \mathcal{F}$, $\mu(A) = 0 \Rightarrow \nu(A) = 0$.

Call two measurable functions X_1, X_2 *equivalent* if $\mu(\{\mathsf{X}_1 \ne \mathsf{X}_2\}) = 0$. Then

Radon-Nikodym Theorem A.30 (*Halmos, p.* 128). *If* $\nu \ll \mu$, *then there exists a nonnegative* $\mathcal{F}$*-measurable function* X *determined up to equivalence, such that for any* $A \in \mathcal{F}$,

$$\nu(A) = \int_A \mathsf{X}\, d\mu.$$

Another way of denoting this is to say that the Radon derivative of ν with respect to μ exists and equals X; that is,

$$\frac{d\nu}{d\mu} = \mathsf{X}.$$

The opposite of continuity is

Definition A.31. *Say that* ν *is singular with respect to* μ, *written* $\mu \perp \nu$ *if there exists* $A \in \mathcal{F}$ *such that*

$$\mu(A) = 0, \quad \nu(A^c) = 0.$$

Lebesgue Decomposition Theorem A.32 *(Halmos, p. 134). For any two measures μ, ν on $\mathcal{F}$, ν can be decomposed into two measures, ν_c, ν_s, in the sense that for every $A \in \mathcal{F}$,*

$$\nu(A) = \nu_c(A) + \nu_s(A)$$

and $\nu_c \ll \mu$, $\nu_s \perp \mu$.

For a σ-finite measure the set of points $\{\omega; \mu(\{\omega\}) > 0\}$ is at most countable. Call ν a *point measure* if there is a countable set $G = \{\omega_j\}$ such that for every $A \in \mathcal{F}$,

$$\nu(A) = \nu(A \cap G).$$

Obviously, any measure ν may be decomposed into $\nu_1 + \nu_p$, where ν_p is a point measure and ν_1 assigns mass zero to any one-point set. Hence, on $(R^{(1)}, \mathcal{B}_1)$ we have the special case of A.32.

Corollary A.33. *A measure ν on $\mathcal{B}_1$ can be written as*

$$\nu = \nu_c + \nu_s + \nu_p,$$

where ν_p is a point measure, $\nu_s \perp l$ but ν_s assigns zero mass to any one-point sets, and $\nu_c \ll l$. [Recall l is Lebesgue measure.]

5. CROSS-PRODUCT SPACES AND THE FUBINI THEOREM

Definition A.34. *Given two spaces Ω_1, Ω_2, their cross product $\Omega_1 \times \Omega_2$ is the set of all ordered pairs $\{(\omega_1, \omega_2); \omega_1 \in \Omega_1, \omega_2 \in \Omega_2\}$. For measurable spaces $(\Omega_1, \mathcal{F}_1)$, $(\Omega_2, \mathcal{F}_2)$, $\mathcal{F}_1 \times \mathcal{F}_2$ is the smallest σ-field containing all sets of the form*

$$\{(\omega_1, \omega_2); \omega_1 \in A_1, \omega_2 \in A_2\},$$

where $A_1 \in \mathcal{F}_1$, $A_2 \in \mathcal{F}_2$. Denote this set by $A_1 \times A_2$.

For a function $\mathsf{X}(\omega_1, \omega_2)$ on $\Omega_1 \times \Omega_2$, its *section at* ω_1 is the function on Ω_2 gotten by holding ω_1 constant and letting ω_2 be the variable. Similarly, if $A \subset \Omega_1 \times \Omega_2$, *its section at* ω_1 is defined as $\{\omega_2; (\omega_1, \omega_2) \in A\}$.

Theorem A.35 *(Halmos, pp. 141ff.). Let X be an $\mathcal{F}_1 \times \mathcal{F}_2$-measurable function; then every section of X is an $\mathcal{F}_2$-measurable function. If $A \in \mathcal{F}_1 \times \mathcal{F}_2$, every section of A is in $\mathcal{F}_2$.*

If we have measures μ_1 on $\mathcal{F}_2$, μ_2 on $\mathcal{F}_2$, then

Theorem A.36 *(Halmos, p. 144). There is a unique measure $\mu_1 \times \mu_2$ on $\mathcal{F}_1 \times \mathcal{F}_2$ such that for every $A_1 \in \mathcal{F}_1$, $A_2 \in \mathcal{F}_2$,*

$$\mu_1 \times \mu_2(A_1 \times A_2) = \mu_1(A_1)\mu_2(A_2).$$

This is called the cross-product measure.

Fubini Theorem A.37 (*Halmos, p.* 148). *Let* X *be* $\mathcal{F}_1 \times \mathcal{F}_2$*-measurable, and*

$$\int |X| \, d(\mu_1 \times \mu_2) < \infty.$$

Then

$$\int X(\omega_1, \omega_2)\mu_1(d\omega_1), \qquad \int X(\omega_1, \omega_2)\mu_2(d\omega_2)$$

are respectively $\mathcal{F}_2$*-and* $\mathcal{F}_1$*-measurable functions, which may be infinite on sets of measure zero, but whose integrals exist. And*

$$\int X \, d(\mu_1 \times \mu_2) = \int \left(\int X \, d\mu_1 \right) d\mu_2$$
$$= \int \left(\int X \, d\mu_2 \right) d\mu_1.$$

Corollary A.38. *If* $A \in \mathcal{F}_1 \times \mathcal{F}_2$ *and* $\mu_1 \times \mu_2(A) = 0$, *then almost every section of* A *has* μ_2 *measure zero.*

This all has fairly obvious extensions to finite cross products $\Omega_1 \times \cdots \times \Omega_n$.

6. THE $L_r(\mu)$ SPACES

These are some well-known results. Let $(\Omega, \mathcal{F}, \mu)$ be a measure space and the functions X, Y be $\mathcal{F}$-measurable:

Schwarz Inequality A.39. *If* $\int |X|^2 \, d\mu$ *and* $\int |Y|^2 \, d\mu$ *are finite, then so is* $\int |XY| \, d\mu$ *and*

$$\left(\int XY \, d\mu \right)^2 \leq \left(\int |X|^2 \, d\mu \right) \left(\int |Y|^2 \, d\mu \right).$$

Definition A.40. $L_r(\mu)$, $r > 0$, *is the class of all* $\mathcal{F}$*-measurable functions* X *such that* $\int |X|^r \, d\mu < \infty$.

Completeness Theorem A.41 (*Halmos, p.* 107). *If* $X_n \in L_r(\mu)$ *and*

$$\int |X_n - X_m|^r \, d\mu \to 0$$

as $m, n \to \infty$ *in any way, then there is a function* $X \in L_r(\mu)$ *such that*

$$\int |X_n - X|^r \, d\mu \to 0.$$

7. TOPOLOGICAL MEASURE SPACES

In this section, unless otherwise stated, assume that Ω has a metric under which it is a separable metric space. Let $\mathcal{C}$ be the class of all open sets in Ω. A measure μ on $\mathcal{F}$ is called *inner regular* if for any $A \in \mathcal{F}$

$$\mu(A) = \sup_C \mu(C),$$

where the sup is over all compact sets $C \subset A$, $C \in \mathcal{F}$. It is called *outer regular* if

$$\mu(A) = \inf_{O} \mu(O),$$

where the inf is over all open sets O such that $A \subset O$, and $O \subset \mathcal{F}$.

Theorem A.42 (*Follows from Halmos, p.* 228). *Any measure on* $\mathcal{F}(\mathcal{C})$ *is both inner and outer regular.*

Theorem A.43 (*Follows from Halmos, p.* 240). *The class of* $\mathcal{F}(\mathcal{C})$*-measurable functions is the smallest class of functions containing all continuous functions on* Ω *and closed under pointwise convergence.*

Theorem A.44 (*Halmos, p.* 241). *If* $\int |\mathsf{X}|\, d\mu < \infty$ *for* X $\mathcal{F}(\mathcal{C})$*-measurable, then for any* $\epsilon > 0$ *there is a continuous function* φ on Ω *such that*

$$\int |\mathsf{X} - \varphi|\, d\mu \le \epsilon.$$

Definition A.45. *Given any measurable space* $(\Omega, \mathcal{F})$ (Ω *not necessarily metric*), *it is called a Borel space if there is* 1–1 *mapping* φ: $\Omega \leftrightarrow E$ *where* $E \in \mathcal{B}_1$ *such that* φ *is* $\mathcal{F}$*-measurable, and* φ^{-1} *is* $\mathcal{B}_1(E)$*-measurable.*

Theorem A.46. *If* Ω *is complete, then* $(\Omega, \mathcal{F}(\mathcal{C}))$ *is a Borel space.*

We prove this in the case that $(\Omega, \mathcal{F}(\mathcal{C}))$ is $(R^{(\infty)}, \mathcal{B}_\infty)$. Actually, since there is a 1–1 continuous mapping between $R^{(1)}$ and (0, 1) it is sufficient to show that

Theorem A.47. $((0, 1)^{(\infty)}, \mathcal{B}_\infty(0, 1))$ *is a Borel space.*

Note. Here $(0, 1)^{(\infty)}$ is the set of all infinite sequences with coordinates in (0, 1) and $\mathcal{B}_\infty(0, 1)$ means $\mathcal{B}_\infty((0, 1))$.

Proof. First we construct the mapping Φ from (0, 1) to $(0, 1)^{(\infty)}$. Every number in (0, 1) has a unique binary expansion $x = .x_1x_2\cdots$ containing an infinite number of zeros. Consider the triangular array

$$\begin{array}{lllll} 1 \\ 2 & 3 \\ 4 & 5 & 6 \\ 7 & 8 & 9 & 10 \\ \cdot & \cdot & \cdot & \cdot & \cdot \\ \cdot & \cdot & \cdot & \cdot & \cdot & \cdot \end{array}$$

Let

$$\Phi(x) = (\Phi_1(x), \Phi_2(x), \ldots),$$

where the nth coordinate is formed by going down the nth column of the array; that is,

$$\Phi_1(x) = .x_1x_2x_4x_7\cdots,$$
$$\Phi_2(x) = .x_3x_5x_8\cdots,$$

and so on. Conversely, if $\mathbf{x} \in (0, 1)^{(\infty)}$, $\mathbf{x} = (x^{(1)}, x^{(2)}, \ldots)$, expand every coordinate as the unique binary decimal having an infinite number of zeros, say

$$x^{(k)} = .x_1^{(k)} x_2^{(k)} \cdots$$

and define $\varphi(\mathbf{x})$ to be the binary decimal whose nth entry is $x_j^{(k)}$ if n appears in the kth column j numbers down. That is,

$$\varphi(\mathbf{x}) = .x_1^{(1)} x_2^{(1)} x_1^{(2)} x_3^{(1)} x_2^{(2)} x_1^{(3)} \cdots$$

Clearly, Φ and φ are inverses of each other, so the mapping is 1–1 and onto. By 2.13, to show Φ $\mathcal{B}_1(0, 1)$-measurable, it is sufficient to show that each $\Phi_k(x)$ is $\mathcal{B}_1(0, 1)$-measurable. Notice that the coordinates $x_1(x), x_2(x), \ldots$ in the decimal expansion of x are measurable functions of x, continuous except at the points which have only a finite number of ones in their expansion (binary rationals). Furthermore, each $\Phi_k(x)$ is a sum of these, for example,

$$\Phi_1(x) = \frac{x_1(x)}{2} + \frac{x_2(x)}{2^2} + \frac{x_4(x)}{2^3} + \cdots$$

Therefore, every $\Phi_k(x)$ is measurable. The proof that $\varphi(\mathbf{x})$ is measurable similarly proceeds from the observation that $x_j^{(k)}(\mathbf{x})$ is a measurable function of $\mathbf{x}$. Q.E.D.

To prove A.46 generally, see Sierpinski, [123a, p. 137], where it is proved that every complete separable metric space is homeomorphic to a subset of $R^{(\infty)}$. (See also p. 206.)

8. EXTENSION ON SEQUENCE SPACE

A.48 (*Proof of Theorem* 2.18). Consider the class $\mathcal{F}_0$ of all finite disjoint unions of finite dimensional rectangles. It is easily verified that $\mathcal{F}_0$ is a field, and that

$$\mathcal{B}_\infty = \mathcal{F}(\mathcal{F}_0).$$

For any set $A \in \mathcal{F}_0$, $A = \bigcup_j S_j$, where the S_j are disjoint rectangles, define

$$\hat{P}(A) = \sum_j \hat{P}(S_j).$$

There is a uniqueness problem: Suppose A can also be represented as $\bigcup_k S_k'$, where the S_k' are also disjoint. We need to know that

$$\sum_k \hat{P}(S_k') = \sum_j \hat{P}(S_j).$$

Write

$$S_k' = S_k' \cap A = \bigcup_j (S_k' \cap S_j).$$

By part (b) of the hypothesis,

$$\hat{P}(S'_k) = \sum_j \hat{P}(S'_k \cap S_j);$$

so

$$\sum_k \hat{P}(S'_k) = \sum_{j,k} \hat{P}(S_j \cap S'_k).$$

By a symmetric argument

$$\sum_j \hat{P}(S_j) = \sum_{j,k} \hat{P}(S_j \cap S'_k).$$

Now we are in a position to apply the Carathéodory extension theorem, if we can prove that $A_n \in \mathcal{F}_0$, $A_n \downarrow \emptyset$ implies $\hat{P}(A_n) \to 0$. To do this, assume that $\lim_n \hat{P}(A_n) = \delta > 0$. By repeating some A_n in the sequence, if necessary, we can assume that

$$A_n = \{\mathbf{x}; (x_1, \ldots, x_n) \in A_n^*\},$$

where A_n^* is a union of disjoint rectangles in $R^{(n)}$. By part (c) of the hypothesis we can find a set $B_n^* \subset A_n^*$ so that B_n^* is a finite union of compact rectangles in $R^{(n)}$, and if

$$B_n = \{\mathbf{x}; (x_1, \ldots, x_n) \in B_n^*\},$$

then

$$\hat{P}(A_n - B_n) \leq \frac{\delta}{2^{n+1}}.$$

Form the sets $C_n = \bigcap_1^n B_k$, and put

$$C_n = \{\mathbf{x}; (x_1, \ldots, x_n) \in C_n^*\}.$$

Then, since the A_n are nonincreasing,

$$\begin{aligned}\hat{P}(A_n - C_n) &\leq \sum_{k=1}^{n} \hat{P}(A_n - B_k)\\ &\leq \sum_{k=1}^{n} \hat{P}(A_k - B_k)\\ &\leq \delta/2.\end{aligned}$$

The conclusion is that $\lim_n \hat{P}(C_k) \geq \delta/2$, and, of course, $C_n \downarrow \emptyset$. Take points $\mathbf{x}^{(1)} \in C_1$, $\mathbf{x}^{(2)} \in C_2, \ldots,$

$$\mathbf{x}^{(k)} = (x_1^{(k)}, x_2^{(k)}, \ldots).$$

For every n,

$$(x_1^{(n)}, \ldots, x_n^{(n)}) \in C_n^*.$$

Take N_1 any ordered infinite subsequence of integers such that $x_1^{(n)} \to x_1 \in C_1^*$ as n runs through N_1. This is certainly possible since $x_1^{(n)} \in C_1^*$ for all n.

Now take $N_2 \subset N_1$ such that

$$(x_1^{(n)}, x_2^{(n)}) \to (x_1, x_2) \in C_2^*$$

as n runs through N_2. Continuing, we construct subsequences

$$N_1 \supset N_2 \supset N_3 \supset \cdots$$

Let n_k be the kth member of N_k, then for every j, $x_j^{(n)} \to x_j$ as n goes to infinity through $\{n_k\}$. Furthermore, the point $(x_1, x_2, \ldots)$ is in C_n for every $n \geq 1$, contradicting $C_n \downarrow \emptyset$.

BIBLIOGRAPHY

[1] BACHELIER, L., "Théorie de la spéculation, Thèse, Paris, 1900," *Ann. Éc. Norm. Sup*. s. 3, **17,** 21–86 (1900).

[2] BAILEY, N. T. J., *The Elements of Stochastic Processes*, John Wiley & Sons, Inc., New York, 1964.

[3] BHARUCHA-REID, A. T., *Elements of the Theory of Markov Processes and Their Applications*, McGraw-Hill Book Co., Inc., New York, 1960.

[4] BIRKHOFF, G. D., "Proof of the ergodic theorem," *Proc. Nat'l. Acad. Sci.* **17,** 656–660 (1931).

[5] ——, "Dynamical systems," *American Math. Society*, New York, 1927, reprinted 1948.

[6] BLACKWELL, D., "A renewal theorem," *Duke Math. J.* **15,** 145–150 (1948).

[7] ——, "Extension of a renewal theorem," *Pacific J. Math.* **3,** 315–320 (1953).

[8] ——, "On a class of probability spaces," *Proc. of the 3rd Berkeley Symp. on Math. Stat. and Prob.* Vol. II., 1–6 (1956).

[9] —— and FREEDMAN, D., "The tail σ-field of a Markov chain and a theorem of Orey," *Ann. Math. Stat.* **35,** No. 3, 1291–1295 (1964).

[10] BLUMENTHAL, R., GETOOR, R., and MCKEAN, H. P., JR., "Markov processes with identical hitting distributions," *Illinois J. Math.* **6,** 402–420 (1962).

[11] BOCHNER, S., *Harmonic Analysis and the Theory of Probability*, University of California Press, Berkeley, 1955.

[12] BREIMAN, L., "On transient Markov chains with application to the uniqueness problem for Markov processes," *Ann. Math. Stat.* **28,** 499–503 (1957).

[13] CAMERON, R. H. and MARTIN, W. T., "Evaluations of various Wiener integrals by use of certain Sturm-Liouville differential equations," *Bull. Amer. Math. Soc.* **51,** 73–90 (1945).

[14] CHOW, Y. S. and ROBBINS, H., "On sums of independent random variables with infinite moments and 'fair' games," *Proc. Nat. Acad. Sci.* **47,** 330–335 (1961).

[15] CHUNG, K. L., "Notes on Markov chains," duplicated notes, *Columbia Graduate Mathematical Statistical Society*, 1951.

[16] ——, *Markov Chains with Stationary Transition Probabilities*, Springer-Verlag, Berlin, 1960.

[17] ——, "The general theory of Markov processes according to Doeblin," *Z. Wahr.* **2,** 230–254 (1964).

[18] —— and FUCHS, W. H. J., "On the distribution of values of sums of random variables," *Mem. Amer. Math. Soc.* No. 6 (1951).

[19] —— and KAC, M., "Remarks on fluctuations of sums of independent random variables," *Mem. Amer. Math. Soc.* No. 6 (1951).

[20] —— and Ornstein, D., "On the recurrence of sums of random variables," *Bull. Amer. Math. Soc.* **68,** 30–32 (1962).

[21] Cramér, H., "On harmonic analysis in certain functional spaces," *Ark. Mat. Astr. Fys.* **28B,** No. 12 (1942).

[22] Darling, D. A., and Siegert, A. J., "The first passage problem for a continuous Markov process," *Ann. Math. Stat.* **24,** 624–639 (1953).

[23] —— and Kac, M., "On occupation times for Markov processes," *Trans. Amer. Math. Soc.* **84,** 444–458 (1957).

[24] Dinges, H., "Ein verallgemeinertes Spiegelungsprinzip für den Prozess der Brownschen Bewegung," *Z. Wahr.* **1,** 177–196 (1962).

[25] Doeblin, W., "Sur les propriétés asymptotiques de mouvement régis par certains types des chaînes simples," *Bull. Math. Soc. Roum. Sci.* **39,** No. 1, 57–115, No. 2, 3–61 (1937).

[26] ——, "Sur certains mouvements aléatoires discontinus," *Skand. Aktuarie-tidskr* **22,** 211–222 (1939).

[27] ——, "Sur les sommes d'un grand nombre de variables aléatoires indépendantes," *Bull. Sci. Math.* **63,** No. 1, 23–32, 35–64 (1939).

[28] ——, "Éléments d'une théorie générale des chaînes simples constante de Markoff," *Ann. Sci. École Norm. Sup.* (3), **57,** 61–111 (1940).

[29] ——, "Sur l'ensemble de puissances d'une loi de probabilité," *Studia Math.* **9,** 71–96 (1940).

[30] Donsker, M., "An invariance principle for certain probability limit theorems," *Mem. Amer. Math. Soc.* No. 6 (1951).

[31] ——, "Justification and extension of Doob's heuristic approach to the Kolmogorov-Smirnov theorems," *Ann. Math. Stat.* **23,** 277–281, (1952).

[32] Doob, J. L., "Regularity properties of certain families of chance variables," *Trans. Amer. Math. Soc.* **47,** 455–486 (1940).

[33] ——, "Topics in the theory of Markov chains," *Trans. Amer. Math. Soc.* **52,** 37–64 (1942).

[34] ——, "Markov chains–denumerable case," *Trans. Amer. Math. Soc.* **58,** 455–473 (1945).

[35] ——, "Asymptotic properties of Markov transition probabilities," *Trans. Amer. Math. Soc.* **63,** 393–421 (1948).

[36] ——, "Renewal theory from the point of view of the theory of probability," *Trans. Amer. Math. Soc.* **63,** 422–438 (1948).

[37] ——, "A heuristic approach to the Kolmogorov-Smirnov theorems," *Ann. Math. Stat.* **20,** 393–403 (1949).

[38] ——, "Continuous parameter martingales," *Proc. 2nd Berkeley Symp. on Math. Stat. and Prob.*, 269–277 (1951).

[39] ——, *Stochastic Processes*, John Wiley & Sons, Inc., New York, 1953.

[40] Dubins, L. and Savage, J. L., *How to Gamble If You Must*, McGraw-Hill Book Co., Inc., New York, 1965.

[41] Dunford, N. and Schwartz, J. T., *Linear Operators, Part I*, Interscience Publishers, Inc., New York, 1958.

[42] Dvoretski, A., Erdös, P., and Kakutani, S., "Nonincrease everywhere of the Brownian motion process," *Proc. 4th Berkeley Symp. on Math. Stat. and Prob.* Vol. II, 103–116 (1961).

[43] DYNKIN, E. B., "Continuous one-dimensional Markov processes," *Dokl. Akad. Nauk.* SSSR **105,** 405–408 (1955).

[44] ——, *Markov processes*, Vols. I, II, Academic Press, Inc., New York, 1965.

[45] EINSTEIN, A., "On the movement of small particles suspended in a stationary liquid demanded by the molecular-kinetic theory of heat," *Ann. d. Physik* **17** (1905). [In *Investigations of the Theory of the Brownian Movement*, Edited by R. Fürth, Dover Publications, Inc., New York, 1956.]

[46] ERDÖS, P., "On the law of the iterated logarithm," *Ann. of Math.* **43,** 419–436 (1942).

[47] —— and KAC, M., "On certain limit theorems of the theory of probability," *Bull. Amer. Math. Soc.* **52,** 292–302 (1946).

[48] ——, "On the number of positive sums of independent random variables," *Bull. Amer. Math. Soc.* **53,** 1011–1020 (1947).

[49] ——, FELLER, W., and POLLARD, H., "A theorem on power series," *Bull. Amer. Math. Soc.* **55,** 201–204, 1949.

[50] FELLER, W., "Zur Theorie der stochastischen Prozesse," *Math. Ann.* **113,** 113–160 (1936).

[51] ——, "On the Kolmogorov–P. Lévy formula for infinitely divisible distribution functions," *Proc. Yugoslav Acad. Sci.* **82,** 95–113 (1937).

[52] ——, "On the integro-differential equations of purely discontinuous Markov processes," *Trans. Am. Math. Soc.* **48,** 488–515 (1940).

[53] ——, "The general form of the so-called law of the iterated logarithm," *Trans. Amer. Math. Soc.* **54,** 373–402 (1943).

[54] ——, "A limit theorem for random variables with infinite moments, *Amer. J. Math.* **68,** 257–262 (1946).

[55] ——, "Fluctuation theory of recurrent events," *Trans. Amer. Math. Soc.* **67,** 98–119 (1949).

[56] ——, "Diffusion processes in genetics," *Proc. 2nd Berkeley Symp. on Math. Stat. and Prob.* 227–246 (1951).

[57] ——, "The parabolic differential equations and the associated semi-groups of transformations, *Ann. of Math.* **55,** 468–519 (1952).

[58] ——, "Diffusion processes in one dimension," *Trans. Amer. Math. Soc.* **77,** 1–31 (1954).

[59] ——, *An Introduction to Probability Theory and Its Applications*, Vol. I, 2nd Ed., Vol. II, John Wiley & Sons, Inc., New York, 1957, 1966.

[60] —— and OREY, S., "A renewal theorem," *J. Math. and Mech.* **10,** 619–624 (1961).

[61] GARSIA, A., "A simple proof of Eberhard Hopf's maximal ergodic theorem," *J. Math. and Mech.* **14,** 381–382 (1965).

[62] GNEDENKO, B. V. and KOLMOGOROV, A. N., *Limit Distributions for Sums of Independent Random Variables*, Addison-Wesley Publishing Co., Inc., Reading, Mass., 1954.

[63] HADLEY, G., *Linear Algebra*, Addison-Wesley Publishing Co., Inc., Reading, Mass., 1961.

[64] HALMOS, P. R., *Measure Theory*, D. Van Nostrand Co., Inc., Princeton, N.J., 1950.

[65] ——, "Lectures on ergodic theory," *Mathematical Society of Japan*, No. 3, 1956.

[66] Harris, T. E., "The existence of stationary measures for certain Markov processes," *Proc. of the 3rd Berkeley Symp. on Math. Stat. and Prob.* Vol. II, 113–124 (1956).

[67] ——, *The Theory of Branching Processes*, Prentice-Hall, Inc., Englewood Cliffs, N.J., 1963.

[68] ——, "First passage and recurrence distributions," *Trans. Amer. Math. Soc.* **73,** 471–486 (1952).

[69] —— and Robbins, H., "Ergodic theory of Markov chains admitting an infinite invariant measure," *Proc. of the National Academy of Sciences* **39,** No. 8, 862–864 (1953).

[70] Herglotz, G. "Über Potenzreihen mit positivem reclen Teil im Einheitskreis," *Ber. Verh. Kgl. Sachs. Ges. Leipzig, Math.-Phys. Kl.* **63,** 501–511 (1911).

[71] Hewitt, E. and Savage, L. J., "Symmetric measures on Cartesian products," *Trans Amer. Math. Soc.* **80,** 470–501 (1955).

[72] Hobson, E. W., *The Theory of Functions of a Real Variable*, 3rd Ed., Vol. I, Cambridge University Press, Cambridge, 1927 (Reprinted by Dover Publications, Inc., New York, 1957).

[73] Hopf, E. *Ergoden Theorie, Ergebnisse der Math.* Vol. 2, J. Springer, Berlin, 1937 (Reprinted by Chelsea Publishing Co., New York, 1948).

[74] Hunt, G., "Some theorems concerning Brownian motion," *Trans. Amer. Math. Soc.* **81,** 294–319 (1956).

[75] Ito, K., "On stochastic processes (I) (Infinitely divisible laws of probability)," *Japan J. Math.* **18,** 261–301 (1942).

[76] —— and McKean, H. P. Jr., *Diffusion Processes and Their Sample Paths*, Academic Press, New York, 1965.

[77] Jessen, B. and Sparre Andersen, E., "On the introduction of measures in infinite product sets," *Danske Vid. Selsk. Mat.-Fys. Medd.* **25,** No. 4 (1948).

[78] —— and Wintner, A., "Distribution functions and the Riemann Zeta function," *Trans. Amer. Math. Soc.* **38,** 48–88 (1935).

[79] Kac, M., "On a characterization of the normal distribution," *Amer. J. of Math.* **61,** 726–728 (1939).

[80] ——, "Random walk and the theory of Brownian motion, *Amer. Math. Monthly* **54,** 369–391 (1949) (Reprinted in [139]).

[81] ——, "On the notion of recurrence in discrete stochastic processes," *Bull. Amer. Math. Soc.* **53,** 1002–1010 (1947).

[82] ——, "On some connections between probability theory and differential and integral equations," *Proc. 2nd Berkeley Symp. on Math. Stat. and Prob.* 189–215 (1951).

[83] ——, "Statistical independence in probability, analysis, and number theory," *Carus Mathematical Monograph*, No. 12, *The Mathematical Association of America*, 1959.

[84] Kallianpur, G. and Robbins, H., "The sequence of sums of independent random variables," *Duke Math. J.* **21,** 285–307 (1954).

[85] Karamata, J., "Neuer Beweis und Verallgemeinerung der Tauberschen Sätze, welche die Laplaceschen Stieltjesschen Transformationen betreffen," *J. für die reine und angewandte Math.* **164,** 27–40 (1931).

[86] KARLIN, S., *A First Course in Stochastic Processes*, Academic Press, New York, 1966.
[87] —— and McGREGOR, J. L., "Representation of a class of stochastic processes," *Proc. Nat'l. Acad. Sci.* **41**, 387–391 (1955).
[88] KHINTCHINE, A., "Über einen Satz der Wahrscheinlichkeitsrechnung," *Fundamenta Math.* **6**, 9–20 (1924).
[89] ——, "Déduction nouvelle d'une formula de M. Paul Lévy," *Bull. Univ. d'Etat Moskou.*, Sér Internat. Sect. A., 1, No. 1, 1–5 (1937).
[90] ——, *Mathematical Foundations of Statistical Mechanics*, Dover Publications, Inc., New York, 1949.
[91] —— and KOLMOGOROV, A., "Über konvergenz von Reihen, deren Glieder durch den Zufall bestimmt werden," *Rec. Math* (*Mat. Sbornik*) **32**, 668–677 (1925).
[92] KOLMOGOROV, A., "Sur la loi forte des grandes nombres," *C. R. Acad. Sci. Paris*, **191**, 910–912 (1930).
[93] ——, "Über die analytischen Methoden in der Wahrscheinlichkeitsrechnung," *Math. Ann.* **104**, 415–458 (1931).
[94] ——, "Sulla forma generale di un processo stocastico omogeneo," *Atti Acad. naz. Lincei Rend. Cl. Sci. Fis. Mat. Nat.* (6), **15**, 805–808, 866–869 (1932).
[95] ——, "Anfangsgründe der Theorie der Markoffschen ketten mit unendich vielen möglichen Zuständen," *Rec. Math. Moscov* (*Mat. Sbornik*), **1** (43), 607–610 (1936).
[96] ——, "Interpolation and extrapolation of stationary random sequences" (Russian), *Izv. Akad. Nauk. SSSR, Ser. Mat.* **5**, 3–114 (1941).
[97] ——, "Stationary sequences in Hilbert space" (Russian), *Bull. Math. Univ. Moscov* **2**, No. 6 (1941).
[98] ——, *Foundations of Probability* (translation), Chelsea Publishing Co., New York, 1950.
[99] LE CAM, L., Mimeographed notes, Statistics Department, Univ. of Calif., Berkeley.
[100] LÉVY, P., "Théorie des erreurs. La loi de Gauss et les lois exceptionalles," *Bull. Soc. Math.* **52**, 49–85 (1924).
[101] ——, "Sur les séries dont les termes sont des variables éventuelles indépendantes," *Studia Math.* **3**, 119–155 (1931).
[102] ——, "Sur les intégrales dont les éléments sont des variables aléatoires indépendantes," *Annali R. Scuola. Sup. Pisa* (2), **3**, 336–337 (1934) and **4**, 217–218 (1935).
[103] ——, *Théorie de l'Addition des Variables Aléatoires*, Bautier-Villars, Paris, 1937.
[104] ——, "Sur certains processus stochastiques homogènes," *Comp. Math.* **7**, 283–339 (1939).
[105] ——, *Processus Stochastiques et Mouvement Brownian*, Gauthier-Villars, Paris, 1948.
[106] LINDEBERG, Y. W., "Eine neue Herleitung des Exponentialgesetzes in der Wahrscheinlichkeiterechnung," *Math. Z.* **15**, 211–225 (1922).
[107] LOÈVE, M., Appendix to [105].
[108] ——, *Probability Theory*, 3rd Ed., D. Van Nostrand Co., Inc., Princeton, N.J., 1963.

[109] LYAPUNOV, A. M., "Nouvelle forme du théorème sur la limite de probabilités, *Mem. Acad. Sci. St. Petersbourg* (8), **12,** No. 5, 1–24 (1901).

[110] MARKOV, A. A., "Extension of the law of large numbers to dependent events" (Russian), *Bull. Soc. Phys. Math. Kazan* (2) **15,** 155–156 (1906).

[111] MARUYAMA, G., "The harmonic analysis of stationary stochastic processes," *Mem. Fac. Sci. Kyusyu Univ.* **A4,** 45–106 (1949).

[—a] MCSHANE, E. J., *Integration*, Princeton University Press, Princeton, N.J., 1944.

[112] MISES, R. v., *Probability, Statistics, and Truth*, William Hodge and Co., 2nd Ed., London, 1957.

[113] NEVEU, J., *Mathematical Foundations of the Calculus of Probability*, Holden-Day, Inc., San Francisco, 1965.

[114] OREY, S., "An ergodic theorem for Markov chains," *Z. Wahr.* **1,** 174–176 (1962).

[115] PALEY, R., WIENER, N., and ZYGMUND, A., "Note on random functions," *Math. Z.*, 647–668 (1933).

[116] POLYA, G., "Über eine Aufgabe der Wahrscheinlichkeitsrechnung betreffend die Irrfahrt in Strassennetz," *Math. Ann.* **89,** 149–160 (1921).

[117] POSPIŠIL, B., "Sur un problème de M. M. S. Berstein et A. Kolmogoroff," *Casopis Pest. Mat. Fys.* **65,** 64–76 (1935–36).

[118] PROKHOROV, Y. V., "Convergence of random processes and limit theorems in probability theory," *Teor. Veroyatnost. i. Primenen* **1,** 177–238 (1956).

[119] RYLL-NARDZEWSKI, C., "Remarks on processes of calls," *Proc. of the 4th Berkeley Symp. on Math. Stat. and Prob.* Vol. II, 455–465 (1961).

[120] SAATY, T. L., *Elements of Queuing Theory, with Applications*, McGraw-Hill Book Co., Inc., New York, 1961.

[121] SAKS, S., "Theory of the integral," *Monografje Mathematyczne*, Tom VII, Warsaw-Lwow (1937).

[122] SHEPP, L. A., "A local limit theorem," *Ann. Math. Stat.* **35,** 419–423 (1964).

[123] SHOHAT, J. A. and TAMARKIN, J. D., "The problem of moments," *Math. Surveys*, No. 1, *Amer. Math. Soc.*, New York (1943).

[—a] SIERPINSKI, W., *General Topology*, University of Toronto Press, Toronto, 1952.

[124] SKOROKHOD, A. V., "Limit theorems for stochastic processes, "*Teor. Veroyatnost. i. Primenen* **1,** 289–319 (1956).

[125] ——, "Limit theorems for stochastic processes with independent increments," *Teor. Veroyatnost. i. Primenen* **2,** 145–177 (1957).

[126] ——, *Studies in the Theory of Random Processes*, Kiev University, 1961; Addison-Wesley Publishing Co., Inc., Reading, Mass., 1965 (translation).

[127] SMITH, W. L., "Renewal theory and its ramifications," *J. Roy. Statist. Soc.* (3), **20,** 243–302 (1958).

[128] SPARRE ANDERSEN, E., "On the fluctuations of sums of random variables," *Math. Scand.* **1,** 163–285 (1953) and **2,** 195–223 (1954).

[129] SPITZER, F., "A combinatorial lemma and its applications to probability theory," *Trans. Amer. Math. Soc.* **82,** 323–339 (1956).

[130] ——, *Principles of random walk*, D. Van Nostrand Co., Inc., Princeton, 1964.

[131] STONE, C. J., "Limit theorems for random walks, birth and death processes, and diffusion processes," *Illinois J. Math.* **7,** 638–660 (1963).

[132] ——, "On characteristic functions and renewal theory," *Trans. Amer. Math. Soc.* **120,** 327–342 (1965).

[133] ——, "A local limit theorem for multidimensional distribution functions," *Ann. Math. Stat.* **36,** 546–551 (1965).

[134] STRASSEN, V., "An invariance principle for the law of the iterated logarithm," *Z. Wahr.* **3,** 211–226 (1964).

[135] ——, "A converse to the law of the iterated logarithm," *Z. Wahr.* **4,** 265–268 (1965).

[136] TROTTER, H. F., "A property of Brownian motion paths," *Illinois J. Math.* **2,** 425–433 (1958).

[137] VILLE, J., *Etude Critique de la Notion de Collectif*, Gauthier-Villars, Paris, 1939.

[138] VOLKONSKIĬ, V. A., "Random substitution of time in strong Markov processes," *Teor. Veroyatnost. i. Primenen* **3,** 332–350 (1958).

[139] WAX, N. (editor), *Selected Papers on Noise and Stochastic Processes*, Dover Publications, Inc., New York, 1954.

[140] WIDDER, D. V., *The Laplace Transform*, Princeton Univ. Press, Princeton, 1941.

[141] WIENER, N., "Differential space," *J. Math. Phys.* **2,** 131–174 (1923).

[142] ——, "Un problème de probabilités énombrables," *Bull. Soc. Math. de France* **52,** 569–578 (1924).

[143] ——, *Extrapolation, Interpolation and Smoothing of Stationary Time Series*, MIT Press and John Wiley & Sons, Inc., New York, 1950 (reprinted from a publication restricted for security reasons in 1942).

[144] YAGLOM, A. M., *An Introduction to the Theory of Stationary Random Functions*, Prentice-Hall, Inc., Englewood Cliffs, N.J., 1962.

INDEX

ABCDE698